ALBUM *of* SCIENCE

The Nineteenth Century

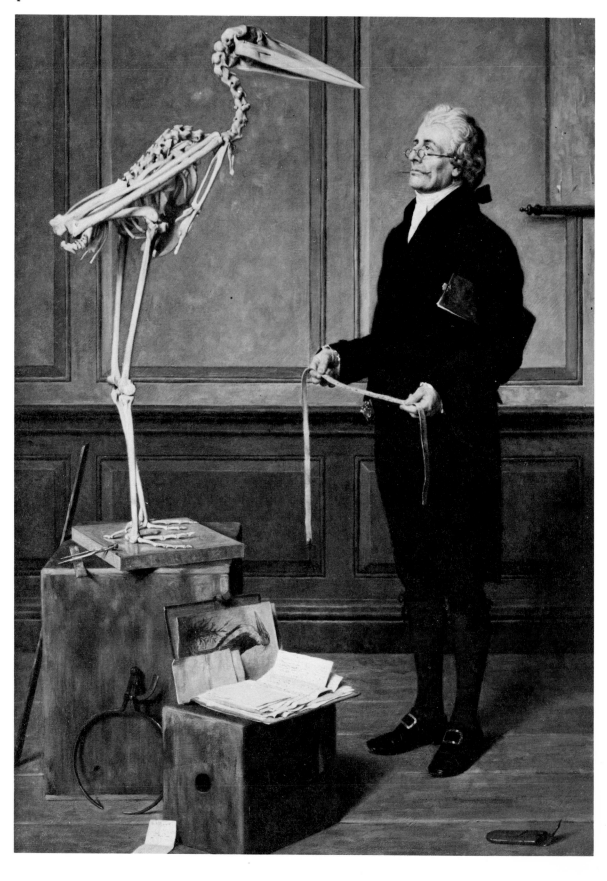

1

ALBUM *of* SCIENCE

The Nineteenth Century

L. PEARCE WILLIAMS

I. B. Cohen, General Editor, Album of Science

Charles Scribner's Sons NEW YORK

THE SCRIBNER PICTORIAL REFERENCE LIBRARY

The visual record of history, though rich and very informative, has been one of the most neglected of the resources available for the study of the past. In large measure this has been because pictorial material has not been presented systematically, in and of itself; it has been seen rather as illustrating or supplementing written texts. The Scribner Pictorial Reference Library fills a need by making accessible many thousands of significant historical pictures, comprehensively indexed and cross-indexed and carefully documenting the past in a way in which words alone cannot do. The library comprises a number of series of pictorial reference volumes covering major fields of knowledge. New volumes are continually being added.

Album of Science
Album of American History
Album of World Civilization
Album of American Culture

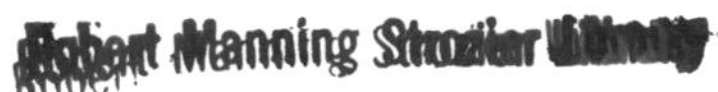

Library of Congress Cataloging in Publication Data

Williams, Leslie Pearce, 1927–
The nineteenth century.

(Album of science)
Bibliography: p.
Includes index.
1. Science—History—Pictorial works. I. Title.
II. Series.
Q125.6.W54 509′.034 77–3907
ISBN 0–684–15047–6

This book published simultaneously in the United States of America and in Canada—

1 3 5 7 9 11 13 15 17 19 M/C 20 18 16 14 12 10 8 6 4 2

Printed in the United States of America

To Sylvia

1. (Frontispiece) *Science is measurement.* "When you can measure what you are speaking about and express it in numbers you know something about it; but when you cannot measure it, when you cannot express it in numbers, your knowledge is of a meagre and unsatisfactory kind; it may be the beginning of knowledge, but you have scarcely, in your thoughts, advanced to the stage of *science*, whatever the matter may be." This statement by Lord Kelvin, made in 1883, expressed the prevalent nineteenth-century attitude toward science. The painting illustrates the puzzlement of the natural scientist, whose subject offers little opportunity for significant measurement. He was faced with the question of whether all of natural history was to be excluded from "science"—or whether the close observer of nature was as near to scientific "truth" as the mathematical analyst.

Contents

Acknowledgments

It is a pleasure to thank all those who helped in making this volume possible. The most important person is the man who conceived the series of which this volume is a part—Professor I. Bernard Cohen of Harvard University. His constant help and useful suggestions have greatly aided me in putting together a kind of book that I had never anticipated writing. It is literally true that without Professor Cohen, this work would never have been composed.

At Cornell, a large number of people have given me the benefit of their expertise and knowledge. Ellen Wells, librarian of the History of Science Collection, and her helpers gave unstintingly of their time to insure that I got just the right illustrations. The reference librarians ordered works not available at Cornell on interlibrary loan and showed me how to find pictures that were particularly difficult to get at. The circulation librarians stretched many rules so that rare books could be taken out to be photographed without calling in Pinkerton agents to protect them. All pictures that carry no special attribution were photographed by the people at the photographic Services at Cornell. My gratitude to them for their constant helpfulness is unbounded. My student aide, Bennie DiNardo, has done invaluable work by checking all the picture attributions for me. A simple thank you seems hardly adequate.

The people at Charles Scribner's Sons were most helpful. Charles Scribner himself gave valuable advice and constant encouragement. Kenneth Heuer was able to suggest many ways to make life easier and the three editors with whom I worked—Jane Anneken, Jack Ennis and Norman Kotker—could not have been nicer or more willing to aid me. I thank them all.

The composition of a book always puts a strain on the family of an author. But few authors' families could have suffered as mine did during the years in which this book took shape. My wife and children were used to manuscripts lying around all over, but hundreds of xeroxes and hundreds of captions were more than they had ever faced. My thankfulness to them for not trampling on them, or losing them, or otherwise scattering them to the winds can hardly be expressed. I can only hope that the finished product will justify their patience with me.

Needless to say, any errors in this work are mine alone.

L.P.W.

Foreword

"The Wonderful Century" was the name given to the nineteenth century by Alfred Russel Wallace. One of the most remarkable features of that century was the emergence of modern science as a powerful national resource and world force, almost as we know it today. Large-scale science, in which research teams worked together with large financial support, first appeared toward the end of the century, in Germany, where the successful production of synthetic dyes rapidly struck tremendous blows at economies that had until then depended heavily on the agricultural production of plants that were sources of dyestuffs, such as madder and indigo. The germ theory of disease revealed at last the cause of many contagious and infectious illnesses and soon afterwards research provided the means for prevention and cure. Industry became increasingly based on science, especially organic and inorganic chemistry, electromagnetism, and animal and plant biology. A science-based technology introduced new forms of human communications—notably the telegraph and telephone—that constituted a revolution in the affairs of men and nations. And, at the same time, the veil of man's ignorance was torn aside to reveal the composition of stars; the origins and descent of man; the evolution of plants and animals; the constitution of matter; the age and formation of the earth; the interconvertibility of natural forces and forms of energy; the laws of electricity and magnetism; and the seemingly all-embracing electromagnetic theory of light.

Although there were many new inventions made during the sixteenth, seventeenth, and eighteenth centuries, it was not until well into the nineteenth century that technology introduced truly significant changes in men's daily lives. It is sometimes difficult for us today to recognize how great a change was produced in nineteenth-century conditions of living by apparently simple innovations based on science, by a shift from inventions based on mechanical ingenuity and gadgetry to those deriving from practical applications of the scientific understanding of nature and a knowledge of nature's laws. A vivid example was given by Wallace, one of the discoverers of the principle of natural selection, who said "one of the most vivid recollections" of his childhood was of "seeing the cook make tinder in the evening, by burning old linen rags, and in the morning, with flint and steel, obtaining the spark which, by careful blowing, spread sufficiently to ignite the thin brimstone match from which a candle was lit and fire secured for the day." Often the tinder would accidently become damp or the flint wear out and then, after repeated failures, fire had to be obtained from a neighbor. In light of this background we can understand why Wallace gave so high a place in the list of marvels of his century to "lucifers"—wooden matches tipped with a phosphorus compound that ignited on being struck, the fruit of the chemist's art. At the time of Wallace's youth, he later recalled, "savages in any part of the world" could still "obtain fire as easily as the most civilized of mankind." These so-called kitchen matches may serve as a symbol of the many ways in which nineteenth-century applied science began to alter almost every possible aspect of traditionally daily life.

As Pearce Williams shows in the following pages, it was in the nineteenth century that science developed into a fully recognized profession, with regular posts and the support system of established research laboratories with paid assistants and technicians. Symbolic of this new status was the conscious invention of the term "scientist" by William Whewell in 1841. As professional and specialized scientific societies came into being, there were also national organizations in which were united science teachers, practicing scientists, and amateurs of science. For this was also the great age of the popularization of science, through magazines and the lecture platform, and it was a time when amateur scientists could still make valuable individual contributions to the advancement of knowledge with microscope and telescope, or by collecting and classifying plants, insects, and minerals; recording meteorological conditions; and so on.

For these reasons the portrayal of nineteenth-century science requires more than a record of significant discoveries. It is the aim of this volume, accordingly, to show the reader science in its many aspects. This is the first of a series of five volumes: it will be followed by others on science in antiquity and the Middle Ages; science in the Renaissance and the seventeenth and eighteenth centuries; the physical sciences in the twentieth century; and the life and health sciences in the twentieth century. In each of them, the reader will be given a "tour" of science, with views of the laboratories, observatories, botanical gardens, menageries, and museums in which scientific research was carried out; the instruction halls for the training of scientists; the instruments of science, and the ways in which they were used; scenes of public lectures and demonstrations; and even examples of public ridicule of science. In short, these volumes are not intended to be pictorial histories of scientific advances (although such materials abound in them), as much as albums that will give a visitor from a distant time, today, a panorama of previous science in its physical, cultural, and social environment.

In addition to drawings, paintings, engravings, etchings, mezzotints, and woodcuts, the nineteenth century specifically made available two new documentary methods: the photograph and the mass-produced lithograph for magazine and newspaper reproduction. This volume is therefore based, in part, on early contemporary photographic illustrations of scientific scenes as well as on the actual photographs used in scientific research—primarily in astronomy and the physical sciences—to record and to give evidence in regions beyond the limits of human vision. The sources used to illustrate this volume thus are themselves a record of the change in the technique and craft of making and reproducing images.

I. Bernard Cohen

A Note on Pictures

The plan of the ALBUM OF SCIENCE has been to reproduce only pictures that are contemporary with the period they illustrate, making each picture valuable as a primary source of information. Although the written sources relevant to the history of science have been published widely, pictorial sources have generally been neglected. The ALBUM OF SCIENCE has been designed to correct this imbalance. The value of the pictorial documentation which it provides has been enhanced by the inclusion of a comprehensive fully cross-referenced index. In some cases the contents of the pictures themselves have been indexed where it was felt that this would provide further information that would be useful. For example, picture number 75, which was included to show Louis Pasteur in his laboratory, is indexed in such a way that a reader looking for a picture of a microscope being used or laboratory instruments is referred to it even though neither topic is specifically mentioned in the text accompanying the picture. The picture is indexed or cross-referenced under the following categories:

apparatus and instruments, scientific, used in chemistry
autoclave, used for sterilizing (left background)
chemistry, laboratories, of Pasteur, Louis
laboratories
microscope, use of
Pasteur, Louis
portraits: Pasteur, Louis
pumps, suction, for filtration, moved by crank mechanism

This sort of index, which has proved of value in various other volumes of the Scribner Pictorial Reference Library, can provide a new and useful approach to the study of the history of science.

A Note on Sources

As this album is intended to be both more than and less than a pictorial history of science in the nineteenth century, the reader should be made aware of what guided me in the selection of the illustrations. What I have attempted to do is to illustrate the major developments of nineteenth-century science within the social context of the period. I have, therefore, generally avoided going to technical scientific sources for illustrations except when these were the only ones available to make a particular point. I have tried to draw illustrations from those sources generally available to the broad public of the nineteenth century, hoping in this fashion to put science into its social context. In some cases, this has led to a distortion of the historical course of scientific history; but since the distortion reveals how contemporaries saw things, I have considered this a valuable insight into the period. Science in the nineteenth century was both popular and increasingly popularly supported. It seems important to me that this popular dimension be fully presented.

L.P.W.

Introduction

The nineteenth century was one of the greatest centuries in the history of Western civilization. It witnessed revolutions in literature and the arts as well as in politics; vast new areas of the earth and of the mind were opened up for exploration; new sciences were created and old sciences rigorously reexamined and burnished until they shone with the same lustre as the new. It was in the nineteenth century that Western civilization completed its conquest of the world, providing, for the first time in human history, a common denominator for all human discourse. A fundamental component of that discourse in the twentieth century is science. What had started out on the periphery of Hellas in early antiquity as a peculiar and idiosyncratic quest for reality, now dominates the world of the human intellect. It was in the nineteenth century that science rose to this position of dominance and one purpose of this volume is to illustrate that process.

One of the themes that runs throughout the century and this volume is that of the growth of the ability literally to see more of reality as the century progressed. In 1800 men sailed the seas, touched the shores of vast continents like Africa and Australia, could distinguish features of the moon and other members of the solar system, and, though blurred and fuzzy, could make out the general form of microorganisms. By 1900 giant telescopes penetrated beyond our own galaxy to other island universes that had been mere luminous blobs for the astronomers of the eighteenth century. Men were able to plumb the depths of the seas and discover the strange forms of life that lived at great depths and pressures. The continents were explored, their flora and fauna described,and the pattern of plant and animal distribution became a vital clue to the understanding of the evolution of life itself. The microscopic world was revealed as one rivaling the ordinary world in the richness and variety of its forms of life. What is more difficult to illustrate is the way the mind's eye penetrated ever deeper into the night surrounding the secrets of matter and its combinations. Here the forms could only be intuited and their reality less easily proven. But even, or perhaps we should say especially, here, the power of the scientific imagination would not be denied. The round, billiard-ball atoms and clumpy molecules of classical atomism gave way to delicately arranged fields of force and molecules of such exquisite symmetry that they rivaled Gothic cathedrals in their construction.

The triumph of science in the nineteenth century did not take place in a social vacuum. If society were to support the scientific enterprise, society must get something for its investment. The payoff in the twentieth century comes from the marriage of science and industry, but this was not so obvious in the nineteenth century. It was not until well past midcentury that it was clearly and generally realized that science lay at the basis of industrial power. When it was realized, the sciences benefited. This benefit came in various forms. Education was reformed to include the sciences, the professional scientist emerged as an ordinary professional man rather than an oddity, and facilities for research were expanded The sciences also benefited from the results of industrial advance. For example, in 1800 a useful telescope could be made by a scientific instrument maker using handicraft methods. The giant telescopes of the end of the century were beyond the handicraft stage. They were the products of heavy industry and relied as much on the new technological results—ball bearings, precision gearing, proper lubrication—as did a steam turbine or a giant milling machine.

A more homely aspect of the public relations of science was its appeal to the general public. By and large, the advances of theoretical physics were beyond the capabilities of the average man to understand, but the new vistas that were being discovered could be appreciated by anyone. Museums, botanical gardens, planetaria and a whole host of other means of exhibiting the new science were created so that the public could appreciate the new discoveries. What better way to recruit new scientists than to excite the young with the wonders of nature? And what better way to justify the expenditure of public funds, than to stimulate the people to enjoy the results. In a way that no other intellectual influence had ever done, science permeated the culture of the nineteenth century. Although creative scientific thought was (and is) immutably elitist, science, as a social phenomenon, was democratic. By the end of the century, if something were not scientific, it was somehow not really respectable. It was in the nineteenth century that science was first used, for example, to sell goods; and the "commercial" with the omnipresent scientist made its appearance.

Finally, the growth of science in the nineteenth century created a new subculture of scientists. Science thrives on communication and criticism and requires social institutons for its health. The number of scientists multiplied rapidly in the nineteenth century and soon outgrew the older scientific institutions. The Royal Society of London and the Paris Academy of Sciences survived, even throve, but they could not meet the new conditions. Larger, less formal bodies were needed both to stimulate the exchange of ideas and to make manifest the growing numbers (and power) of scientists. National boundaries were increasingly irrelevant to science, and scientific projects often required international sponsors. The result was the creation of new bodies—the national associations for the advancement of science—and the growth of informal international symposia and meetings. Such bodies and occasions offered the opportunity for vigorous and stimulating intellectual exchange, as well as the prospect for social relaxation and a vacation from everyday tasks.

There is an obvious way to illustrate the progress of science in the nineteenth century. If we begin with the instruments and what could be seen and handled scientifically at the beginning of the century, it is fairly easy to trace the course of science by focusing on improvements in what could be seen and measured. Thus we shall begin with the galactic heavens and simply move downward toward the earth—through the solar system, the earth's atmosphere, and to the earth itself. Then we shall look at the biosphere—from man, through animals and plants, to microorganisms. Finally, there is the realm of the invisible—that of molecules and atoms and imponderable "fluids" such as heat and light and electricity and magnetism. Although their constituents are invisible, these substances or forces manifest themselves either to the imagination or to the senses in ways that permit of their scientific treatment. If we proceed in this way, we will be able to catch a glimpse of the path that science took during the last century.

An album of science is just that: a series of snapshots of the scientific family of the nineteenth century with a focus on the things that interested that family. It is intended to be enjoyed as well as to inform and when the reader has closed the album, he will have some sense of what the life of science and of the scientist was like in this golden age.

I

The Environment of Science

Philosophy and the World of Science

At the beginning of the nineteenth century, everyone could agree that science was the study of nature but that would have been about as far as agreement would have reached. Did this "study" mean the observation of natural facts as, for example, the collection and tabulation of meteorological data? Or did science mean the interpretation of such natural facts according to theological or metaphysical preconceptions? Or, was the whole thrust of science intended precisely to free men from theology and metaphysics and confine scientific reality to quantitative and mathematical truths? Various answers were given to these questions and these answers provided the bases for different philosophies of science that were held by different people throughout the nineteenth century. Moreover, the philosophies of science that attempted to define what the proper sphere of science was had important effects on the activities of scientists themselves. People tended to look for the facts that their own passionately held philosophy suggested were important to find and to ignore those areas that were ruled out by that philosophy. In the eighteenth century, the French physicist Charles Coulomb had carefully measured the forces of attraction and repulsion exhibited by static electricity and magnetism. These forces were found to be different and Coulomb's followers attributed them to different kinds of material particles. By accepting as fact the conjecture that forces must emanate from basic particles—a philosophical assumption justified by an appeal to the Newtonian system of physics—these scientists ruled out the possibility of the transformation of electrical and magnetic forces into one another. They did not look for the magnetic effect of an electrical current after the invention of the voltaic pile in 1800 because their philosophy of science had eliminated that possibility. It is no coincidence that this effect was discovered in 1820 by Hans Christian Oersted, an ardent disciple of a rival philosophy, that of Immanuel Kant, in which forces, not particles, were considered to be the ultimate reality. One of the corollaries of this philosophy was that the forces of nature were transformable into one another.

A basic part of the intellectual environment of science in the nineteenth century, then, was the philosophical dimension and this tended to differ along national lines. The English, French, and Germans looked at nature in different ways because they came to nature with different philosophical preconceptions. The leaders in science at the beginning of the nineteenth century, the French, were the strongest proponents of the doctrine that nature was essentially and fundamentally mathematical. The great French mathematicians of the eighteenth century and early nineteenth century—d'Alembert, Lagrange, and Laplace, to mention only the more famous—had provided French science with a mathematical instrument of great subtlety and power. They expected their followers to use it, almost exclusively, for the investigation and expression of the laws of nature. If a physical hypothesis could not be put in mathematical form, it almost necessarily followed that the hypothesis was not worth taking seriously.

The English held to a somewhat less rigorous view. To be sure, Isaac Newton had established the whole mathematical tradition with the publication of his *Philosophiae Naturalis Principia Mathematica* (Mathematical Principles of Natural Philosophy). In English scientific thought in the nineteenth century a strong tradition of empiricism persisted. Its patron saint Sir Francis Bacon was believed to have lost his life in the cause of experimental science as the result of trying to freeze a chicken with snow. Contact with nature as well as abstraction from it was certainly an equally valid way of knowing nature. Furthermore, most of the "scientists"* in England at the beginning of the century were amateurs whose knowledge of mathematics was sketchy or nonexistent. They wished to advance science by keeping voluminous records of barometric pressure, temperature and rainfall, or by reporting interesting and striking natural phenomena. Their subject was natural history or, if carried to a higher plane, natural philosophy, and close observation, common sense and a dread of premature generalization were the qualities necessary for success. The very idea of a formal philosophy of science struck people as slightly absurd. Nature was there to be observed and discovered. Why make a big metaphysical fuss about it?

A different philosophical tradition was in vogue in the states of Germany at the beginning of the nineteenth century. There Kant, in particular, had awakened the Germans to the power of metaphysics.

The *Critique of Pure Reason* challenged the very foundations of the Age of Reason in both England and France. Reason, Kant attempted to prove in particularly dense Germanic phrases, is not unlimited and omnicompetent. It had its limits and to ignore these was to push reason beyond its ability to comprehend the world. In the process of describing the limits of Reason, Kant also leveled a devastating attack at the very foundations of contemporary science. He tried to show that things such as space, time, matter, and causality were not simple givens upon which classical physics could rest securely, but highly complex aspects of the mind and the way in which the mind apprehends nature. Common sense, here, could lead only to scientific disaster and a whole generation of Germans eagerly embraced the implications of Kantian philosophy to build a new vision of physical reality. These nature philosophers saw nature as an organic whole that was destroyed by the contemporary practice of analysis. They challenged the atomic doctrine and insisted instead on a kind of primitive field theory that emphasized the interconnectedness of natural phenomena and processes. In this approach they resembled the Romantic artists of the day for, like the Romantics, they were driven by the desire to encompass the cosmos in their theories.

These disparate philosophies of science led their adherents quite literally to see nature in different ways. They were, therefore, important aspects of the environment of science at the beginning of the nineteenth century. As the century advanced, these differences became less obvious but they did not disappear. At the end of the century, the great revolution in physics that began with the Special Theory of Relativity was caused, at least in part, by the fact that Albert Einstein had been influenced by a number of unorthodox ideas in the philosophy of science, including those of the Austrian physicist Ernst Mach.

The general philosophical tendencies of the French, English, and Germans were not always purely expressed. A metaphysical German could and did use mathematics to express his metaphysics; empirical Englishmen could worry over the place of God in a determinist universe; and even positivist French scientists could occasionally build metaphysical models to enable them to write differential equations. Yet, by and large, there were philosophical differences that did follow national lines and it is important to keep them in mind. These differences did affect the way science was done in the nineteenth century.

* The word was not coined until the middle of the nineteenth century when science had become a recognized profession.

2. *The theory and practice of the pendulum.* In 1851, Léon Foucault illustrated the rotation of the earth by means of a large pendulum, thereby focusing attention on the theory of the pendulum. This tableau, which appeared soon afterward and is based on Foucault's work, reveals some typically French attitudes toward science and nature. The "Philosopher" pays no attention to the beauty of the scene, seeing neither the vegetation nor the pretty girl in the swing. What he observes is abstracted to a geometrical problem, free from passion and life. The girl's response to his mathematical analysis is "True, but who cares?" To the philosopher, science *is* measurement; to the young girl, it is irrelevant because it omits life. 2

3

PUNCH'S PENCILLINGS.—No. XXXVI.

SOCIAL MISERIES.—No. 7.

METAPHYSICS.

"What you say about Corporeity is all very well, but it presupposes the idea of—(hic)—absolute spirituality and transcendental—(hic)—perfection—(hic); b'sides, it's incompatible—(hic)—with the def'nition of space."—(Hic.)

"Well!—don't—go—old fellow. Have some m-m-m-ore—g-g-g-r-o-g-grog."—(Hic.)

3. *Kant, metaphysics and the English.* The difficult philosophy of Immanuel Kant had considerable influence on science in nineteenth-century Germany. In England, however, it ran head-on into the common-sense school that dismissed metaphysics as so much empty verbiage. One British reaction was to ridicule it, as the satirical weekly *Punch* does here.

4. *A French cartoonist defines the fundamental concepts of physics.* The French cartoonist Cham uses scientific terms to parody contemporary human foibles. A "solid body" (a) earns its owner 100,000 francs from a desperate anatomist; smugglers of spirits (liquid body, b) are caught at the customs bar; gendarmes are defied by poachers (refractory body, c); fat men cannot fit into small carriages (impenetrability, d); hussars consume enormous quantities of alcohol (porosity, e); women faint with the "vapors" (passage of a body into a state of vapor, f); a dragoon forces back the crowd at the funeral of a distinguished person, thus punning on the words *pompe foulant*, which mean both "force pump" and "crowded funeral" (g); soldiers clear out a theater presenting a play that has been prohibited by the authorities (method of creating a vacuum, h).

4a-h (left to right, top to bottom)

(Corps solide.)

(Corps liquide.)

(Corps réfractaire.)

(Impénétrabilité des corps.)

(Porosité.)

(Passage d'un corps à l'état de vapeur.)

(Pompe foulante.)

(Manière de faire le vide.)

5

6

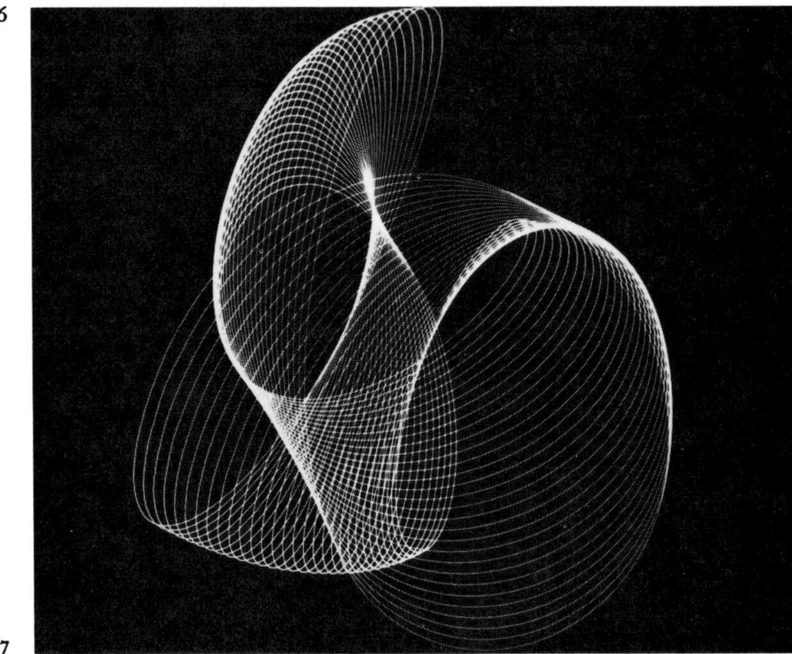

7

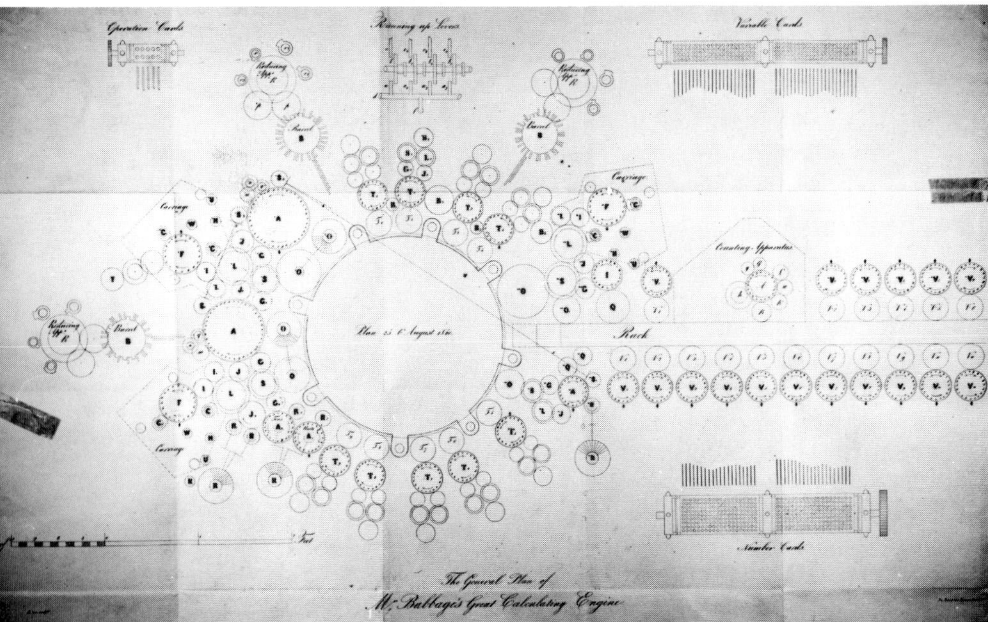

5. *The photoratiograph.* The photoratiograph was a machine invented to illustrate the beauty of complex mathematical curves. It is the ancestor of a modern toy, the Spirograph, that permits children to make intricate mathematical designs. The machine and the curves it produced were graphic illustrations of the intimate relations between mechanics and mathematics.

6. *A harmonogram produced by the photoratiograph.* The beauty of mathematical figures such as these provided evidence for those who argued that mathematics did not exile beauty from the world. No conflict need exist between the admirers of natural forms and the partisans of mathematics and mathematical methods.

7. *Charles Babbage's calculating engine.* Charles Babbage (1792–1871) devoted his life to the creation and perfection of his "calculating engine," the first really sophisticated computer. This layout drawing gives some indication of its complexity. The "calculating engine" carried out the various operations demanded of it by strictly mechanical means, through the use of differential gears and gear trains that "calculated" by turning specified amounts in each operation. Results could be "stored" in a mechanical memory gearbox. The results of succeeding operations could be recalled by connecting the memory box to the operational train at whatever point desired.

8

8. *Model of Babbage's engine.* The engine was built before machine tools permitted the close and standard machining of metal parts. Each part, therefore, had to be finished by hand. Since the accuracy of the engine depended upon the accuracy of the machining, almost impossible demands were made by Babbage on his workmen. Not surprisingly, the engine was never completed.

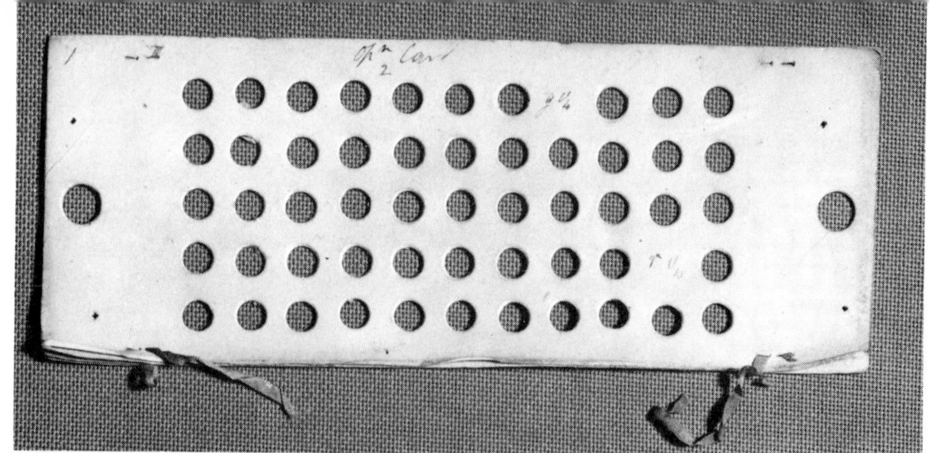

9

9. *Punch card for performing an operation on Babbage's engine.* Punch cards for directing machines were not invented by Babbage, but he was the first to use them as an aid to calculation. The holes in the card permitted spring-loaded rods to pass through them. These rods, in turn, activated certain gears and wheels that performed the required operation. What the machine did depended upon the pattern of holes, much as is the case with a modern IBM punch card that electronically activates a computer.

10. *Thought contemplates infinity.* The role of philosophy in science was particularly strong in Germany, and it led Germans to seek a universal meaning in the cosmos. Here the problem is put succinctly, with just enough *Weltschmerz* to appeal to the German soul. Man, naked and alone (and sexually bowdlerized) leans against a sphinx that presumably could reveal the secrets of the universe if it wanted to. Gazing into infinite space, man seems lost. Science clearly is more than common, everyday experience; it involves the question of man's place and purpose in the universe. That is the "riddle of the universe," whose solution the sphinx may never disclose. The mood of the picture runs distinctly counter to the more obvious equation of science and progress that was commonplace.

10

The Scientific Society

The social world of science changed dramatically in the nineteenth century. The great scientists of the seventeenth and the eighteenth centuries who had created and consolidated the scientific revolution had tended to work in isolation and with the most primitive apparatus. One need only call to mind Descartes meditating in his winter quarters as a soldier or Newton at his farm in Woolsthorpe.

They reflected their age in what they believed to be the proper objectives of natural philosophy. What both Descartes and Newton sought was a fundamental insight into the nature of reality that would enable them to unlock the mysteries of nature. In both cases, the result was a wide-ranging system that pretended to contain that key by which nature's treasure chest of truth could be opened. It is hard to imagine how either Descartes or Newton could have used collaborators. In spite of fairly continuous correspondence with other savants, each remained a solitary genius looking for a new world view; and world views, as Einstein in the twentieth century was to illustrate again, are intensely personal and created only by intensely personal visions of physical reality.

By the nineteenth century, the scientific scene had changed greatly. The giants of the seventeenth century had created a new world view and the *Principia* had provided the necessary mathematical and philosophical apparatus to explain it in detail. Indeed, that was what science was all about at the beginning of the nineteenth century; the large picture had been sketched in and what was necessary now was to fill in the details. Details, as such, should not be dismissed as negligible. They were not only the ultimate proof of the correctness of the system as a whole, but it was detail that offered the hope of ameliorating the condition of mankind. Lavoisier could provide a general theory of the nature of acids but it was Berthollet's discovery of the bleaching power of hypochloric acid that revolutionized the bleaching of clothes in France. Great doctors from Hippocrates and Galen in antiquity to Boerhaave and Morgagni in the eighteenth century could speculate on the causes of disease, but it was William Jenner's attention to detail in his observation of the immunological potential of cowpox that virtually eliminated smallpox as a periodic scourge.

When it comes to details, the social organization of science becomes a prominent factor in the advance of science. Not everyone can create a world view, but almost everyone can help to observe, measure, and define the details that world views imply or that are necessary for the validation and application of a general theory. But, to find psychological and professional satisfaction in the delineation of detail, it is vitally important for those involved in this task to communicate easily and frequently with one another. The excitement of science of this kind comes from speaking with others who, like oneself, are discovering and describing pieces of the same jigsaw puzzle and seeing where the pieces fit in. The older scientific organizations were incapable of fulfilling this need. By the beginning of the nineteenth century, such bodies as the Royal Society of London or the Academy of Sciences of Paris had become too narrowly based to satisfy the social requirements of the new scientists. The Royal Society had, during the eighteenth century, become a social club in which birth and station were more important than scientific achievement. Many of the fellows had only the slightest acquaintance with science and used their F.R.S. as social cachet. The Academy of Sciences, on the other hand, had been turned into a rigorously professional society with such high standards for admission that only the top scientists of France could belong. Lesser lights had to be content with their position on the periphery of the scientific world. There was no place, in either of these bodies, for the scientist of the middling sort who, nevertheless, was becoming increasingly important as science and its applications assumed greater importance in the overall scheme of nineteenth-century things.

It was to meet the needs of the growing number of scientific practitioners that the national associations for the advancement of science, or their equivalents, were founded. The initiative, not surprisingly, came from Germany where there was no national unity or national organization which could bring together scientists from the various German states. Princely courts, of course, had always patronized science and the arts, but that circle, even when institutionalized into such a grandiose society as the Berlin Academy of Sciences, was too small to include all German natural philosophers. The German scientists first assembled as a group in 1822 in Leipzig and the Deutsches Naturforschers Versammlung (German Scientific Association) became a popular and an annual convention during which German scientists could exchange both social and scientific intelligence. In 1831 the British Association for the Advancement of Science was founded and rapidly assumed an important place in the scientific life of Great Britain. The American Association for the Advancement of Science was founded in 1842.

The French case is interesting because it is anomalous. The French Association for the Advancement of Science did not hold its first meeting until 1872, a generation after other associations. The need for some kind of scientific colloquy that would permit the informal exchange of scientific views was as great in France as elsewhere for there were scientific congresses held in the period before the French Association was formally created. These congresses were on an *ad hoc* basis, however, and never attracted the kind of public and official attention that the other associations did. This was because the Academy of Sciences in Paris held a near monopoly on science and bled off the natural leaders of rival organizations. Paris was the lodestar for all French scientists. They came to the capital and, if they were good enough, they entered the scientific establishment of the Academy where communication was easy because the Academy met frequently to discuss scientific matters. Anyone not at Paris or not in the Academy was simply not of stature sufficient to make communication desirable. Such an attitude, which reflected the older situation of science, was inimical to the new. More than one observer of the contemporary nineteenth-century scientific scene attributed the rise of German science and the relative decline of French science to the fact that while France had great individual scientists there were few of the rank and file who could fill in the all-important details. German science, on the other hand, made special provisions for such private soldiers in its scientific army. The failure of French science to organize on a broader basis merely reflected the failure of the French to realize that the environment of science had changed. It was a failure for which the French were to pay by loss of economic and military power by the end of the century.

The new associations of science were more than organizations for the easing of communications between scientists. They also reflected the new importance of science in the nineteenth century. Science had long been viewed as a means for the amelioration of man's lot, but it was not until the nineteenth century that that dream had started to become a reality. The industrial revolution radically altered the exploitation of natural resources and science appeared to offer a rational approach to maximizing this exploitation. National welfare, it was felt, depended upon the encouragement of science and of scientists but most national governments were leery of supporting something they did not understand. Hence the national associations of science took it upon themselves to preach the gospel of science to the public in the hope of gaining public support of science. They were eminently successful, at least, in interesting the public. Science became the fashion and the meetings of the national associations tended to be jammed. Thus a forum of considerable importance was created. It was no accident that Thomas Henry Huxley chose to debate the Darwinian position with Bishop Wilberforce before the British Association when it met at Oxford. Huxley knew that he would reach a wide audience and that his remarks would be amply and fully covered in the press.

The new associations also marked the growth of science as a profession. The older societies had recognized and inducted genius into their midst. The new ones were content with competence. They made mere interest in and some facility with science the only requirement for membership. It was in the nineteenth century that science became a profession in which one could earn one's livelihood. The need for scientists was reflected in the fact that women gradually began to enter scientific fields and were accepted. Because science was a new profession, unlike medicine and law where prejudices died hard, it became attractive to women whose talents often entitled them to a place of honor.

11. *The eighteenth-century meeting place of the Royal Society.* The Royal Society was the oldest extant scientific society in the nineteenth century. From 1710 until 1782, its members had met in a house in Crane Court designed by Sir Christopher Wren. The meeting hall looked more like a salon than a place of serious scientific business. The portraits on the walls reflected the extrascientific interests of the Fellows. Mary, Queen of Scots and the Earl of Bedford had little to do with the scientific questions discussed when Isaac Newton and his successors presided over the Society. The room was small, and it obviously could not serve as a vital center for English science.

12. *The Royal Society's meeting room in Burlington House.* In 1857, the Royal Society moved to Burlington House, Piccadilly, where it remained for the rest of the century. The atmosphere of the room reflected the fact that, by then, the Royal Society had become a professional group of scientists. The walls were hung with the portraits of scientific notables, among whom may be distinguished Isaac Newton, Humphry Davy and Christopher Wren. The room was designed for serious business. The Fellows occupied the banks of the lecture room from which they could see and hear about the scientific novelties brought before them by their President.

13

13. *A meeting of German scientists.* It was in Germany that the social base of science was broadened to include all those interested in the advancement of science. Now those investigators who had not been elected to one of the prestigious academies or societies could meet with their more illustrious colleagues and discuss the burning scientific issues of the day. The tone of such meetings was relaxed and informal; indeed, some observers, like the creator of this cartoon, felt that conviviality was the major reason for the meetings.

14. *The British Association at Bath.* The professionalization of science brought scientists a responsibility to explain to the public what they were doing and how their work affected the ordinary citizen. The British Association for the Advancement of Science (founded 1831) attempted to fulfill both these obligations by holding annual meetings in the cities of Britain. When the BAAS met at Bath in 1846, Sir Charles Lyell, the most famous geologist of the day, devoted his presidential address to the geology of Bath and its environs. His reputation and his subject attracted a crowd that filled the main theater in Bath to capacity. The citizens of that city came away enlightened and impressed with the new estate of scientists in the realm.

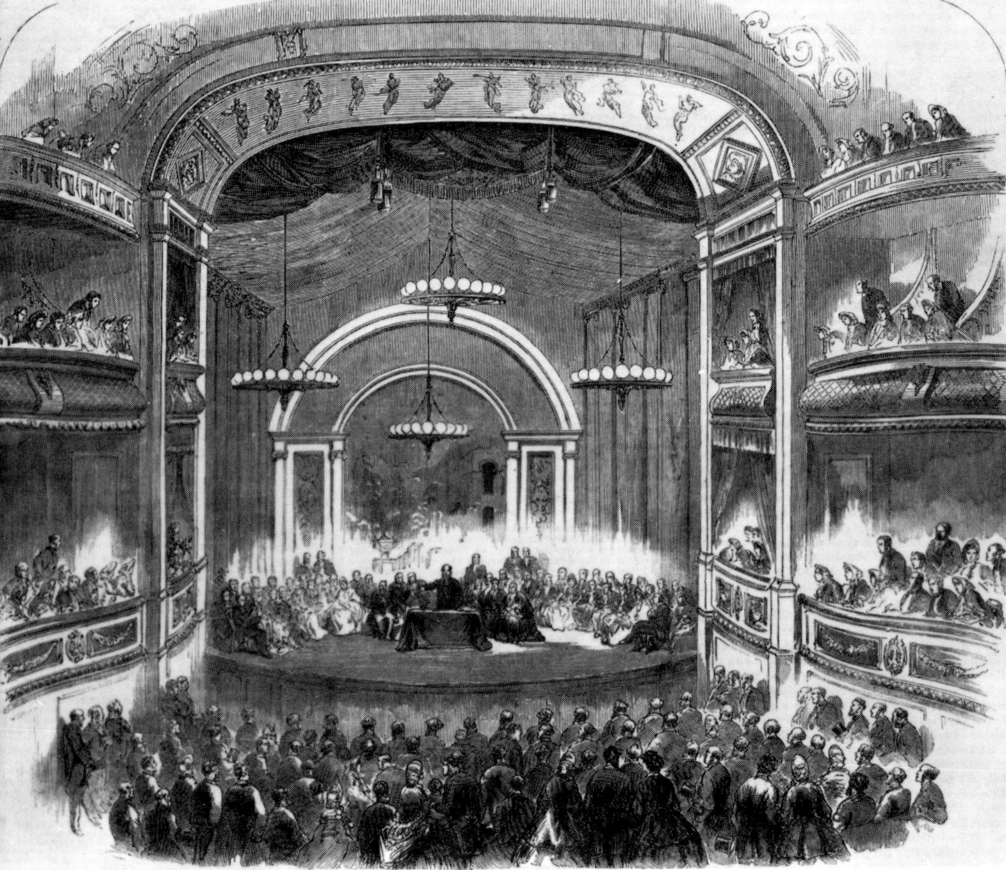

14

15

15. *Science and pleasure: The British Assocition meeting at Brighton.* In this engraving, the artist has captured many of the moods and attitudes of a meeting of the BAAS. British science was obviously in good hands when it was expounded by so solid a gentleman as the one shown here addressing his wife and daughter, and by so earnest a man as the one on the left, who is explaining some obscure point to a monocled dandy. The main figure is a splendid example of what Victorian womanhood was supposed to be. She had dutifully come to the meeting and sketched some point of interest. The outing is certainly edifying, which undoubtedly gave it added value in the minds of those who looked at the picture.

16. *Meeting of the BAAS at Newcastle-on-Tyne.* Attendance at the inaugural address of the annual meeting was one of the social high points of the provincial town in which the session was held. This engraving of the 1863 meeting well illustrates the vast numbers who were drawn to the affair. Newspapers in Newcastle proudly reported that the vast hall was filled to capacity—4,000 persons.

17. *The French scientific scene: the 1847 Congress at Tours.* Meetings such as this were sporadic in France until the foundation of the French Association for the Advancement of Science in 1872. They were justified, according to the President of the Congress, as scientific jousts in which ideas could confront one another. Only some 700 persons attended; and the correspondent of *L'Illustration,* in which this view of the Congress appeared, concluded that it was a great waste of time.

16

17

18

DAUGHTERS OF SCIENCE.

British Association (to her sister of another land). "WE, AT LEAST, CAN MEET WITH NEIGHBOURLY CORDIALITY."

18. *Scientific hands across the sea.* The congresses and meetings of the various national associations for the advancement of science had always stressed the international aspect of science, which rose above national politics or quarrels. In 1899, relations between France and England were strained by the Dreyfus affair, which the English considered a tragic miscarriage of justice, and by imperial rivalries overseas. Nevertheless, the French and the British AAS exchanged visits in that year, with an official French delegation arriving at Dover in the middle of September. The meetings were cordial and science was, in fact, able to lift its followers above the passions of the day.

19

19. *The two girl graduates in natural science at Oxford.* The new professionalism of science attracted women of talent who found other, older professions closed to them. Such women were considered curiosities in the Victorian world and were greeted with some amazement by the popular press. The Misses Pollard (left) and Kirkaldy (right) were able to enter science because of their formal education, both having received their degrees from Oxford. They were the first women to work in the anatomical department of the University Museum, a fact reported with slight shock in the press.

20. *A German woman in science.* Fraulein Marie von Chauvin could not use the German universities as a route to a career in science. She was largely self-taught, having had the good fortune to count some eminent zoologists among her friends. Her work on embryological development in salamanders was recognized as first-rate by her male colleagues.

21. *Mrs. Ayrton lecturing to the Society of Electrical Engineers.* The original caption of this picture read: "The unusual spectacle of a lady addressing a scientific body on a highly abstruse subject and elucidating a much-vexed problem was presented at a meeting last week of the Society of Electrical Engineers." That the woman, a Mrs. Ayrton, could speak intelligently on the subject of the hissing of the electric arc in large arc lamps seemed almost to upset the rational basis of nature and was, therefore, deemed to be eminently newsworthy. There could be no doubt of her command of her subject, supported as she was by graphs and an assistant with two pointers to help illustrate her text.

20

21

The Observation of Nature

The great Newtonian synthesis, achieved by Newton and his followers in the eighteenth century, had provided a solid and firm framework of theoretical physics that claimed to account for most of the basic forces at work in the world. The most dramatic success of Newtonian physics, of course, was its solution of the problem of the motion of the planets in the solar system. The hypothesis or principle of universal attraction, could be formulated exactly and mathematically as $F = \frac{km_1m_2}{r^2}$, where F is the force of gravity, k is the gravitational constant, m_1 and m_2 are the two masses gravitating towards one another, and r is the distance between their centers of gravity. This principle appeared to account for the motions of the heavenly bodies, particularly the planets, although it was recognized that perturbations caused by the actions of planets upon one another would throw the calculations off. Just how far off could be determined only by accurate observations. There had been some advances made in telescopic accuracy in the eighteenth century, but it was in the nineteenth that real precision was achieved.

The Greenwich Observatory and the Paris Obsefvatory had been built when the scientific revolution was in full swing. They had been quite advanced for their time, but they were, basically, primitive observation posts with relatively primitive instruments. Neither was placed particularly well with regard to astronomical observation for both were either in or very near a large city. When London and Paris were illuminated, first with gaslight and later with electricity, and also became subject to urban smog, the two observatories became obsolete. Nevertheless, they both did yeoman service in the early years of the century. Small telescopes could be housed in them and observations of stellar and planetary transits could be made and recorded with fair accuracy.

Observatories proliferated in the nineteenth century for many reasons. The most obvious is the fact that any given observatory can observe only certain portions of the sky. The southern hemisphere, for example, remained unmapped until observatories were set up in South Africa and Australia. Observatories were also matters of national and local prestige. To have an observatory was to be part of the mainstream of Western culture and those peoples that doubted their modernity eagerly sought the external signs of it. Russia and the United States were both on the periphery of the Western world and both made serious efforts to prove that their peoples were not barbarians. The great Pulkovo Observatory was founded in 1839 near Saint Petersburg at a time of intense political repression in Russia. Harvard College Observatory was founded in the same year. The one was a state institution whose renown could be expected to buttress an otherwise obscurantist regime. The other was an appendage to the oldest American educational institution of higher learning and represented the high-water mark of American science. It served as an inspiration to newer and rawer communities that wished to preserve or attain a basic level of scientific culture. Cincinnati founded an observatory that could compete with Harvard and that was an object of great civic pride. Private observatories, like the Warner Observatory in Rochester, New York, and a number of similar places in England, were pointed to by neighbors as evidence of scientific cosmopolitanism in the provinces. Presence of an observatory was concrete evidence of modernity.

The work of most of these smaller and provincial observatories was pedestrian at best, and negligible at worst. Yet there were times when they contributed significantly to the body of facts concerning the heavens. The appearance of comets and eclipses called forth masses of data. For the better observation of the latter, it was necessary to send out expeditions if the path of the eclipse did not coincide with an established observatory. So, temporary observatories were created for specific astronomical events. The data were recorded and the observatory then abandoned. The sheer number of facts that were recorded by all the observers in the nineteenth century was, if one may be pardoned the use of the word, astronomical. Positions and motions were recorded with infinite care and increasing precision. As we shall see, such observations could lead to important new discoveries that had escaped less systematic exploration of the heavens.

The heavens were not the only object that could be observed and measured. The weather had always been a good topic for conversation and even for science before the nineteenth century. Country gentlemen in England and on the Continent had kept detailed records of daily temperatures and barometric pressure in the eighteenth century in the hope of discovering some meteorological pattern that would permit accurate prediction. The hope had proven vain but that did not discourage those who felt that even the weather must come under the rule of natural law. Since the weather seemed to be made high up in the atmosphere, the thing to do was to get up there and keep records. Thus was born the meteorological observatory, placed on mountain peaks and subject to extremes of wind and temperature. The proliferation of such meteorological observation posts did not provide a scientific basis for meteorology but it did permit the early detection and reporting of weather conditions that could be inimical to ocean commerce or agriculture. With the invention of the telegraph, such meteorological bad news could be reported to those involved and some precautionary measures might then be taken.

Closely allied to the meteorological station was the magnetic observatory which kept tabs on and measurements of the fluctuations of the earth's magnetic field. There was an obvious utilitarian aspect in that the magnetic compass was the most important navigational instrument and that it reacted to the terrestrial magnetic field. But there was also the simple desire to observe an intrinsically interesting aspect of the earth. Since a single magnetic observatory would be relatively useless, there was a concerted attempt to set up a network of such stations from whose data the activity of the earth's magnetic field as a whole could be deduced. The initiative came from Germany where C. F. Gauss and W. Weber created the first magnetic observatory. It spread rapidly throughout the Western world until, by the end of the century, the terrestrial magnetic lines of force could be mapped with some accuracy.

The various observatories were set up to measure permanent aspects of nature. The desire to observe and measure, however, extended to all kinds of phenomena. What was the speed of a cannon ball? Apparatus could be (and was) devised to measure it. What was the temperature and air pressure in the upper atmosphere? Balloons could ascend with scientists and instruments to measure both. What was the speed of light? What forms of life lived at the depths of the sea? What kinds of plants lived in the Andes? What forms of mankind inhabited central Africa? All these were questions of fact that the scientists of the nineteenth century thought well worth asking, in spite, sometimes, of the incredible difficulties involved in finding the answers. By the end of the century, most of the visible environment had been subjected to observation and, where possible, measurement.

22. *The meridian telescope and mural circle in the Paris Observatory.* The telescope is in the foreground, the mural circle behind it. Both instruments date from the early nineteenth century and were used to determine stellar positions. The meridian telescope is fixed in the plane of the meridian (the plane that passes through the polestar perpendicular to the terrestrial equator at the point of observation). It is used to observe the precise time of a star's passage through the meridian plane. This measures what astronomers call the right ascension of the star. The mural circle measures the altitude of the star (or declination, its distance from the equator). Both instruments are refinements of those used since the sixteenth century.

23

23. *The transit instrument at the Paris Observatory.* The transit telescope is a later nineteenth-century modification of the meridian telescope and reflects the new metalworking technology. It performed the functions of both the meridian telescope and the mural circle, permitting the astronomer to determine right ascension and declination from a single observation. It is equipped with accurately machined gauges so that the readings, though still subject to human error, are free from errors in manufacture or calibration.

24. *Reflecting equatorial telescope by John Browning.* The evolution of the telescope as a precision instrument is almost complete in this example, made by John Browning in the 1870s. The equatorial mount, driven by finely machined clockwork, permits the telescope to be "fixed" on a star or planet and to turn automatically with it. The astronomer may observe over time without making constant adjustments, or—more imporant—a photographic plate may be exposed while the telescope remains focused on the celestial object. This instrument was available at a cost low enough that the amateur astronomer could afford it.

24

25. *Greenwich Observatory in the nineteenth century.* The Greenwich Observatory was designed and built by Sir Christopher Wren in the seventeenth century. It is the locus of the Greenwich meridian, the 0 point on the meridian scale. The observatory, a pleasant boat ride from the city of London, was a tourist attraction in the nineteenth century, as it is now. On July 28, 1851, when there was a total eclipse of the sun, crowds flocked to Greenwich, where telescopes had been set up to observe the phenomenon.

25

27

28

26. *The Paris Observatory.* The Paris Observatory, like the one at Greenwich, was designed and built in the seventeenth century. Unlike the Greenwich Observatory, it was in a large city and, by the midnineteenth century, was almost useless for accurate astronomical work. It was the first observatory to fall victim to the effects of modern urban lighting and smog.

27. *John Herschel's observatory at "Feldhausen," Cape of Good Hope, South Africa.* Observatories in Europe could be used only to observe stars in the northern hemisphere. That left half of the celestial sphere unmapped. In 1833, John Herschel, son of the famous astronomer Sir William Herschel, set out from England to observe the nebulae of the southern hemisphere. It took him five years to catalog them all. The telescope he used, designed by his father, was a reflector with an 18.25-inch mirror and a focal length of 20 feet. The results of his observations were published in 1847 and added significantly to knowledge of the nature and distribution of nebulae.

28. *Pulkovo Observatory, Saint Petersburg.* National observatories were to the nineteenth century what academies of science had been to the eighteenth—essential institutions of national and cultural prestige. The Pulkovo Observatory, opened in 1839 under the personal supervision of the great Danish-born astronomer F. G. W. Struve, became one of the great observatories of Europe under Struve, his son Otto, and his grandson Hermann. It was destroyed during the Nazi siege of Leningrad in World War II.

29. *Warner Observatory, Rochester, New York.* National prestige, civic pride, and individual enterprise all contributed to the bringing of scientific astronomy to America. In 1883, H. H. Warner of Rochester, New York, built a private observatory on the south side of East Avenue. The site was described in the history of the observatory as "one of the most beautiful and fashionable streets" of Rochester and seemed a most unlikely spot for a scientifically productive observatory. The superb equipment included a 16-inch refractor, then the third largest in the United States. Although the observatory cost a fortune, no significant astronomical work was ever done there.

30. *Cross section of the Royal Observatory, Potsdam.* The professionalization of astronomy in the nineteenth century led to the appearance of professional architects who specialized in observatories. Unlike the Paris and Greenwich observatories, which were delightful town houses equipped with domes or windows for observing, the new observatories were austere structures in which the housing of instruments, not the comfort or convenience of observers, was paramount. Note the massive foundations, which ensured steadiness, and the subordination of all details to the requirements of the dome. This was clearly a no-nonsense structure.

31. *The great equatorial of the Royal Observatory, Greenwich.* The older observatories found themselves falling behind newer ones, which could take immediate advantage of new technological possibilities and instrumental improvements. To survive, they had to expand. The great equatorial was completed in 1860 and incorporated into the Greenwich Observatory, where, because of proximity to London, the "seeing" was increasingly poor. Within twenty years, the observatory was obsolete.

29

30

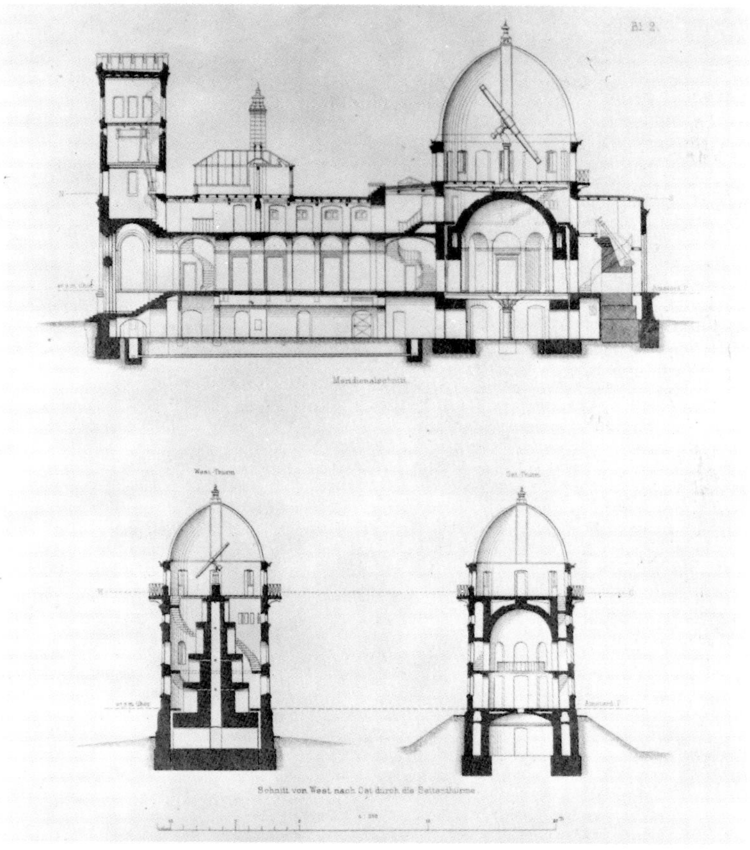

31

32

32. *The astronomers waiting for the solar eclipse of 1871.* Some astronomical events can be observed only in out-of-the-way places where temporary observatories have to be set up. In 1871, there was a total eclipse of the sun that could best be observed in India. A British team under Norman Lockyer traveled there to observe it, and set up instruments in an old fort. Lockyer is seated on the left, using one of the expedition's spectrometers while another member observes the sun through the second spectrometer.

33

33. *The meteorological observatory on the Puy de Dôme.* Observatories displaced castles as the most prominent buildings on the mountains of Europe in the nineteenth century but, surprisingly, they were not astronomical observatories. They were, rather, meteorological observation posts, for it was felt that weather could best be studied at a high altitude. Astronomical observatories were not located on mountains until the end of the century. The meteorological observatory on top of the Puy de Dôme in France beautifully illustrates the trend. Placing a weather station there was appropriate, for it was on the Puy de Dôme that Pascal's prediction that the mercury in a barometer would fall as the climber carrying it ascended was proved correct.

34

34. *The new meteorological observatory at Ben Nevis, Scotland (1883).* It was about 1883 that Mark Twain said that everyone spoke of the weather, but no one did anything about it. The weather had been observed methodically by amateurs for a century before scientific observatories were established to keep really accurate records. The station was placed on the top of Ben Nevis because the Ben was the highest mountain (4,400 feet) in the British Isles. High-altitude weather, it was assumed, determined general weather patterns. The observations made at Ben Nevis could be telegraphed to the rest of the world, so that ocean voyagers and other interested parties could be advised of what weather seemed to lie ahead. Although the weather could not be changed, one at least could be warned of what was coming.

35. *The thermometer at Ben Nevis.* The instruments for weather observations remained simple and traditional. The thermometer was a simple mercury bulb type. The most complicated part was the cage necessary to protect the thermometer from the debris whipped up during the gales that swept across the mountaintop.

36. *The barometer at Ben Nevis.* The cairn of stones provides eloquent testimony to conditions at the weather station. The wind was only one of the elements that made life difficult. Meteorologists frequently complained that the mist was too dense for them to record their instrument readings with pencils; the pencils became so damp they simply smudged the paper.

35

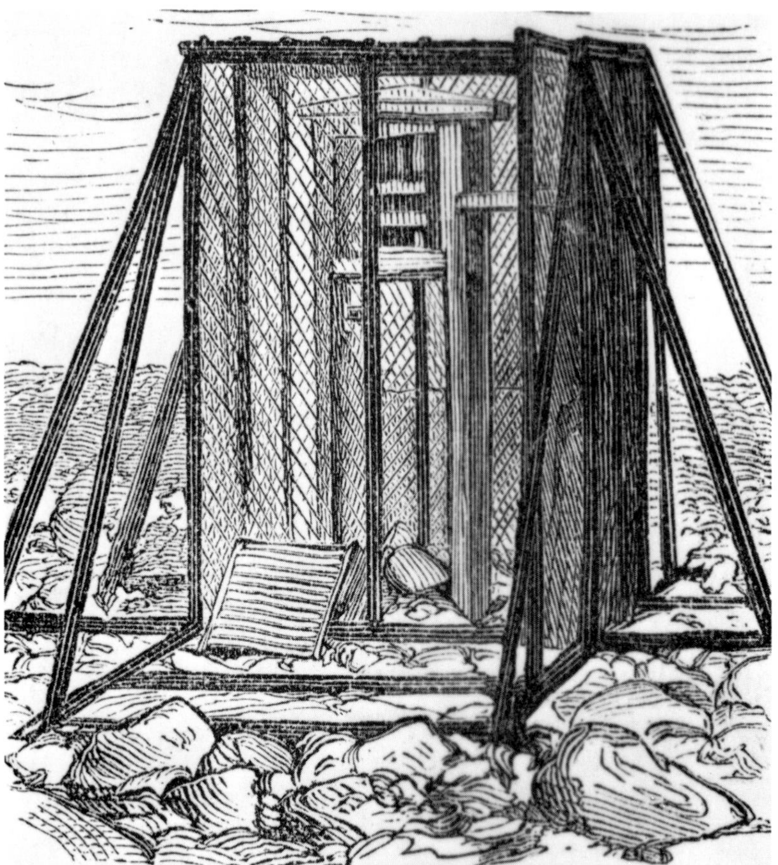

36

37. *The Ben Nevis Observatory in winter.* Conditions were often extreme. In midwinter it was not unusual to have to dig out, literally, through feet of snow. Then the rounds had to be made to record the readings of the instruments.

38. *The interior of the observatory.* Despite the extreme conditions outside, life was tolerable inside: there is an almost club-like atmosphere in the hut. The fact that it is a scientific center is betrayed only by the presence of the telegraph under the clock.

39. *The Greenwich Magnetic and Meteorological Observatory.* The Greenwich Magnetic and Meteorological Observatory was set up by George Airy, the Astronomer Royal, who had a passion for measuring everything that could be measured. Airy's aim was to reduce as many atmospheric phenomena as possible to precise quantities. Readings from a thermometer, a barometer, a hygrometer, and an anemometer were taken every hour. The dew point was determined four times a day; wind direction and velocity were carefully plotted; the amount of cloud cover was estimated and the kinds of clouds recorded; rainbows were noted; and daily rainfall was measured. Most important, however, were the readings of terrestrial magnetism and atmospheric electricity. No iron was used anywhere, and the main building was laid out in the form of a cross with three magnets occupying the south, east, and west arms. An 80-foot mast in front of the building served as a probe for atmospheric electricity. From it a wire ran to a carefully insulated apparatus that held a variety of electrical measuring instruments.

37

38

39

40. *The declination magnet.* This magnet was placed in the south arm of the building, where an observer could observe both the polestar and the magnet. The magnet was suspended by a thread, *e*, around which it could move. A copper bar, *i*, served to damp the swings of the magnet and hold it along a terrestrial line of force. The observations of declination were made through *b* and *c*, a lens and a cross hair. The instrument was of remarkable sensitivity.

41. *The horizontal magnet.* The horizontal magnet was placed in the east wing of the building, at right angles to the magnetic meridian. Two filaments, *ee*, supported the magnet, again damped by a copper bar sheath, *i*. This suspension turned the magnet into a torsion scale capable of reflecting minute variations in terrestrial magnetism in the horizontal plane. These minute variations could be detected by viewing a scale placed on the south wall of the eastern wing and reflected in a mirror, *b*.

42. *The vertical magnet.* This magnet was placed in the west wing and registered changes in the vertical component of the earth's magnetic field. The magnet was pivoted in the center and allowed to turn on this pivot. A mirror, *b*, magnified the small variations that occurred, so that they could be observed through a telescope placed some distance away.

40

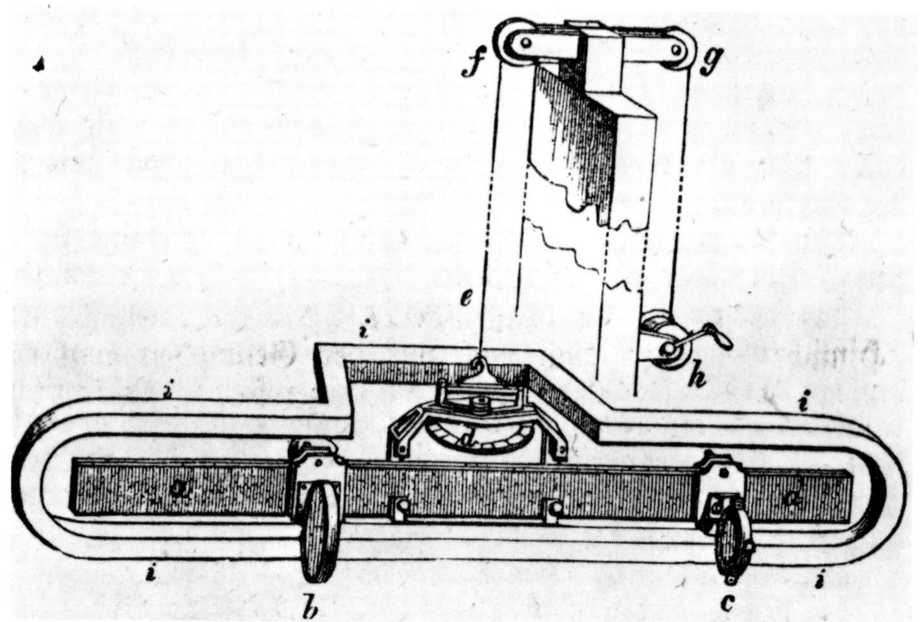

41

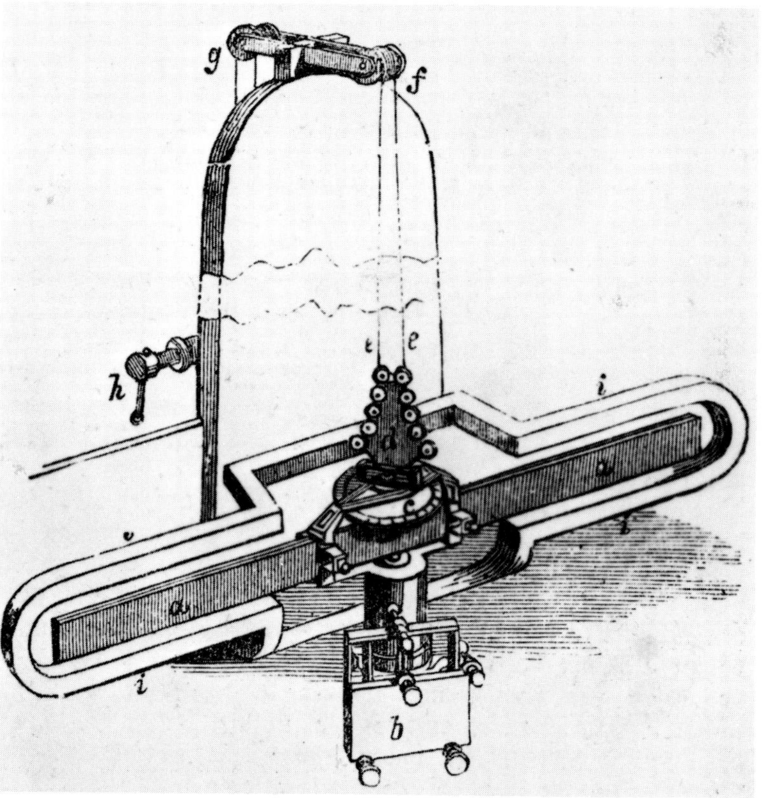

42

43. *Apparatus for the study of atmospheric electricity.* This series of instruments was hung from the outside of the observatory by a glass rod, *cc*, which guaranteed that it was insulated. A hook, *a*, was attached to a wire that ran to the top of the 80-foot mast. Electrical influences were then conducted through the glass rod to a copper bar, *g*, to which the various instruments were attached. A very delicate gold-leaf electrometer, *h*, had a dry cell on each side of the leaves to provide a weak electric field that reacted immediately to atmospheric electricity. A similar instrument, *l* (at right), was somewhat less sensitive to changes in electrical potential. Between *h* and *k* was *j*, an ordinary torsion electrometer. The length of electrical sparks was measured by *k*, presumably during thunderstorms. A straw electrometer, *m*, permitted frequent readings of atmospheric electricity and could be used to measure electrical quantities as well.

44. *Measuring the speed of a cannonball.* The incredible velocity and magnetizing property of electric current was utilized to measure the speed of a cannonball. In the background are two upright "screens," consisting of electrical wires connected to the shed. Within the shed the wires from the first screen are wound around a bobbin. A pendulum bob of soft iron is held magnetically against the bobbin until the cannonball smashes the wire net of the screen and disrupts the current. The pendulum then begins to descend along a quadrant scale. When the cannonball strikes the second screen, it breaks a current holding a knife switch open. That switch now closes, magnetizing the quadrant scale and making the pendulum stop immediately. Since the period of the pendulum is easily measured, the distance of the full swing can be read off the quadrant, and the distance between the two screens is known, the velocity of the cannonball is easily computed.

43

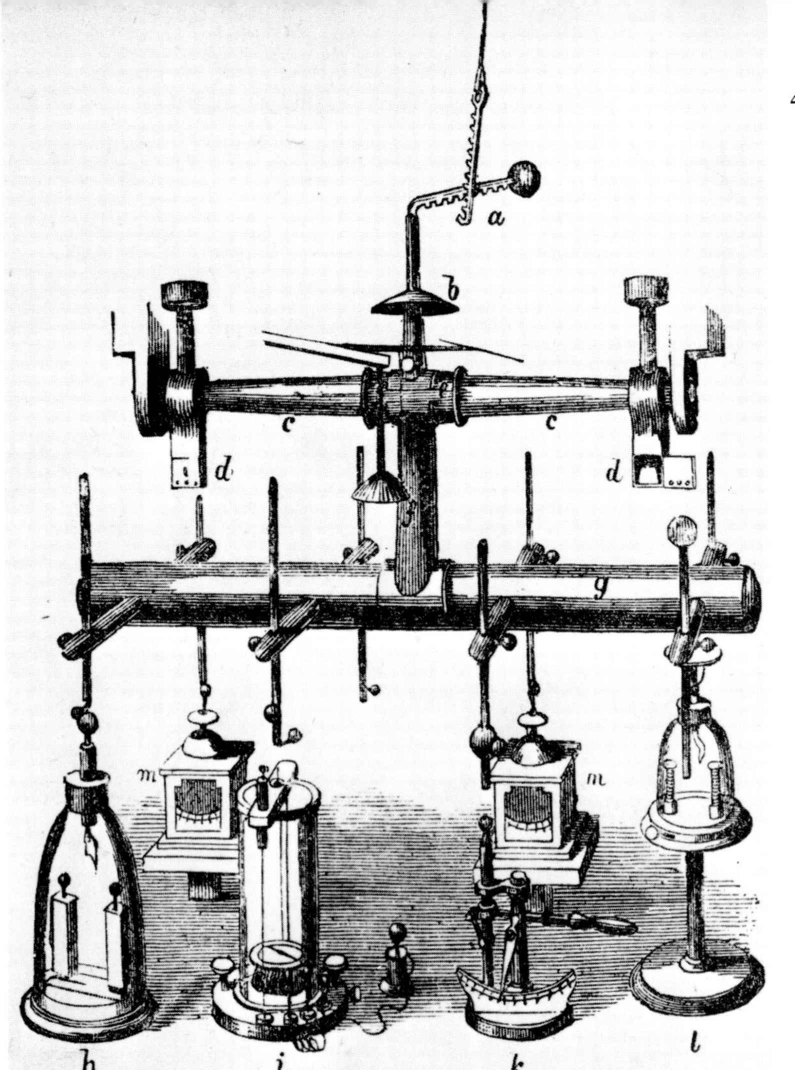

44

45. *Measuring wind velocity on Mount Washington, New Hampshire.* Not all dynamic measurements could be taken while wearing a top hat and sitting in a comfortable hut. The winds on Mount Washington often reach 80 miles per hour, and they are best measured from the position illustrated.

46. *The Greely Arctic Expedition: the farthest point north.* Measurements are very difficult to take in certain locations. In 1883, an expedition under the command of Lt. Adolphus Greely, U.S. Army, penetrated to latitude 83 degrees, 24 minutes, north on the Greenland coast, close to the North Pole, where it was hard to take a sighting of the sun. This was the farthest north that Europeans had yet reached, but it was still almost 400 miles from the North Pole.

45

46

47. *The scientific balloon ascent by Joseph Gay-Lussac and Jean-Baptiste Biot in 1804.* The hot-air balloon was invented in the eighteenth century but not used for scientific purposes until the nineteenth. In 1803, an amateur physicist reported that the earth's magnetic field and the current from a voltaic pile both decreased rapidly in intensity as altitude increased. This seemed odd, so Biot and Gay-Lussac, two young and ardent scientists, were commissioned by the Paris Academy of Sciences to investigate. On August 24, 1804, the first truly scientific balloon expedition set forth to observe the physics of the upper atmosphere. At 13,000 feet (4,000 meters) they found that neither of the reported effects was true.

Our illustration demonstrates what happened when nineteenth-century illustrators, ignorant of the sciences, tried to depict scientific events. Gay-Lussac, on the left, is shown holding a glass flask, presumably because his name was associated with gases. Biot, on the right, holds a combination thermometer and barometer. Nowhere to be seen are the pivoted magnet and the voltaic pile whose actions were the main reason for the ascent.

48. *Gay-Lussac's balloon ascent of September 16, 1804.* Shortly after his ascent with Biot, Gay-Lussac made another expedition into the upper atmosphere. This painting records his departure from the Luxembourg Gardens in Paris. The only piece of scientific apparatus visible is a mercury barometer, although the purpose of the trip was to collect samples of air at different heights. Gay-Lussac reached a height of 7,016 meters (*ca.* 22,500 feet) and discovered, to his great discomfort, that the temperature decreased as altitude increased. During his six hours aloft he almost froze to death.

47

48

49. *An English balloon expedition in midcentury*. Little progress in ballooning took place in the half-century between Gay-Lussac's voyage and the one whose beginning is depicted here. Nor did the scientific results improve. In this ascension of 1852, precisely the same observations of temperature, relative humidity, and barometric pressure were made that had been recorded by Gay-Lussac. The only difference between the two expeditions was that the photographic camera had been invented, so the later trip could be captured accurately on a photographic plate. (This engraving was made from the daguerreotype.) The instruments shown appear to be a barometer, a thermometer, and, in the background, a compound pendulum for measuring the force of gravity. The decoration of the gondola, with fringe and tassels, is typically Victorian.

50. *A projected balloon expedition to the North Pole (1890)*. By the end of the nineteenth century, the advance of technology permitted considerable modification of the earlier hot-air balloon, with its pendant basket. In 1890, an expedition was fitted out that utilized the latest in balloon techniques and equipment. The balloon had a volume of 15,000 cubic meters when filled with hydrogen. This provided a lifting force of 16,500 kilograms (18.1 tons). The gondola was no mere wicker affair, but a secure observation platform and a mobile home.

49

50

51

52

51. *Living quarters of the gondola.* The inside of the gondola was a regular scientific laboratory. There was room for the scientists and for their instruments, and for sled dogs that also served as experimental subjects. There was even a small pot-bellied stove in the corner—a bit daring, considering its proximity to 15,000 cubic meters of hydrogen. With such a sophisticated setup, the observation and measurement of phenomena in the upper atmosphere became simple and routine.

52. *Measurement of the speed of light.* The speed of light was increasingly recognized as a basic constant in the science of the nineteenth century, and determining it accurately became a matter of some concern. In this French experiment, light passed through a rapidly revolving toothed wheel at the observatory at Montlhéry to a mirror set up at the Tour de Montlhéry, 23 kilometers away, then traveled back to the observatory. The distance from the observatory to the tower was known with great precision, since it had been used for the determination of the meter. By varying the speed of rotation of the toothed wheel, the "echo" from the original signal could be blocked by one of the teeth of the wheel. Knowing the speed of the wheel, it was then a simple matter to determine the speed of light. This method yielded the figure of 300,400 kilometers per second.

53. *Apparatus for collecting deep-sea specimens.* The depth of the sea, as well as the heights of the atmosphere, were investigated in the nineteenth century. At the end of the century, an artist depicted the instruments available for collecting data from the seas and their bottoms.

1. A net with swinging glass door.
2. Deep-sea sounding apparatus.
3. Purse net for collecting specimens at a given depth.
4. Deep-sea weir.
5. Deep-sea dredge for bottom-dwelling creatures.
6. Deep-sea sounding cylinder equipped to penetrate the bottom and secure a core. The "rings" in the foreground are detachable ballast used to drive the cylinder into the bottom. They fell off on impact.

54. *Undersea photography.* Man has never been content to send recording instruments into inaccessible places. He wants to see for himself. The invention of the camera and of equipment permitting human survival at great heights or in great depths of the sea was necessary for the unhampered examination and recording of the terrestrial environment. The problem of taking photographs at considerable depths in the ocean was solved by the apparatus shown here. A barrel was filled with oxygen under pressure. A bell jar atop it contained a spirit lamp into which magnesium dust was blown by a current of oxygen. The burning magnesium produced enough light for photography.

53

54

The Contemplation of Nature

A skeptical historian once wrote that historians have attributed everything from the pyramids of Egypt to the French Revolution to the rise of the middle class which must, by now, have risen out of sight. It is true that the middle class has been used to explain almost every aspect of modern life and we shall have to appeal to it once again here. It really was the middle class that found science exciting and liberating in the nineteenth century and wanted, somehow, to be able to contemplate the world that science revealed. We may ignore here those members of the bourgeoisie who were interested in science because it was useful and could help build private fortunes through its application to industry and agriculture. What we are after are the institutions that were created or enlarged and improved in the nineteenth century and that made the world of science available to those with the leisure and the desire to look.

The most prominent public exhibition of nature at the beginning of the century was the Muséum d'Histoire Naturelle in Paris, of which the Jardin des Plantes was the most popular part. The Muséum had been created by the revolutionary government out of the old Jardin du Roi and had been intended quite explicitly to bring Parisians and other Frenchmen and women into direct contact with the beneficence and harmony of nature. This contact would then, it was hoped, awaken the natural virtue inherent in every heart and thereby lay the groundwork for the Republic of Virtue that the Revolution was intended to usher in. The ideology may have failed, but the Muséum prospered. Here was recreated for urban dwellers a botanical survey of the world, insofar as plants could be acclimatized to Paris. Where acclimatization was impossible, the climate of the natural habitat was reproduced by building special buildings to house special collections. A Parisian could transport himself to Brazil or to Java simply by passing through a door. Or, he could wander through beautiful woods and admire shrubs and flowers from distant lands.

Sentiment led one to the Jardin des Plantes; curiosity took one to the other parts of the museum or to other collections. The nineteenth century witnessed the creation of enormous showrooms where specimens of the various objects to be found in nature were collected and carefully labelled. For us who are familiar with the picturesque displays of modern museums it may seem difficult to imagine that someone could have found enjoyment in walking down an aisle, surrounded by bare bones or minerals, each with its card describing its species and its provenance. Yet, mineralogical and zoological collections drew considerable numbers of people to them and municipalities often paid considerable sums for the kernel of a good collection. Obviously those who came to look at the objects felt that they were gaining something by their inspection.

Later in the century, the attempt was made to bring living nature to an urban public by recreating the total environment within the walls of a museum or botanical or zoological garden. The aquarium built for the Le Havre Maritime Exhibition is an excellent example of this trend. A visit did allow visitors to feel that they had made contact with living nature, however briefly.

The creation of public places where the interested public could view natural history created a whole new industry that was to have some scientific repercussions. The popularity of a natural historical museum could be greatly enhanced by the addition of rare and beautiful specimens from distant and exotic places. High prices were charged for such specimens and specialists emerged whose major talent was their ability to trap or collect novelties from all corners of the globe. One of these trappers was Alfred Russel Wallace whose experiences in the field led him to suggest a theory of evolution through natural selection in a letter to Charles Darwin. It was this communication that induced Darwin to write *On the Origin of Species.*

55. *An appeal from science.* The growing importance of science in the culture of Europe was becoming evident by the second half of the nineteenth century. It seemed only proper to suggest establishing a science museum where good burghers and their children could gaze upon the achievements of this "new" field. The view of science held by the layman is beautifully illustrated in this *Punch* cartoon. Science is depicted as a bespectacled bluestocking, looking a bit disordered and frumpy, with an electric light in her hair to crown her glory.

56. *The botanical garden of the Museum of Natural History, Paris.* The problem of bringing science to the people had long attracted the attention of scientists who felt that appreciation of nature ranked among the natural rights of man. One solution was the botanical garden, which, as a journalist wrote in 1847 of the one at Paris, "pleases everyone. It charms all ages and attracts all classes of society. The small child can run through it without danger and amuse himself in the middle of a thousand wonders. The adolescent, the mature adult, the aged can all find solitary byways proper for dreams or meditation. There, too, are innumerable objects for study and always enough shade, depending on the hour and the place." One could absorb botany and fall in love with plants almost without thinking. The practical people of the nineteenth century were attracted by a place that was educational as well as pleasant.

55

AN APPEAL FROM SCIENCE.

"Am I not worthy of as much Consideration as Music and Geology? Why should not *I* have a Museum?

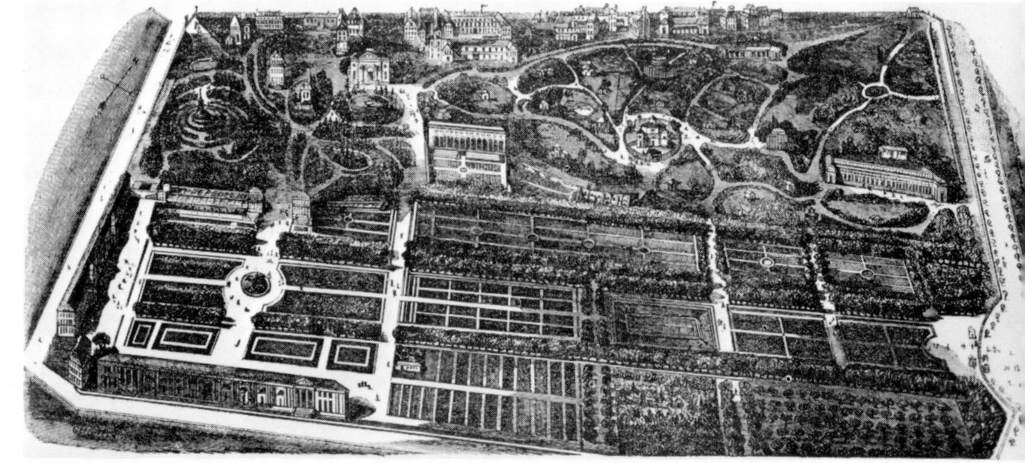

56

57. *The gallery of comparative anatomy in the botanical garden.* For those who wished to confront nature on an intellectual level rather than on a casual promenade, there were museums displaying the results of scientific investigations. Here, in the comparative anatomy gallery, one could see the changes nature had wrought on certain basic skeletal forms. Little attempt was made to lead the visitor into the subject. It seemed sufficient to put him in the middle of a group of skeletons and allow him to decide what they were all about. This approach probably piqued, but left unsatisfied, the curiosity of the majority of those who passed through the gallery.

58. *The natural history gallery in the botanical garden.* This gallery looks a bit like a contemporary living room. Stuffed birds and animals provide the decor for glass cases containing other specimens. The public was admitted only on Tuesdays and Fridays, the other days being reserved for serious scientists. Here one scientist appears to be explaining a point of zoology to an interested family.

59. *The acclimatization garden in the Bois de Boulogne.* Here, in a marvelous iron and glass building, nature could be brought indoors and controlled. Not every Frenchman could visit the tropics, but the tropics could be recreated in Paris through the wonders of modern science and technology. The iron grillwork and the vast glass expanses were made possible by modern industry. The garden served as a monument both to the plant world and to the ingenuity of nineteenth-century Western man.

57

58

59

60. *The gallery of comparative anatomy at the School of Medicine, Paris.* This collection was a far more serious affair than the public galleries of the Museum of Natural History. Created in 1845, it served as a teaching aid for the study of comparative anatomy. More than mere curiosity was involved here; the medical students were expected to master the subject by using the anatomical examples that had been brought together here. This was a strictly professional museum.

60

61

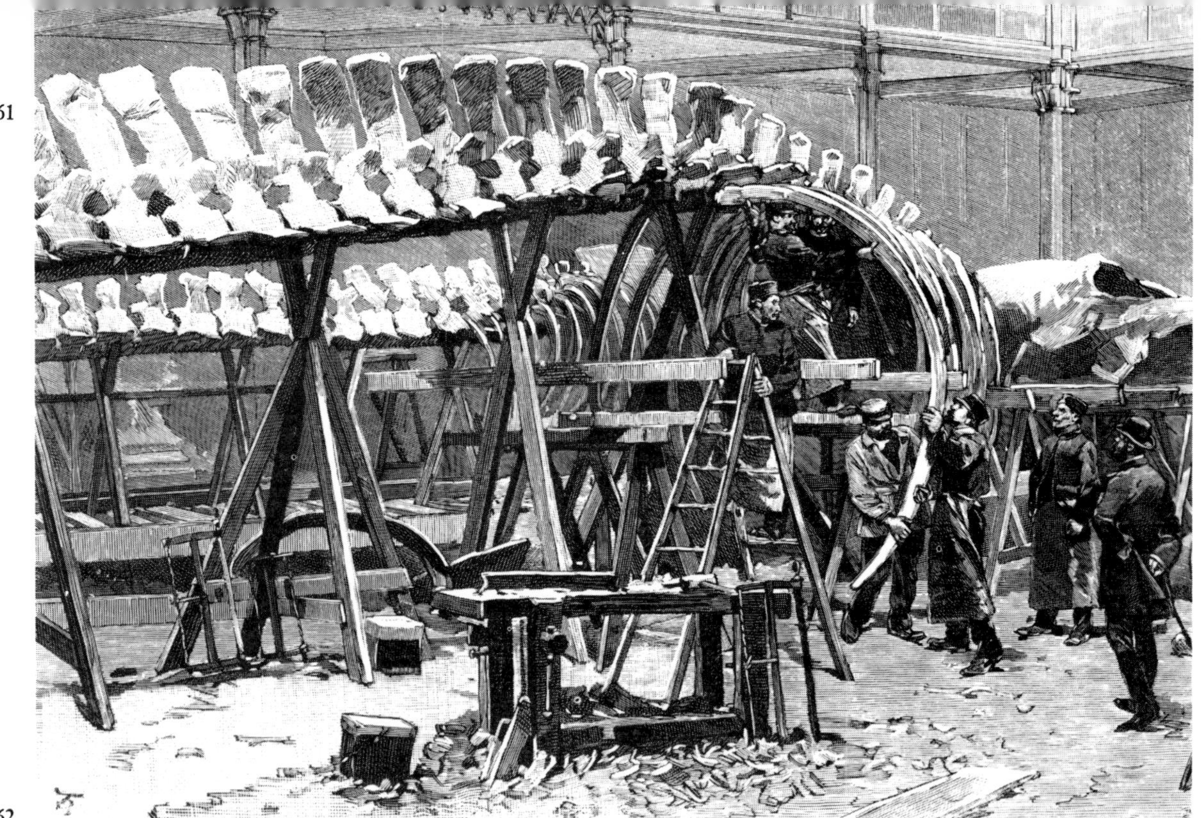

62

61. *Mounting a whale skeleton for the Museum of Natural History, Paris.* One of the appeals of a museum was that it brought the uncommon into public view. And, of course, the more spectacular the uncommon exhibit was, the more the public liked it. The skeleton of a whale was guaranteed to be a sensation, so the new gallery opened in 1889 featured not one, but six, such skeletons. Mounting them required almost as much ingenuity and technical skill as assembling a nineteenth-century railway station. Iron rods and buttresses were skillfully inserted into the bones to give them a "lifelike" appearance, and the exhibit was a smashing success.

62. *The mineralogical gallery at the Museum of Natural History, Paris.* The educational stamina of people in the nineteenth century was extraordinary. The mineralogical gallery was a popular place, in spite of the fact that it simply exhibited tens of thousands of mineral specimens. The modern mind reels at the prospect of passing down those rows of neatly labeled rocks, patiently examining each, firm in the knowledge that one is, somehow, bettering oneself.

63. *The aquarium at Le Havre.* Live specimens clearly are more interesting to the public than the dead husks of animals or inanimate minerals. Hence the attempt was made to bring living nature to popular view. The aquarium constructed for the Havre Maritime Exhibition in 1868 was a superb example of what could be done along these lines. The whole building was designed to resemble Fingal's Cave in the Scottish Hebrides. The columns were imitations of the columnar basalt that is a striking feature of the original, and verisimilitude was increased by building outdoor pools in which seals cavorted and above which seagulls wheeled.

64. *The interior of the aquarium.* The motif of Fingal's Cave was continued inside the building and lent a note of authenticity to the exhibits. Besides the tanks with their living specimens, as shown here, there were sea pools with water of different temperatures; these permitted the exhibition of almost every known species of crustaceous, molluscan, radiate, and vertebrate sea animal; of polyps, zoophytes, sea reptiles, and sea insects; and of a great variety of seaweeds and sea plants. The final touch of verisimilitude was the sea sand that covered the floor.

65. OVERLEAF: *The Microscopical Aquarium, Berlin.* The Microscopical Aquarium in Berlin was created by the city and located in an old mint. The main "reading room," where the public could inspect specimens, was well lighted by windows in the dome. Off this main room were smaller rooms containing special collections. Some 50 microscopes were available for use by the general public. Attendants demonstrated the use of the microscope and helped amateurs to identify specimens.

63

64

The Study of Nature: The Laboratory

Collections of plants, animals, minerals, and microorganisms did more than titillate and amuse the public; they also served as essential data for the progress of science. Jean-Baptiste Lamarck, for example, was led to his evolutionary ideas while engaged in the Herculean task of cataloguing the invertebrate collection of the Muséum d'Histoire Naturelle. But the observation and contemplation of natural objects could answer only a few of the questions that could be put to nature. She was more likely to respond when tortured either by the application of forces not usually found in nature or, more likely, by isolating some aspect of natural phenomena and dealing with it exclusively. Both means are artificial and require experiment rather than observation. The great galvanic pile that Napoleon sponsored for the École Polytechnique permitted Gay-Lussac and Thenard to use electrical currents that were otherwise unattainable in their experiments. Similarly, the laboratory founded by Justus Liebig at Giessen in 1826 enabled him to teach his students chemistry by example rather than by rote. Substances could be isolated and then put together in various combinations, many unnatural, so that their interrelations could be observed and described. The laboratory became the center for those sciences that permitted questions to be put to nature under carefully controlled circumstances which then yielded fairly clear-cut answers. Laboratories, of course, had existed before the nineteenth century, but one of the novelties of the nineteenth century was the extension of the laboratory as a research tool to sciences previously not deemed susceptible of laboratory study. Foremost among these sciences was that of biology—the word itself dates from the first year of the century. Living matter had long been considered different from ordinary matter and, therefore, not capable of being studied in the same way. Nineteenth-century scientists proved this view wrong. Magendie in France, Purkinje in Bohemia, and Johannes Müller in Germany all showed that living tissues could be subjected to experimental study and that conclusions could be drawn from experiments on such tissues that were as reliable and solid as those drawn by the chemists from their experiments. Laboratories of animal and vegetable physiology became important and essential adjuncts of universities and hospitals where these subjects were taught and investigated.

The laboratory was more than a research tool. It became an invaluable aid to the teaching of the sciences. Liebig was the first chemist systematically to use a laboratory for the formation of future chemists. The idea spread and laboratories became as common as classrooms for they were, in fact, as necessary. Their dissemination, however, was uneven. Germany led in building and equipping them and soon profited from this foresight by taking the scientific lead. France, in spite of the example of the École Polytechnique and Magendie's work, lagged behind. Research laboratories were often miserably equipped and teaching laboratories remained rare. England also failed to understand the importance of the laboratory for both research and teaching. Fortunately, the Royal Institution of Great Britain provided superb facilities for a long chain of original investigators, beginning with Humphry Davy in the early 1800's and continuing through Faraday into the late nineteenth century. But, generally, English students had few opportunities to do original research until they had been able to create their own laboratories.

In the early years of the century, a laboratory could be created by a single individual with a rather modest expenditure of private funds. John Dalton, after all, was able to come up with the atomic theory using handmade and extremely crude apparatus. As the century progressed, it became increasingly difficult for someone to do original scientific research without relying on some kind of outside help. Equipment became more sophisticated and more expensive; the industrial revolution made it possible to build larger and more intricate instruments or machines without which pressures, temperatures, or other conditions necessary for probing nature's mysteries could not be obtained. One need only look at the equipment in the laboratory of the École Normale Supérieure (Fig. 76) or that used by Pictet to liquefy oxygen to recognize that the industrialization of the West had an immediate impact upon science.

The scientist himself became institutionalized, depending for the very instruments of his trade upon large subsidization from some outside source. In most cases this source was a university, for science grew primarily within the walls of academe. The creation of the Cavendish Laboratory at Cambridge in 1873 and the Clarendon Laboratory at Oxford in 1872 is typical of this aspect of the institutionalization of the laboratory. Industry also began to use laboratories as their value became clear. Again, it was Germany that led the world in attaching research and testing laboratories to industrial plants for quality control and for the creation of new products. German leadership in the manufacture of aniline dyes was the direct result of its investment in chemical laboratories oriented towards the aniline dye industry. By the end of the century, the picture of the scientist had altered rather radically from what it had been when the century began. The lonely investigator, like Dalton or Ampère or Davy, ingeniously constructing his own simple apparatus and establishing new truths unaided had given way to the industrial or academic entrepreneur in a white coat surrounded by bizarre and mysterious instruments. Such a man spoke with almost priestly authority on matters of life and death (if he were Pasteur) or on the qualities of a product being sold to the public. Science had become the ultimate arbiter and scientists had become a new estate.

66

67

66. *The naturalist.* For the amateur, not all scientific pleasures were to be found in public museums or institutions. A sizable portion of the "informed public" kept private collections. The subject of this picture collected butterflies; Charles Darwin, more professionally, collected beetles. Such collections brought the curious amateur into contact with at least one part of nature.

67. *The great electrical battery at the Ecole Polytechnique.* At the beginning of the nineteenth century, current electricity was the focus of scientific interest. The voltaic pile that produced it was the most important new scientific tool of the era. Soon it was realized that the bigger the pile, the more spectacular its effects. Really large piles were beyond the resources of the individual scientist, so outside help was necessary. In 1813, Napoleon had this huge battery constructed to ensure French supremacy in the new galvanic science. Though it cost a small fortune, it paid off only in prestige, producing not a single important scientific discovery.

68

69

70

68. *Liebig's chemical laboratory in Giessen.* In 1842, Justus Liebig made a laboratory an essential part of the teaching of chemists at the University of Giessen. Students were actively involved in research from the moment they were introduced to chemistry. Lectures provided the theory, but only the actual manipulation of chemical apparatus could convey the essence of the science. Liebig's laboratory is strikingly modern. Apparatus similar to that shown here was in use until after World War II, when it was displaced by electronic gadgets. Only the alembic on the table at the right looks out of place; it was made obsolete by the Liebig condenser.

69. *Liebig in his laboratory.* Liebig conducted his own research in a corner of the laboratory. He was, therefore, constantly available to his students, who would follow his work and then return to their own. They literally learned at the side of their master.

70. *The northern entrance to Liebig's laboratory, Munich.* Chemical laboratories, unlike astronomical observatories, do not dictate forms to the architect. A chemical laboratory can look like a classical temple or like a Renaissance palazzo. When Liebig accepted an appointment at the University of Munich, the city built a laboratory for him that, complete with frescoes, blended into Munich's architectural style. Chemists soon demanded more austere and impressive monuments to their science.

71. *The laboratory of plant physiology at the Museum of Natural History, Paris.* One triumph of nineteenth-century science was to widen enormously the scope of the experimental study of nature. Living organisms such as plants could be studied scientifically in laboratories. One of the earliest of these was the laboratory of plant physiology at the Museum of Natural History in Paris, which was created in 1873. The laboratory itself was basically a chemical laboratory, but it was set up so that plants could easily become the objects of experimental study.

72. *Early photograph of laboratory equipment (ca. 1851).* The photographic camera was one of the most important inventions of the nineteenth century. This early photograph has preserved the image of the standard equipment in a simple laboratory, whose objects have taken on the quality of a still life.

73. *The chemical laboratory at the Prefecture of Police, Paris.* Chemistry was probably the most useful of the sciences in the nineteenth century, and its utility was soon recognized. On May 22, 1881, the Paris police opened their own laboratory—to be used not for the apprehension of criminals but for the detection of additives and harmful ingredients in food sold in the Paris markets. The laboratory had become part of the machinery of government.

71

72

73

74

74. *The laboratory on board HMS Challenger.* Not all subjects for study could be brought into the laboratory. Sometimes the laboratory had to go to the subject, as was the case with the great oceanographic voyage of HMS *Challenger*. On December 21, 1872, the *Challenger* left Plymouth harbor on the most ambitious scientific exploration yet planned: to map the depths of the seas and to report on the flora and fauna encountered. The laboratory, though simple, was adequate to the task; from it came volumes of reports on the discovery, and the anatomy and physiology, of countless new organisms.

75

75. *Louis Pasteur in his laboratory in Paris.* The fact that human diseases could be studied and attacked in the laboratory brought both science and the laboratory to the center of nineteenth-century awareness. Louis Pasteur did not create the medical laboratory, but his work caught the public imagination. He became a secular saint and the laboratory became the new cathedral of the religion of science.

76

76. *The laboratory of the École Normale Supérieure, Paris.* The laboratory developed with the rest of nineteenth-century society, drawing particularly upon new industrial techniques. In 1873, the French chemists Henri Sainte-Claire Deville and his associate Henri Debray were chosen by the government to prepare the platinum that was to serve as the new standard for the meter. The purification of the metal and its reduction to a liquid required intense heat. This was provided by the burning of hydrogen in oxygen, both gases being introduced into the crucible through the tubes visible on the lid. The operation was more like a process encountered in a steel mill than in a place of scientific research. The pouring of the platinum was a public occasion. Adolphe Thiers, President of the Republic, stands at left center, shielding his eyes with a piece of dark glass.

77

78

77. *The liquefaction and solidification of hydrogen.* The union of science and industry is well illustrated here. At Geneva, in 1873, Raoul Pictet liquefied the permanent gases. His laboratory is indistinguishable from a factory; the power that runs the industrial mills is here connected to a powerful compressor. This experiment not only provided final proof of the generality of the laws of matter by showing that the "permanent" gases could be reduced to liquid and solid states; it also inaugurated the age of refrigeration, which was to revolutionize the food supply of the Western world.

78. *The first commercial advertisement using science.* The new central role that laboratory science played in modern society was graphically illustrated in this advertisement, the first to draw upon the prestige of science. Cadbury's Cocoa Essence was not only a treat for the taste buds; it was also "scientifically determined" to be nutritionally valuable, containing "flesh-forming ingredients." The analysis that accompanied the advertisement explained precisely how Cadbury's Cocoa Essence could contribute to one's general health. An entry in the laboratory notebook in the foreground testifies to the absolute purity of this cocoa. Who could doubt these claims, when they were made by such a scientifically imposing figure as the gentleman depicted here?

The Teaching of Science

The new class of scientific men was created in the nineteenth century by formal education. By and large, the older type of self-educated investigator of nature tended to disappear to be replaced by the person with a doctorate from a respectable (usually German) university.

France was the first Western country to attempt a massive, national reform of education that included the sciences as an essential element of instruction. During the French Revolution, central schools, in which mathematics, physics, chemistry, and natural history were offered, were set up in every department of France. The sciences had been introduced into the curriculum for ideological as well as practical reasons, it being hoped that scientific method would lead to the realization that liberty and the Republic were both in accord with natural law. They were therefore eliminated during the Napoleonic reform of education with only mathematics being retained. Similarly, the École Polytechnique, founded in 1794 and soon to become the foremost institution for the teaching of the highest reaches of theoretical science, was "reformed" by Napoleon into a practical school for military engineers. Even so, it was the foremost institution for scientific education in the world in the first quarter of the nineteenth century and served as the model for such schools as the U.S. Military Academy, the Massachusetts Institute of Technology, and the host of polytechnic schools that sprang up in Germany and other countries.

The lead in scientific education passed rapidly from France to Germany. It is not at all clear why this was so. German universities did have the ability to assimilate the sciences and make them part of their regular course of instruction. Possibly this happened because the universities in Germany were essentially run by their faculties and were able to introduce scientific education more easily than were those in France which were controlled directly by the state. In England the sciences languished at the great universities of Oxford and Cambridge largely because these institutions had become finishing schools for the aristocracy and the upper middle class. Science smacked of the workshop and was, therefore, *infra dig.* In any case, it was the German university that led the world in the nineteenth century in training future scientists and, by midcentury, it was standard practice for the budding scientist to go to a German university for the Ph.D. which made him into a professional.

The very professionalization of science tended to hinder its growth in places like England, which clung to the ideal of liberal education. Even after it became clear that scientists were made, not born, the English were reluctant to provide the kind of advanced and specialized training that marked their German contemporaries' education.

University education produced research scientists. What about the practical applied scientist? Where was he to come from? The answers, once again, differed from country to country. In this area, the United States began to move ahead with the passage of the Morrill Land Grant Act in 1862 which provided grants of federal land to colleges or universities that would undertake serious training in the agricultural and mechanical arts. In France, the École Centrale des Arts was founded to create practical engineers but one school clearly could not do the job. In England, there was a strong but short-lived movement to educate artisans through Mechanics' Institutes which would teach skilled workers the scientific principles underlying their traditional practices. The same approach was attempted in France where workers could attend lectures at the Conservatoire des Arts et Métiers to learn the basic principles of machinery and the fundamental laws of nature upon which the arts were based. Both movements failed, largely because industrialists did not pay off for such technical education and because workers would have to be extraordinarily devoted to their tasks to undertake the serious study of nature after a full day in the shops. In Germany, the rank and file of technically trained people was created by popular education. Prussia had free and universal elementary education long before any other Western country. From this group came those who went on to the Technical High Schools that produced engineers and other skilled technicians for German industry. The result was the clear superiority of German industry and science by the end of the century. The Germans (and the Americans) were the first to realize that scientific and technical expertise were the foundations of material power.

Not all teaching of science took place in the universities or before workingmen's audiences. There was a large and interested audience for the popular exposition of science to be found in the middle classes whose new wealth and leisure were important aspects of nineteenth-century society. There grew up in all countries in the Western world a new kind of science teaching aimed at this group. Faraday at the Royal Institution was a master of the art of the popular scientific lecture, and his popularity helped to keep the Royal Institution solvent during years of financial difficulty. More importantly, the growth of popular science had a striking effect upon the general mores of the West. It was a cornerstone of the new materialism that was, increasingly, a fundamental characteristic of Western civilization. The conversion of a large and politically dominant segment of the population of Western countries to a rational and scientific culture was accomplished during the 1800's. What could be measured and counted and weighed was what was real; everything else was mere tradition or sentiment. The greatness of a country no longer was to be seen in its art or its culture or its humanitarianism, but in the number of tons of steel it poured per annum.

79

80

81

79. *A boys' school in Berlin.* The advance of science left the social institutions required for its support far behind. Only the select few could work in Liebig's laboratory, yet society needed increasing numbers of scientists. The schools that ought to have provided them remained as they had been for a hundred years. In this German public school, an abacus aids in calculation; a blackboard permits demonstration of simple examples. The rest is in the teacher's head.

80. *Greenwich Royal Hospital School: a lesson on steam machinery (1848).* The need for specialized education had been recognized early in some quarters. The Royal Navy had set up educational facilities to train officers at Greenwich. The school followed the principle that underlay Liebig's laboratory: that the best practical instruction comes from working with basic instruments of a science. Here the steam engine and its applications were studied by inspecting and analyzing actual steam engines.

81. *The class in mathematics and optics at Greenwich.* Some studies remained academic and, as had been traditional, were learned from books and lectures. Here only a few visual aids distinguish this class from one in Latin or Greek. An optical system is diagramed on the wall at the end of the room; flanking it are some simple geometrical figures. For the rest, it was study and memorize.

82. *The schoolmaster of the future: an English view.* It is bizarre that the wedding of science and technology was not universally recognized in England, where it had first occurred. Throughout the nineteenth century, science retained an exotic and ethereal aura, and seemed to be removed from the practical concerns of the day. In this cartoon, young boys of the poorer classes, it is suggested, would be wasting their time on the "impractical" sciences; they should be taught a trade.

83. *The physics lecture hall at the new Collège Chaptal, Paris.* The new science would not be denied, and institutions had to be changed in order to accommodate it. The Collège Chaptal, named after Napoleon's great Minister of the Interior and one of the first French industrial chemists, was built to house 1,000 students and cost 4 million francs. It contained rooms like the physics lecture hall, in which physical laws could be demonstrated before hundreds of students. In this sense, it was more efficient than Liebig's laboratory, yet it lacked that sense of personal involvement that made Liebig's laboratory such an exciting place.

82

Trigonometry
Geometry
Geology
Ethnology
THE SCHOOLMASTER OF THE FUTURE.
(And the sooner we get him the better.)
British Workman. "BOTHER YOUR 'OLOGIES AND 'OMETRIES, LET *ME* TEACH HIM SOMETHING USEFUL!"

83

84

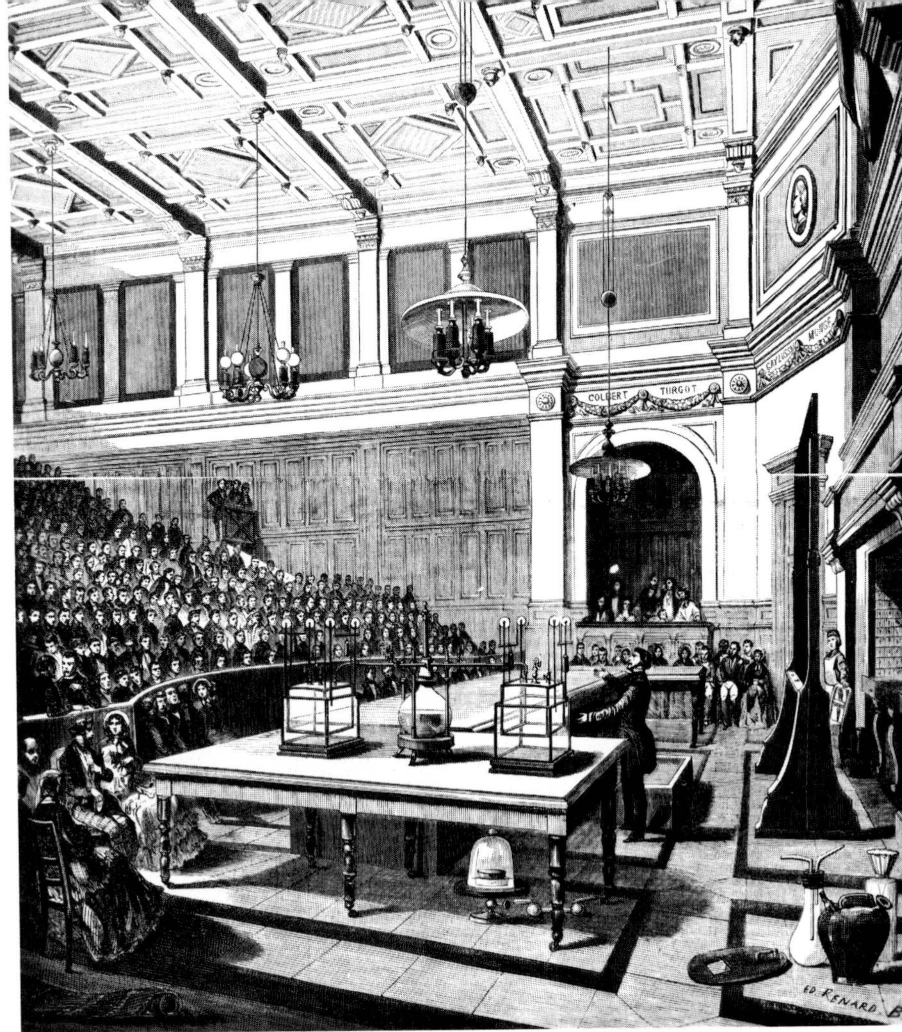

84. *The lecture hall and public course for workers at the Conservatory of Arts and Trades, Paris (1850).* The advance of scientific technology had obvious implications for the status and condition of skilled workers. If they could learn some basic scientific principles, they might be in the best position to suggest significant, and profitable, improvements in industrial processes. Hence, in England, Germany, and France, free public courses in elementary science were instituted. We may conclude that the lectures were well attended, if this illustration represents a typical one. One point is worth noting: the bourgeois members of the audience, in the front row, are carefully segregated from the working class. And this was only two years after the great social revolution of 1848!

85

Scientific Researches! — New Discoveries in PNEUMATICKS ! — or — an Experimental Lecture on the Powers of Air —

85. *New discoveries in pneumatics: a lecture at the Royal Institution of Great Britian.* The "informed public" wanted to keep abreast of the newest scientific discoveries. For them to do so, new institutions had to be created. Foremost among these in the nineteenth century was the Royal Institution of Great Britain, founded by Count Rumford in 1799. The English caricaturist James Gillray here depicts a public lecture given there. The Professor of Physics, Thomas Garnett, is administering nitrous oxide (laughing gas) to a volunteer. Behind Garnett stands the young Humphry Davy, who had made his scientific reputation with his monograph on nitrous oxide, while Count Rumford smiles benignly near the doorway.

86

86. *Michael Faraday lecturing to the royal princes at the Royal Institution.* The Victorians demanded more decorum from their lecturers than Regency society did. Michael Faraday's lecture was suitable for Prince Albert, Victoria's consort, and for their young sons. Faraday was one of the great popularizers of science in England, and lectures like this did much to assure the kind of public support that kept English science first-rate.

87

87. *Perrini's planetarium.* R. Perrini of Hampstead Road, London, was a private citizen who recognized the general public's thirst for scientific knowledge and catered to it. His planetarium consisted of a gaslit sun around which the planets moved in ellipses. The constellations appeared in the background. The machinery that moved the planets could be slowed down or speeded up to taste. The various "houses" of the zodiac on the outside provided inspiration to both the astronomer and the astrologer.

88

89

88. *A phrenological lecture.* The pseudo sciences throve in the new popular scientific atmosphere. Phrenology was particularly popular, for it offered a quick and easy method of character and personality analysis. The topic was made for caricature; in the example given here, the members of the audience have skulls weirder than many of those in the "professor's" collection of specimens.

89. *The illustrated scientific lecture.* Scientific and technological advances helped to inform the interested public. The carbon-arc lamp and photography made the slide lecture possible. Whereas before, one could look at specimens only individually, as in the Microscopical Aquarium at Berlin, now one could, through the use of slides, view a specimen and hear a lecture about it simultaneously. (The slide projector has not really changed in 100 years.)

90

91

90. *The great electric induction coil at the Polytechnic Institution.* Size and power were central to the nineteenth-century concept of machines, and exerted a great attraction upon the general public. So, when Professor Pepper of the Polytechnic Institution of London exhibited his gigantic induction coil, people flocked to see it in action. It was an impressive show. The coil was 9 feet, 10 inches, long and 2 feet in diameter. Its core of thin, soft iron rods weighed 123 pounds; 3,770 yards of copper wire made up the primary coil and 150 miles of copper wire composed the secondary coil. The current for the primary coil came from 40 wet cells. The spark from this giant, 29 inches long, could perforate a glass plate 5 inches thick.

91. *Physics for the people.* The "informed public" could go to lectures or visit museums. The lowest classes could not, so science went to them. Here, on the Champs Elysées in Paris in 1843, a "physicist" demonstrates the laws of electricity to passersby. With an assistant to turn the crank on a static-electricity generator, and with an adequate supply of Leyden jars, a good popular show could be mounted. When the hat was passed, the popular "scientist" undoubtedly reaped an adequate reward.

92

SCIENCE.
Professor Parallax (*enthusiastically*). "Oh! My dear Mrs. S., if You can manage to Stoop down, here is 'Capella' shown most Beautifully!!"
[*But by this time, it being a fine frosty night, poor* Mrs. Spudgrove, *having seen the Moon, and Jupiter and his Satellites, and Saturn, and Double Stars, and no end of Nebulæ, had had almost enough of it!*

93

92. *The amateur scientist.* The thrills of science were an inspiration to many nonscientists who could afford the necessary equipment. Such people tended to become fanatics and social menaces, as this 1868 cartoon from *Punch* illustrates.

93. *Parisians observe the eclipse of July 28, 1851.* By the mid-nineteenth century, eclipses in Europe were matters of only mild public curiosity. Almost everyone knew what to expect, and the entrepreneur selling time on a telescope drew only a small crowd. Europeans had become scientifically sophisticated.

94

94. *Eclipse of the moon at Constantinople, 1877.* Europeans considered science as their own monopoly. This sense of superiority was buttressed by engravings such as the one shown here. Although the Turks were still a European political power, they obviously were not part of the same civilization. Their firing muskets in an attempt to frighten away the evil spirit swallowing the moon brought smiles of pity and mild contempt to the faces of the "informed public" of Europe.

II

The World of Nineteenth-Century Science

Observing the Heavens

Astronomy in the nineteenth century was essentially a science of more precise measurement and of optical exploration. There were no great new theories of cosmogony or cosmology that could be based solidly upon the astronomical facts. Such theories had to wait for the twentieth century. Instead, what astronomers did was to perfect their instruments, perfect their methods of observation, introduce such novelties and fundamental aids as photography, and vastly extend the range of astronomical observation. The result was to consolidate the older Newtonian system of the world, measure and comprehend some of the basic characteristics of the stellar system of our galaxy, and create problems for later astronomers and cosmologists to wrestle with.

In the two centuries since Galileo had first directed a telescope at the heavens, only one really major improvement had been made in the optical system of a refractor. In the eighteenth century, an achromatic lens combining crown and flint glass was perfected. This lens permitted astronomers to avoid chromatic aberration and thereby gain a sharper image. Unfortunately, it was impossible to cast flint glass into large lenses, so refracting telescopes were forced to remain in the 4-inch range. It was not until the beginning of the nineteenth century that this restriction was removed by rapid cooling of the flint glass which provided larger lumps of relatively homogenous structure. Joseph Fraunhofer used this technique to build the 9½-inch refractor for Dorpat Observatory, which was, for many years, the largest refractor in the world.

The reflecting telescope had been invented by Newton and appeared to provide a way of avoiding chromatic aberration since the angle of reflection is the same for all colored rays. Again, however, there was a technical barrier to widespread use of the reflector. The mirrors had to be made out of speculum metal, a combination of copper and tin, which tarnished rapidly. Large mirrors also easily deformed from their own weight and this, plus the necessity of shutting down and repolishing after brief periods of viewing, prevented reflecting telescopes from displacing refractors.

Given these technical problems, the main progress in astronomy in the early years of the nineteenth century was made by increasing the precision of the observations and by extending these observations to every part of the celestial sphere. The increase in precision came from new techniques of observation and from the mathematical analysis of the origin of errors. Gauss's and Laplace's working out of the least-mean-square method of error correction provided a new dimension in the analysis of observational facts. New and better timing devices, ultimately using electricity, made it possible to refine standard astronomical observations of stellar and planetary positions to great precision. Such work produced important results. By keeping careful track of stellar positions over long periods of time, it was finally possible to detect stellar parallax in 1838 thus providing solid proof, if any were still required, of the earth's passage around the sun. More importantly, it provided a first measurement of astronomical distances beyond the solar system. Even the proper motion of the sun and of the solar system itself could be detected as it moved through the galaxy. Within the solar system, the new precision permitted the perturbations of the orbit of Uranus to be used to calculate the position of a new planet, Neptune. That discovery was the most widely publicized astronomical event of the century and served to gain popular acceptance of the Newtonian system of the world.

The seemingly pedestrian task of mapping the heavens put considerable pressure on astronomical instrument makers for there was the universal desire to build larger instruments to reach farther into space. When larger lenses could be made, they were, but this, in turn, strained other technological aspects of telescope construction. Larger lenses had longer focal lengths which required larger tubes to enclose them, which, because of their weight, caused serious difficulties in constructing properly accurate and strong driving mechanisms to keep telescopes pointed at their objects. These problems were exactly the same as those faced in industry, where power and strong materials were in constant demand, and the solutions came from industry. The telescopes of the late nineteenth century were large machines made possible by the evolution of heavy industry. They were driven by motors that applied science had discovered and designed. In this sense, the penetration of the heavens was linked intimately to the growth of Western industrialization.

The major breakthrough in sheer power came with the development of giant reflecting telescopes. If one were willing to take the time to polish and repolish a speculum mirror, quite impressive results could be gained from a large reflector. William Herschel was able to see and draw nebulae at the beginning of the century and William Parsons, the third earl of Rosse, by the late 1840's, was able to distinguish the spiral structure of some of these nebulae. Few astronomers were willing, however, to devote so much time to the pure mechanics of telescope care and reflectors did not become popular until after Justus Liebig had discovered the process for silvering glass. From midcentury on, the reflecting telescope with silvered-glass mirror became the primary instrument for the probing of the deeper reaches of space. The results were striking. All one need do is compare the drawings of nebulae made by Herschel at the beginning of the century with the photographs taken at the end to appreciate how much more of the universe an astronomer could see when the nineteenth century ended. Theories of cosmogony and cosmology to explain what could be seen could hardly have arisen before the objects themselves came into man's ken.

The final improvement in what could be "seen" was, of course, the photographic camera, which could make visible on a photographic plate objects that the naked eye could not see at all. Photography came of age by the end of the first half of the nineteenth century and was almost immediately applied to astronomy. Time exposures required tracking mechanisms of exquisite accuracy and when these were made available, astronomy profited. A photograph could be studied at leisure and often revealed details that were not available through any other method of observation. The application of the camera to the astronomical spectroscope, finally, made possible a new science, astrophysics.

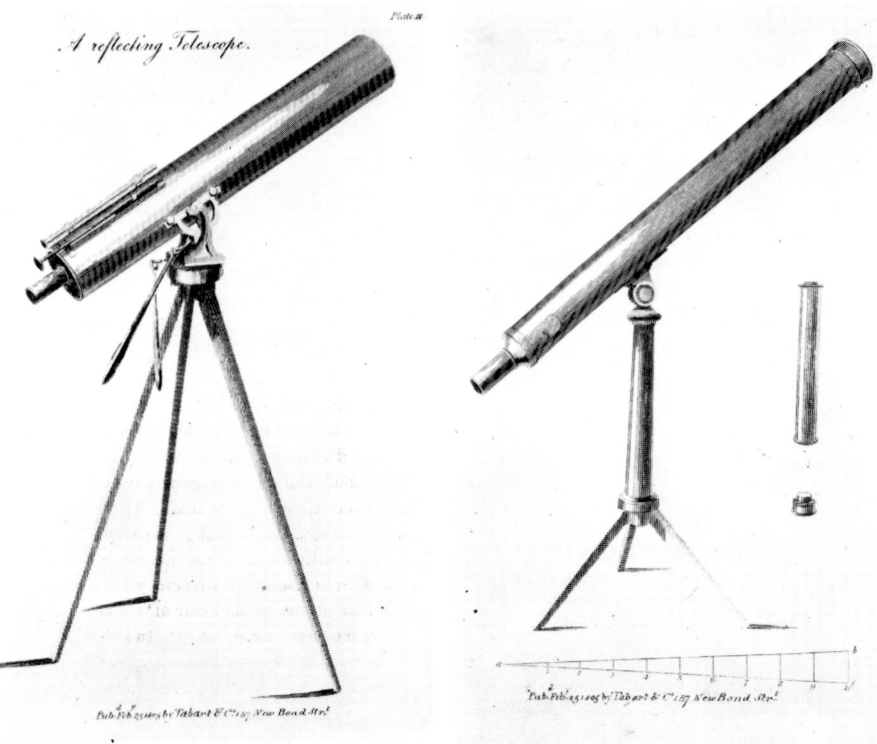

95

96

95. *A reflecting telescope, 1805.* At the beginning of the nineteenth century, telescopes were still mainly the products of artisans working in small workshops. Their size was restricted by the metallurgical techniques of the day, the ability to cast glass, and the power of the available driving mechanism. Most telescopes, therefore, were like this one—small and moved by hand. They could be used with some precision, but the size of the field and their magnifying power were not very impressive. Although reflecting telescopes had the advantage of avoiding chromatic aberration, they were restricted in size by the difficulty of making large mirrors from speculum metal, an alloy of copper and tin.

96. *A refracting achromatic telescope.* Small refractors, such as this one, were the most popular astronomical instruments at the end of the eighteenth century. John Dollond's optical work had eliminated chromatic aberration in small telescopes so that the image was sharp and clear.

97. *John Herschel's 20-foot reflector.* This was the telescope Herschel took to South Africa to observe the heavens in the southern hemisphere. Herschel's sketch reveals the clumsy mounting apparatus necessary to give the telescope mobility. There was no tracking mechanism; adjustments had to be made by hand, using various cranks.

97

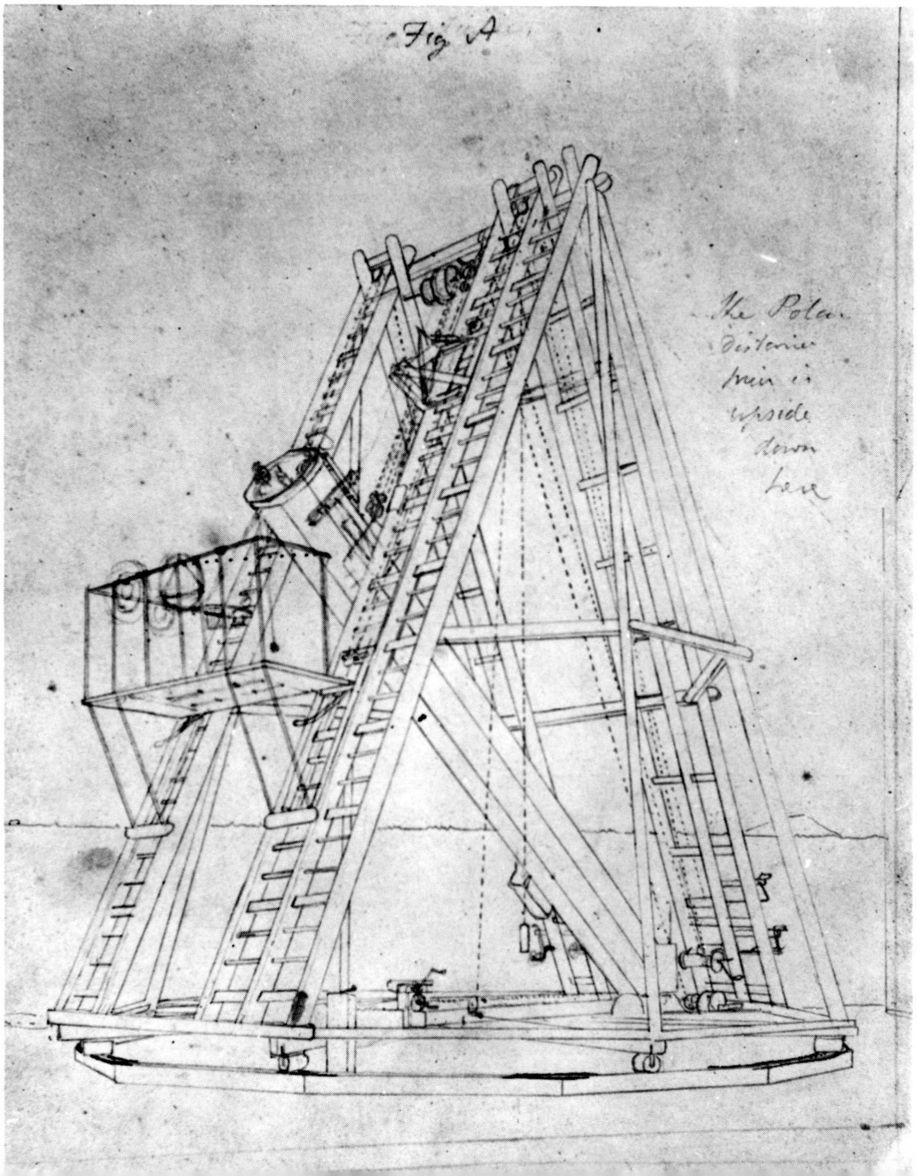

98

99

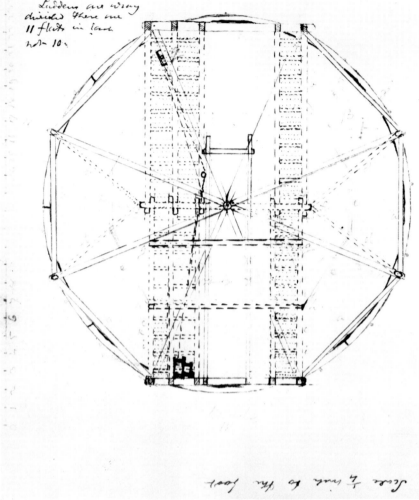

98. *Top view of the scaffolding for the 20-foot reflector.*

99. *Photograph of the 40-foot reflector at Slough.* The greatest telescope of the early nineteenth century was the 40-foot reflector made by William Herschel, John Herschel's father. It reached the limit of telescope construction, given the techniques of the time. The mirror, of speculum metal, was 4 feet in diameter and allowed the elder Herschel to make the first systematic observations of nebulae. Because its mirrors tarnished rapidly, the telescope was not practical for extended observations. This is one of the first photographs ever taken. John Herschel made it just before the telescope was dismantled in 1839.

100. *The great telescope of the Earl of Rosse (1844).* The Earl of Rosse's telescope, in Ireland, was the last large reflector to use speculum metal. It was truly gigantic: the mirror was 6 feet in diameter and 5.5 inches thick, and it weighed 3.75 tons. To mount this monster required a stone castle-like structure, which was built with walls parallel to the meridian. The telescope rested on a universal joint and could be raised and lowered by two men using pulleys.

101. *Mechanism for raising and lowering the Rosse telescope.* The mechanism shown here permitted the telescope to be raised and lowered in the plane of the meridian. Because very little side-to-side motion was possible, observations were restricted to a narrow portion of the sky.

100

101

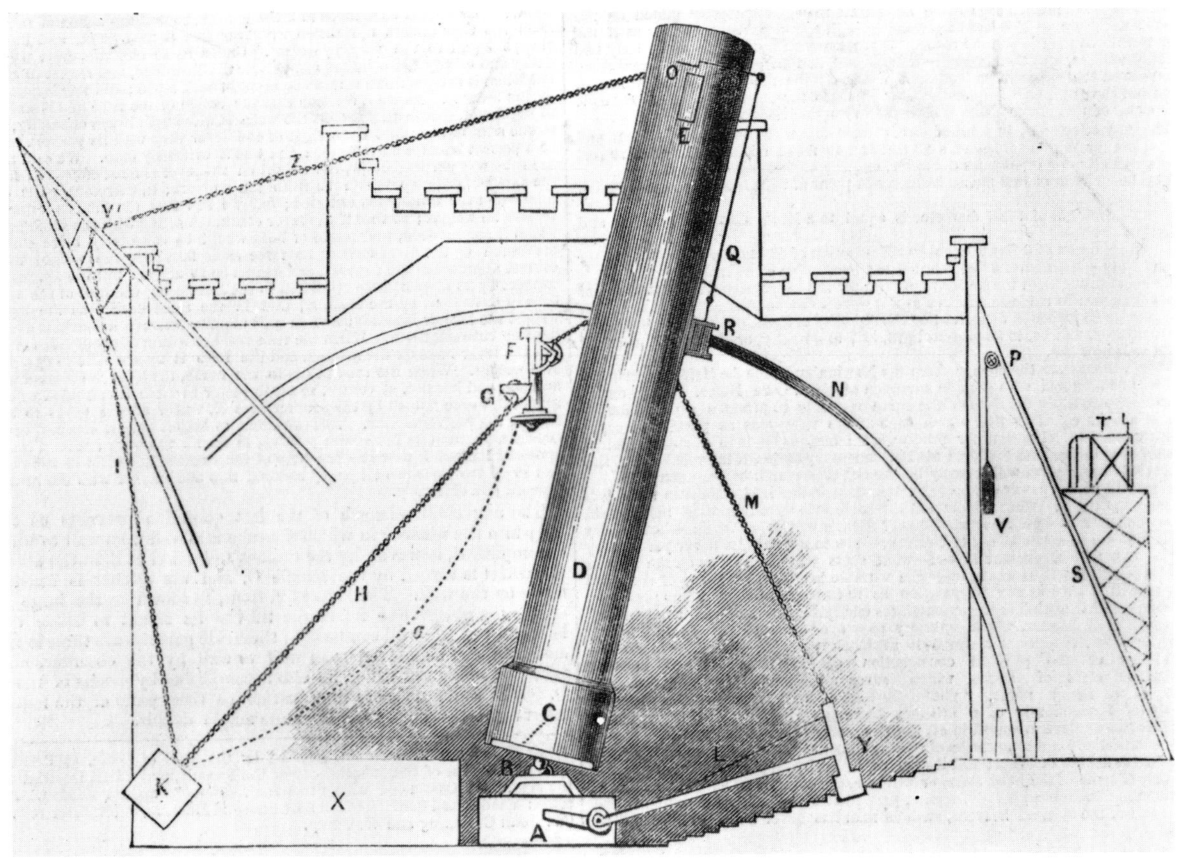

102

103

102. *The giant refractor at Wandsworth Common (1852).* This was the largest achromatic refractor of English design and construction until that time. It was conceived by a country clergyman, Mr. Craig, vicar of Leamington, designed by the engineer William Gravatt, and financed by Lord Spencer. The telescope was 85 feet long, and its tube was built like a riveted boiler for a steam engine. The mounting was ingenious and simple: the tube was suspended in a large iron ring hanging from a tower that, by its sheer mass, was completely vibrationless. The eyepiece end could be cranked up and down by means of a windlass, and the whole telescope pivoted on the ring and could be moved by pushing it along a track.

103. *The refractor at the Midland Observatory.* An amateur astronomer, Mr. Lawson, of Bath, left his excellent collection of astronomical instruments to a committee at Nottingham, provided money could be raised to house them in a proper observatory. Lawson's refractor, shown here, represents a somewhat more sophisticated instrument than the Craig telescope. Its optical properties were no better; but it could be maneuvered with considerably greater ease and precision, which made it altogether a more elegant astronomical system than its larger sister at Leamington.

104. *The grand telescope at the Paris Observatory (1875.)* The new reflecting telescope of the Paris Observatory, representing the latest in optical and metal technology, was officially inaugurated on October 7, 1875. The process of silvering glass had been invented by Justus Liebig in the 1850's and applied to telescope mirrors by Léon Foucault in 1857. Such mirrors did not tarnish easily and were some four times lighter than those made of metal—this one weighed 500 kilograms. The Paris telescope was the result of advanced industrial technique: its parts

continued on page 68

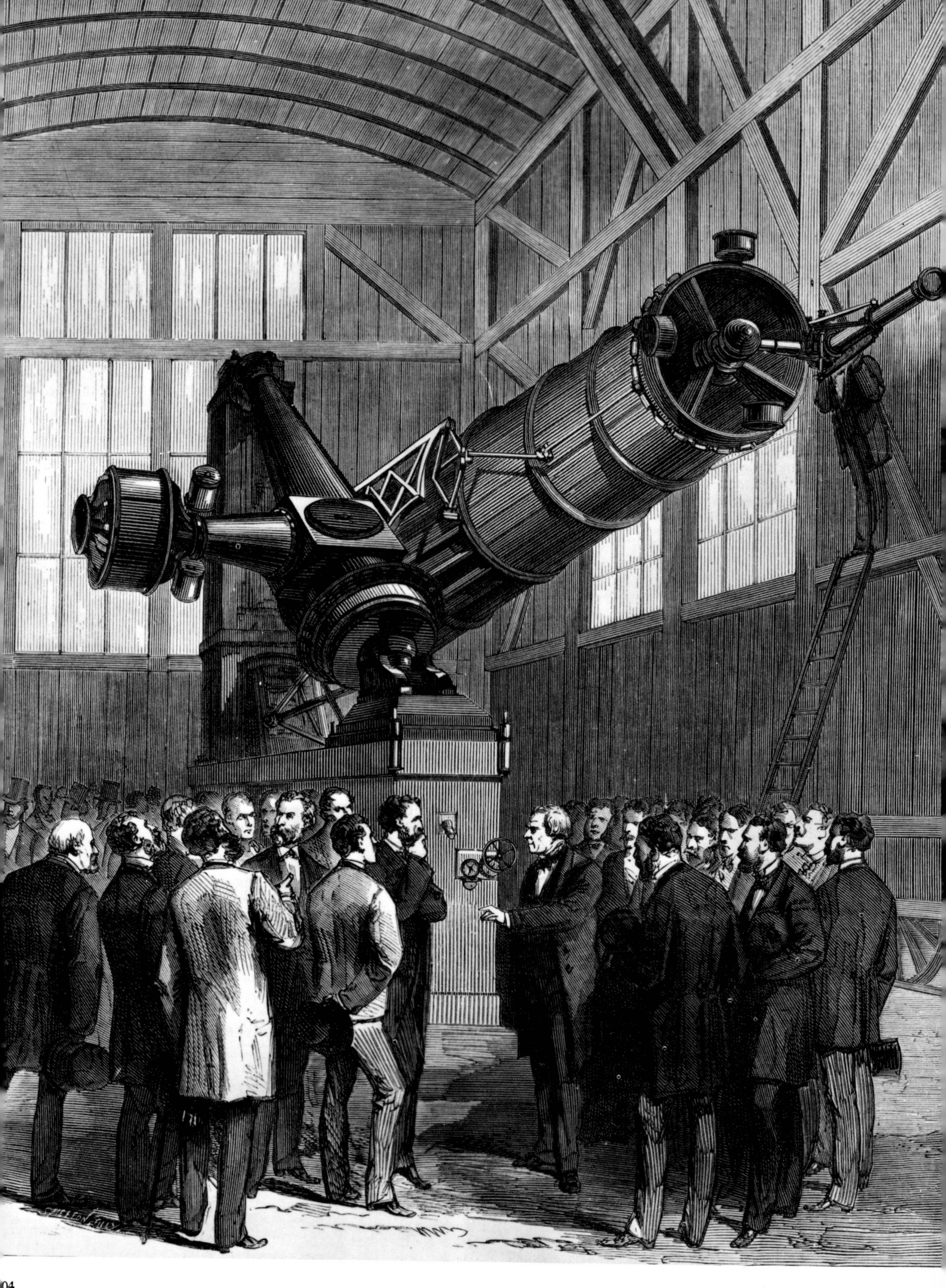

continued from page 66
were carefully and closely machined, and the clockwork that turned it was precise. The whole was housed in an iron shed to protect it from bombs—a lesson learned during the siege of Paris in 1870–1871. The shed moved on a track so that the telescope could be covered and uncovered easily.

105. *Father Secchi's photographic telescope.* The ultimate refinement of the telescope was its combination with photography to yield permanent and accurate records of the heavens. Father Angelo Secchi, at the Vatican Observatory, was one of the first to attach a camera to his telescope in order to record sunspots. The image of the sun was projected on the plate labeled *OQ* in the picture reproduced here. There it could be examined or recorded on a photographic plate.

106. *Clockwork for Secchi's phototelescope.* It was essential that the image of the sun not move on the photographic plate if Secchi's photographs were to be of value. To ensure immobility, accurate machinery that compensated for the movement of the earth was absolutely essential. The clockwork shown is the sort required for a small and simple telescope.

105

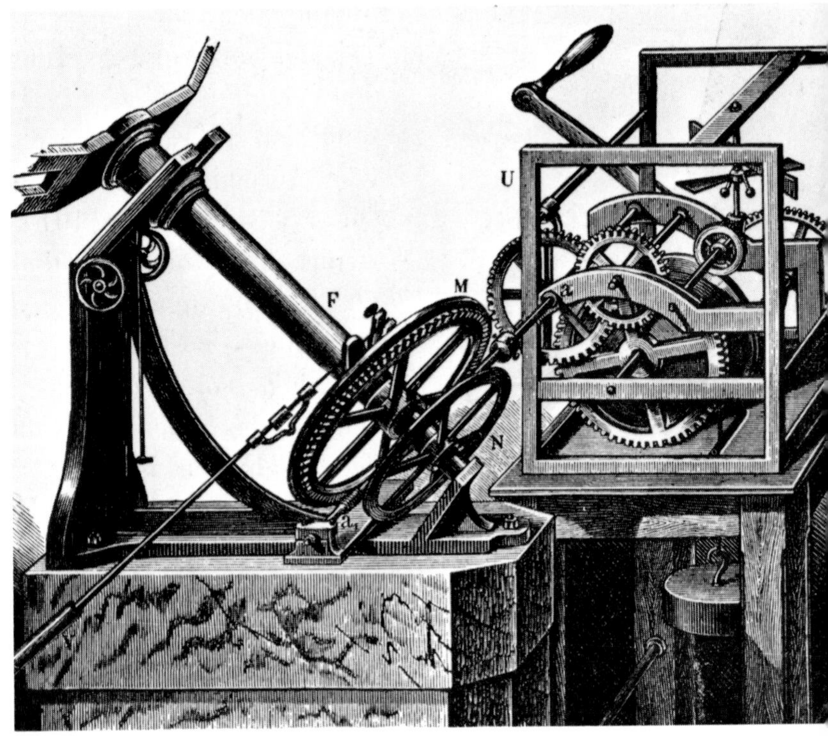

106

107. *An early photograph of the moon.* This photograph of the moon was taken by J. A. Whipple at the Harvard College Observatory on February 26, 1852, with a 15-inch refractor. It shows what a difference photographs of celestial objects could make. The visible detail is clearly fixed and can be interpreted rather easily.

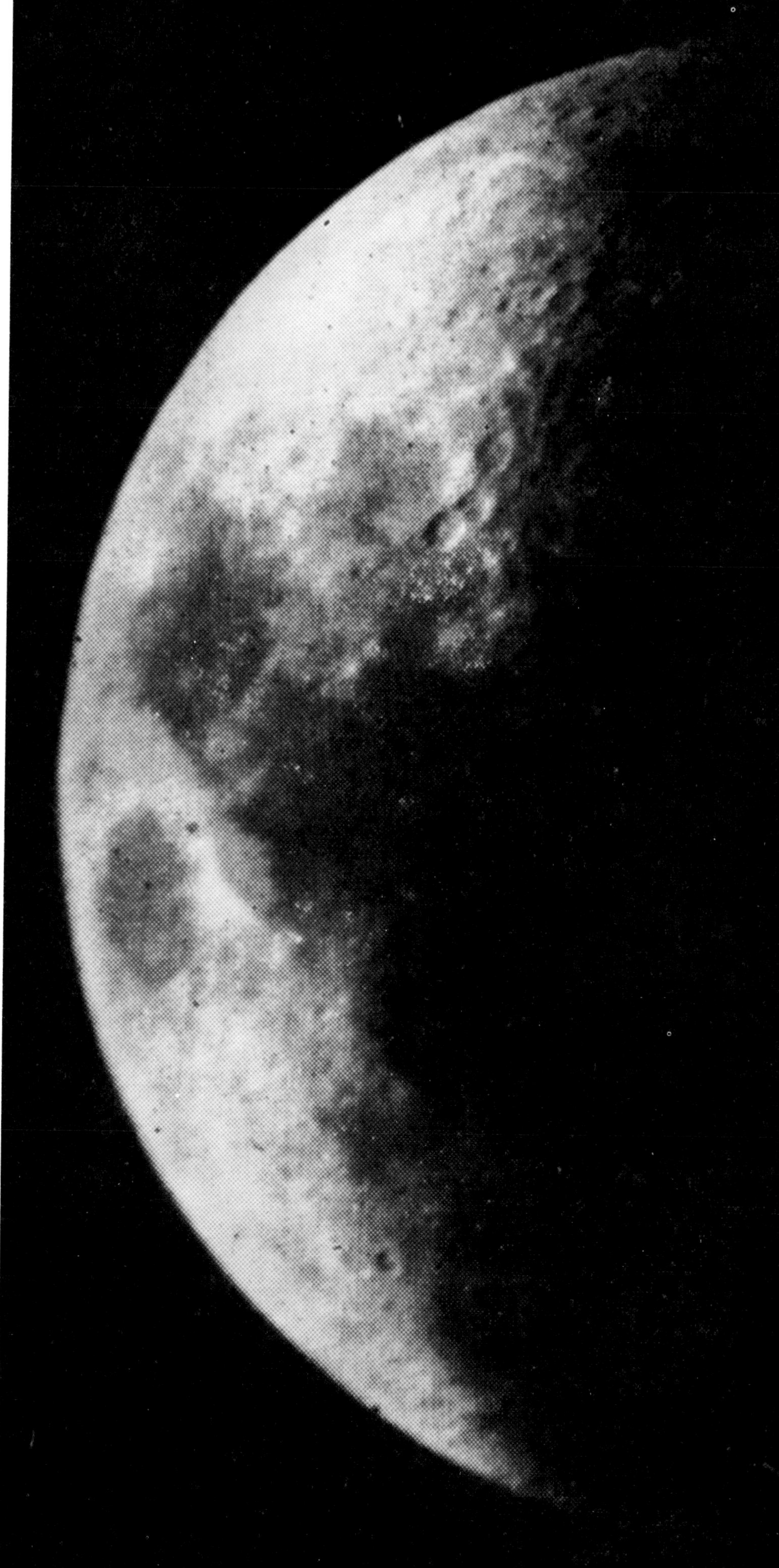

108

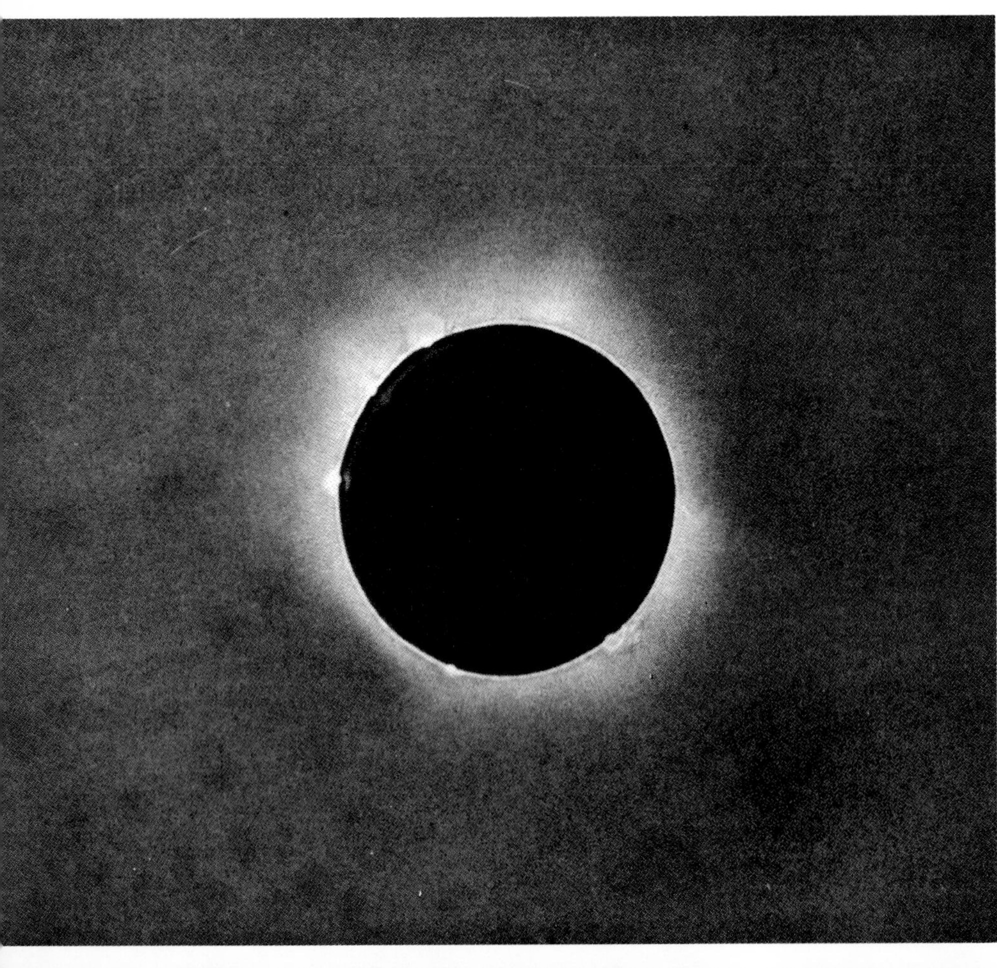

108. *The first successful daguerreotype of a solar eclipse.* This photograph was made on July 28, 1851, at Königsberg. Copies, in which the solar flares are apparent, were used throughout the century to illustrate the phenomenon (See Figure 181).

109. *An early photograph of a double star.* The most valuable astronomical photographs were made after it was realized that long time exposures could reveal things that were difficult or impossible for the eye to see. This shot, made by the wet-plate method in 1857, shows one such sight—two images of the components of Mizar.

110. *The first photograph of the Orion Nebula.* Sometimes, it was not immediately obvious that the camera was superior to the naked eye. Compare this photograph, taken by Henry Draper on September 30, 1880, at an exposure of 51 minutes, with John Herschel's 1834 drawing of the Orion Nebula (Figure 116).

109

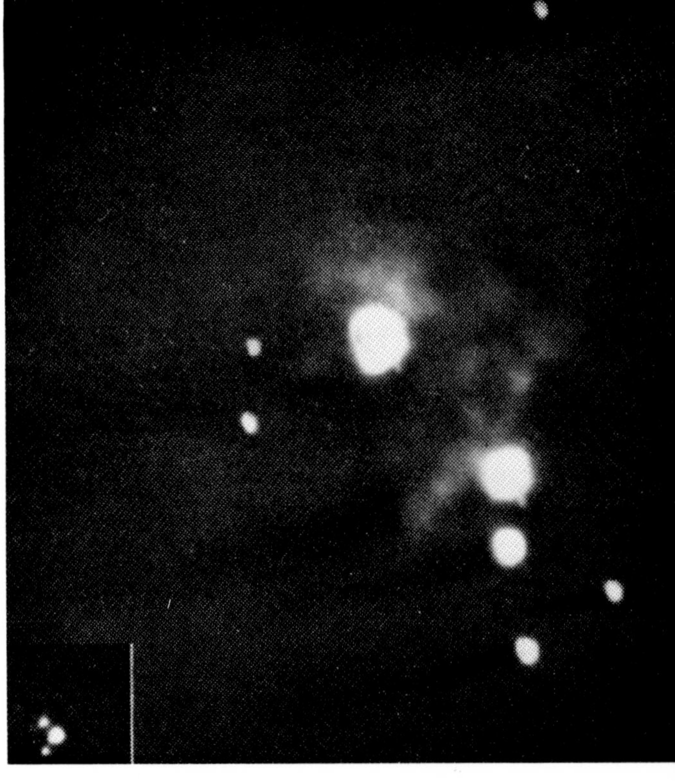

110

Stars and Galaxies

As has already been mentioned, the major accomplishment of sidereal astronomy in the nineteenth century was the mapping of the heavens. By the end of the century, this task had been done sufficiently well for astronomers to desist for the moment and try to make sense of what they had found. It might be well to sum up what was known here.

The distribution of stars in space was known to a high degree of accuracy. From this knowledge, the proper motion of stars could be found with relative ease simply by comparing observations made over the course of years. Double stars had been detected and, in some cases, their actual orbits mapped. The principle of gravitation was thereby proven to be truly universal as Newton had suggested. The larger structure of systems of stars could only be guessed at. Herschel had already realized that the sun was part of a galaxy whose rough shape he could make out by carefully observing the distribution of stars in space. The Milky Way was that part of our galaxy in which this stellar distribution was densest and from which it could be concluded that the galaxy was lenticular in shape. The great mystery for nineteenth-century astronomy was the nature of the nebulae. Were they, as Herschel suspected, large blobs of luminous gases within our galaxy? Or, as photographic definition improved and intimate nebular structure became visible (at least in some cases), was it not possible to speculate and believe that nebulae were things apart from our galaxy—even galaxies of their own so distant as to be almost unimaginable? The questions were of fundamental importance but they were not to be answered in the 1800's. The mystery of the nebulae was to be solved only in the twentieth century.

111

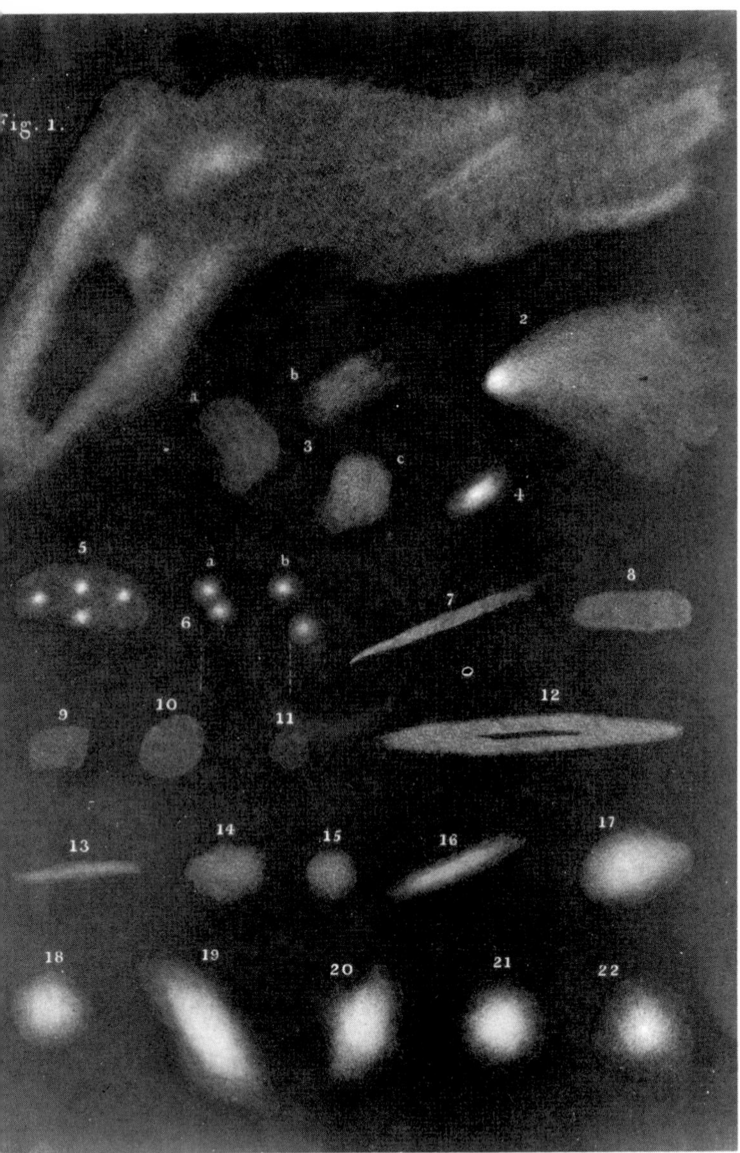

112

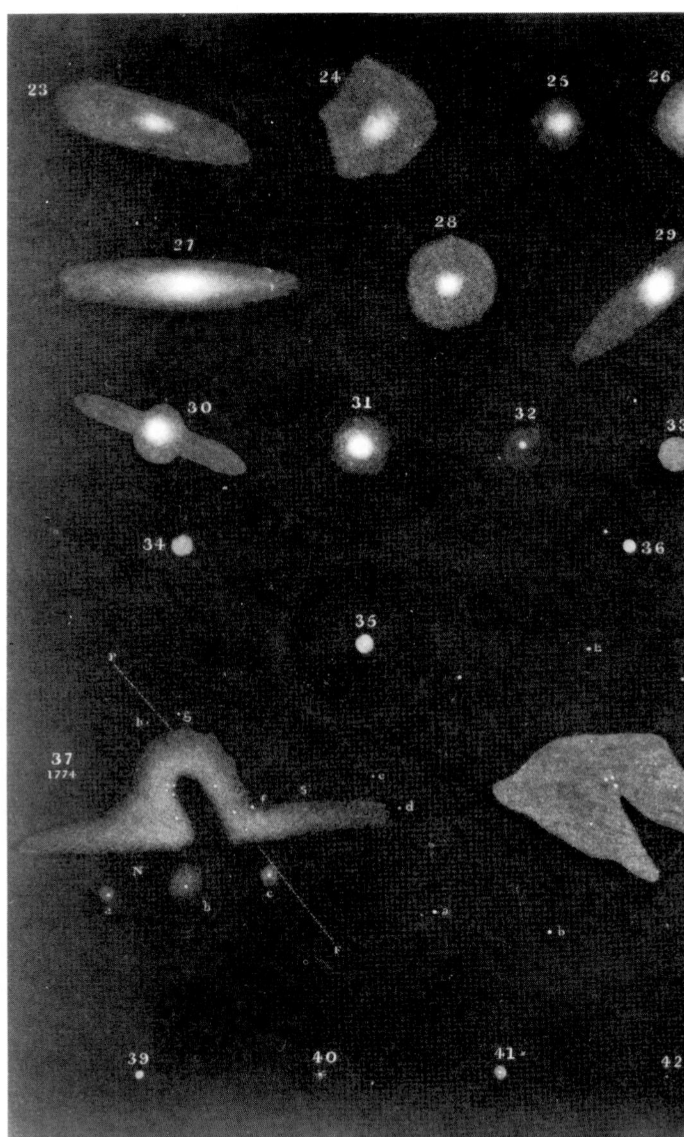

111. *William Herschel's drawings of nebulae.* Using his great 40-foot reflecting telescope, Sir William Herschel was able to detect and draw what he called "diffusions of nebulosity." In 1811, he cataloged a number of new "nebulosities" and published two plates containing drawings of 42 of them. This plate records 22 very vague nebulae, which Herschel considered to be stars in their earliest form.

112. *Developed nebulae as drawn by Sir William Herschel.* The condensation of "diffused nebulosity" into stars is illustrated in this plate. Numbers 29, 30, and 31 represent nebulae that were depicted many times in the nineteenth century.

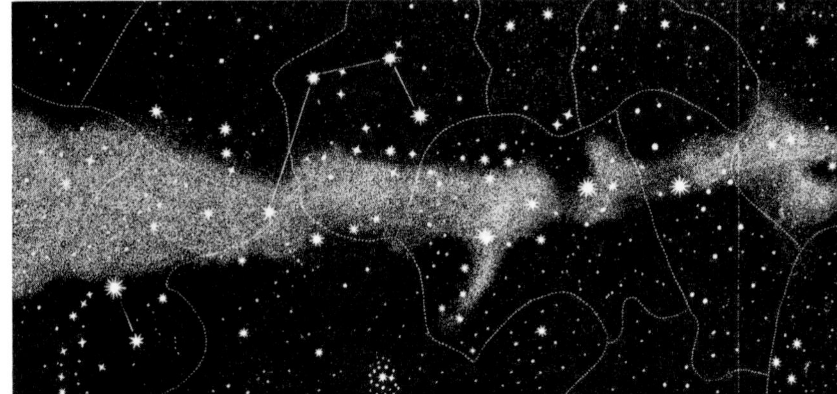

113

113. *The Milky Way—north and south.* William Herschel was the first to suggest that the earth was part of a galaxy that appears to us, on the earth, as the Milky Way. This illustration shows the view of our galaxy, from the inside looking out, that was available in the midnineteenth century. The section visible from the northern hemisphere appears at the top, that from the southern hemisphere at the bottom.

114. *The Clouds of Magellan: the Great Cloud (Nubecula Major).* When John Herschel mapped the skies of the southern hemisphere, one of the most striking features he observed was the Clouds of Magellan. The Great Cloud appeared to be a system of stars that had condensed out of the nebulosity. The picture of the Great Cloud is surrounded by the names of neighboring stars.

115. *The Clouds of Magellan: The Small Cloud (Nubecula Minor).* The Small Cloud appeared to illustrate the condensation of a single star from a cloudy mass of luminous material. This, of course, upheld William Herschel's theory of nebulae.

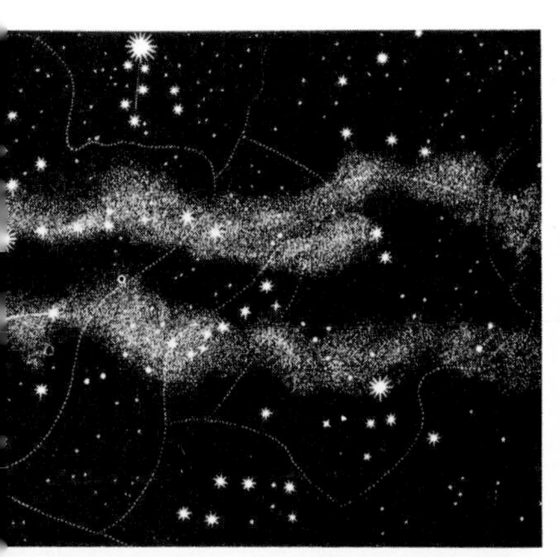

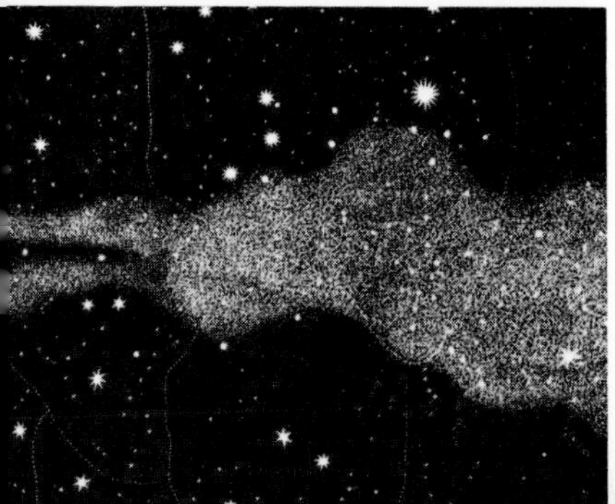

114

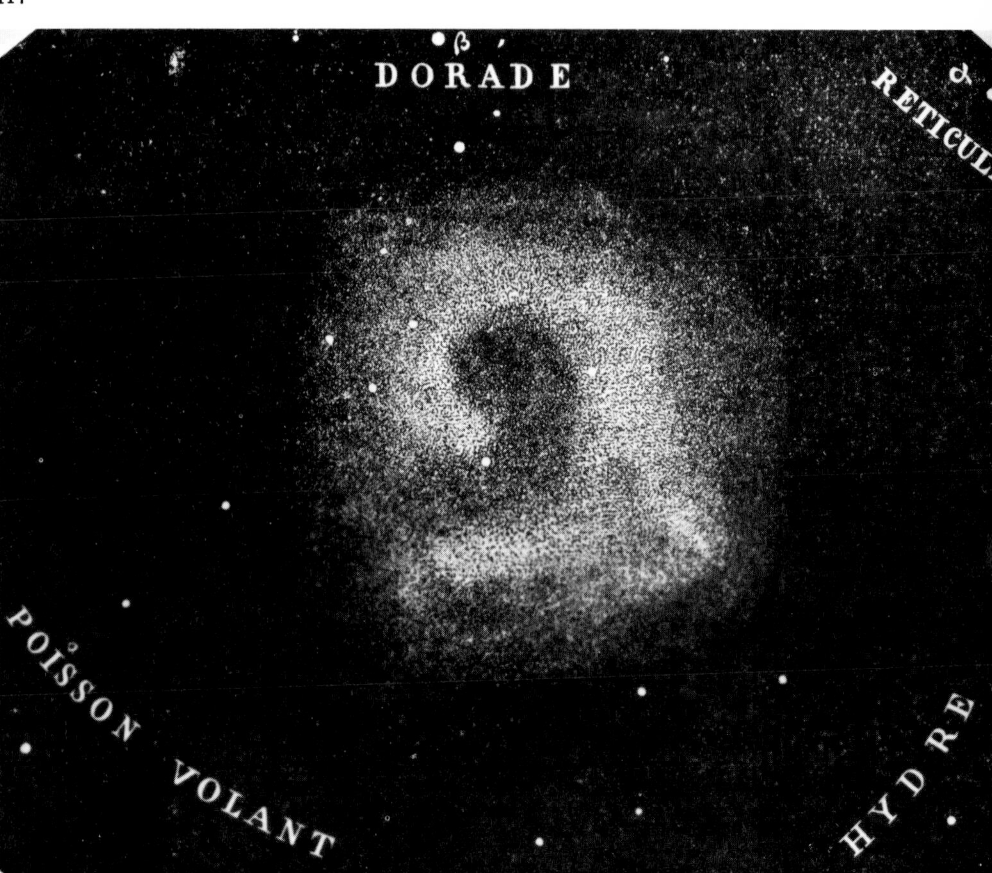

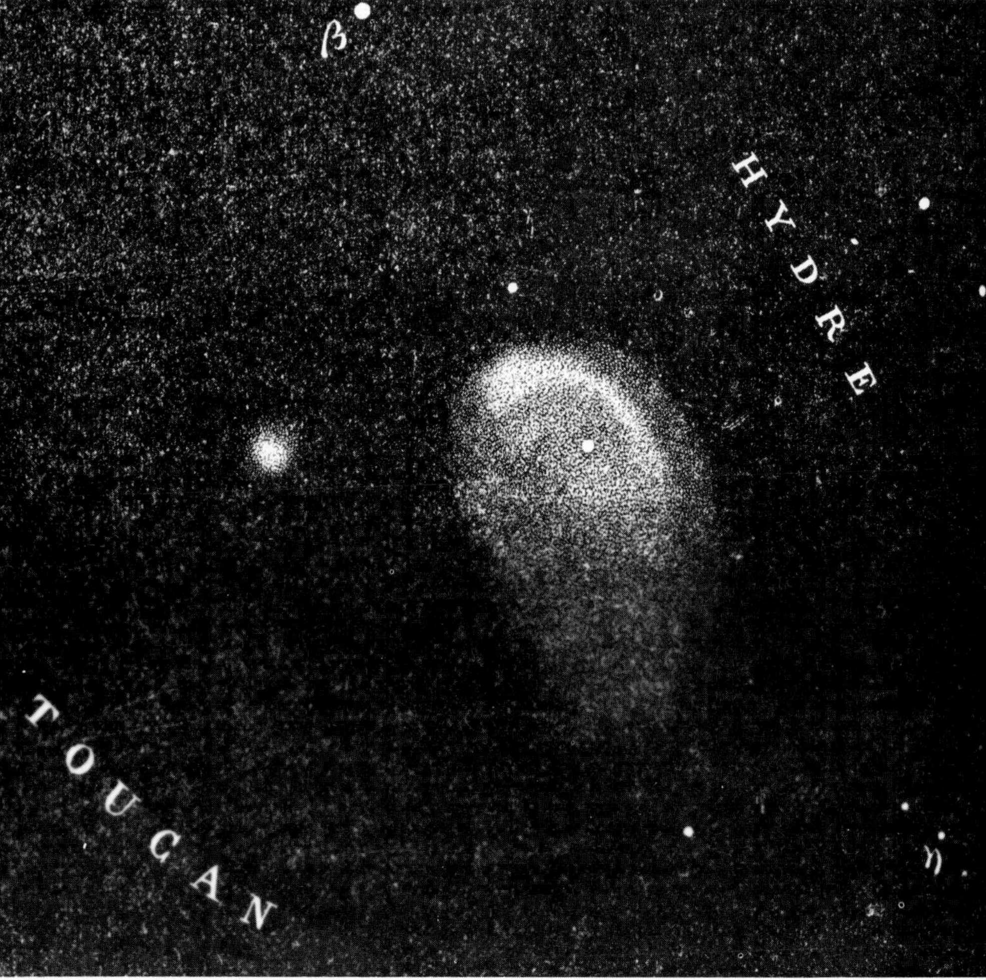

115

116

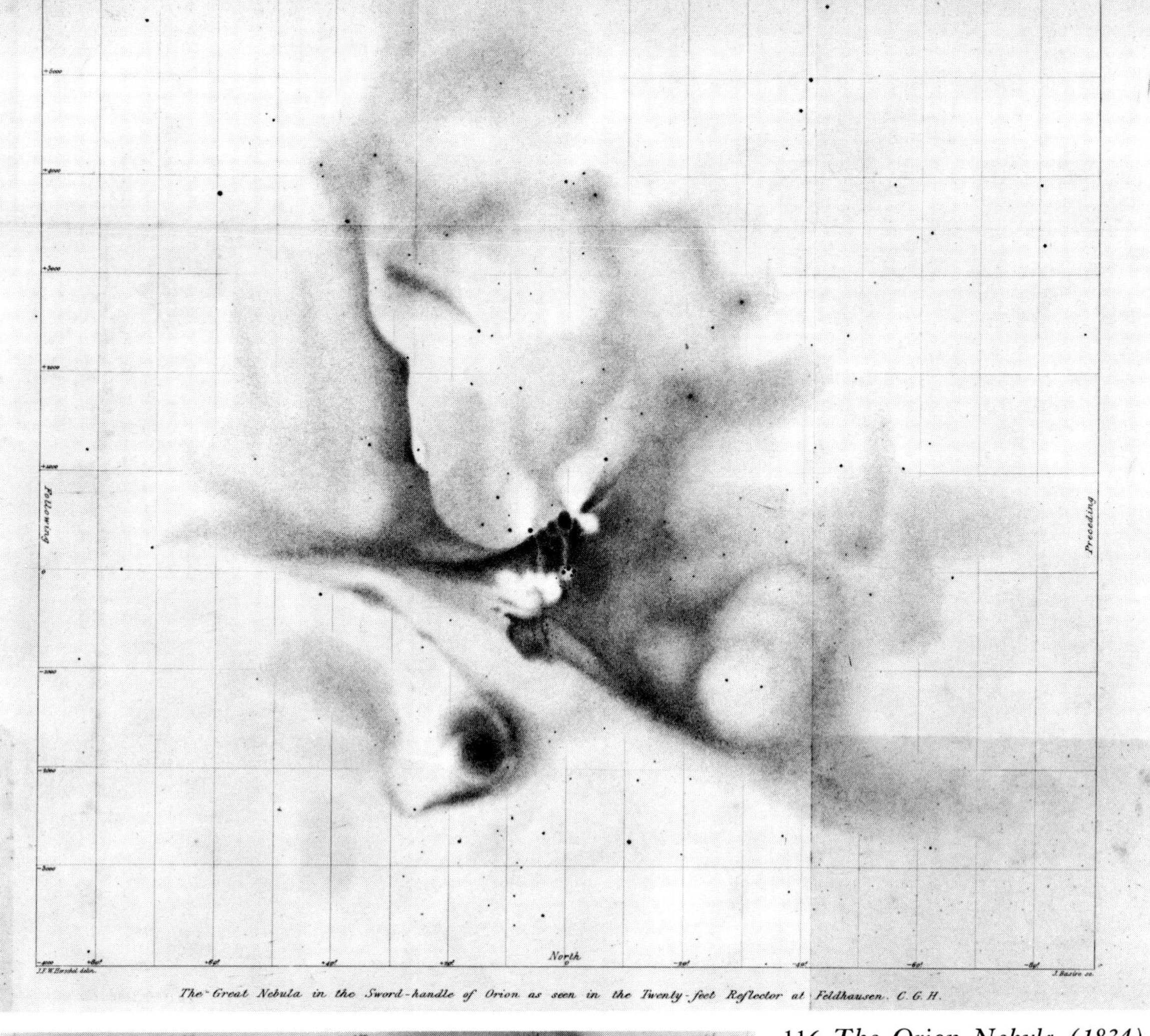

Following
Preceding
North
The Great Nebula in the Sword-handle of Orion as seen in the Twenty-feet Reflector at Feldhausen. C. G. H.

117

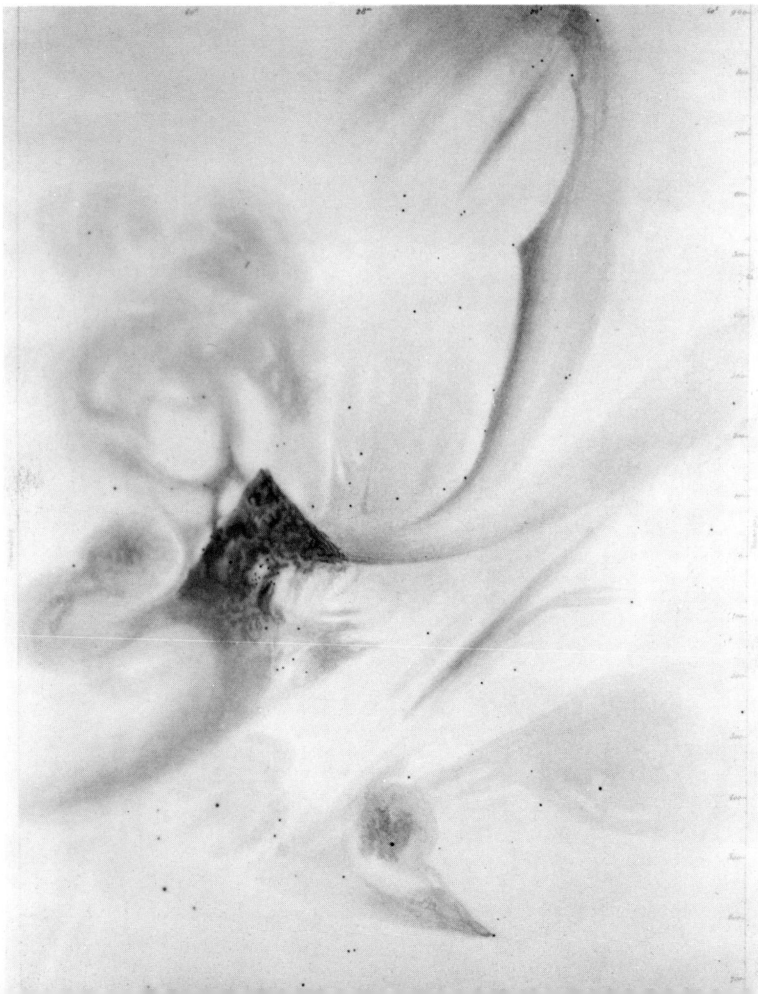

116. *The Orion Nebula (1834).* The great nebula in Orion had occupied the attention of Sir William Herschel, and it continued to fascinate his son. In this dramatic and hauntingly beautiful drawing, John Herschel depicted what appeared to be a cosmic process in full development.

117. *The great nebula in Orion, as observed and drawn in 1867.* G. P. Bond, using the refractor at Harvard College Observatory, sketched the Orion Nebula in 1867. There had been some improvement in definition in the 30-odd years since Herschel had recorded his observations.

118. *A photograph of the great nebula in Orion.* The advent of astronomical photography made it possible to "see" details the human eye could not detect. This photograph was made in the 1890's by the Englishman Isaac Roberts, one of the last independent astronomical amateurs.

119

120

119. *The great nebula in Orion.* The camera and the 40-inch refractor combined to produce this magnificent photograph. It was taken at the Yerkes Observatory in 1901 at an exposure of 3 hours.

120. *The ring nebula Messier 57 Lyrae.* This figure should be compared with the globular nebulae pictured in Figure 112. It shows clearly how much more could be seen with greater precision of detail by the latter part of the nineteenth century.

121. *The ring nebula in Lyra, photographed from the Lick Observatory.* This and Figure 122 illustrate the peak of astronomical viewing power in the nineteenth century. The telescope at the Lick Observatory in California, a 36-inch refractor, was a superb instrument; this photograph reveals its ability to capture the details of an astronomical object. The modern astronomer looking at this picture had to propose hypotheses quite different from those Sir William Herschel had suggested on the basis of what he could see in 1811.

122. *The spiral nebula M51 Canum Venaticorum.* The difference that the camera and a high-powered telescope made in what could be seen in the farthest reaches of the heavens is graphically demonstrated by this photograph. Compare it with the "nebulosities" drawn by Herschel (Figure 116).

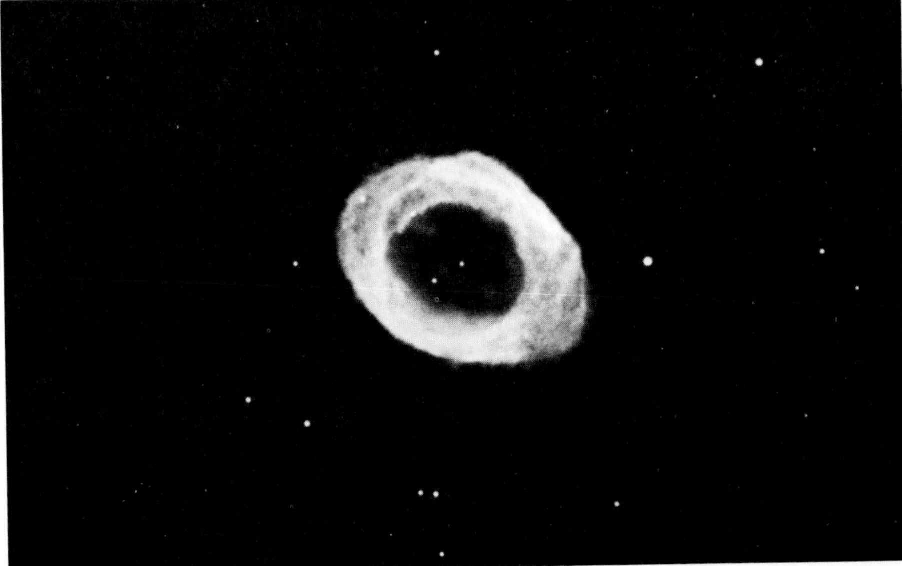

121

Comets

In terms of popular appeal, comets were clearly the most spectacular astronomical bodies. By the nineteenth century, they had lost their status as ill omens, but they still fascinated those who observed them. Two basic questions about comets demanded attention. Were they, as they seemed, wanderers from outer space that intruded into the solar system or were they just rather eccentric members of the family of planets and asteroids? Secondly, what were they? The first question was to be answered in the nineteenth century, the second is still not settled.

The first comet to be assigned a regular and predictable course was that whose orbit Edmund Halley correctly computed in the seventeenth century. Encke's comet was shown to follow an elliptical path in 1823 (by Johann Franz Encke) and so joined Halley's comet as a permanent member of the solar system. During the nineteenth century, measurements made on cometary paths tended to prove that these paths were elliptical, even though the ellipses may have been extremely elongated or eccentric. In those few cases where the elements of the orbit indicated either a hyperbolic or parabolic path, which would have meant that the comet came from and returned to outer space, it could be shown that these paths were the results of gravitational encounters with members of the solar system. The nineteenth century concluded, therefore, that comets were part of the solar system, distinguished from other bodies by their peculiar paths and peculiar composition.

The question of what a comet is was a more difficult one to answer. By their very nature, comets are difficult to observe accurately. They are naturally "fuzzy" and give ample scope for the imagination. The "structure" of the head of Donati's comet, as represented in Figures 127-130, gives some idea of what comet-watchers saw. From their drawings, it was impossible to derive any firm conclusions although the fact that comets are often accompanied by meteor showers led many to assume that they were merely a peculiar form of meteorite. What was peculiar about them was first detected through spectroscopic observations. Donati, in 1864, first examined a comet with a spectroscope. Later observations indicated that comets contain hydrocarbons, cyanogen, and carbon monoxide. How these substances are formed and how they become associated with a celestial object remained mysteries for the nineteenth century.

123

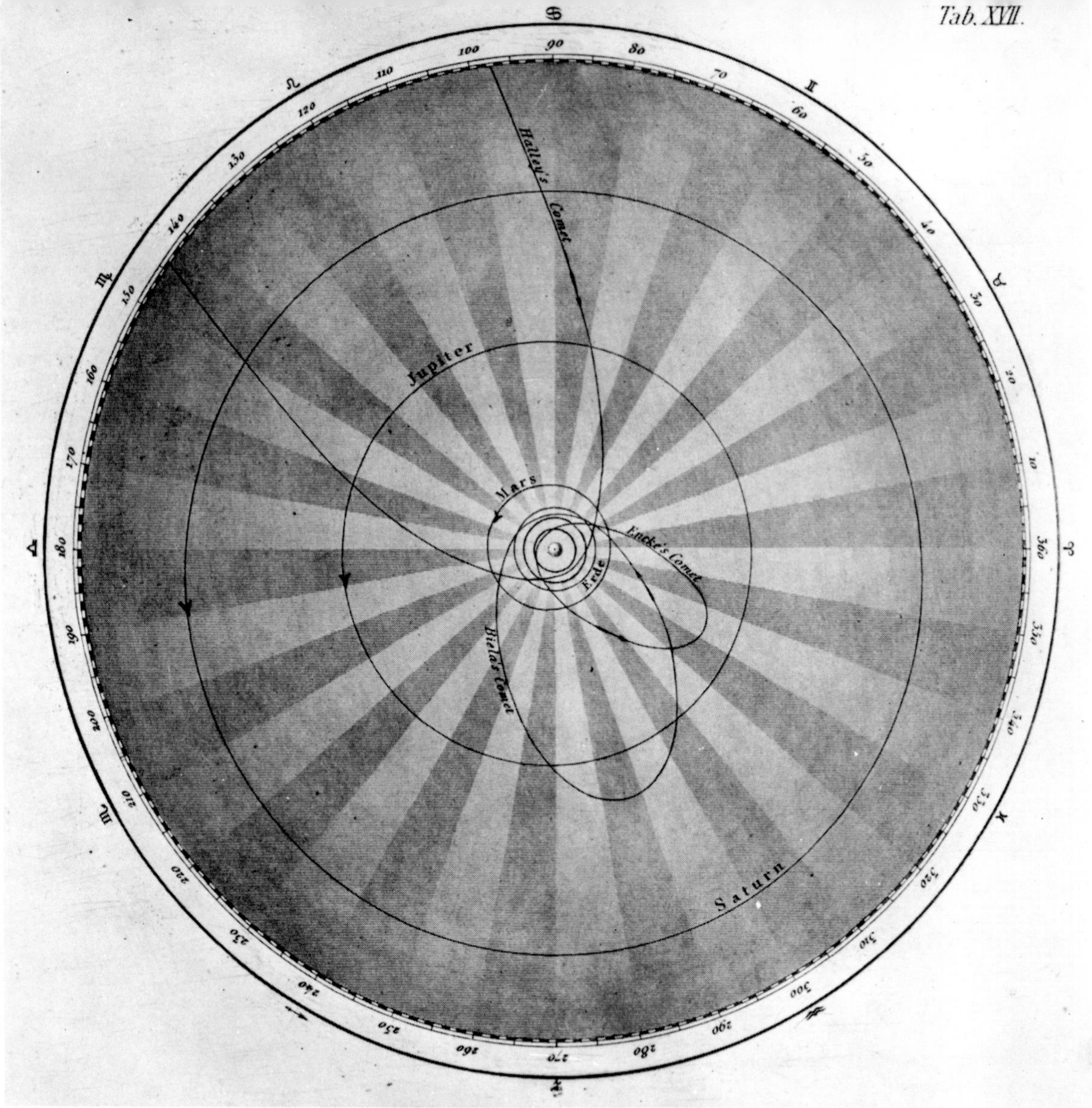

123. *The paths of comets.* Between the nebular heavens and the solar system were the comets, which had fascinated and terrified men since the dawn of time. Some of their mystery was dissipated for the astronomers of the nineteenth century. The paths of comets, like those of the planets, eventually could be reduced to tidy little diagrams like this one, which was drawn in 1875 and shows the paths of Halley's, Encke's, and Biela's comets.

124

124. *The great comet of 1811.* For most people, comets were awesome sights; and the comet of 1811 was particularly striking. There were still many who felt that it presaged disaster, and the defeat of Napoleon's Grand Army in Russia in 1812 must have served to confirm the superstitious in that belief.

125

126

125. *Comets as seen in Spain.* By the nineteenth century, Spain had become an intellectual backwater of Europe. This figure illustrates Spain's backwardness in two ways. First, the *Semanario pintoresco* ("Pictorial Weekly"), in which this picture appeared, used woodcuts, while the illustrated journals of the day in France, England, and Germany used steel engravings to produce their illustrations. Second, the situation portrayed in the picture is impossible. No comets would be visible so close to the sun as the two shown here are, nor were two comets of such size ever seen in such proximity to one another, going in opposite directions around the sun. Apparently scientific accuracy was not a requirement for the "informed public" of Spain.

126. *Donati's Comet.* In 1858, Donati's Comet appeared and was keenly observed by astronomers throughout the world. This picture shows how it appeared to the naked eye in Cambridge, England.

127. *Views of the head of Donati's Comet.* The nature of comets fascinated the astronomers of the nineteenth century. There was, therefore, a good deal of attention focused on the head of the comet, which appeared to be the most important part. These drawings of a comet's head, made between September 21 and October 5, 1858, at the Cambridge Observatory, are extremely difficult to interpret and provide few clues to the ultimate solution of the cometary problem.

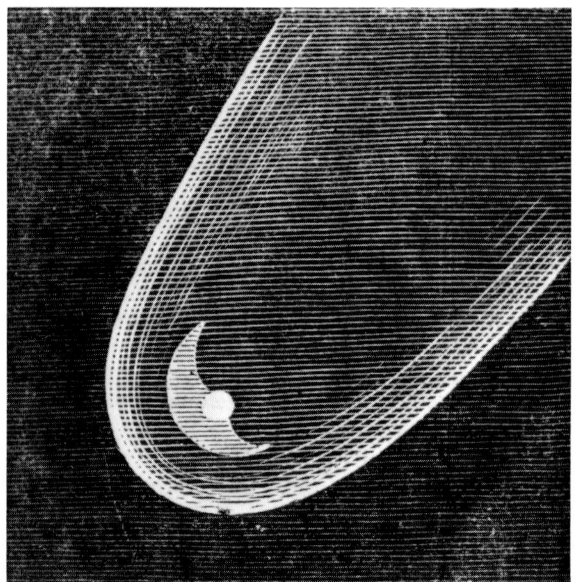

Sept. 21, 8h p.m.

Sept. 27, 7h. p.m.

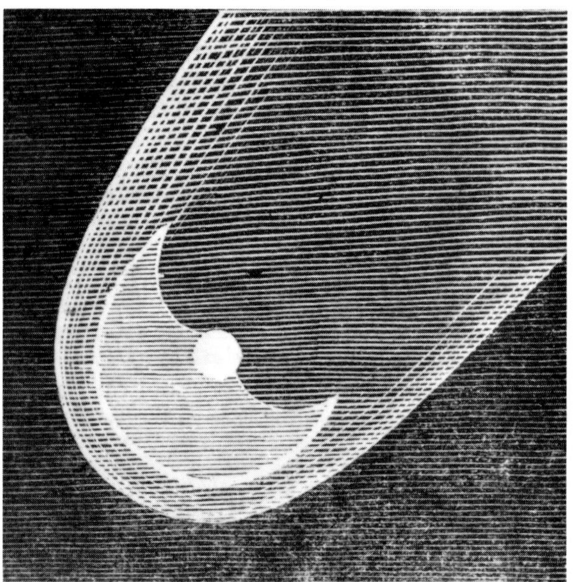

Sept. 24, 8h. p.m.

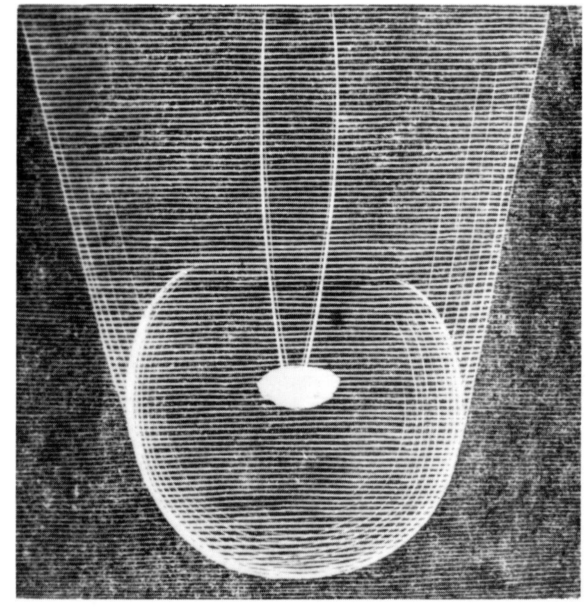

Sept. 30, 8h. p.m.

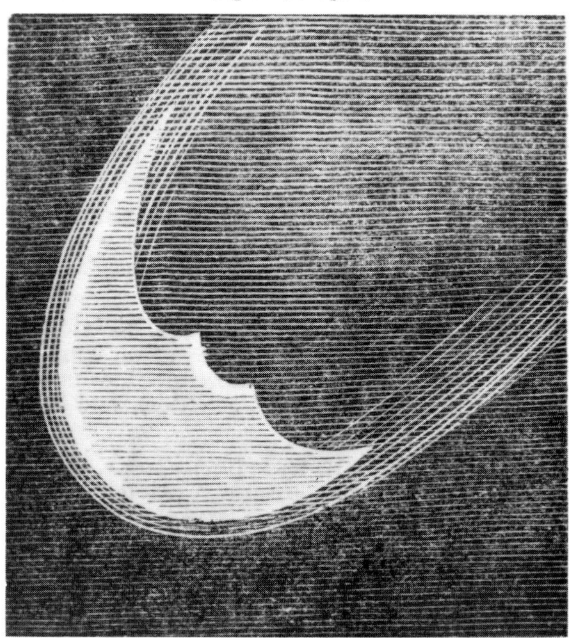

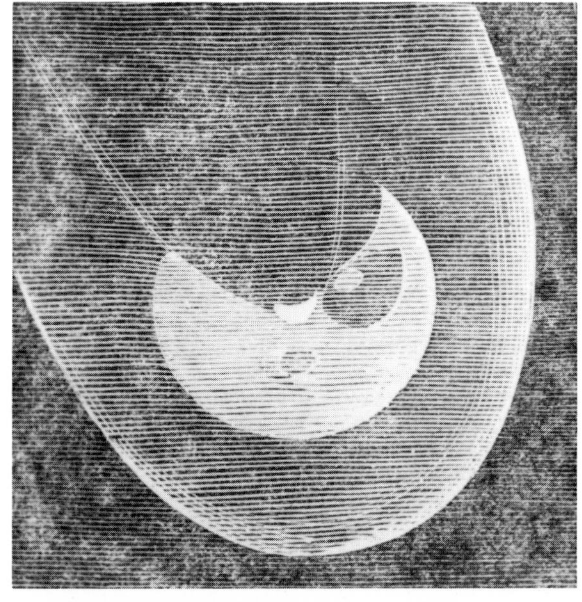

128

128. *Donati's Comet as seen from Slater's Observatory*. Using a 20-foot refractor, Mr. Slater of Euston Road, London, observed and drew the head of the comet on October 1, 1858. His comet seems easier to understand, for the head appears to be solid and surrounded by luminous material. It differs so greatly from the drawings made at Cambridge that one wonders whether the two observatories were viewing the same object.

129. *Donati's Comet*. The most systematic account of Donati's Comet was that published by G. P. Bond of the Harvard Observatory. The greatest care was taken with the engravings to ensure their accuracy. They provide the best account of a comet's "natural history" in its passage through the solar system. The comet is shown here as it enters the solar system.

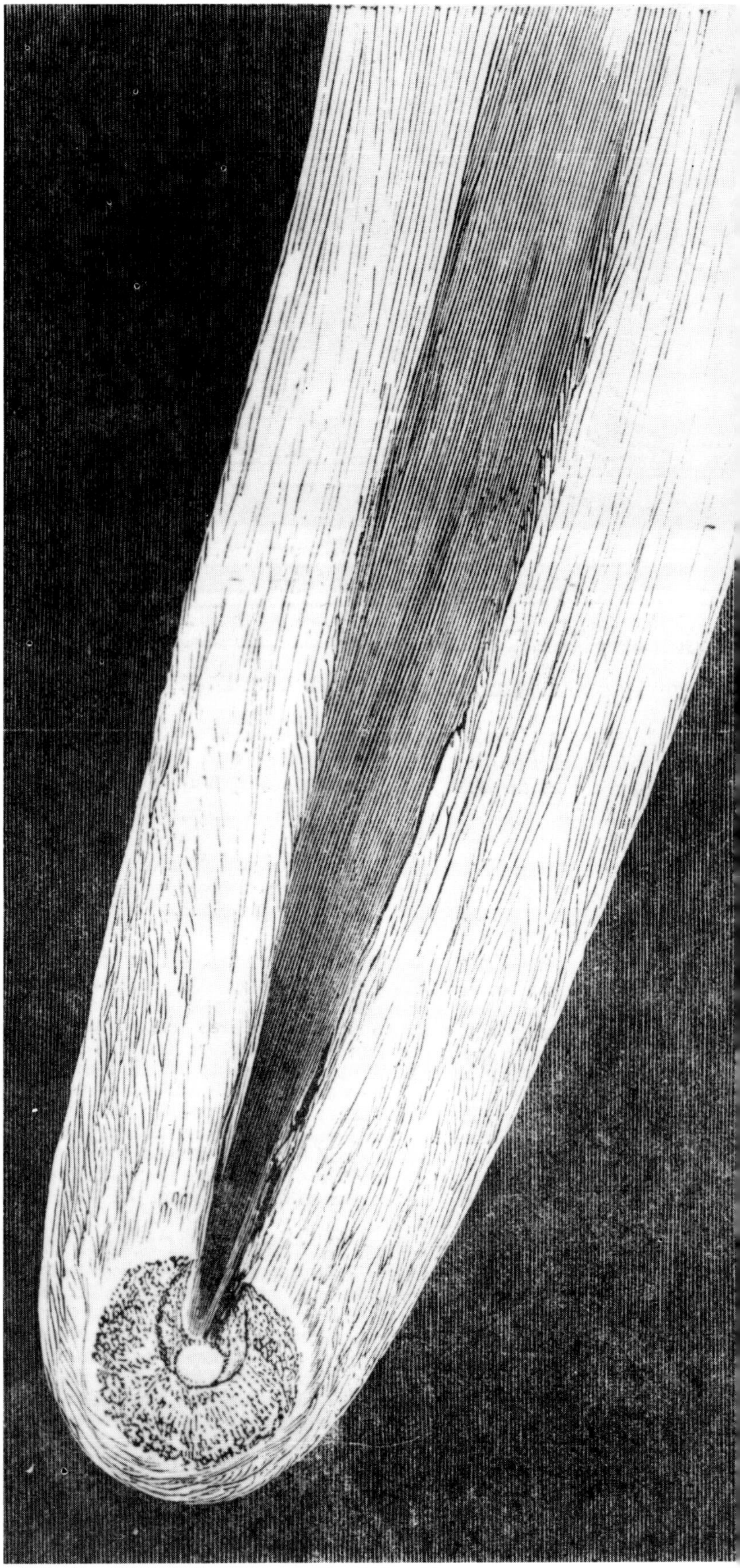

129

130

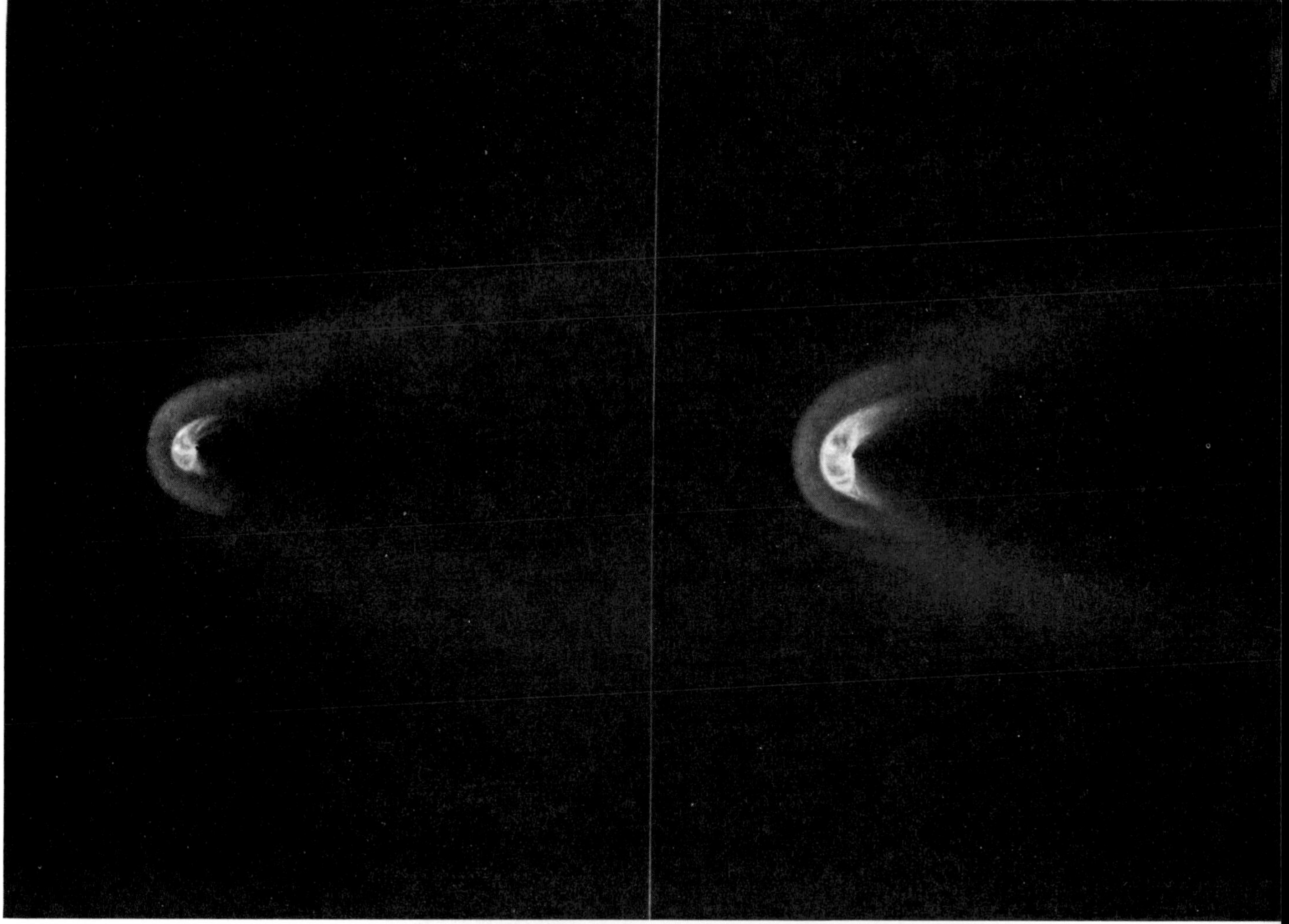

130. *The head of Donati's Comet.* The bizarre appearance of the head of the comet is illustrated here. It would certainly seem to be something quite different from the solid body represented by Slater in Figure 128.

131. *The comet—a French fantasy.* Donati's Comet stirred the imagination, and this French allegory was one of the results. It is, to put it mildly, puzzling to the modern viewer. What a comet has to do with winemaking or why its appearance should be accompanied by bibulous *putti* is clear probably only to the Gallic mind.

131

The Solar System and Transits of Venus

The great triumph of the eighteenth century had been the application of Newton's laws to the planets and their orbits. This, together with the measurement of the absolute distance of the sun from the earth at the time of the transits of Venus in 1761 and 1769, appeared to settle the dynamics of the solar system once and for all. But, as instruments and observations improved in the nineteenth century, doubts began to creep in. The solar parallax (half the angular diameter of the earth as it would appear from a position at the center of the sun) that had been calculated from the transits of Venus was called into doubt and since this figure is fundamental to the calculation of astronomical distances, it cast all other dimensions of the solar system into the doubtful column. These doubts troubled only the professional astronomer for they had little or no effect on the qualitative representation of the solar system. This, at least, appeared to have been laid out, once and for all, and the many planetaria and orreries and other representations of the solar system were presented confidently to the public as the true picture of their immediate astronomical environment.

The opportunity for redetermining the solar parallax offered itself with the 1874 transit of the planet Venus across the disk of the sun. Even with the improved instruments that had become available, there was considerable concern among astronomers about their ability to profit from the transits. The eighteenth-century transits had revealed an observational problem that had injected a large possibility for error in the recording of the raw data. When Venus made contact with the limb of the sun, a crucial point in the observations, the contact was not a clean one. Instead of remaining a distinct, circular disk, Venus appeared to flow into the limb as it approached contact and this made it exceedingly difficult to determine the exact time of contact. Similarly, when Venus lost contact with the limb and moved over the surface of the sun, there was another teardrop effect which blurred the moment of loss of contact. The problem appeared to be caused by the failure of the eye to detect the point of contact and two main efforts were directed toward eliminating this source of error. If astronomers could practice on some kind of a model, then they could gain sufficient experience to make the observation properly and accurately. Models were invented to simulate the transit and astronomers assiduously worked on accustoming their eyes to the peculiar problem involved. At the same time, photography was called upon to assist the astronomer. The power of the light source (the sun) guaranteed the ability to make fast exposures, so that all that was required was to invent something like a photographic machine gun. That was exactly what was done, the result being a camera that worked on exactly the same principle as the recently invented Gatling gun. What practice and ingenuity could do had been done by 1874.

Like the eighteenth-century transits, this astronomical event was an international extravaganza. The more observations that could be made from widely separated observation posts, the better the results ought to be and 1874 saw the scattering of astronomers all over the globe. International cooperation was sought and achieved. Observatories were set up, equipped with the most modern instruments ready for the event. An enormous amount of data was collected from both the transit of 1874 and that of 1882. The results from just the German expeditions took twenty years to reduce and were published in five large volumes. But the actual solar parallaxes that emerged from the different national efforts were disappointing. They varied from

8.76″ to 8.88″ and astronomers had to wait for the twentieth century for a really accurate determination of the solar parallax (8.794″).

The failure to achieve the precision that had been hoped for came after a half-century of increasing confidence in the ability to make fine astronomical measurements. This ability, indeed, had led to one of the more dramatic astronomical events of the century. Uranus had been discovered by William Herschel in the eighteenth century and its orbit had been the object of considerable observational and computational attention. Early in the nineteenth century, Alexis Bouvard, Laplace's computational assistant, had computed tables for Uranus that differed from earlier data that had been accumulated when Uranus was considered a star. There was increasing suspicion that the perturbations that this indicated meant that the motion of Uranus was being disturbed by a new, unknown planet. In 1842-1843, J. C. Adams, then a mathematics student at Cambridge University tackled the problem and in 1845 suggested to G. B. Airy, the astronomer royal, where the unknown planet should be. Airy and the director of the Cambridge Observatory, to whom Adams had also addressed his results, felt that no mere stripling like Adams could possibly have found something that they had missed, and so neither took the time or the trouble to look. In the meantime, Urbain Jean Le Verrier had taken up the problem in Paris and, by 1846, could publish the elements of the orbit of the new planet. Again, no astronomer took the time to look where Le Verrier suggested until after he had written to the Berlin Observatory in September with a specific request to look where the planet was supposed to be. This time, Galle, the Berlin astronomer, looked; and there was Neptune. The discovery of Neptune made a deep impression on the world of the mid-nineteenth century. That a man could discover a new celestial object by simply making computations at a desk revealed the power and the truth of the Newtonian world machine for all to see. Here was the true enlightenment, to be compared with the old traditions of superstition and obscurantism. Le Verrier was lionized and science was given an enormous boost as *the* proper approach to truth.

The transits of Venus and the discovery of Neptune impressed the general public, even if this public did not quite understand how or why the results were obtained. But, a subject that could be counted upon to excite everyone, layman and astronomer alike, was that of the inhabitability of the other members of the solar system. Conjectures about the plurality of worlds were, of course, not new; Giordano Bruno had used the idea and was burned at the stake in 1600 for heresy. But now, in the nineteenth century, astronomers had the instruments available to discover the truth of the doctrine and the informed public eagerly awaited their verdict.

The first results of the new instruments were disappointing. The closest candidate for harboring extraterrestrial life was the moon and lunar life had been vigorously supported by William Herschel in the eighteenth century. In 1816 clouds were reported on the moon and one astronomer even claimed to have recognized fortifications and other buildings among the lunar craters. With the new refractors, things appeared to be somewhat different. In 1834 Bessel showed that there could be no lunar atmosphere sufficient to support life, and, later in the century, it was demonstrated from the kinetic theory of gases that all oxygen and water vapor must long since have diffused into space from the lunar surface. The moon was out as a home for extraterrestrial life.

Jupiter and Saturn also appeared unlikely candidates for life, given their enormous size. In the course of the century, however, it was possible to refine observations so that some details could be seen and it was also possible to travel in the imagination to these planets. The public could be entertained by "views" from these celestial points, even if everyone recognized that there probably was no one there actually to do the viewing. The real breakthrough came with the close study of Mars. Earlier telescopes had been unable to detect more than the grossest features of the red planet. The ice cap could be seen, and some vague kind of markings made out. In 1859, during the opposition of Mars, the Italian Jesuit astronomer Pietro Secchi, announced that he had seen two "canali" that appeared to be permanent features of the Martian landscape. "Canali" does not necessarily mean canals in Italian, but the term stuck and in 1877, Secchi's compatriot, Giovanni Schiaparelli, took advantage of the new opposition of Mars to study its surface closely. The canals were reported in rather great number and became the objects of passionate debate and mapping. Canals, of course, implied canal builders and Mars watchers were carried away by their eagerness to map the achievements of the Martians. The American Percival Lowell founded the Lowell Observatory in Arizona for precisely this purpose in 1894 and enthusiastically added to the details. The knowledge that we are not alone in the universe obviously comforted Westerners whose sense of cosmic isolation was increasing as traditional religious ideas crumbled before the assault of reason and science. This would seem to account for the enthusiasm with which descriptions of the Martian surface were greeted in the popular press. The belief in Martians was to remain strong until well into the twentieth century.

The only other member of the solar system to attract more than professional attention was Venus. Like Mars, it was a candidate for habitation although its temperature militated against the appearance of life as we know it on earth. Nevertheless, it was scrutinized with some care. Its history in the nineteenth century was the opposite of that of Mars. The features that earlier astronomers had thought they had seen were discovered to be illusions. Venus was recognized as shrouded in a thick cloud cover that did not permit direct observation of the surface. No one really took Lowell's discovery of Venusian canals seriously since, in this case, only he could see them.

By the end of the nineteenth century, the planetary members of the solar system were familiar to all Westerners through the popular press. Professional astronomers had mapped their orbits and as much of their surfaces as could be studied. All mysteries but one had been removed from the planets as physical bodies. The one that remained was the peculiar nature of the orbit of Mercury. Its "perturbations" led Le Verrier and his followers to suggest another new planet between Mercury and the sun that no one could detect. It was not until Einstein's general theory of relativity appeared in the middle of World War I that this anomaly was explained.

132

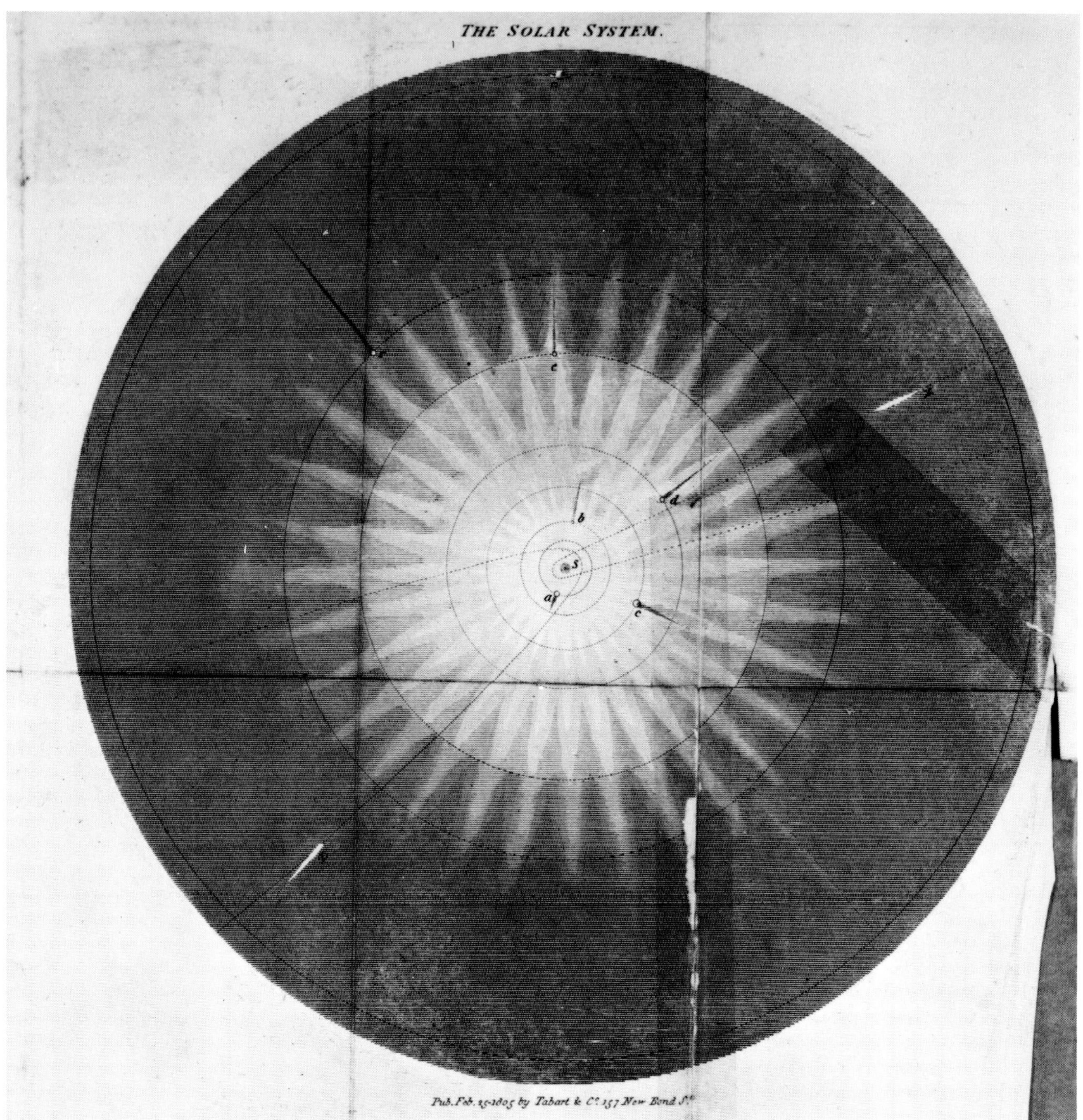

132. *The solar system.* Comets were interlopers in the solar system whose nature could only be guessed at. The planets and their satellites, on the other hand, were well known (it was thought) by the beginning of the nineteenth century. When a scheme such as this was published in 1805, it represented solid knowledge of the Newtonian heavens.

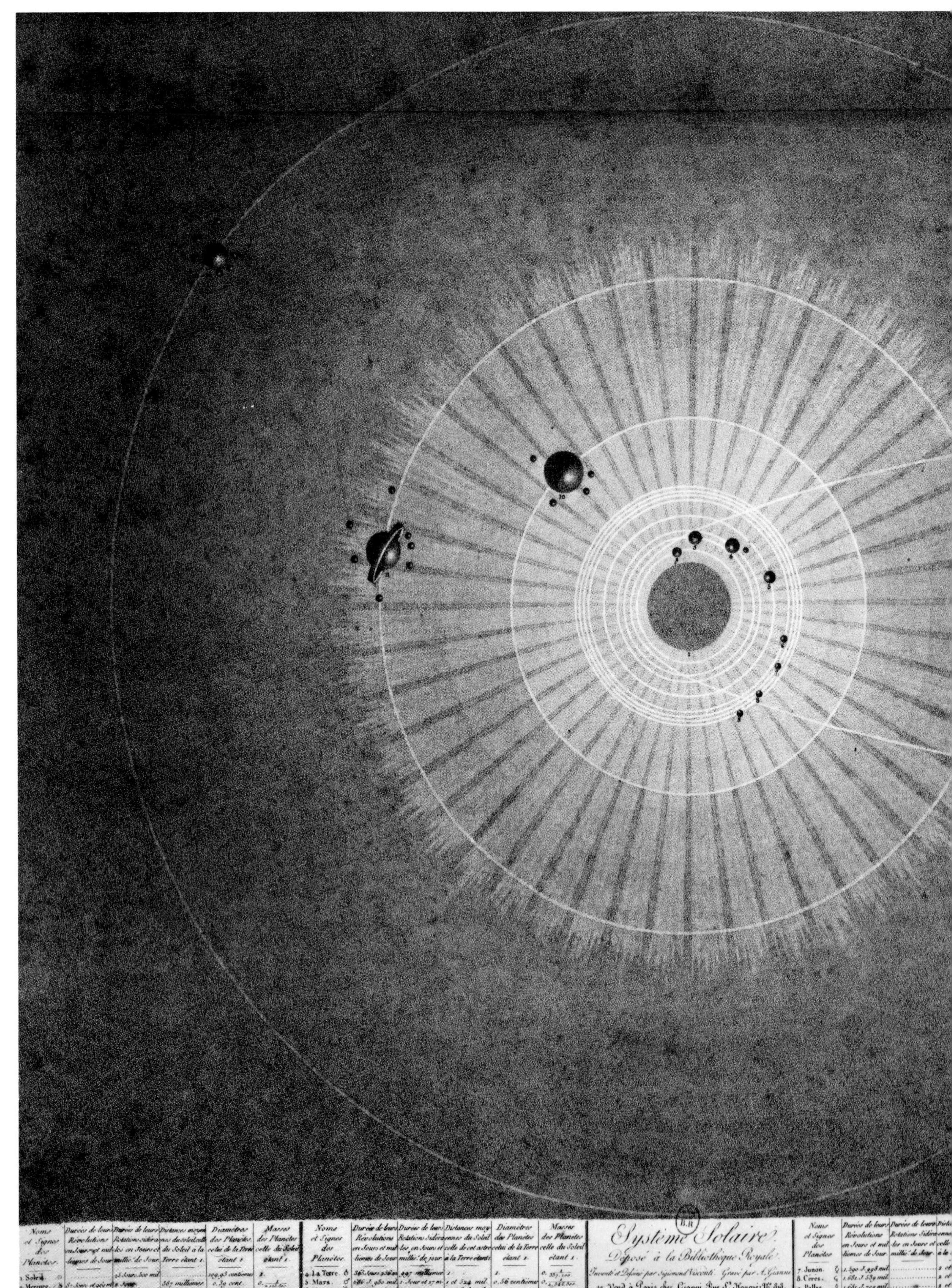
B.R
Système Solaire
Déposé à la Bibliothèque Royale
Inventé et Dessiné par Sigismond Visconti. Gravé par A. Gianni
se Vend à Paris, chez Gianni, Rue St. Honoré No. 313.

Noms et Signes des Planètes.	Durée de leurs Révolutions en Jours et millièmes de Jour.	Durée de leurs Rotations Sidérales en Jours et mil.^s de Jour	Distances moyennes du Soleil celle de cet astre à la Terre étant 1.	Diamètres des Planètes celui de la Terre étant 1.	Masses des Planètes celle du Soleil étant 1.
10. Jupiter. ♃	4,332. 596 mil.	414 millièmes.	5 et 203 mil.	11 et 56 cent.	0, 1/1067
11. Saturne. ♄	10,758. 970 mil.	426 millièmes.	9 et 539 mil.	9 et 61 cent.	0, 1/[illegible]
12. Uranus. ♅	30,688. 713 mil.		19 et 183 mil.	4 et 26 cent.	0, 1/[illegible]
13. Comète.					

134

135

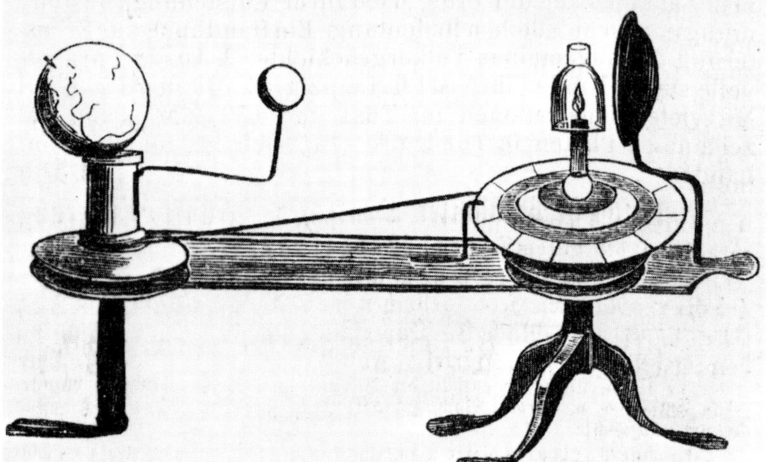

133. *A detailed Newtonian guide to the planets.* This engraving, after a drawing by Sigismond Visconti, represents the solar system as it was known when the nineteenth century began. All the planets and their satellites are represented. Note that the scale is much distorted; the sun is far too small.

134. *David Richter's planetocometarium.* The "fine structure" of the solar system was discovered in the nineteenth century, as illustrated by the orrery shown here. Between the orbits of Mars and Jupiter are the orbits of 36 asteroids. Eleven cometary orbits also are represented.

135. *The earth-moon-sun system.* Interest in the solar system inspired the building of simple models that could be used to demonstrate astronomical phenomena. This "Tellurium," as it was called, illustrated the phases of the moon, the seasons, and day and night.

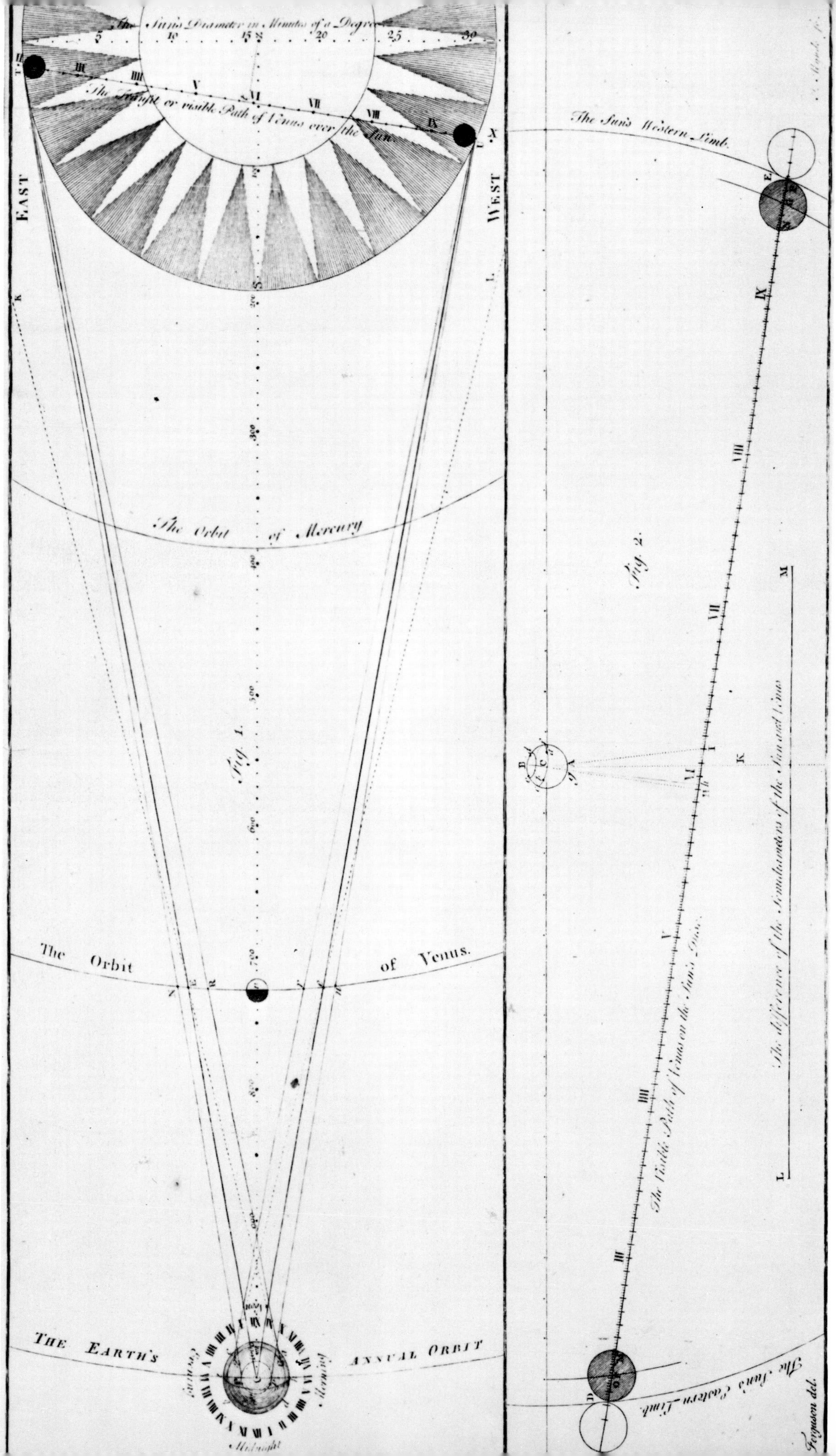

The Sun's Diameter in Minutes of a Degree

The Transit or visible Path of Venus over the Sun.

EAST

WEST

The Orbit of Mercury

Fig. 1.

The Orbit of Venus.

THE EARTH'S ANNUAL ORBIT

Midnight

The Sun's Western Limb.

Fig. 2.

The Visible Path of Venus on the Sun's Disc.

The difference of the Semidiameters of the Sun and Venus

The Sun's Eastern Limb.

Ferguson del.

137

136. *The transit of Venus.* One of the more important relationships in the solar system was that involving the transit of Venus across the disk of the sun. The transit had been used in the eighteenth century to discover the solar distance, as these diagrams illustrate. From observations taken at widely separated parts of the earth, recording the exact time Venus entered and left the sun's disk, it was theoretically possible to calculate, by using simple geometry, the absolute distance of the sun from the earth.

137. *The transit of Venus (1874).* The 1874 transit of Venus was anticipated as a great occasion of international scientific cooperation. A *Punch* drawing captured this international flavor rather nicely.

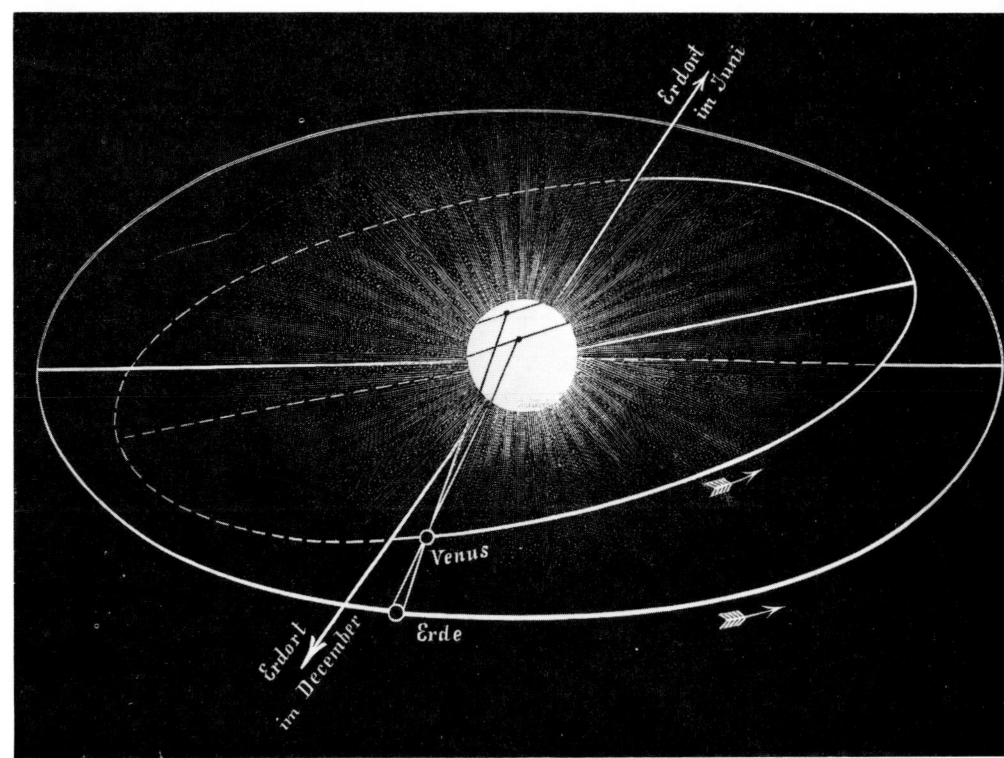

138

138. *An overview of the transit of Venus.* The geometry of the transit was clearly represented in this German drawing.

139

140

139. *Training for observing the transit.* During the eighteenth-century transits of Venus, observations of the exact times of Venus' entrance into and exit from the sun's disk were extremely difficult to make. Sir G. B. Airy, the Astronomer Royal of England, devised the apparatus shown here to enable astronomers to practice observing the phenomenon. An appropriately sized black disk sunk in a glass plate was passed at exactly the right speed over a brightly illuminated field of the same angular diameter as the sun's disk, thus making all the phases of the simulated transit occur in the proper order and at the proper rate. With enough practice on the model, it was hoped, the actual observations would be accurate.

140. *A French model of the transit.* The transit of 1874 proved to be a bit of a disappointment. Even with practice on models, it was difficult to observe the exact moment of contact between Venus and the sun's limb. A French commission was appointed to study the problem before the next transit, in 1882. This commission determined that the problem was one of diffraction of light around the image of Venus. A new model was then created to train astronomers to compensate for the diffraction effects. In the model, the disk of Venus was pulled across a round space illuminated by a powerful electromagnetic arc lamp. The Venus disk and the "sun" were both connected to electrical apparatus so that their actual contact was recorded immediately.

141. *The observer's station for viewing the model of the transit of Venus.* The observation post was 100 meters from the model of Venus and the sun. Here astronomers could practice recording the exact time that the Venus disk and the "sun" touched and then evaluate their findings by comparing their figures with those given by the electrical signals resulting from the contact.

142. *The photographic revolver.* Some of the observational problems attendant on the transit of Venus could be eliminated, or at least minimized, by using a camera. Astronomers had to find a way to take a large number of pictures in a short time. The photographic revolver was a solution. A piece of film was cut into the shape of a doughnut and placed in a magazine that could be turned by a handle. Exposures could be made simply by turning the magazine. Here a heliostatic mirror is shown reflecting the image of the sun into the photographic revolver. The result was the first "movie."

143a, b, c. *Exploded view of the magazine of the photographic revolver.* Figure 143a shows the butt plate of the magazine that holds the film. The little rectangles, marked *F*, are the shutters that, when opened, expose the film. Figure 143b shows how the revolver is loaded with the doughnut-shaped film. Figure 143c is a side view of the magazine.

142

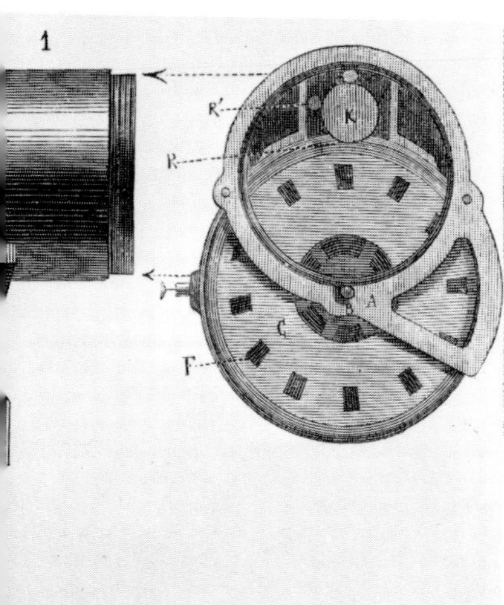

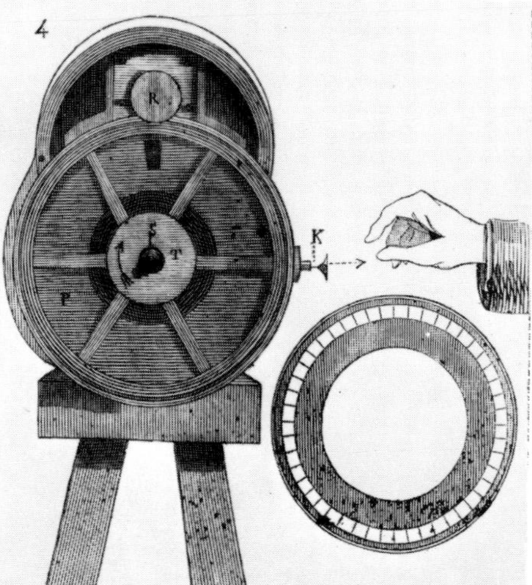

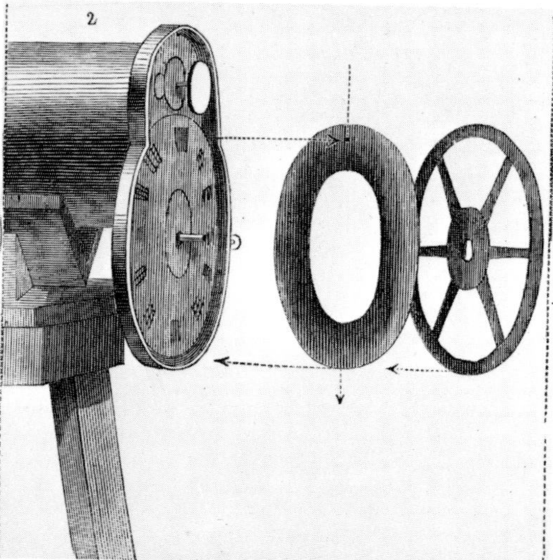

143
a, b, c

144

145

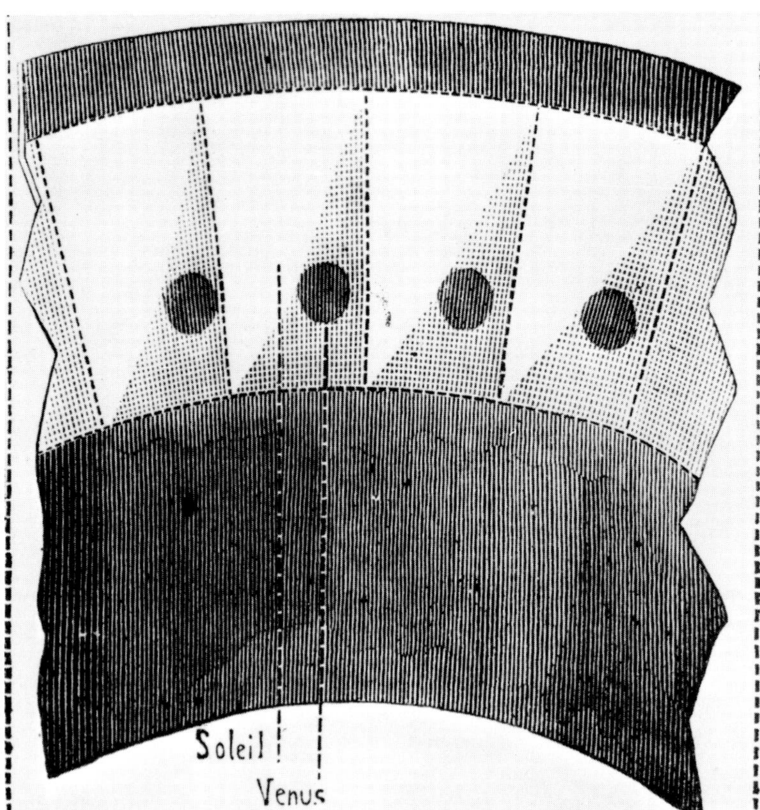

144. *Photographs of the contact of Venus and the sun taken with the photographic revolver.* A section of the photographic revolver's doughnut-shaped film shows Venus' gradual penetration of the sun's disk. Attaching the revolver to an electrical mechanism that would record the exact time of exposure of each "shot" of the revolver was all that was needed to measure the time of the transit accurately.

145. *A transit station in Honolulu.* The more observations that were taken at different points on the earth's surface, the greater was the possibility of obtaining accurate calculations of the solar distance. This station at Honolulu was typical of the makeshift observatories created to observe the 1874 transit all over the world.

146

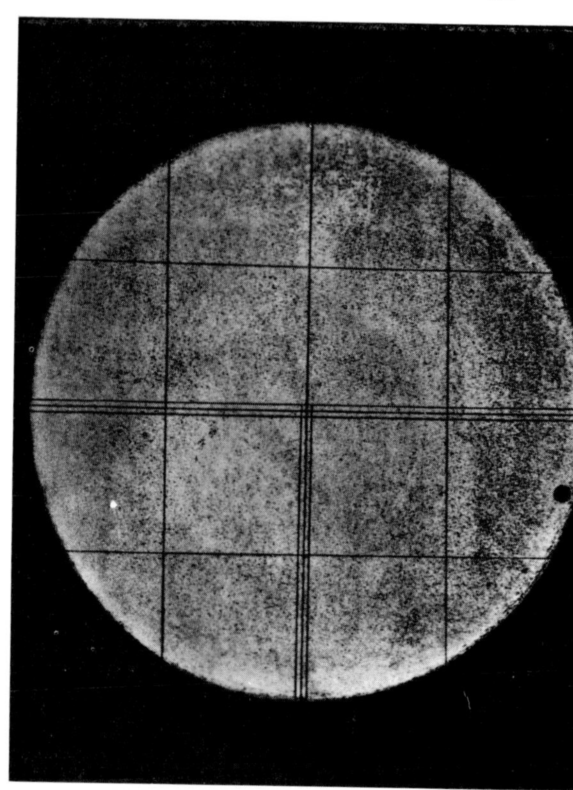

146. *The French station in China.* This was a far more substantial observatory than the one in Honolulu. Note particularly the quality of the telescopes and the fact that electrical power has been brought in to ensure accuracy.

147. *The transit of Venus.* The passage of Venus across the sun's disk in 1882 is here recorded by photography. The grid system helped make it possible for astronomers to define the times of their observation accurately.

148

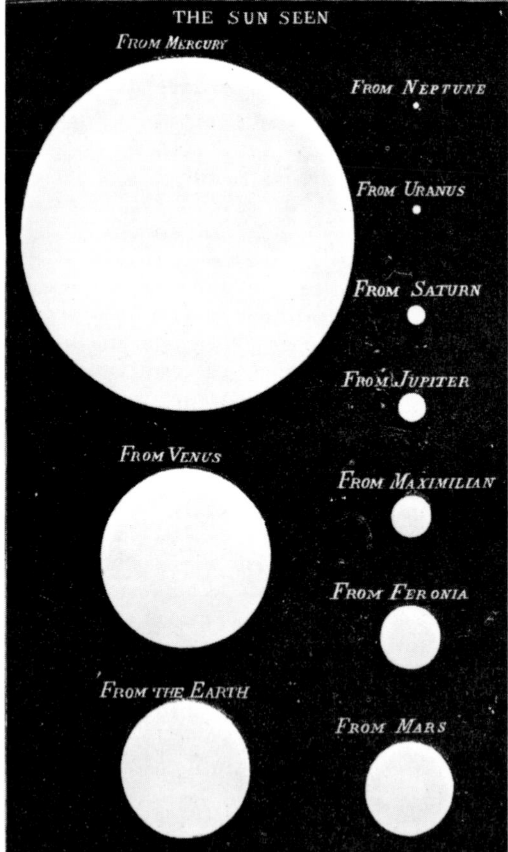

149

148. *The sun as seen from different planets.* The sun is the dominant member of the solar system, and it was of some interest to depict how it must have looked from the various planets and from the asteroids Maximilian and Feronia.

149. *The evolution of the solar system.* The sun not only dominated the solar system; it was also considered to have given birth to it. Or, perhaps to put it more accurately, it was believed that the sun and the planets had condensed out of a volume of primordial gas, as shown here.

150. *The sun.* The sun is not an easy body to observe with the naked eye, and it is difficult to detect elements of its structure. Here a picture of the sun, dating from 1802, makes little or no advance over what Galileo reported in the seventeenth century. The letter *a* indicates a sunspot.

150

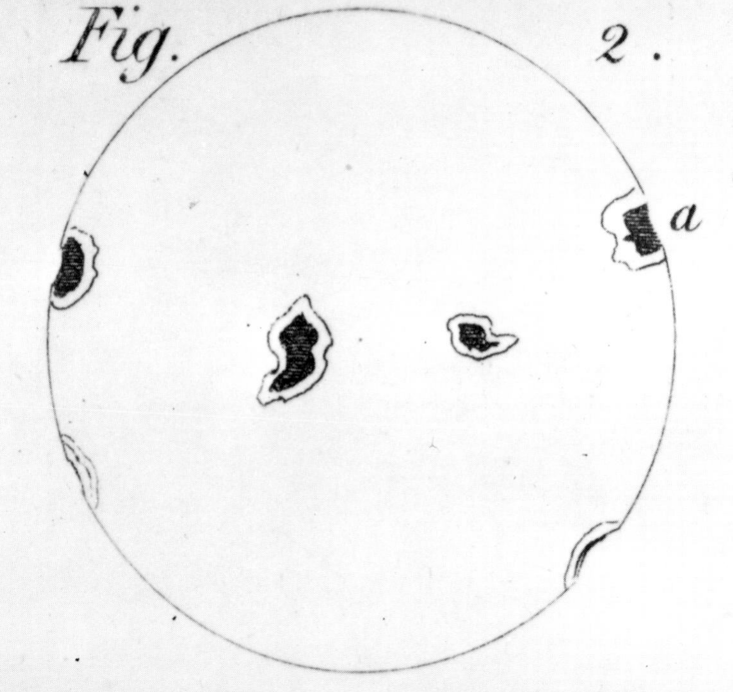

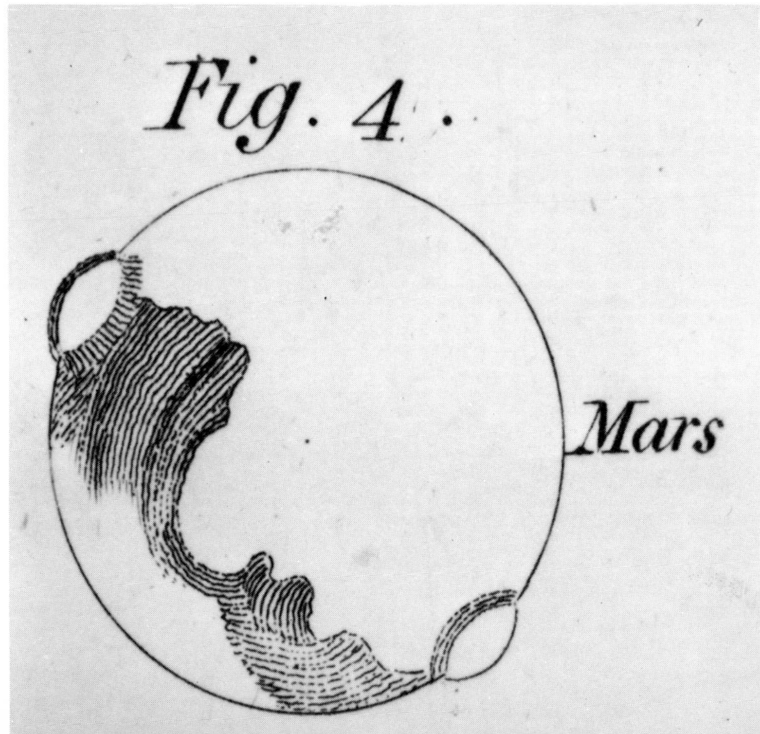

151. *Mars*. This view of Mars, made in 1802, exaggerates the polar caps. The shadings would appear to indicate surface features much like terrestrial continents.

151

152

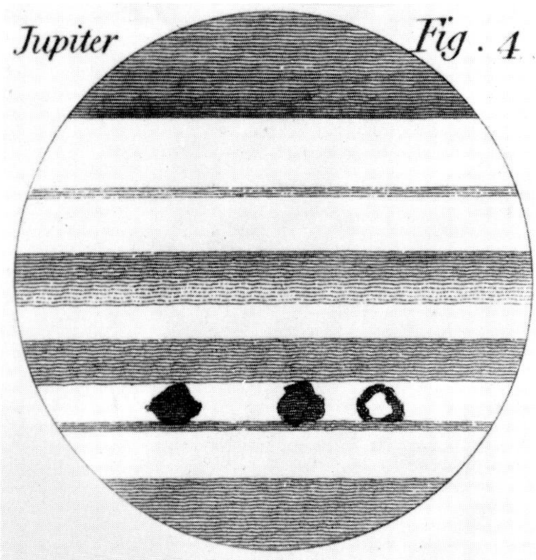

153

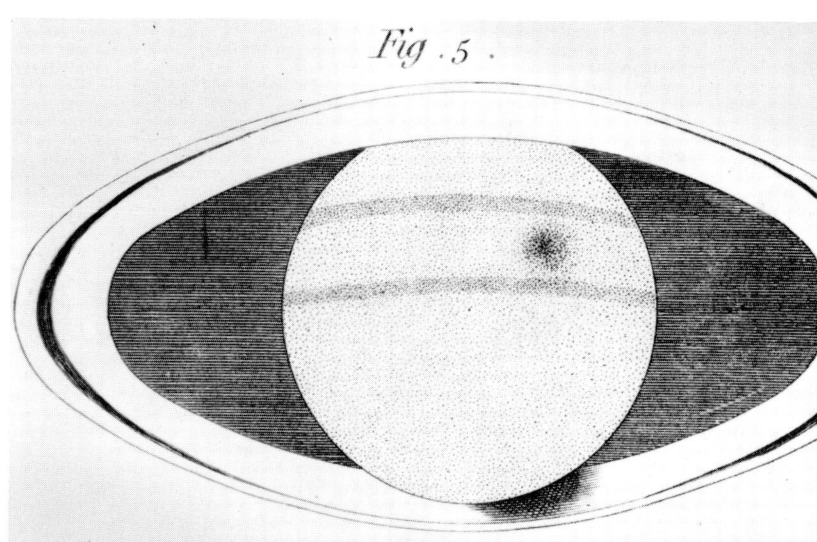

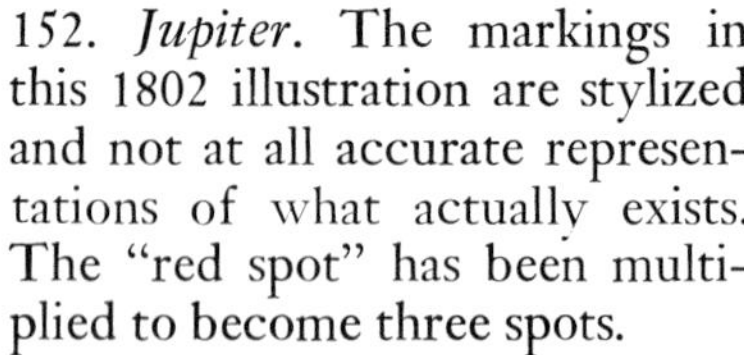

152. *Jupiter*. The markings in this 1802 illustration are stylized and not at all accurate representations of what actually exists. The "red spot" has been multiplied to become three spots.

153. *Saturn (1802)*. Here only two rings are visible and surface markings are indicated only in the vaguest way.

154. *The moons of Jupiter and practical affairs (1802)*. The moons of Jupiter provided a handy celestial timepiece that could be used for determining longitude on the earth. The times at which one of these moons passed behind Jupiter could be accurately observed and computed for an observatory such as the one at Greenwich. These observations and calculations could be reduced to tabular form and made available to travelers. Then, if someone—say, in the woods of Canada—observed the time of passage of the moon into Jupiter's shadow, and compared that time with the time at Greenwich, it was a simple matter to determine just where on the earth he was. All that need be done was to determine the difference in times. Let us assume it was six hours. The whole earth turns in 24 hours, so he was 6/24ths of the way around the globe, which equals 90 degrees west of Greenwich.

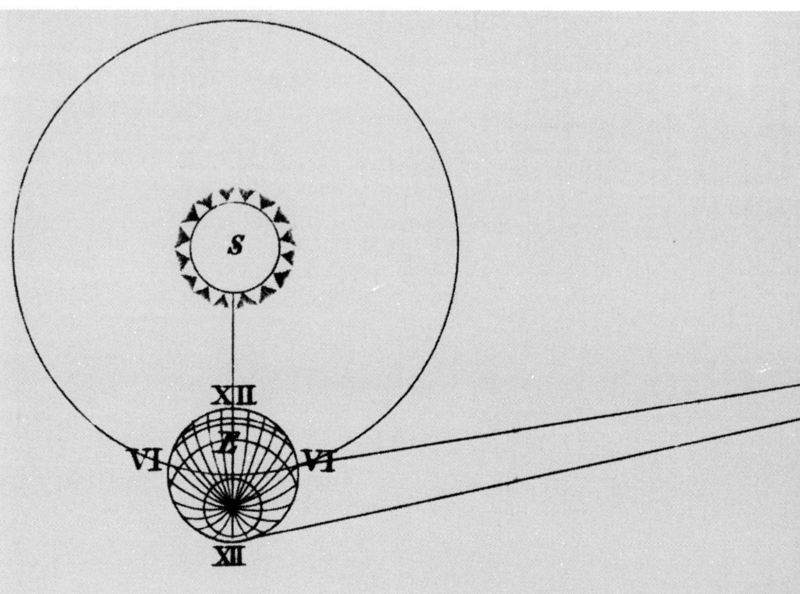

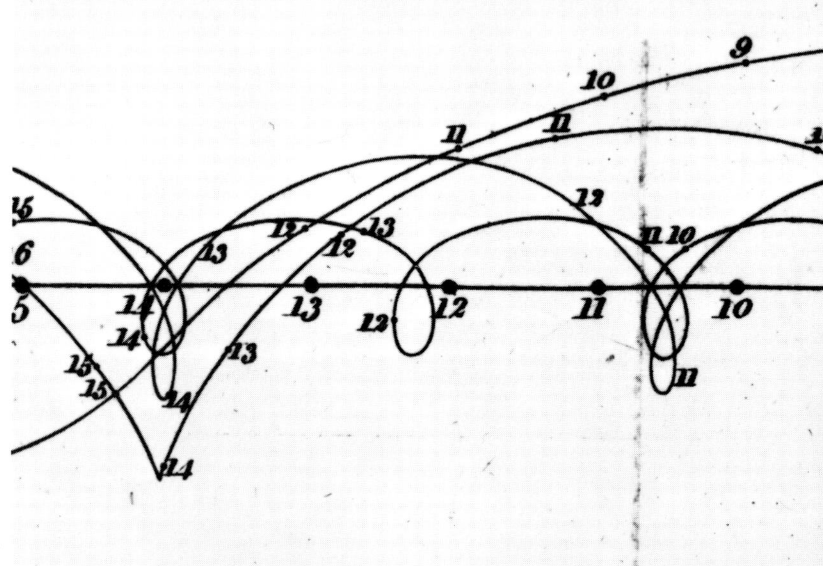

154

155

156

155. *Paths of the satellites of Jupiter.* By the beginning of the nineteenth century, it was possible to determine very complicated celestial orbits by means of accurate telescopic observations. The figure here shows the various routes taken by the moons of Jupiter (labeled Sat. 1, Sat. 2, etc.) as the planet moves along the straight line of its orbit.

156. *The discovery of Neptune.* One of the most exciting astronomical events of the nineteenth century was the discovery of a new planet — Neptune — by the French astronomer Urbain Le Verrier. The English astronomer John Couch Adams had calculated its position from perturbations in the orbit of Uranus but could not prevail upon British astronomers to look for it. A nationalistic French cartoon shows Adams, under a British flag, sighting Le Verrier's discovery in the scientific press, while the French astronomer actually sights the planet.

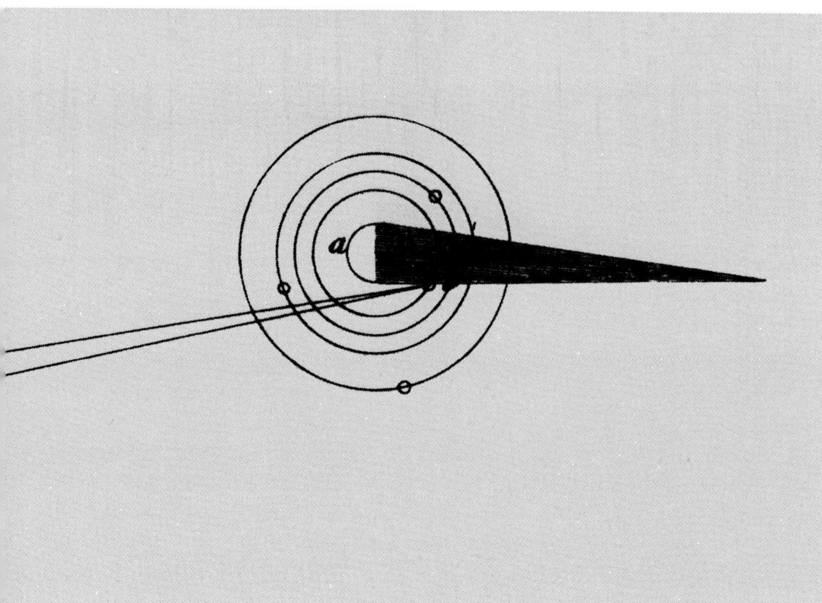

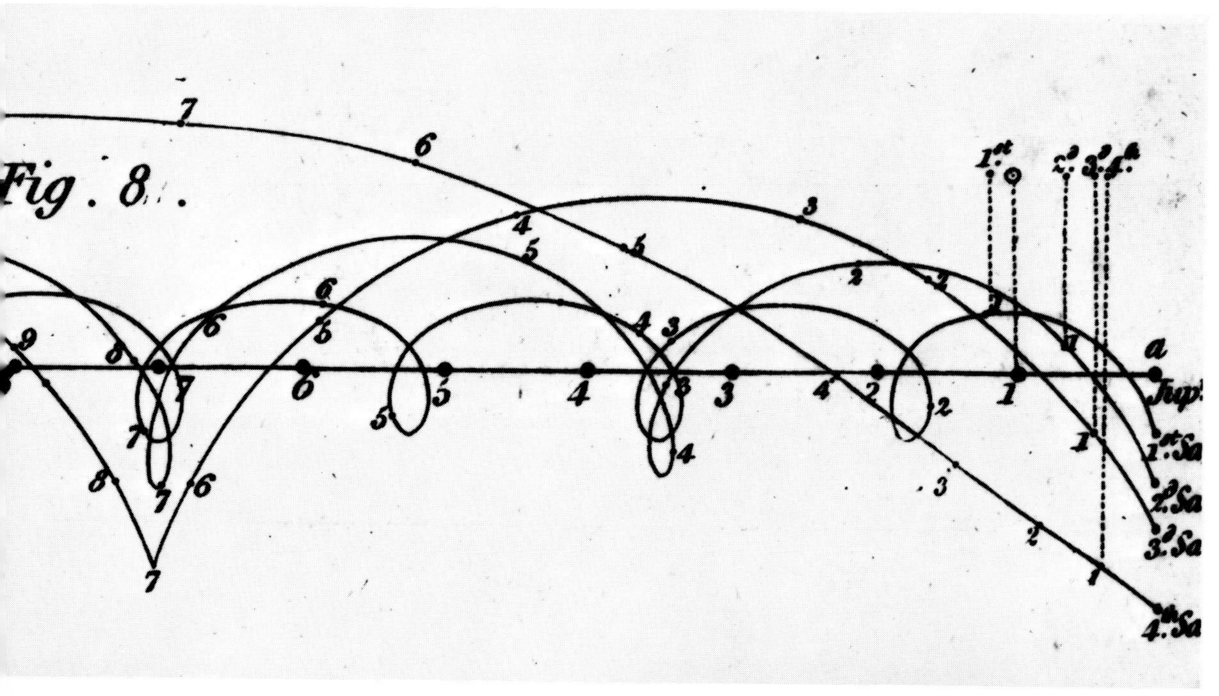

157

157. *Mr. Adams looking for Le Verrier's planet.* In this rather cruel caricature, the French cartoonist Cham sarcastically disposed of Adams' claim to be the codiscoverer of Neptune. His suggestion that Adams had no idea of where to look was hardly fair, for Adams had stated rather precisely where Neptune must be.

158. *Saturn (1805).* Saturn and its satellites are depicted fairly accurately in this early nineteenth-century engraving. The more complicated structure of the ring is hinted at by the existence of a band between the light inner and the dark outer segments.

159. *Imaginary views of Saturn's ring from Saturn.* The eye of the imagination could supplement what the telescope could reveal. Here an artist depicts Saturn's "ring," known by the late nineteenth century to be composed of dust and small particles, as it might have appeared from Saturn under different conditions. At the top is the ring at midnight, in summer; in the middle, at midday in winter; and, at the bottom, at sunset in summer. All views are from the northern hemisphere.

158

160

160. *Saturn's ring over a twenty-year period.* The ring did not always look the same. Part of the reason for this was that Saturn was viewed in different attitudes, so that the perspective changed. Part undoubtedly was due to changes in the distribution of material in the ring. Also, the surface markings of Saturn changed. With the advent of newer telescopes and professional observers, they were no longer seen as simple, geometrical stripes. By 1890, the dimensions of the ring could be measured with some accuracy: its diameter was 176,000 miles, its width was 30,000 miles, and its thickness 40–50 miles.

161. *Jupiter and four satellites (1805).* Jupiter is poorly, almost schematically, represented here. The bands are completely regular and the "red spot" is not shown, possibly because it was on the other side of the planet when the observation was made. Only four satellites are depicted.

162. *Jupiter (1843).* Knowledge of the planet had not improved greatly when this British engraving was made, showing Jupiter on October 21, 1843. The bands are less regular and the large "hump" in the middle band is the "red spot." Only three satellites are shown.

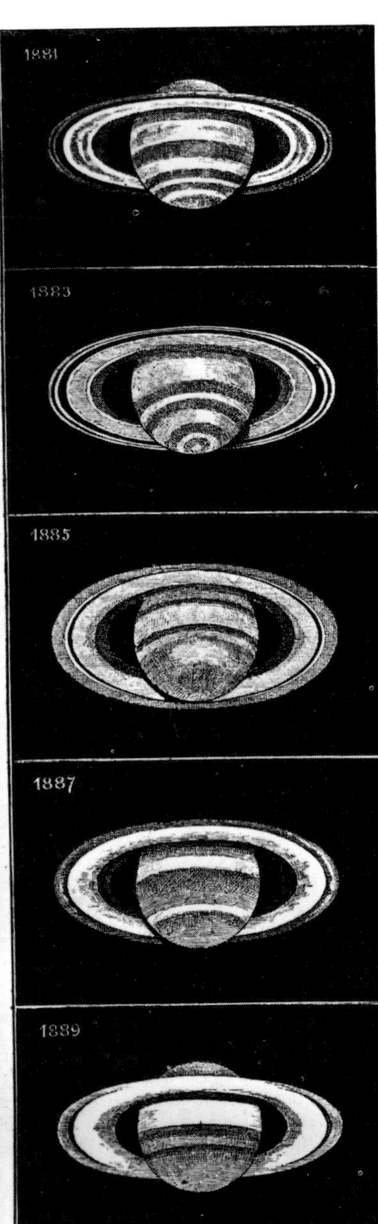

161

162

163. *Jupiter (1889).* By 1889, the telescopic definition of the surface markings of Jupiter was striking. The "red spot" is clearly visible in the top and bottom photographs. The "bands" have been resolved into more complicated structures. Publication of photographs such as these eliminated the distortions that were bound to appear in drawings and engravings. The telescope used, the Lick Observatory refractor, guaranteed the highest optical quality then possible.

164. *Mars (1805).* The planet is only vaguely depicted. The one polar ice cap is reasonably well defined, but the only other feature discernible is the strange serpentine band that "flows" from it.

165. *Mars (1862).* By 1862, greater surface detail was visible. The polar ice cap stands out, and there appears to be a large continent occupying the major part of the northern hemisphere. The observation was made by John Philips, Professor of Geology at Oxford, who was an amateur astronomer.

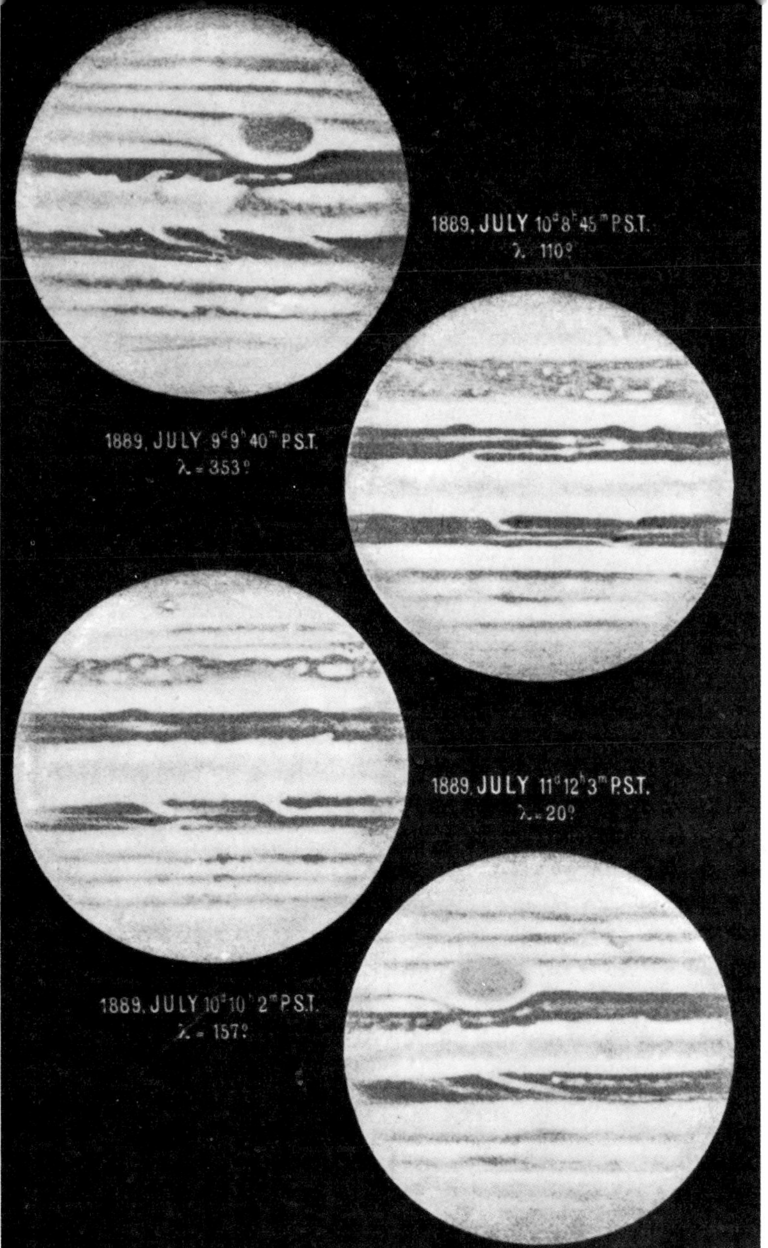

163

164

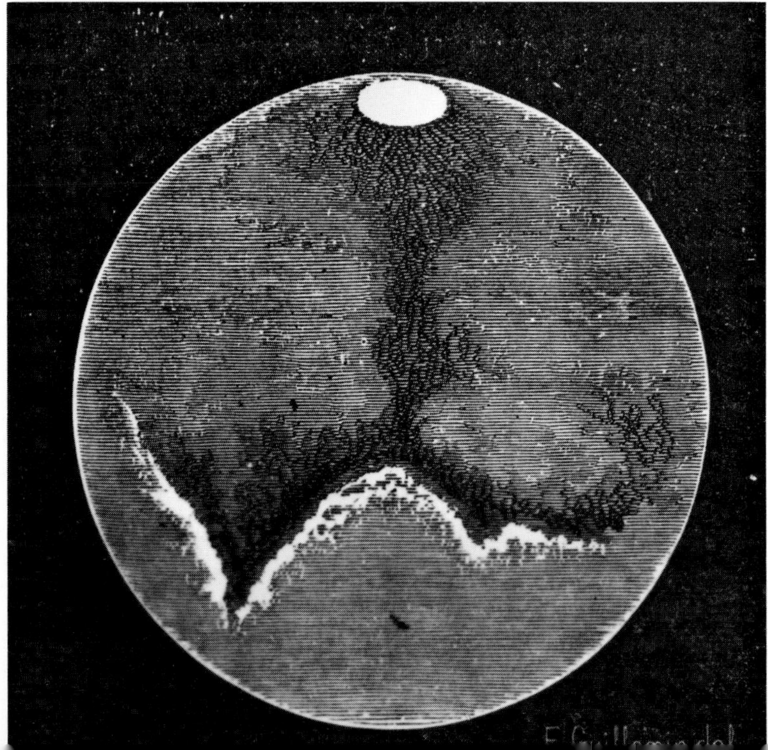

165

PHÉNOMÈEES OBSERVÉS SUR LA PLANÈTE MARS

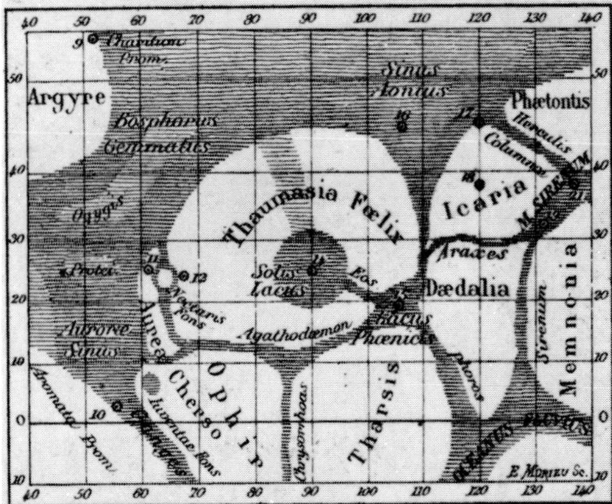

Fig. 1. — Le Lac du Soleil en 1877.

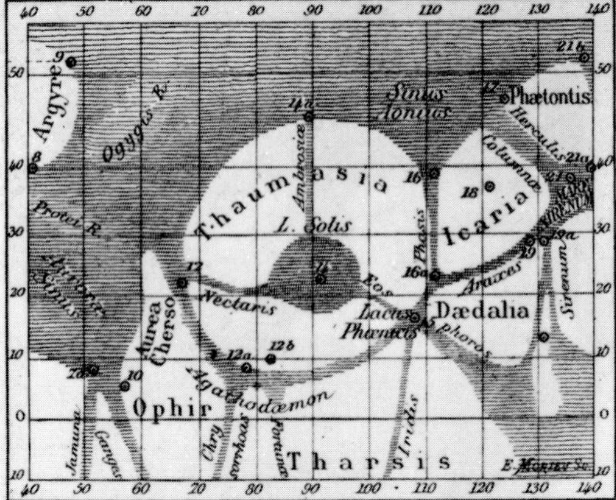

Fig. 2. — La même région en 1879.

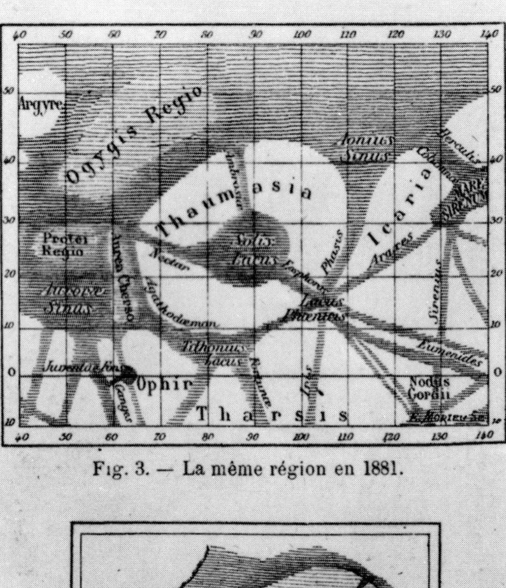

Fig. 3. — La même région en 1881.

Fih. 4. — La même région en 1890.

166. *The surface of Mars.* Mars captured the imagination of the late nineteenth-century public, largely through the efforts of the Italian astronomer Giovanni Schiaparelli, the French popularizer Camille Flammarion, and the American astronomer Percival Lowell. Flammarion was convinced that life and intelligence were scattered throughout the universe, and his observations of Mars strengthened his belief. These engravings, showing one Martian region from 1877 to 1890, indicated the presence of "canals" that hinted at the probability of intelligent life. The "canals," like the natural features observable on Mars, were given names drawn from antiquity. The region shown here appears also at the upper right of the map of Mars, Figure 170.

167a, b, c. *Surface changes on Mars.* These views, which Flammarion believed supported his faith in the existence of life on Mars, appeared to him to be changes occurring over time in the planet's river beds or canal beds. The region shown here can be seen, somewhat differently, at about longitude 320, just south of the Martian equator, in Figure 170.

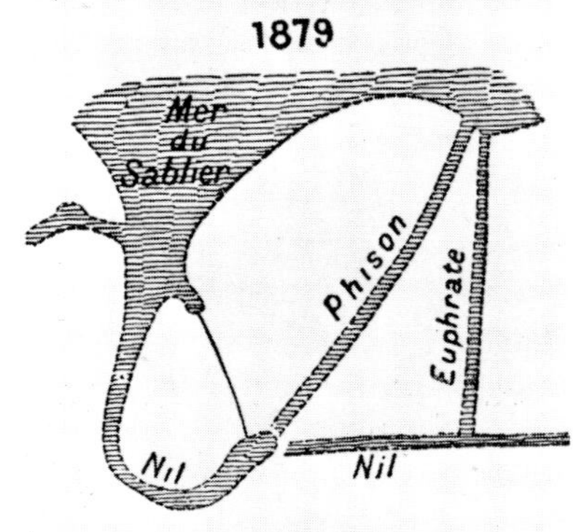

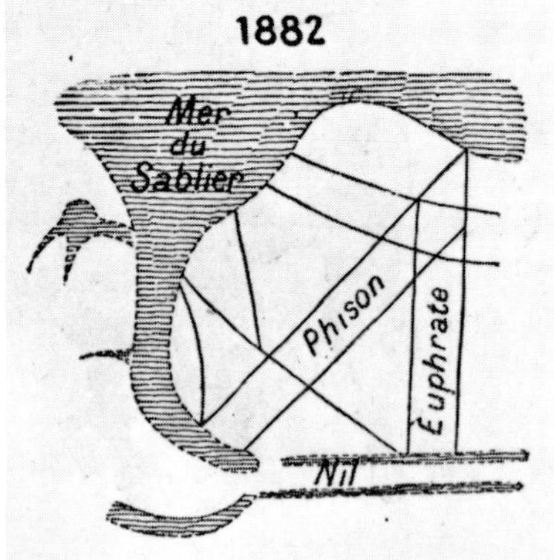

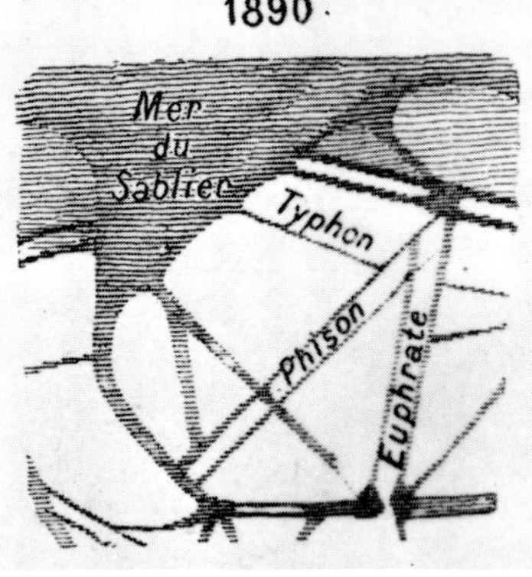

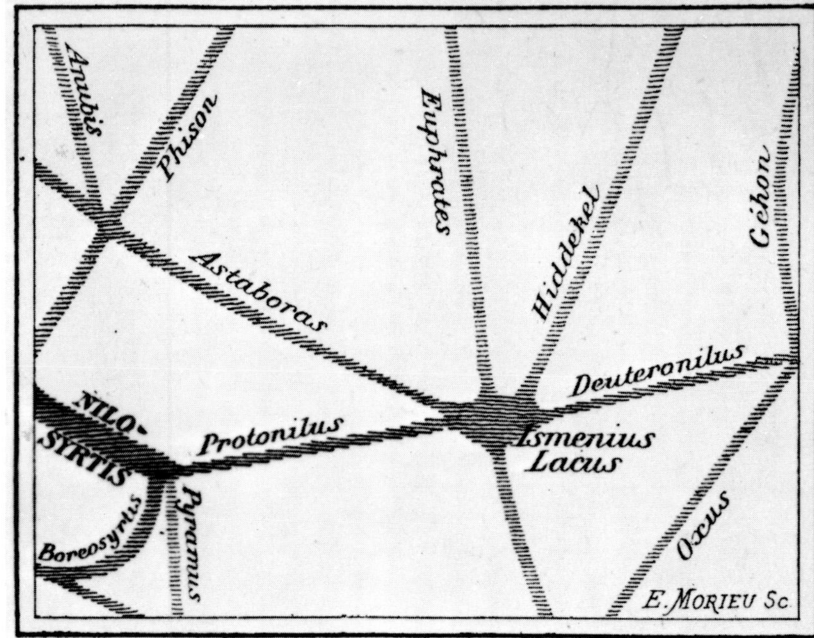

168

169

168. *The intersection of canals.* This view seemed to lead almost inescapably to the supposition that intelligent life existed on Mars. It clearly represents a network of canals whose geometrical form could hardly be due to chance. The practice of giving the names of ancient rivers to these "canals" also subtly supported Flammarion's theory. Labeling a feature "Euphrates" made it difficult *not* to think of it as a river. In Figure 170, the "lake" shown here, Lacus Ismenus, can be seen at 40 degrees south, meridian 330.

169. *Mars (1888).* The detail that could be "seen" by 1888 was striking. The overall geometric grid of canals seemed difficult to explain without the hypothesis of intelligent life on Mars.

170. *A map of Mars.* By the late nineteenth century, it was possible to map details on Mars with almost terrestrial accuracy. Every feature was named, and there was a familiarity to the Martian landscape that seemed to guarantee its reality.

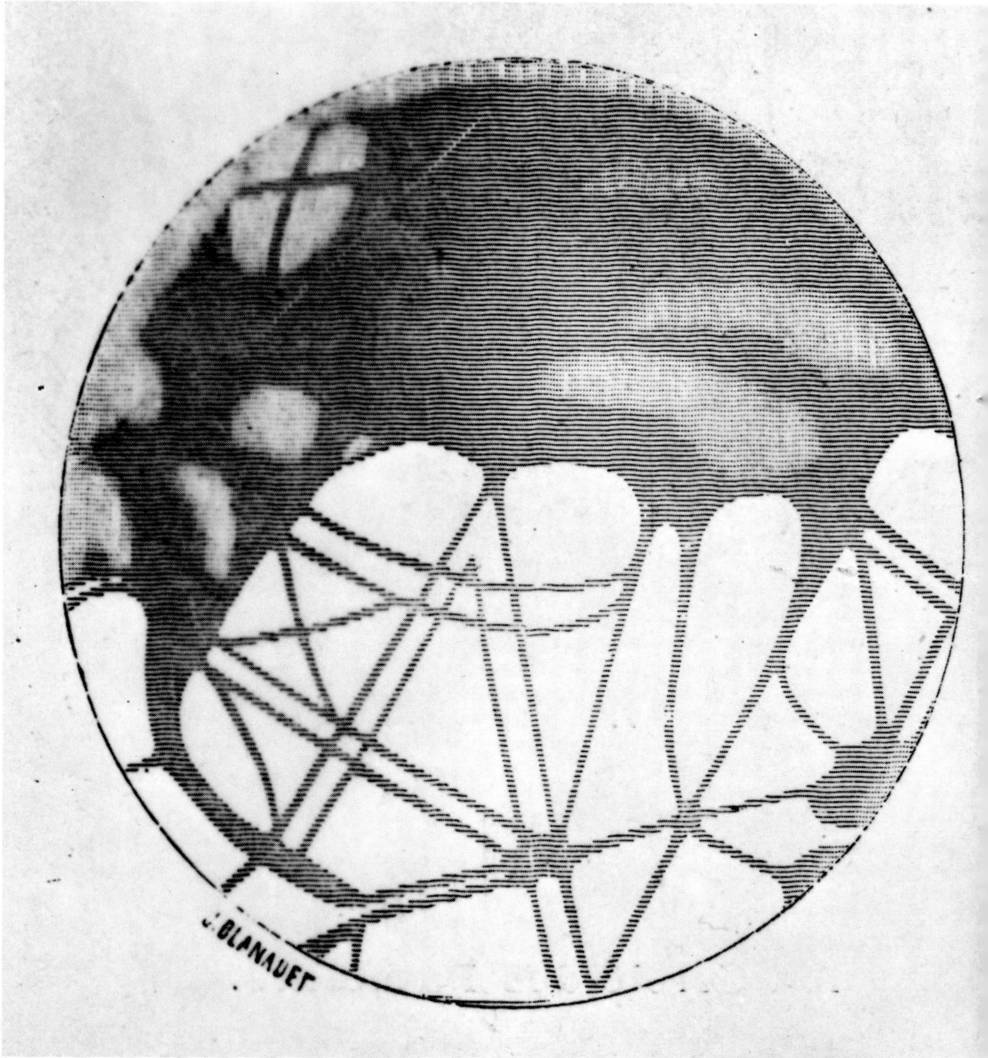

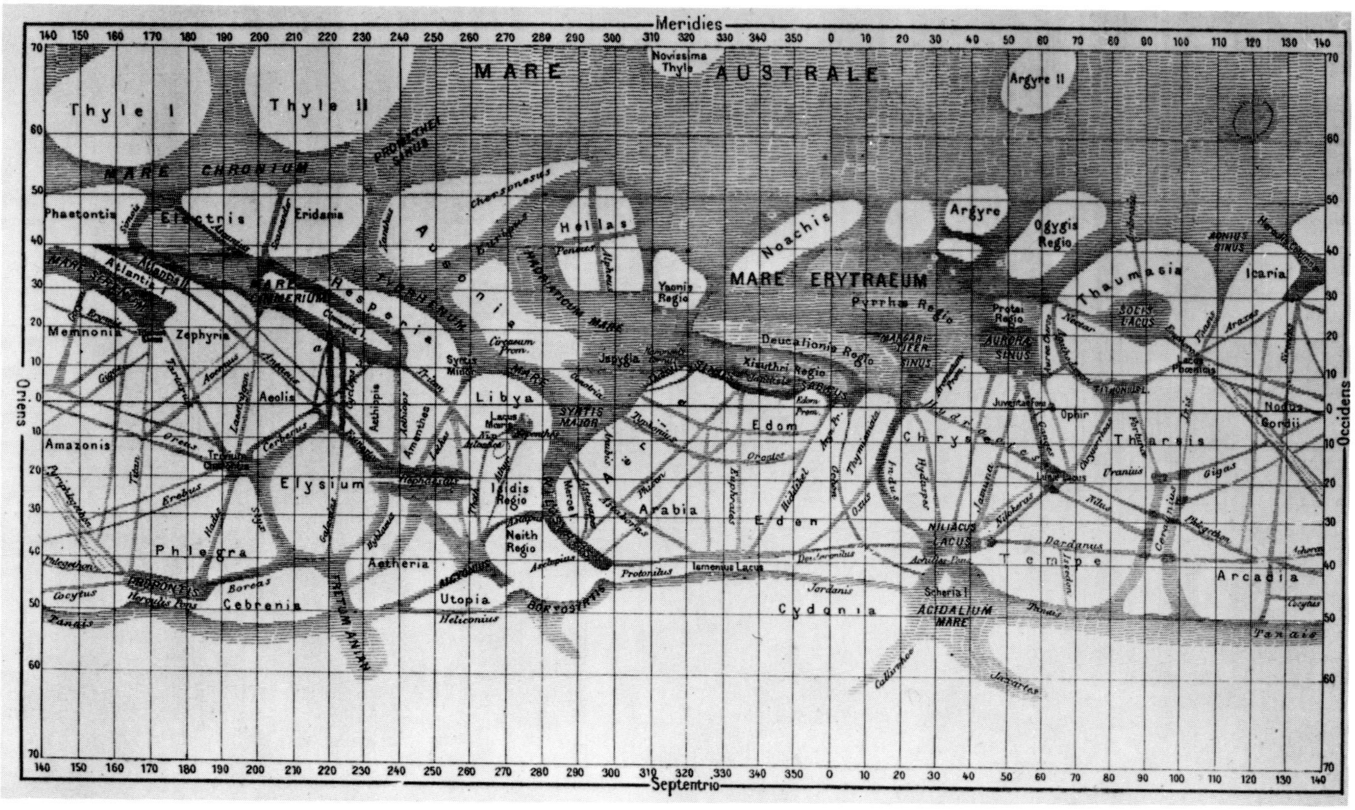

170

171

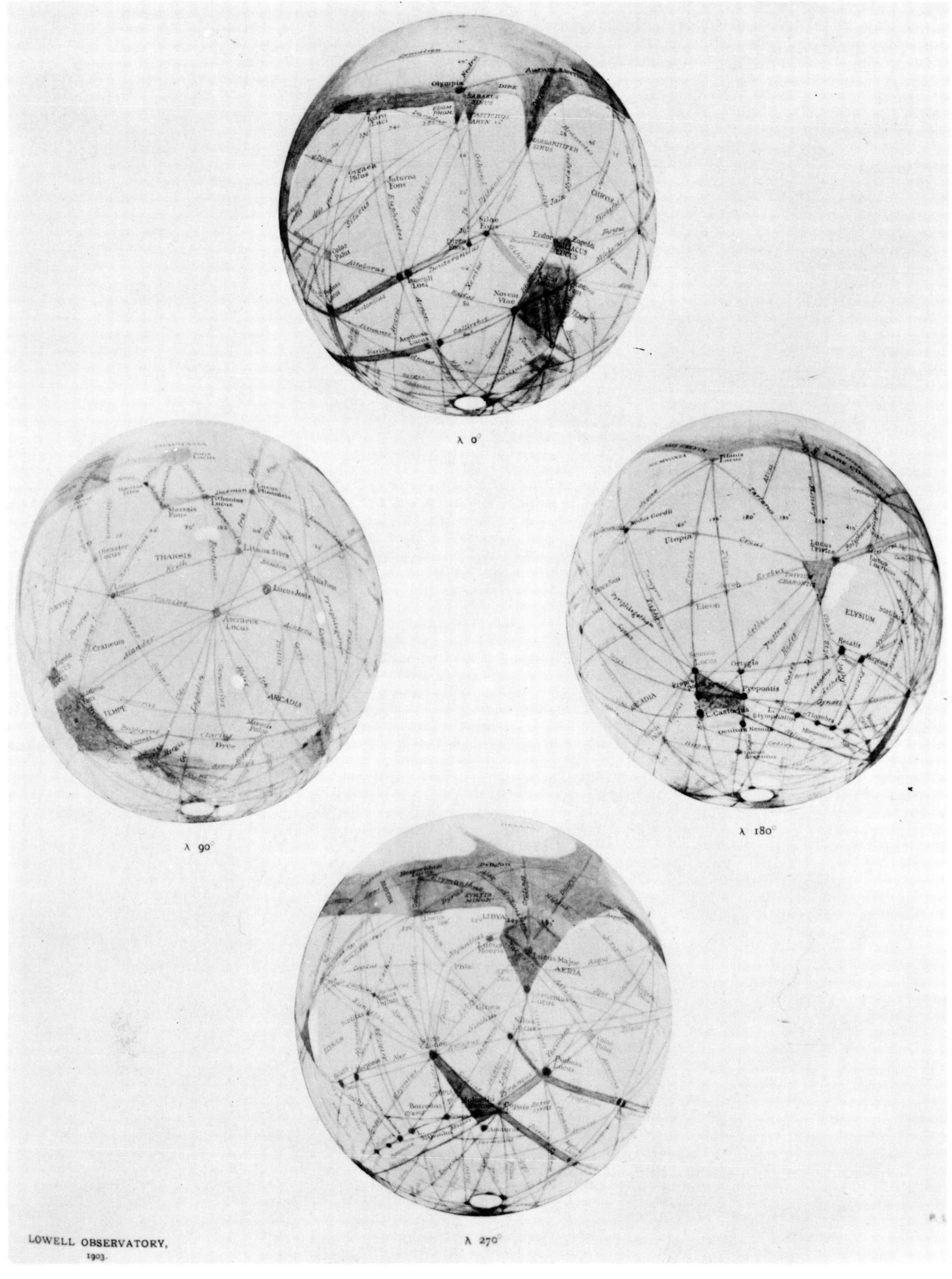
λ 0°
λ 90°
λ 180°
λ 270°
LOWELL OBSERVATORY,
1903.
THARSIS
TEMPE
ARCADIA
ELYSIUM
Utopia
Propontis
AERIA
LIBYA

171. *Believing is seeing.* The foremost exponent of the canal system on Mars was Percival Lowell. From the Lowell Observatory, which he founded in Flagstaff, Arizona, he observed and drew the Mars he was convinced was really there.

172. *Skepticism and seeing.* E.-M. Antoniadi did not believe in the canals or in Martians. What he observed confirmed his skepticism. Only the true believer could find evidence for Lowell's theories in these pictures.

173. *Mars on camera.* The great advantage of photography is that the camera has no preconceived ideas that can mislead it. This photograph, taken in 1909 with the Yerkes Observatory's 40-inch refractor, appears to confirm Antoniadi's doubts. Lowell's "map" is impossible to detect. And yet, might not the slight motion of the camera, or changing conditions in the atmosphere, blur the fine detail of the canals in a long time exposure? The answer was not to be known until the twentieth century.

172

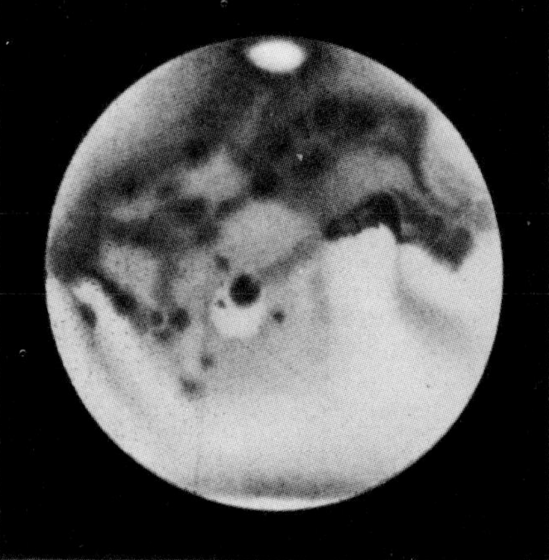

6 octobre. $\omega = 121°$; $\varphi = -21°$.

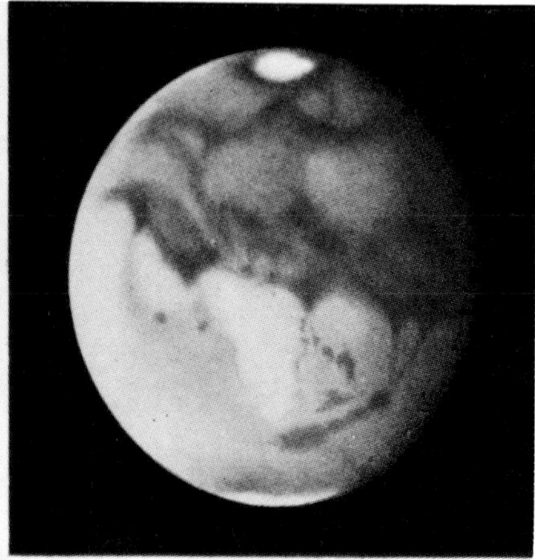

5 novembre. $\omega = 197°$; $\varphi = -24°$.

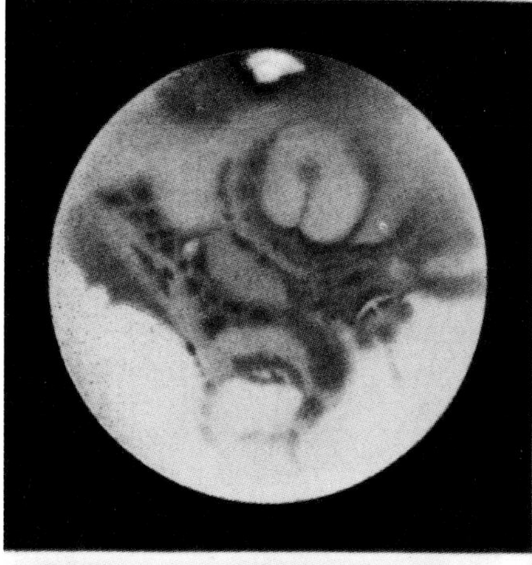

20 septembre. $\omega = 279°$; $\varphi = -20°$.

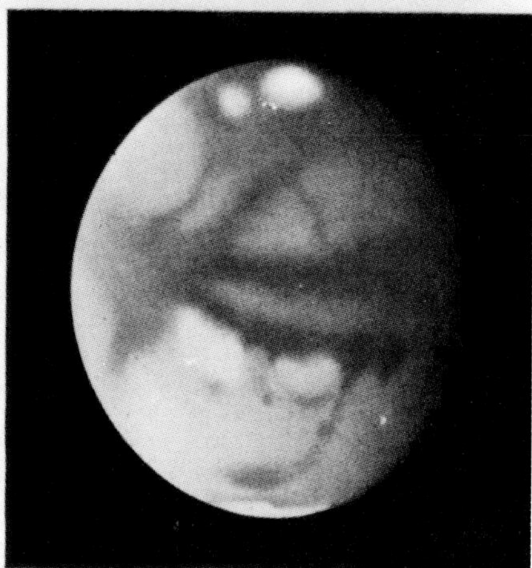

27 novembre. $\omega = 346°$; $\varphi = -25°$.

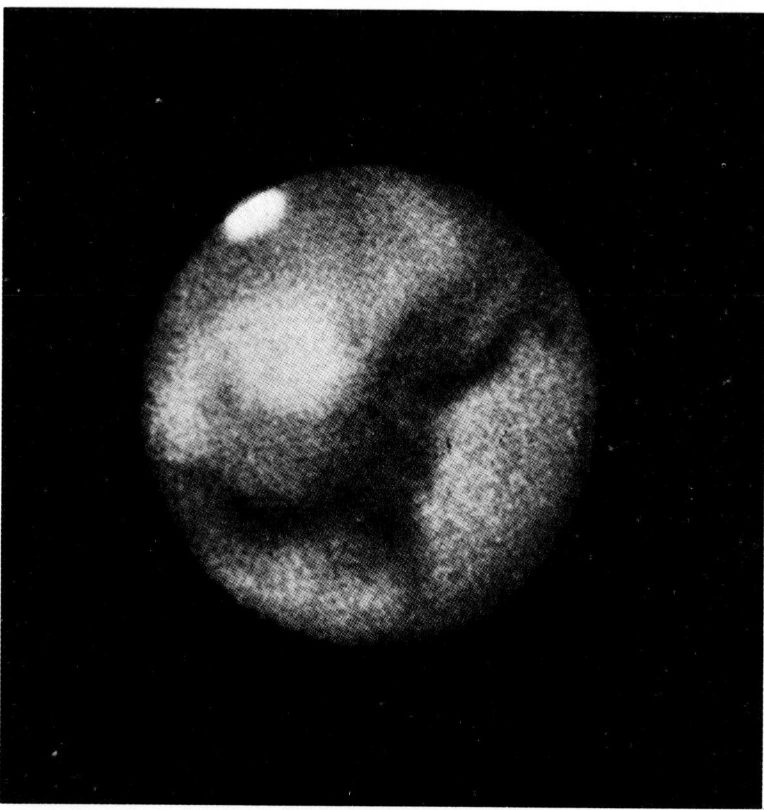

173

174

174. *Venus (1805).* This crude engraving represents Venus during its "half-moon" phase and depicts two surface features. One appears to be a polar ice cap, and the other is merely a dark streak of undetermined origin.

175. *The "surface features" of Venus.* Venus is a particularly difficult planet to observe, for when it is closest to the earth, the hemisphere facing us is dark. It can be observed full-face only when it is at a very great distance from the earth. The figure here is based on an eighteenth-century drawing that shows the phases of Venus and the "continents" of the planet. It was not improved upon throughout the nineteenth century.

176. *The "canals" of Venus.* Percival Lowell could find canals everywhere. This is his view of Venus in 1896. On it he sketched a complicated system of very large canals, thereby contradicting the idea that Venus was covered with clouds.

177. *Venus and Halley's Comet (1910).* Photographs seemed to confirm the theory that Venus was cloud-covered. At least, there are no visible surface markings in this picture taken during the passage of Halley's Comet.

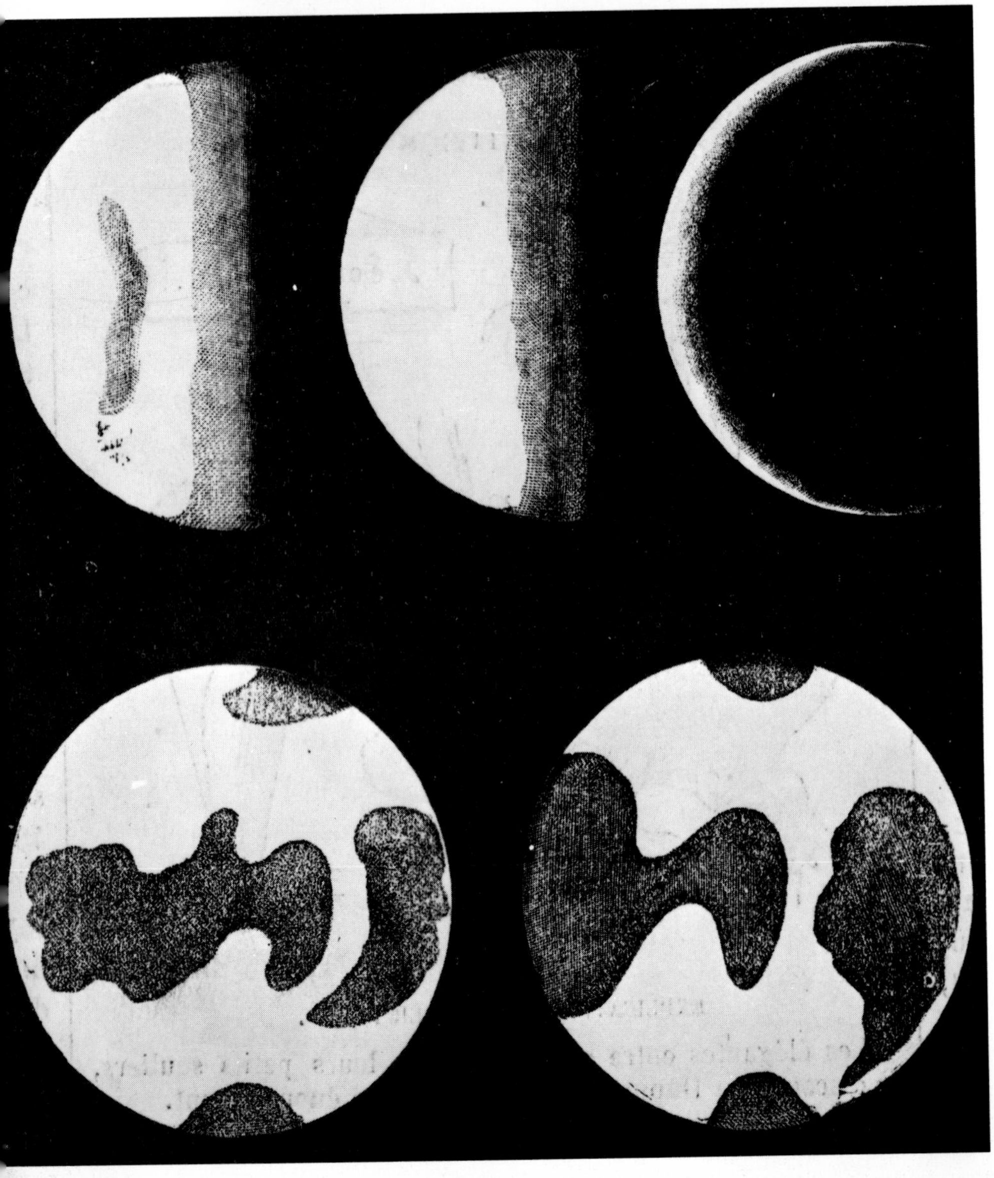

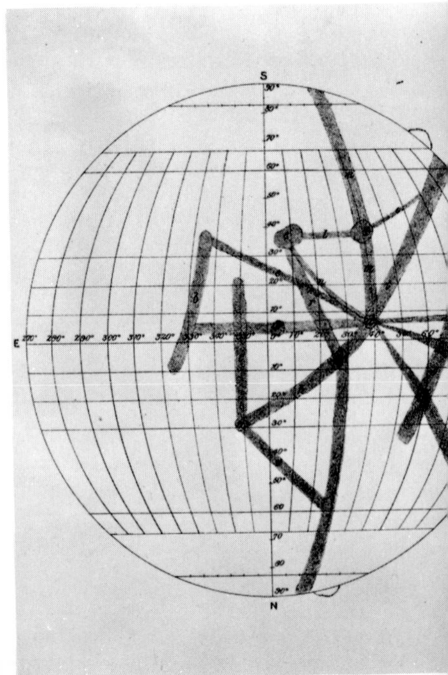

175

176

177

The Sun

Progress in observation and understanding of the sun was less dramatic than other advances made in astronomy in the nineteenth century. Basically, this was the result of observations being far in advance of any possible theory to explain them, solar processes not yielding to theory until the advent of quantum mechanics in the twentieth century.

Two aspects of the sun fascinated both astronomers and the general public. Sunspots, discovered by Galileo, still puzzled observers and the spectacular coronal displays made visible by eclipses could not help but impress witnesses. About all that could be done in the nineteenth century, however, was to observe these solar features and report on them. The one real advance was the discovery by Schwabe, extended by Wolf, in the early 1850's, that sunspots showed a periodic variation. This indicated that they were not merely occasional aspects of the solar atmosphere like clouds on the earth, but must be the results of processes within the sun itself. It was soon found that the variations in the sunspot activity of the sun coincided with magnetic and auroral variations on the earth, although why this should be so eluded astronomers and physicists of the time.

The spectacle of a solar eclipse still could awe even the most rational Westerner and terrify the uninitiated. Eclipses were avidly followed and what was seen could be reported by photograph and was, indeed, spectacular. Solar flares extending for miles into space indicated the power of the solar forces but the source of this power remained a mystery.

After 1859, with the enunciation of the law of Kirchhoff and Bunsen, the sun became susceptible to chemical analysis through spectroscopy. This led to the discovery of helium, first on the sun, and then on the earth; but it did little to explain to physicists of the nineteenth century how the sun worked. Such clues from astrophysics were to be of value only in the twentieth century.

178. *The sun (1805).* This drawing is roughly contemporaneous with that in Figure 150. But in this view, based on a drawing by a professional astronomer, the sunspots are well done and give some hint of their structure.

179. *The first daguerreotype of the sun (1845).* This is the first daguerreotype of the sun, taken by Hippolyte Fizeau and Léon Foucault on April 2, 1845. Their greatest difficulty was in getting sufficiently short exposure times. Using an exposure of 1/60 of a second, they were able to obtain this picture, which contains two sunspot groups.

178

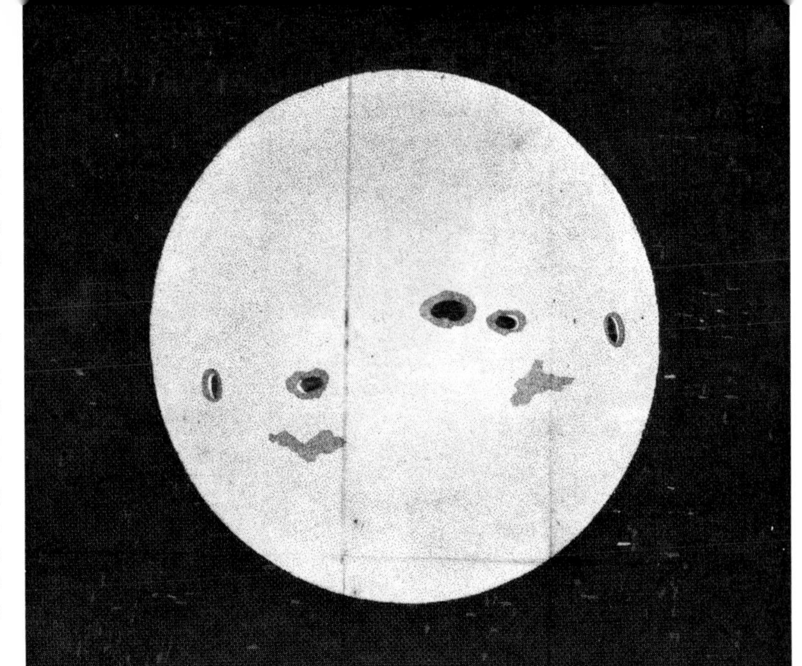

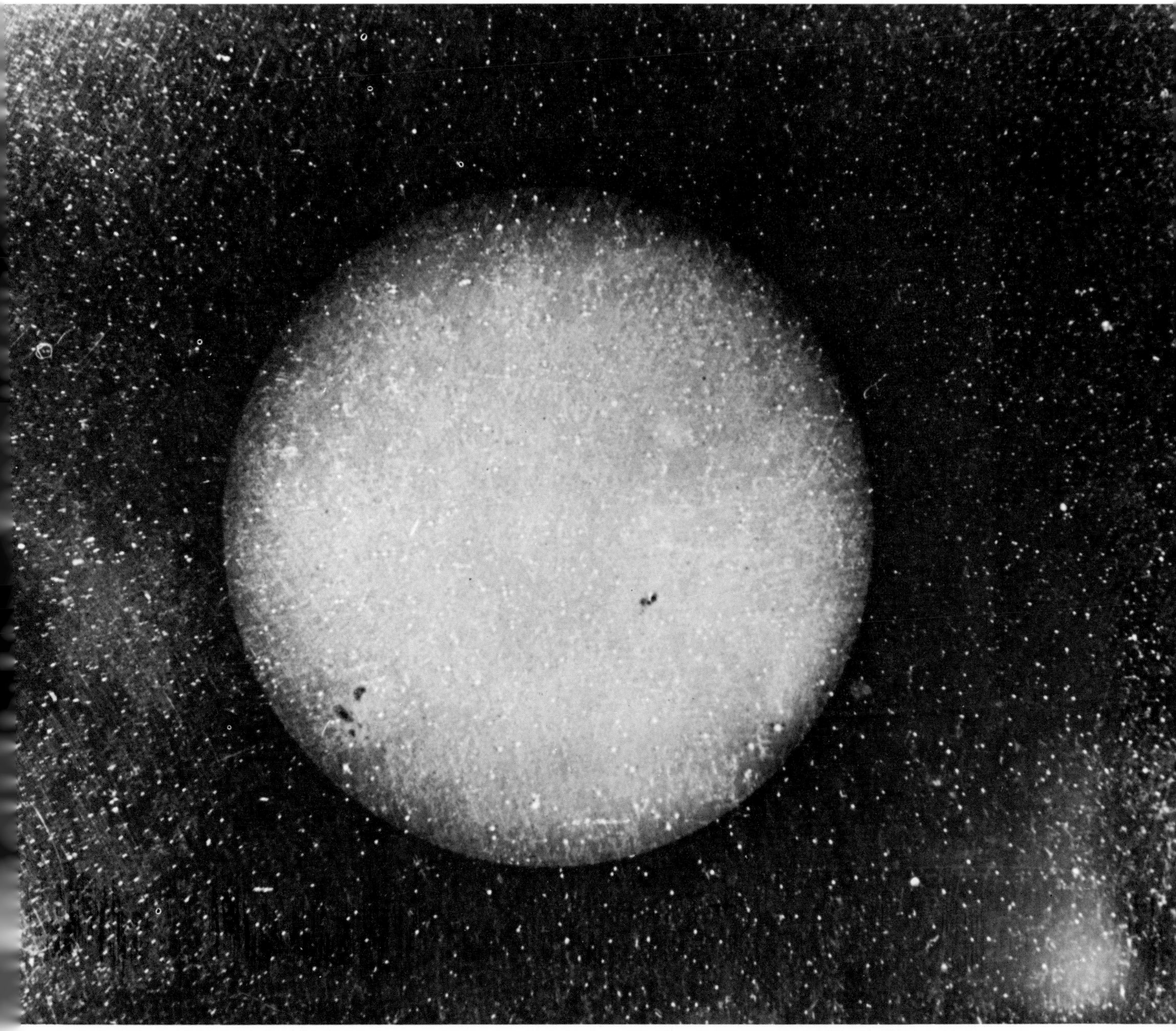

179

180

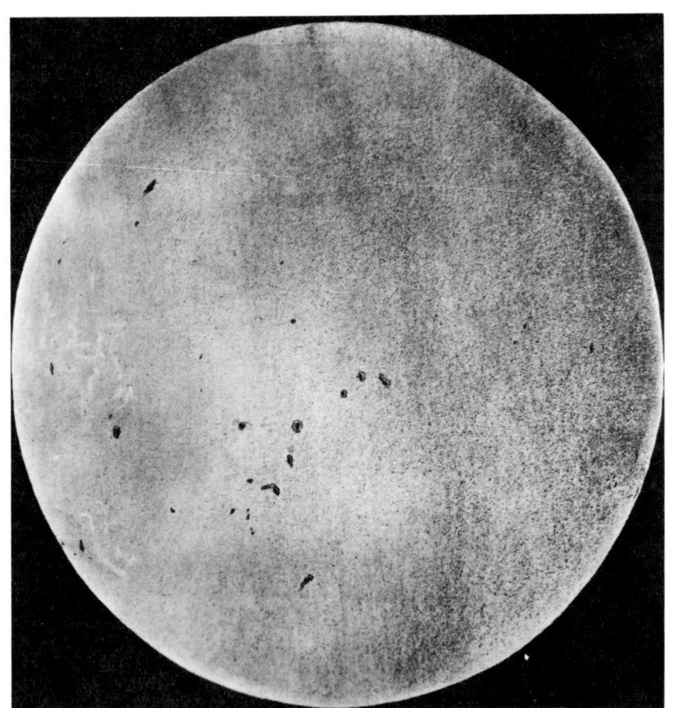

181

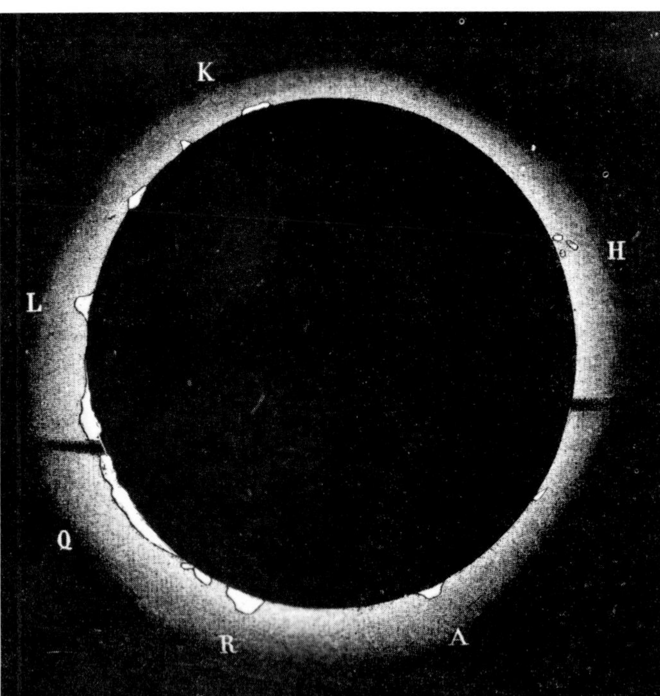

182

180. *A heliograph (1884).* Photographs of the sun were the easiest of all astronomical pictures to take. The exposure time required was extremely short, and excellent results could easily be obtained. The sunspots appear much smaller in comparison with earlier pictures. One suspects some exaggeration in the earlier views, although sunspots did vary considerably in size over time.

181. *An early photograph of a solar eclipse (1860).* The camera was particularly good at capturing the evanescent coronal display during solar eclipses. Salient features, here marked by letters, represent solar flares.

182. *Solar flares.* By means of the spectroscope, solar flares could be directly observed even when there was no eclipse. These prominences were drawn from spectroscopic observations made on July 23, 1871.

183. *A gigantic sunspot.* Sunspots were among the earliest features to be observed on the sun. This one, observed on October 14, 1883, was seven times the size of the earth and was visible to the naked eye.

184. *Wilson's theory of sunspots illustrated.* In 1774, Alexander Wilson suggested that sunspots, marked *A* and *B* in the picture, were really cavities in the luminous envelope of the sun. Thus, the dark area, Wilson hypothesized, was the real surface of the sun, which we can glimpse through the surrounding, blazing gases. It was this theory that underlay nineteenth-century speculation that life existed on the sun.

183

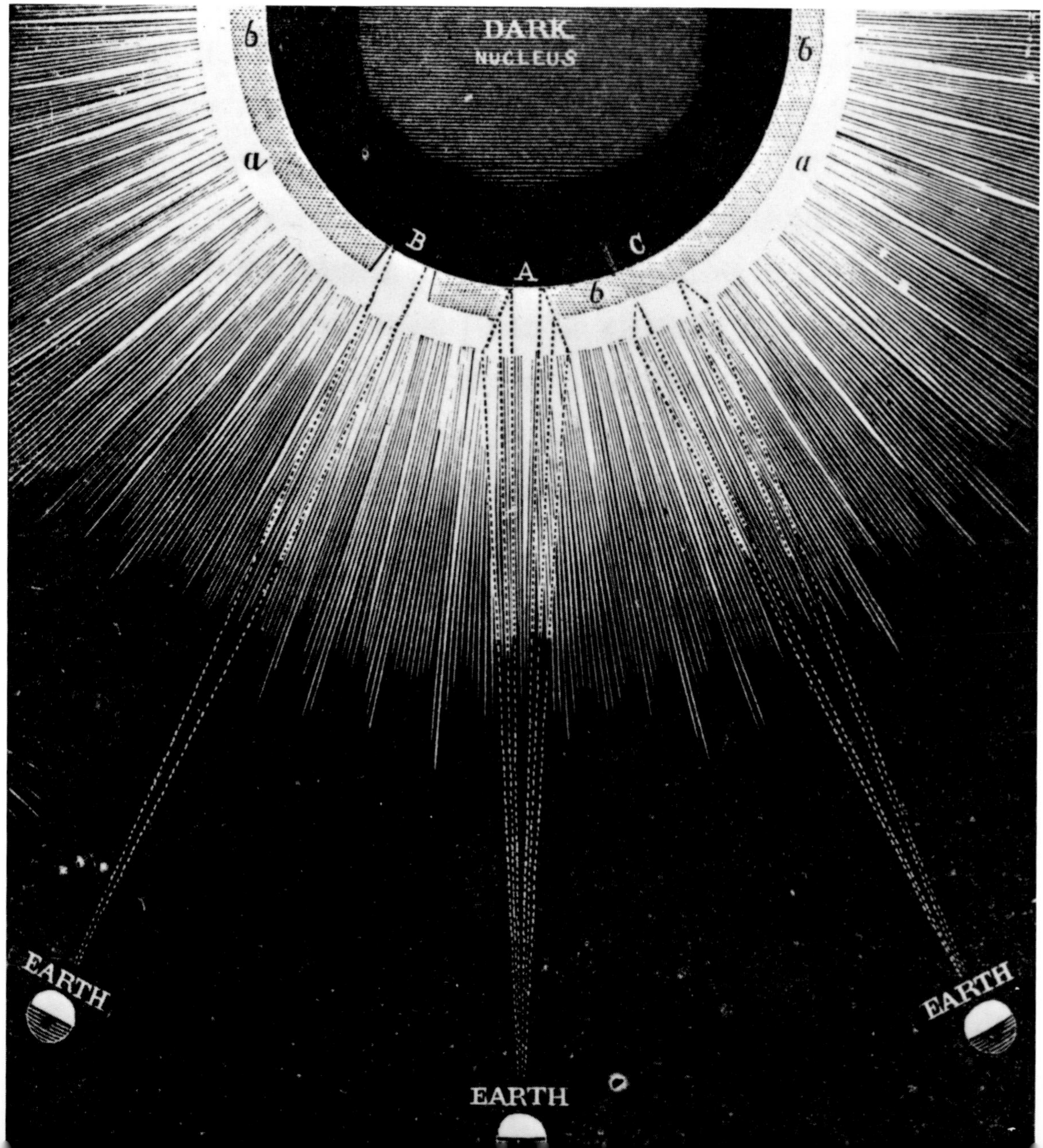

184

The Moon

After the realization that our nearest celestial neighbor was uninhabited, popular interest in it flagged, although it never vanished. The moon was still the obvious object upon which to turn a telescope when engaged in amateur stargazing. Because of its difficult orbit, the moon retained interest for mathematical astronomers who were able, over the course of the nineteenth century, to devise a lunar theory of considerable accuracy. This theory, in turn, paid off in terrestrial navigation in a century which first mapped the earth's seas accurately and completely.

For the rest, the main task facing nineteenth-century selenographers was the accurate mapping of the detail of the lunar surface. This was done, in large part, because this surface was so obtrusively there, and also because only through accurate mapping of lunar surface detail could there be any hope of detecting changes on the moon's surface. The work was incredibly laborious, each feature being carefully drawn on a grid system and topography discovered by repeated observations of the same feature under different conditions of light and shade. The advent of astronomical photography greatly lightened this task. By the end of the century, the moon was no longer an object of mystery or romance, but a well-mapped hemisphere. About the only question that a nineteenth-century observer could ask about the moon's surface that could not be answered by looking at a photograph of it, was what did the other side of the moon look like? Again, the answer was not to come until the twentieth century.

185

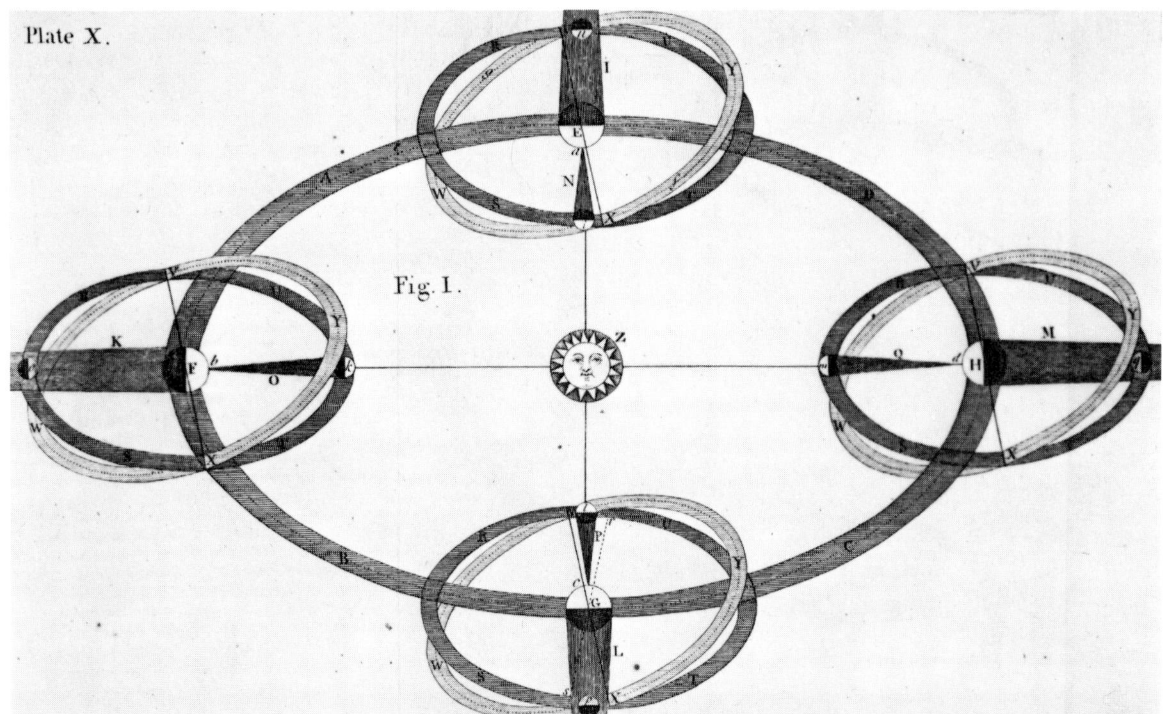

186

185. *The moon-sun-earth system.* Our closest neighbor is the moon. Its position in relation to the sun and the Earth gives rise to the phases of the moon and to both lunar and solar eclipses, as illustrated here.

186. *The solar eclipse of 1858.* The eclipse of March 15, 1858, is graphically depicted here. The path of totality ran across northern Europe. In this engraving the representations of Europe and Africa are reversed, and seen as mirror images of other maps.

187

187. *The moon and the "informed public."* Scientific knowledge of the earth's satellite had not penetrated very far below the scientific upper crust. The casual observer obviously did not know, nor did he care, how far away the moon was. It was nice simply to get a close-up look at it.

188. *Mayer's map of the moon.* The best overall view of the moon at the beginning of the nineteenth century was this one, which had been prepared by Tobias Mayer in 1775.

Tob. Mayer, del.

Phot. u. Druck v. W. Hoffmann, Dresden.

188

189. *A popular telescopic view of the moon.* Contrast this effort with Mayer's map (Figure 188). There are gross errors here that cry out for correction. Since the moon always presents the same face to the Earth, all pictures of the full moon show the same surface and, logically, should have little variation.

190. *A relief model of the moon.* This model was constructed by Thomas Dickert and exhibited at Bonn in 1854. Since the other side of the moon had never been seen, it was only a hemisphere.

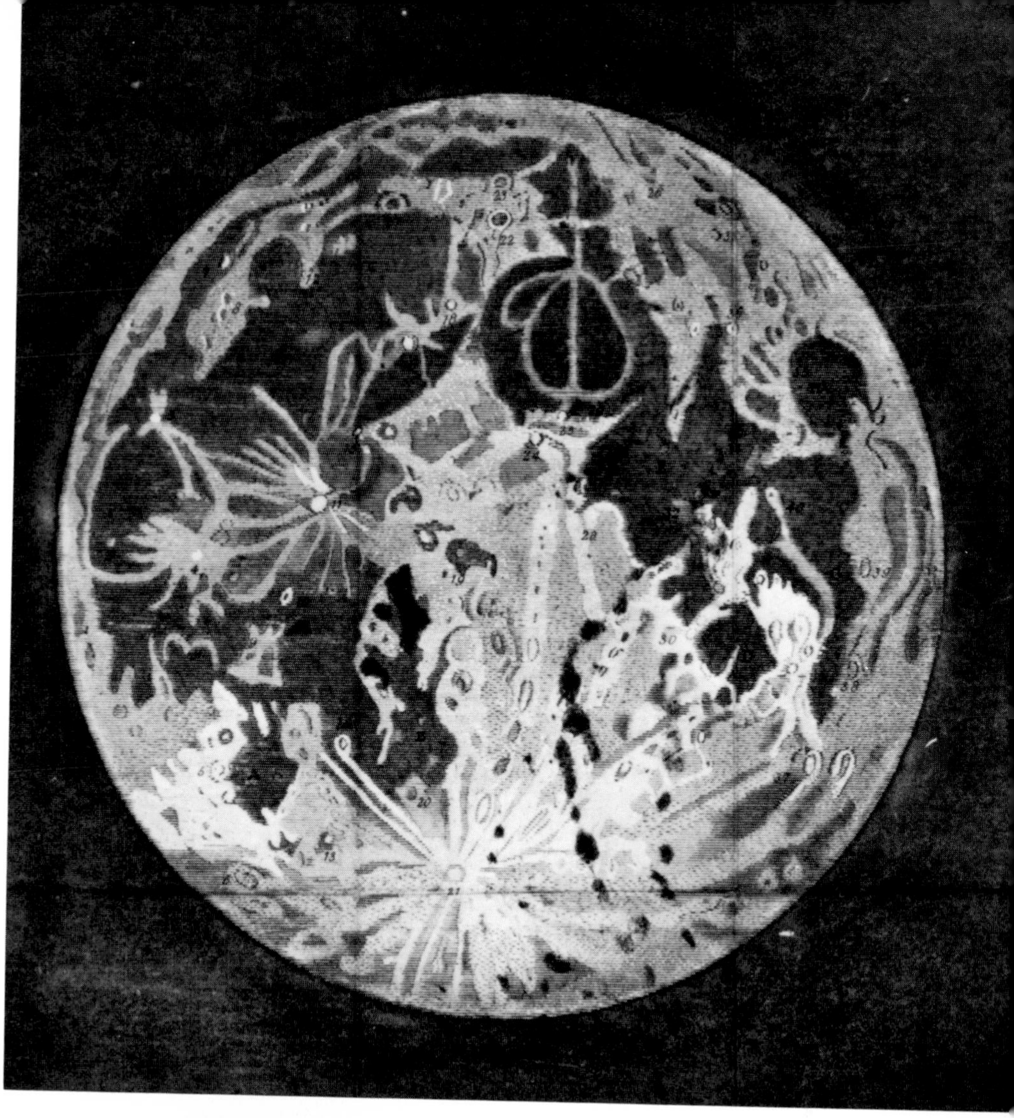

189

190

191

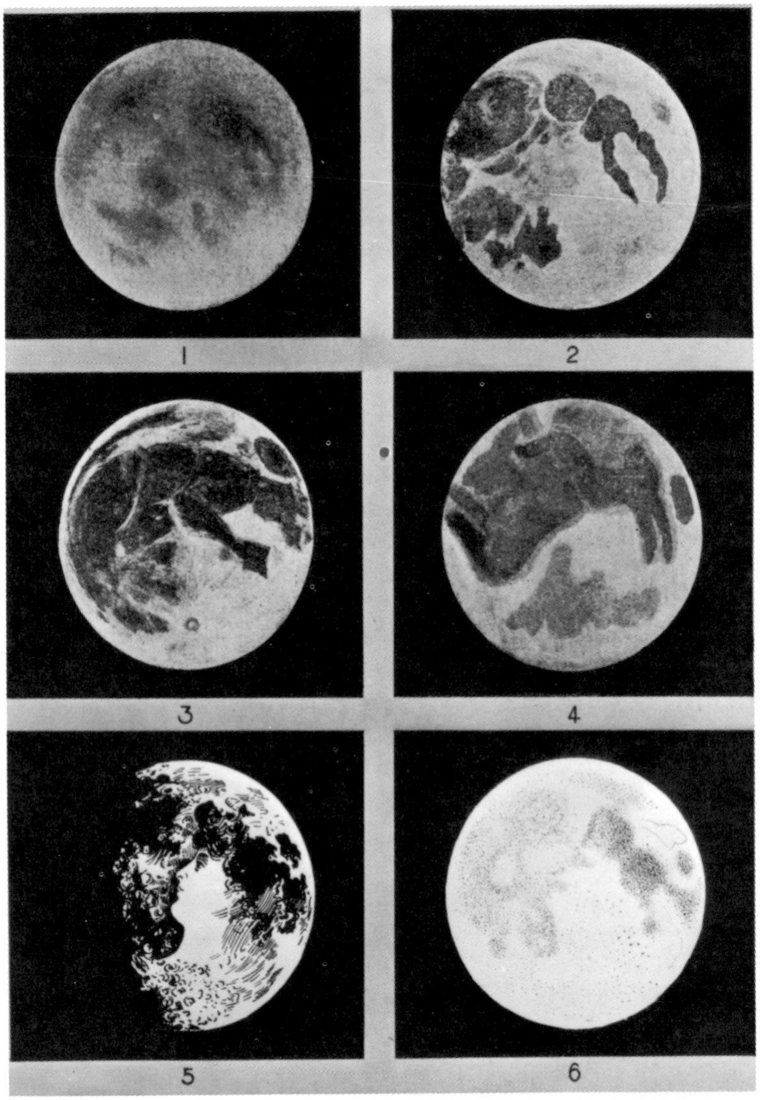

192

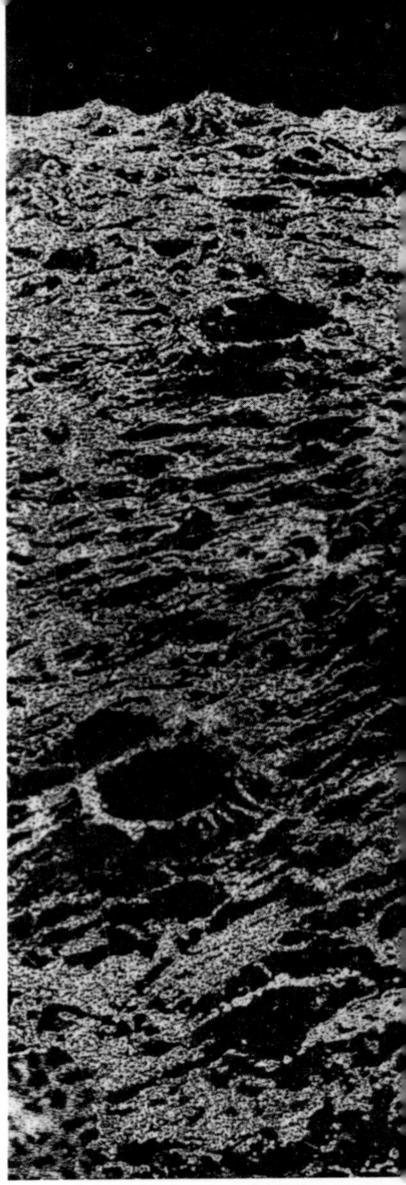

191. *Lunar fancies.* People have been seeing images in the moon since the dawn of time. Here are six that were current and popular in the nineteenth century: 1. The man in the moon. 2. The crab. 3. Girl reading. 4. The donkey. 5. The lady. 6. An astronomer's drawing.

192. *A moon crater.* This model represents what a typical moon crater was supposed to look like. The crater was characterized by a central peak and a circumferential wall.

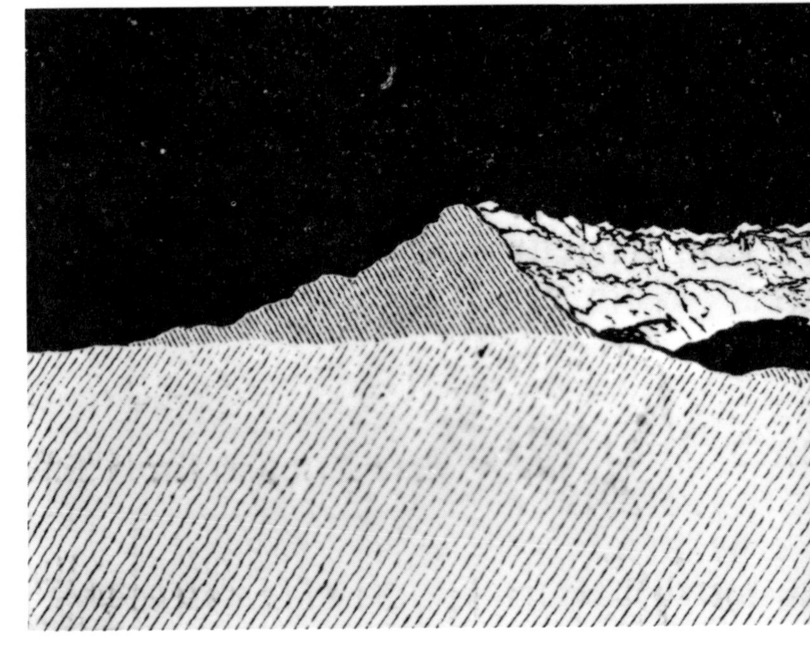

193

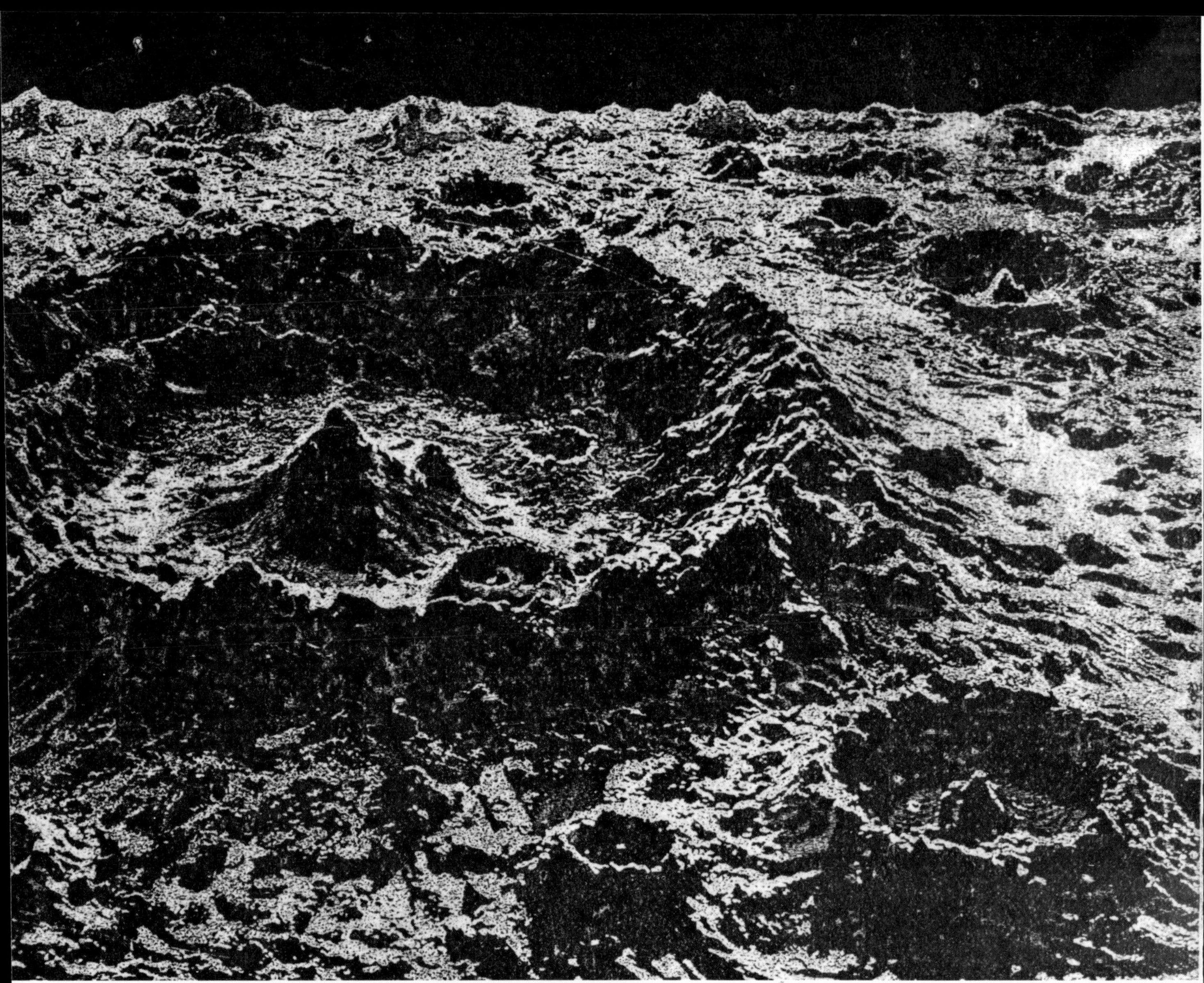

193. *Theory of lunar crater formation.* The crater was assumed to be of volcanic origin. As shown, the central cone was supposed to have been built by direct lava flow; the circumferential wall was considered the result of material that had been hurled into space by volcanic action and had fallen some distance away. The distance between the cone and the wall in the crater Archimedes, some 5.5 miles, could be accounted for by the low lunar gravity.

194. *A lunar landscape.* James Nasmyth, the inventor of the steam pile driver, was an avid observer of the moon. Here, dating from 1885, is his model of the isolated lunar mountain, Pico, as it would appear to an observer on the moon.

195. *Eclipse of the sun, as seen from the moon.* Nasmyth's imagination produced this highly romanticized painting of an eclipse of the sun by the Earth. A beautiful sight, it must have thrilled its nineteenth-century viewers.

194

19

196

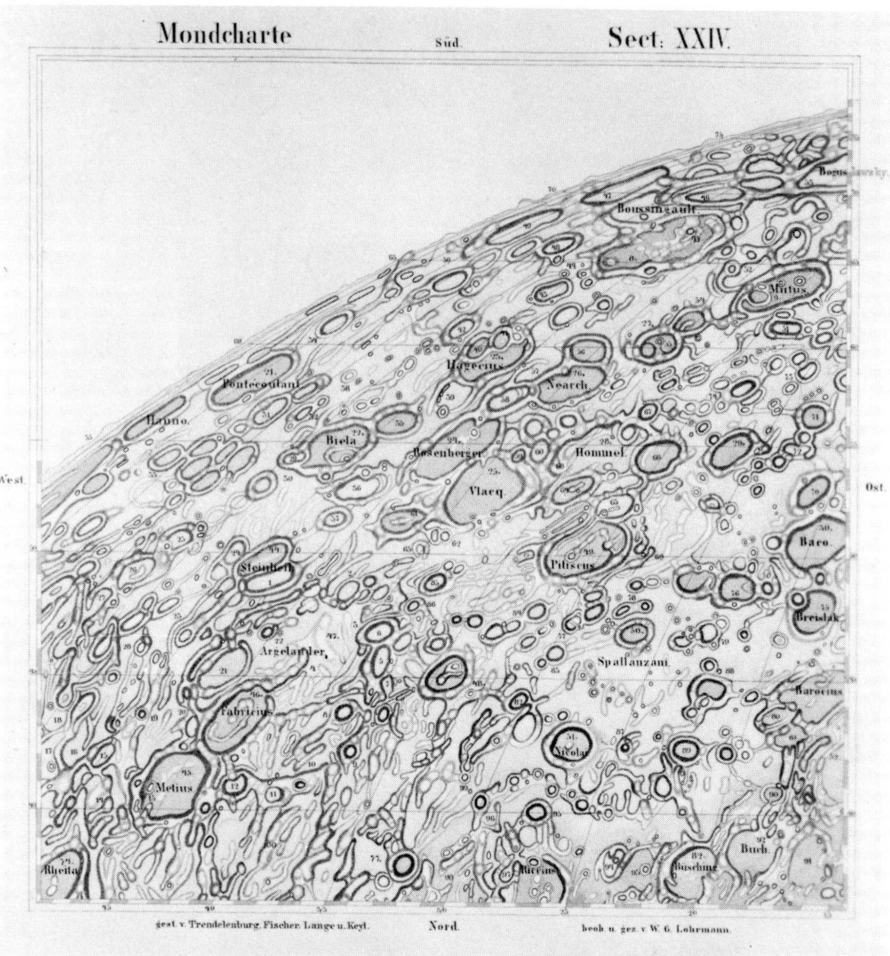

197

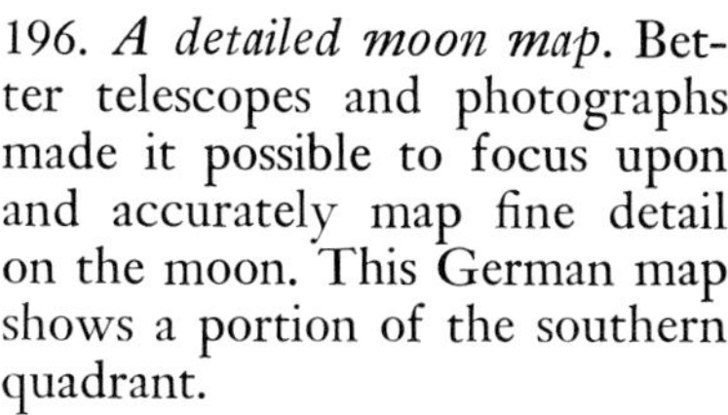

196. *A detailed moon map.* Better telescopes and photographs made it possible to focus upon and accurately map fine detail on the moon. This German map shows a portion of the southern quadrant.

197. *Photograph of the lunar surface.* This photograph, taken at the end of the nineteenth century at the Paris Observatory, shows how much detail could be clearly seen on the moon's surface. Lunar fancies (Figure 191) were difficult to maintain in the face of these lunar facts.

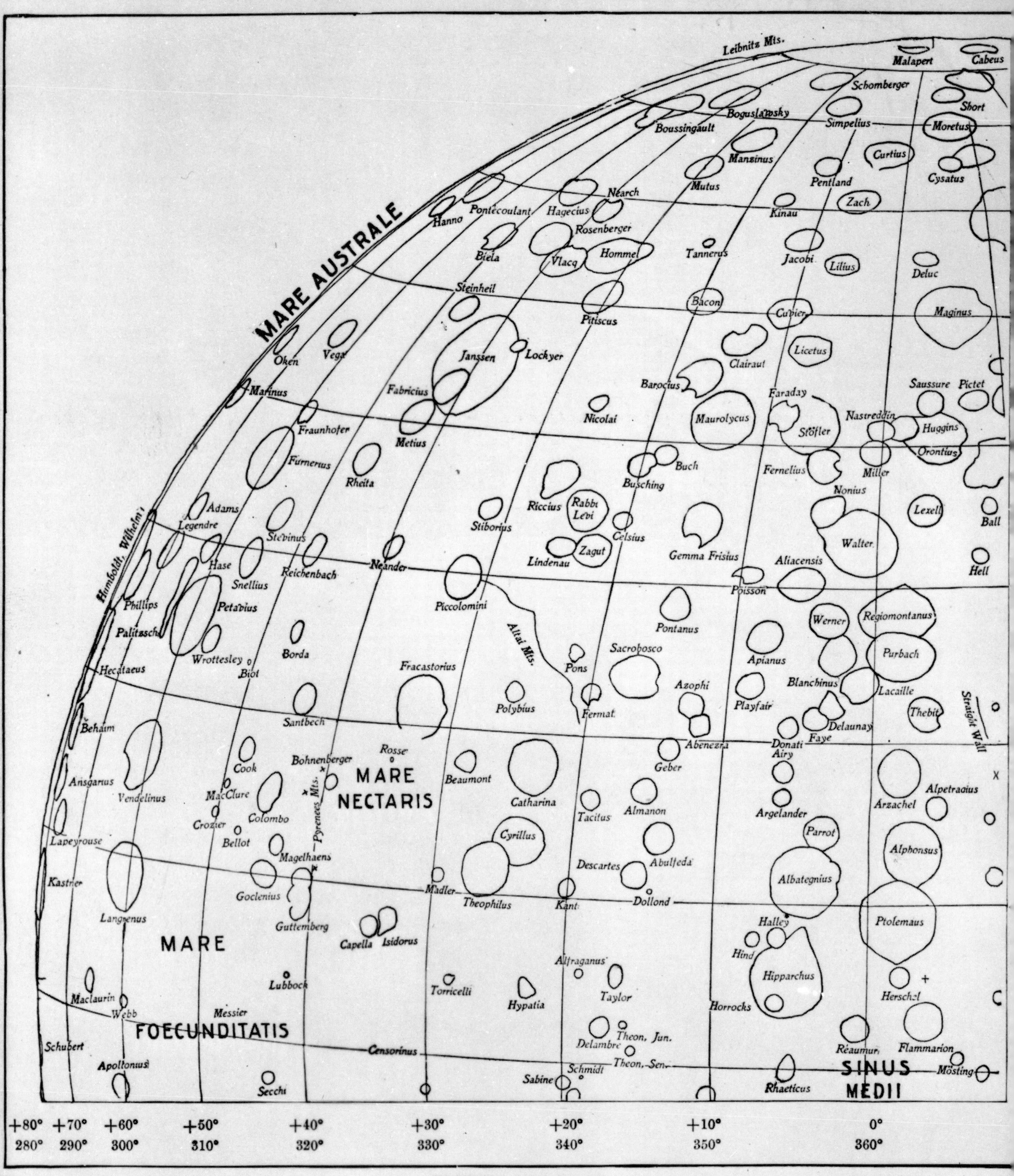

198a, b. *The moon mapped* (1903). This map, shown here and on the following two pages, is based upon a careful photographic survey of the entire face of the moon. It provided an accurate record of the moon's major features.

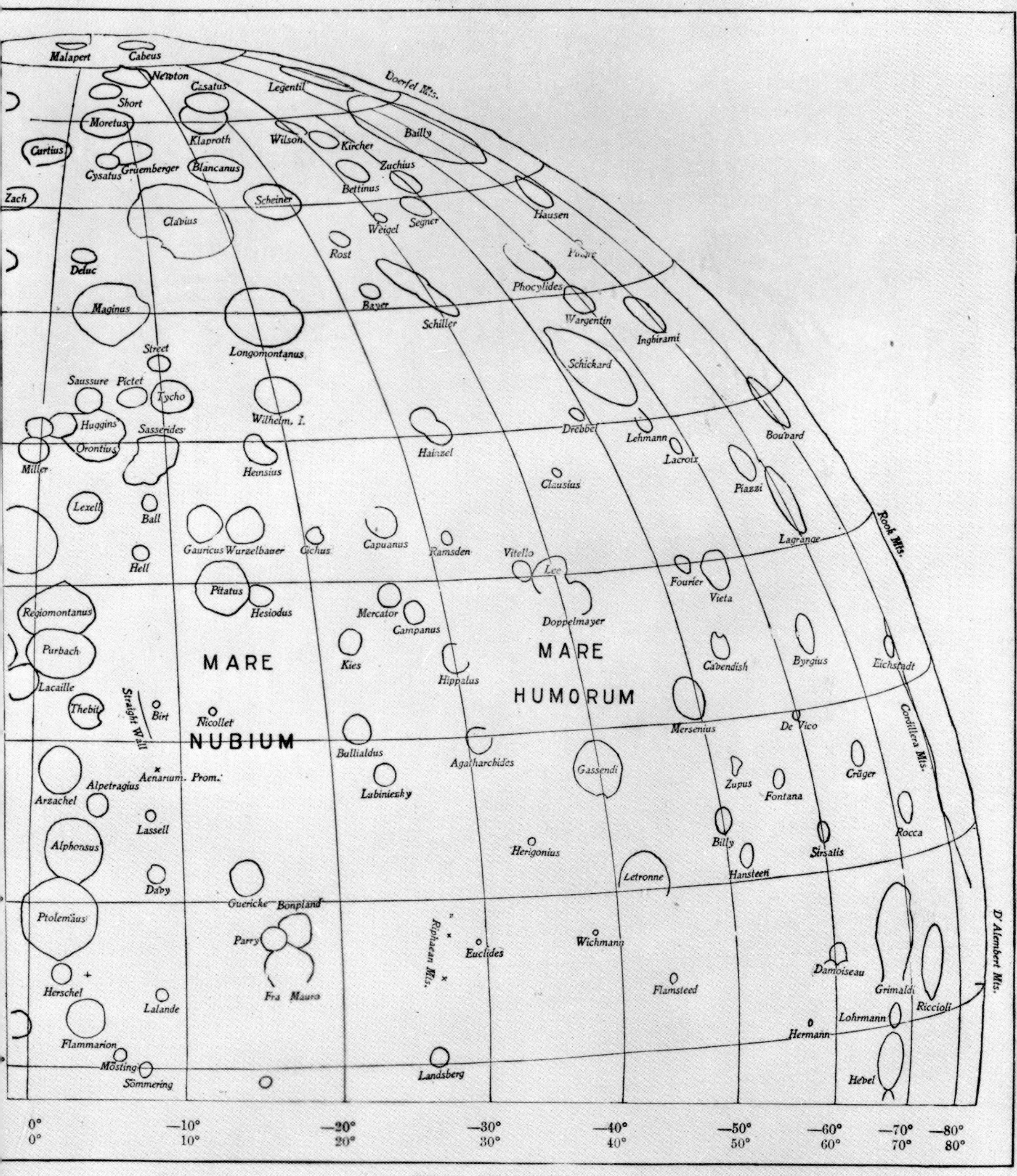

THIRD QUADRANT

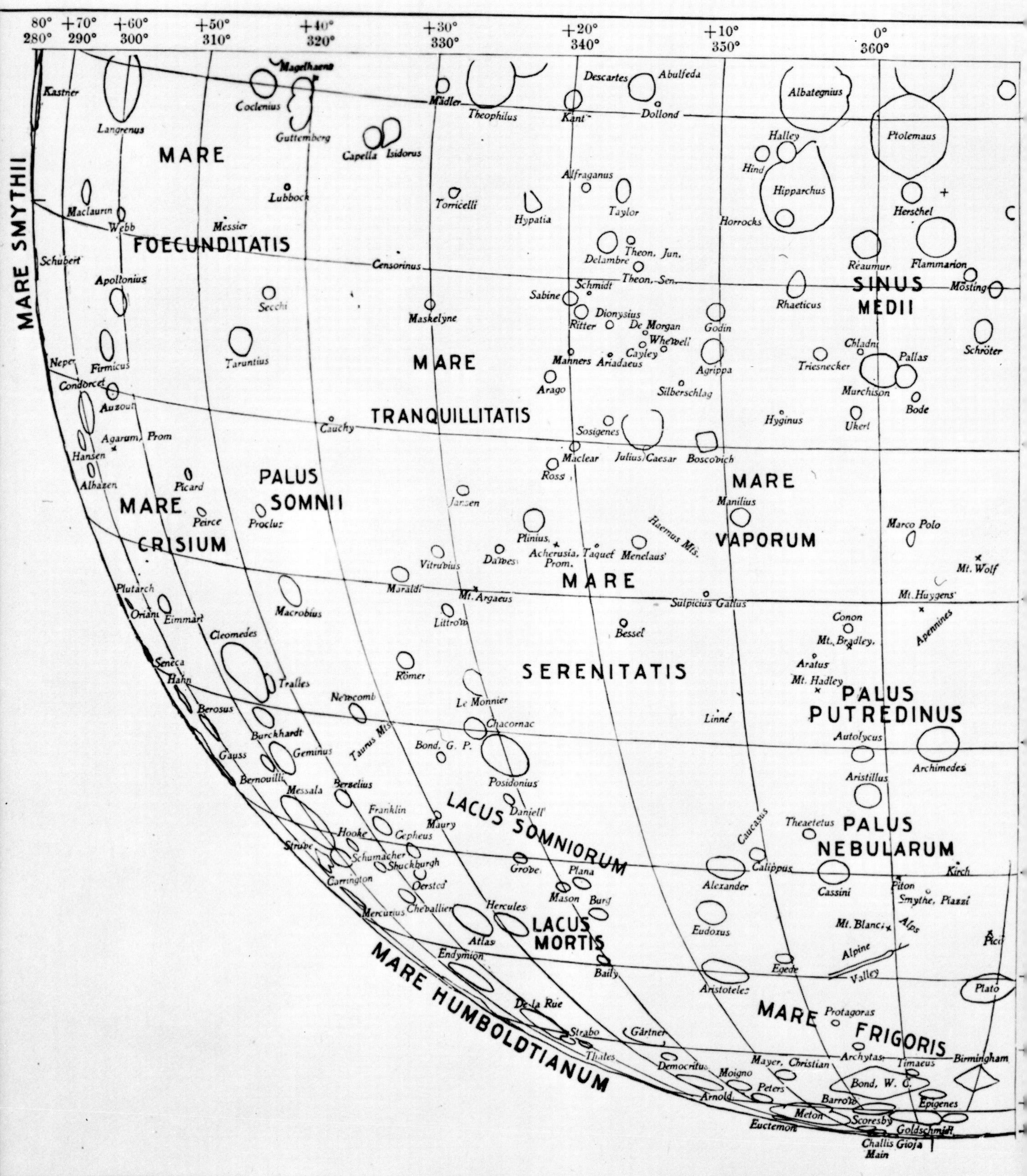

198c, d. *The moon mapped* (1903).

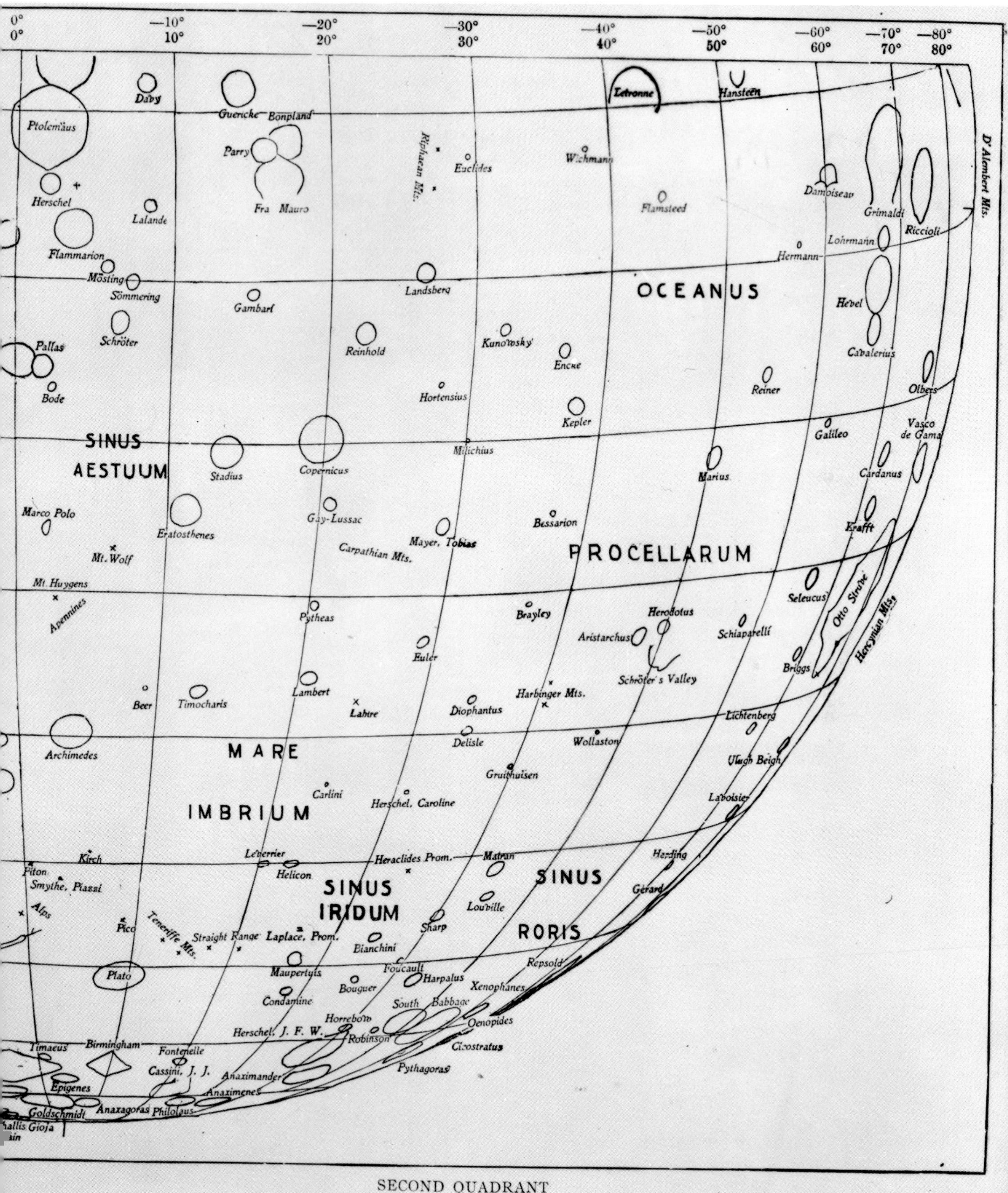

SECOND QUADRANT

The Earth

Just to consider the earth a planet like the other planets in the solar system was to be a bit daring when the nineteenth century began. To view its origins in the condensation of solid matter out of primordial gases smacked of heresy. There was, after all, an account of the origin and basic geological nature of the earth that was central to Western civilization, enshrined by tradition, and literally sacred. The Book of Genesis provided a sufficient account of the origin and development of the earth and to challenge this was to strike at the very foundations of religion. When those foundations were sapped, it was widely believed, upheavals such as the French Revolution occurred and most scientists were a bit leery of contributing to another such cataclysm.

One could play it safe by leaving aside basic and fundamental questions of the origin of the earth and of the geological results of the Noachian flood and still do respectable scientific work. The earth and its associated phenomena could be described without fear of offending anyone's religious sensibilities. So the astronomical status of the earth could be viewed with some calm. The earth *is* a planet, no matter what else it might be, and there was no harm in illustrating the earth's place in the solar system. The discovery of stellar parallax in 1838 proved to even the most religiously devout that the earth was not at the center of the universe and that the command in the Book of Joshua, "Sun, stand thou still," must have been meant as a metaphor. That it was the earth that rotated, and not the sun that went around the earth, was demonstrated conclusively by Foucault's pendulum (1851). All very bland stuff, but the thin edge of the wedge in establishing the earth as just another natural object, devoid of any divine importance in that respect.

199. *The origin of the earth.* The earth's origin is here imagined according to the Kant-Laplace nebular hypothesis—that it was, in the beginning, a whirling mass of gas that, over eons, condensed into its present state.

200 a-f. *The earth as seen from other planets.* In the nineteenth century, it was finally realized that the earth was merely another heavenly body like the other planets. As a result, there was fascination with what the earth looked like from elsewhere in the solar system. The artist here has shown the earth from Mercury (*a*); Venus (*b*); the moon (*c*); Mars (*d*), the earth being the large "star" in the middle; Jupiter (*e*), during a transit of the earth, which is shown as a black spot on the sun's disk; and Saturn (*f*), from which the earth would have been visible only through a telescope.

200a
200b
200c
200d
200e
200f

201

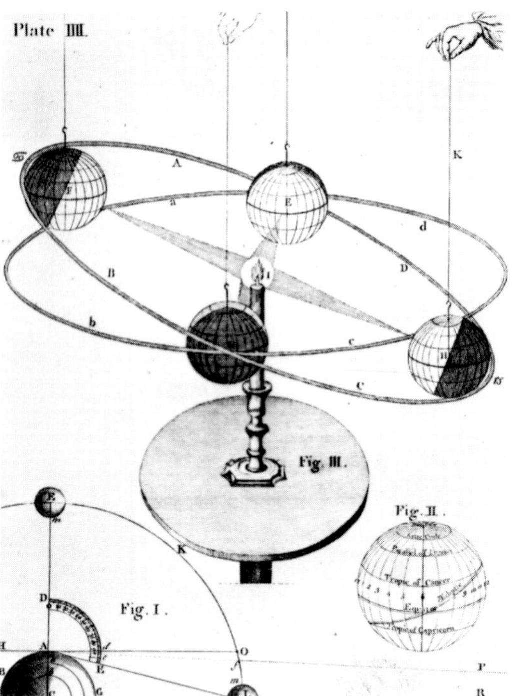

202

201. *The Seasons.* A late eighteenth-century illustration of the way the seasons are produced on the earth.

202. *A model of the earth.* James Wyld, a Member of Parliament, was stimulated by the Great Exhibition of 1851 to construct this model, opened to public view in that same year. It was built on a scale of 10 miles to the inch horizontally and 6 miles to the inch vertically. Topographical features—mountains, rivers, valleys—were made of plaster and carefully fitted to the *inside* of a sphere, thus making them more easily visible.

203. *Proof of the rotation of the earth.* In 1851, Léon Foucault, taking advantage of the long-known fact that a pendulum vibrates in a constant plane, constructed an enormous pendulum to illustrate the rotation of the earth. The pendulum was suspended from the dome of the Pantheon in Paris and vibrated in its plane as the earth turned beneath it. The plane of the pendulum appeared to turn through 360 degrees in 24 hours. The Foucault pendulum, although extremely simple, drew sizable crowds. Here was dramatic and simple proof that the earth *did* turn, despite the evidence of the senses to the contrary.

The Earth's Surface

Attention to the earth as a planet suggested consideration of the earth as a whole, not just regional studies of parts of it. Once again, it was possible to study terrestrial phenomena without hypothesizing about ultimate causes or getting into religious disputes. The man who almost single-handedly made this planetary view of the earth both popular and scientifically respectable was Alexander von Humboldt (1769-1859). Humboldt was an indefatigable traveller who personally explored some of the most inaccessible parts of the globe. Throughout his long career he was guided by a vision that was expressed in his great work, *Kosmos*, which remained unfinished at his death. To Humboldt, the earth was like a living organism. Seas and atmosphere and magnetic field and mountains and plants and animals were all connected together in one gigantic ecological whole. To isolate one aspect of this complex organism and study it alone was to miss the larger truth of the whole and so Humboldt stressed the importance of seeing the earth whole and working out the interrelations of global phenomena. The currents of the sea must be mapped (the Humboldt current is named after him) and the mountains of the world scaled and barometric and magnetic observations taken from their summits. The vast ocean of the atmosphere must be explored and the meteorological events in it classified and analyzed. The magnetic field must be mapped (Humboldt brought back important magnetic observations from South America). Temperatures and barometric pressures all over the world must be measured and correlated for from these correlations Humboldt was convinced would come knowledge of the factors that make the world's weather. The aurora borealis must be studied for it must contain vital clues on terrestrial dynamics. Glaciers must be climbed and probed; volcanoes plumbed and probed to discover the secrets of the earth's crust. From all these observations and expeditions, Humboldt was convinced, would come a description of the earth that would enable us to appreciate it as an entity in itself. And from this global view would emerge a new appreciation of the earth and all its progeny—animal, vegetable, and mineral alike. This was a noble and grandiose vision. It inspired two generations of explorers who now sought intellectual riches rather than Inca gold or Eastern spices.

204. OVERLEAF: *The earth at mid-century*. The nineteenth century was the first century to know the terrestrial globe as a whole. By mid-century, all the continents were known and the major landmasses had been accurately mapped. On the sides of the map are the relative heights of the known mountains in the western and eastern hemispheres. Only the interiors of Africa and South America are relatively blank.

204

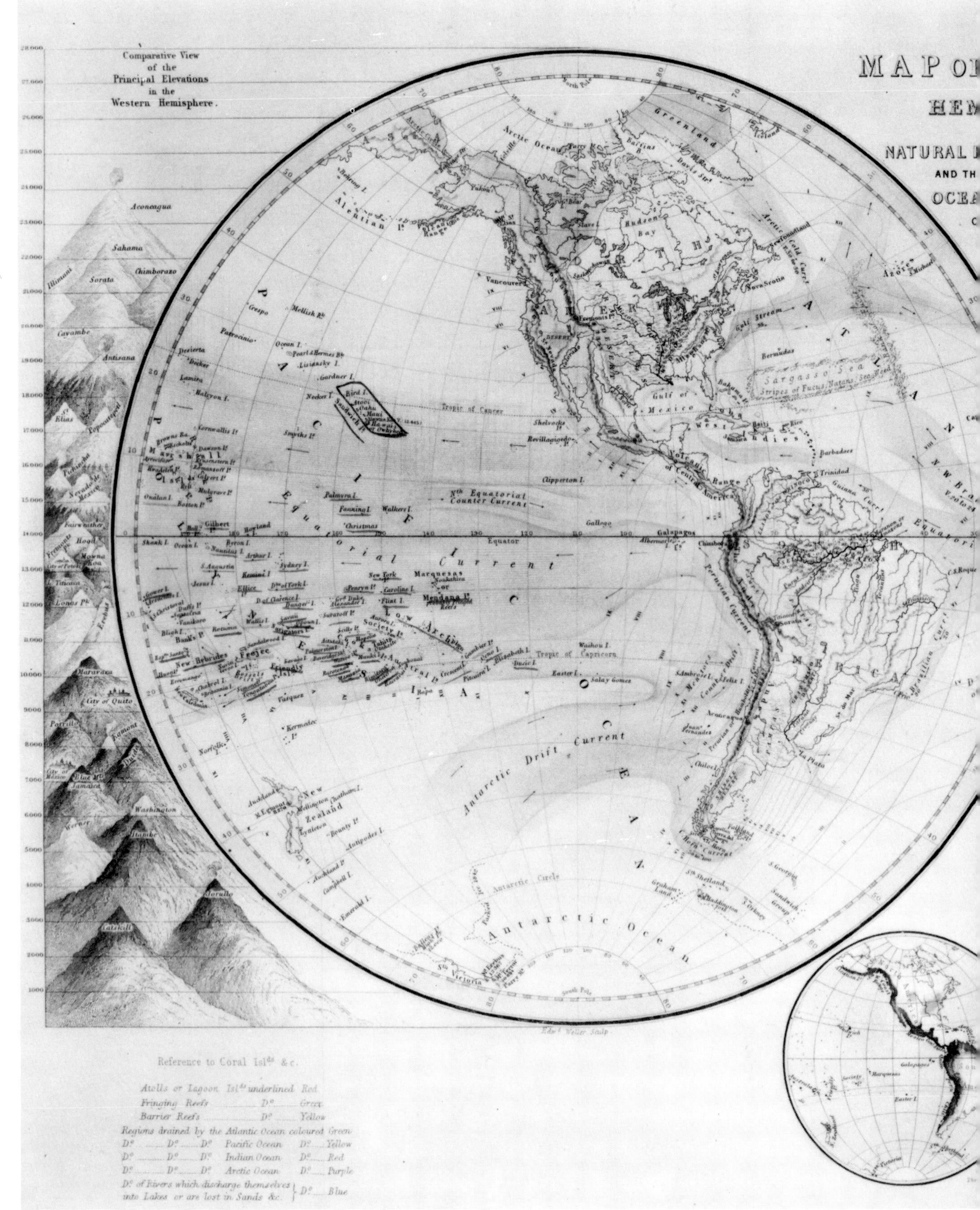

Comparative View of the Principal Elevations in the Western Hemisphere.
Aconcagua
Sahama
Chimborazo
Sorata
Illimani
Cayambe
Antisana
Popocatepetl
Elias
Orizaba
Fairweather
Hood
Mowna Roa
Titicaca
Longs Pk
Erebus
City of Quito
Perote
City of Mexico
Blue Mts Jamaica
Washington
Itambe
Jorullo
Catskill
Arctic Ocean
Greenland
Baffins B.
Davis Str.
Iceland
Aleutian Is.
Hudson Bay
Vancouver
Newfoundland
Nova Scotia
Azores
Gulf Stream
Bermudas
Sargasso Sea
Gulf of Mexico
Tropic of Cancer
West Indies
Barbadoes
Trinidad
Equator
Galapagos
Chimborazo
Equatorial Current
Nth Equatorial Counter Current
Marquesas
Low Archipelago
Society Is.
New Hebrides
Feejee
Friendly Is.
Tropic of Capricorn
Easter I.
Peruvian Current
La Plata
Antarctic Drift Current
New Zealand
Falkland Is.
Cape Horn
Sth Shetland
S. Georgia
Antarctic Circle
Antarctic Ocean
South Pole
Edwd Weller Sculp.
MAP O
NATURAL
OCEA
Reference to Coral Islds &c.
Atolls or Lagoon Islds underlined Red
Fringing Reefs Do Green
Barrier Reefs Do Yellow
Regions drained by the Atlantic Ocean coloured Green
Do Do Do Pacific Ocean Do Yellow
Do Do Do Indian Ocean Do Red
Do Do Do Arctic Ocean Do Purple
Do of Rivers which discharge themselves into Lakes or are lost in Sands &c. Do Blue

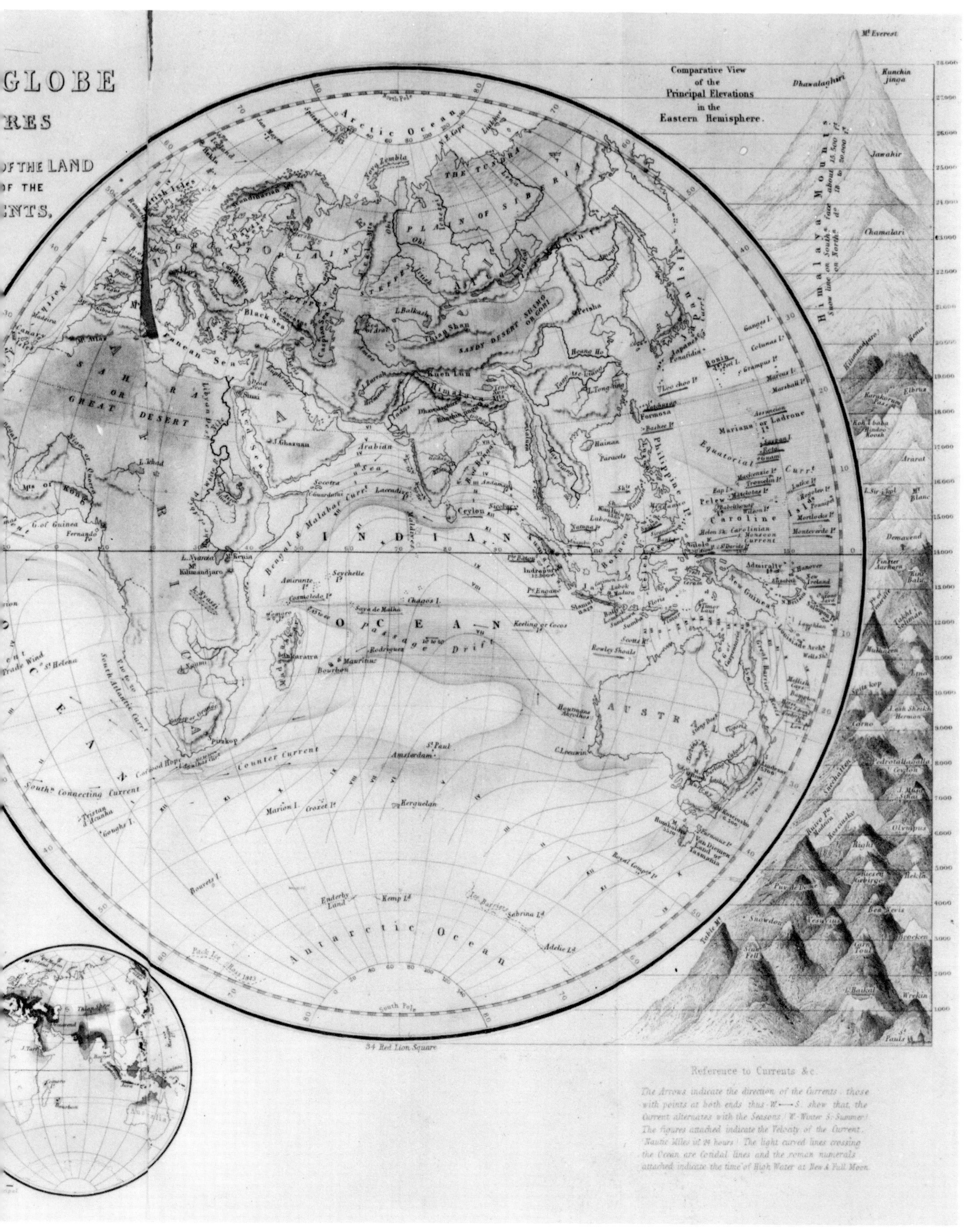
GLOBE
RES
OF THE LAND
OF THE
ENTS.
Comparative View
of the
Principal Elevations
in the
Eastern Hemisphere.
Arctic Ocean
INDIAN OCEAN
Antarctic Ocean
Mt Everest
Counter Current
South Connecting Current
34 Red Lion Square
Reference to Currents &c.

205

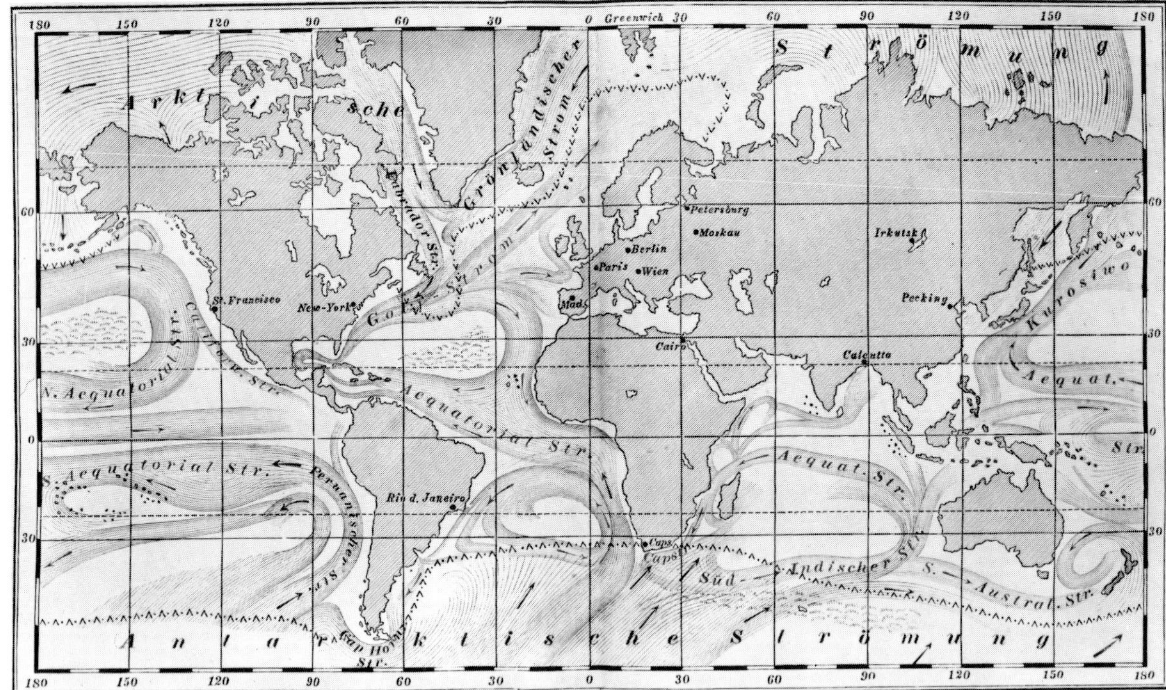

205. *The seven seas.* The seas, so important for nineteenth-century travel and commerce, were charted and mapped as never before. The great ocean currents were carefully noted and the ocean system of water movements, as a whole, was laid bare.

206

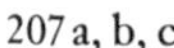

207a, b, c

206. *The atmosphere and its effects.* Meteorology was a nineteenth-century science. Here, a popular schema depicts the manifold meteorological phenomena, each of them numbered: 1. effects of tempestuous winds on land; 2. effects of tempestuous winds at sea: the maelstrom; 3. waterspouts; 4. fog; 5. stratus clouds; 6. cumulus clouds; 7. cirrus clouds; 8. nimbus or rain clouds; 9. cirro-cumulus clouds; 10. rain; 11. snow; 12. perpetual snow; 13. glaciers; 14. aurora borealis; 15. rainbow; 16. halo; 17. mirage; 18. parhelid or mock suns; 19. zodiacal light; 20. ignis fatuus or "will-with-a-wisp"; 21. lightning; 22. lightning conductor; 23. falling stars; 24. aerolites.

207 a, b, c. *The classification of clouds.* In 1803, Luke Howard published the paper "On the Modification of Clouds and on the Principles of Their Production, Suspension and Destruction" in *Philosophical Magazine.* It provided the basic classification and nomenclature for one of the most important meteorological objects. The figures reproduced here are from this essay and identify clouds as follows:

a. Top: Cirrus; middle: Cumulus; bottom, "Stratus occupying a valley at sun-set, in the midst of which is supposed a spot of higher ground with trees."

b. Top: Cirrocumulus; middle: a light and a dark cirrostratus; bottom: mixed and distinct cumulostratus.

c. Approach of a shower, "the superior sheet stretching in different parts to windward, and cumuli advancing towards and entering the mass, the whole of which constitutes the nimbus."

208. *The amateur meteorologist.* Meteorology was the perfect science for the dilettante. Meteorological observations could be taken anywhere, and recorded and compiled voluminously. Here two hikers use a portable mercury barometer to determine atmospheric pressure.

209. *Snow crystals.* Of all meteorological productions, the most beautiful when examined closely is snow. Snowflakes, as illustrated here, are dazzling in their variety and symmetry, and represent both the order and the complexity of nature.

210. *The earth's magnetic field.* Some of the symmetries and complexities of the earth's properties were hidden from direct view and had to be sought out. The earth's magnetic field exerted a visible effect on the traveler's compass, but its subtlety and delicacy could be appreciated only when, as in this illustration, it was carefully observed and measured over long periods of time. This is a picture of the first magnetic observatory, set up by Carl Friedrich Gauss and Wilhelm Weber at Göttingen in the 1830's.

209

210

211

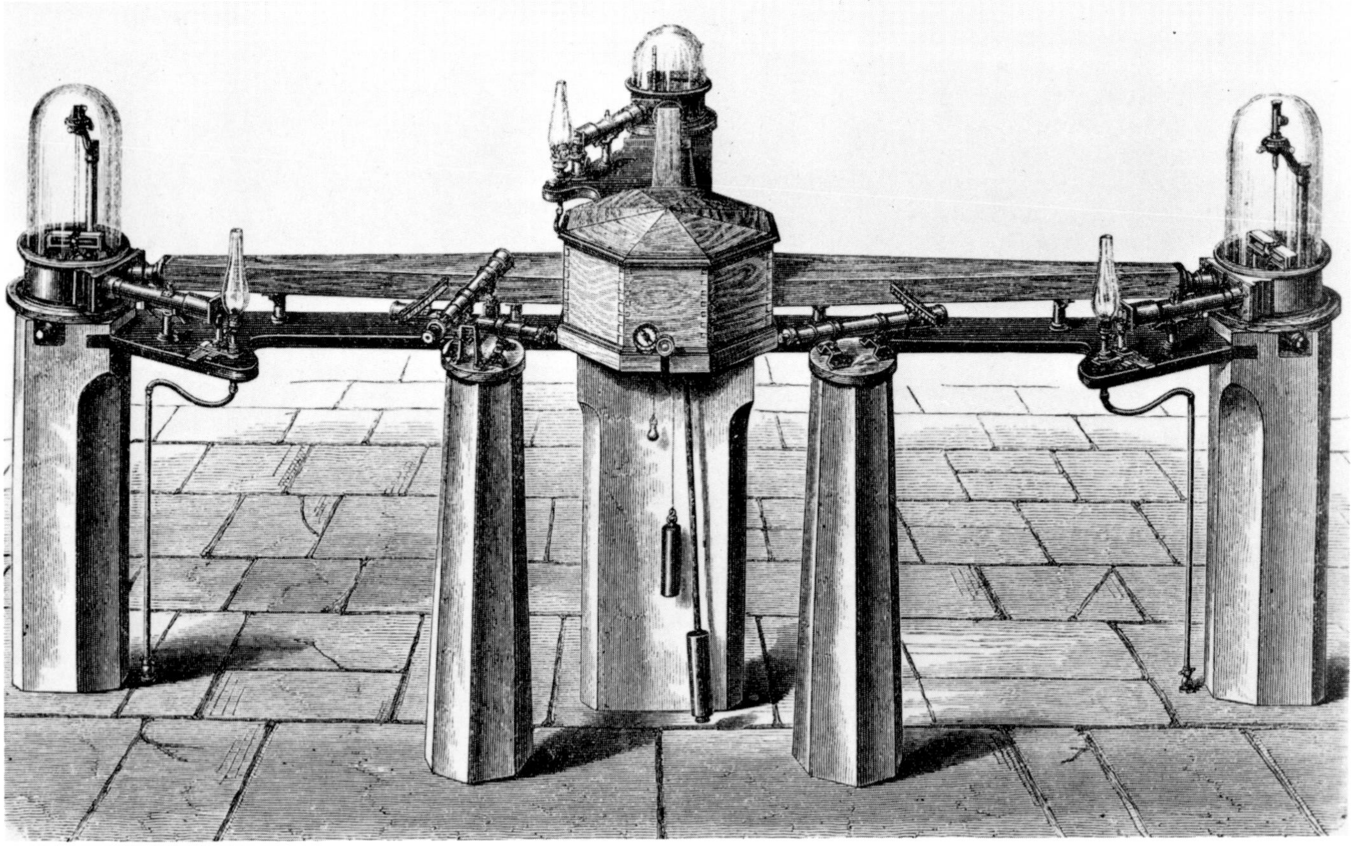

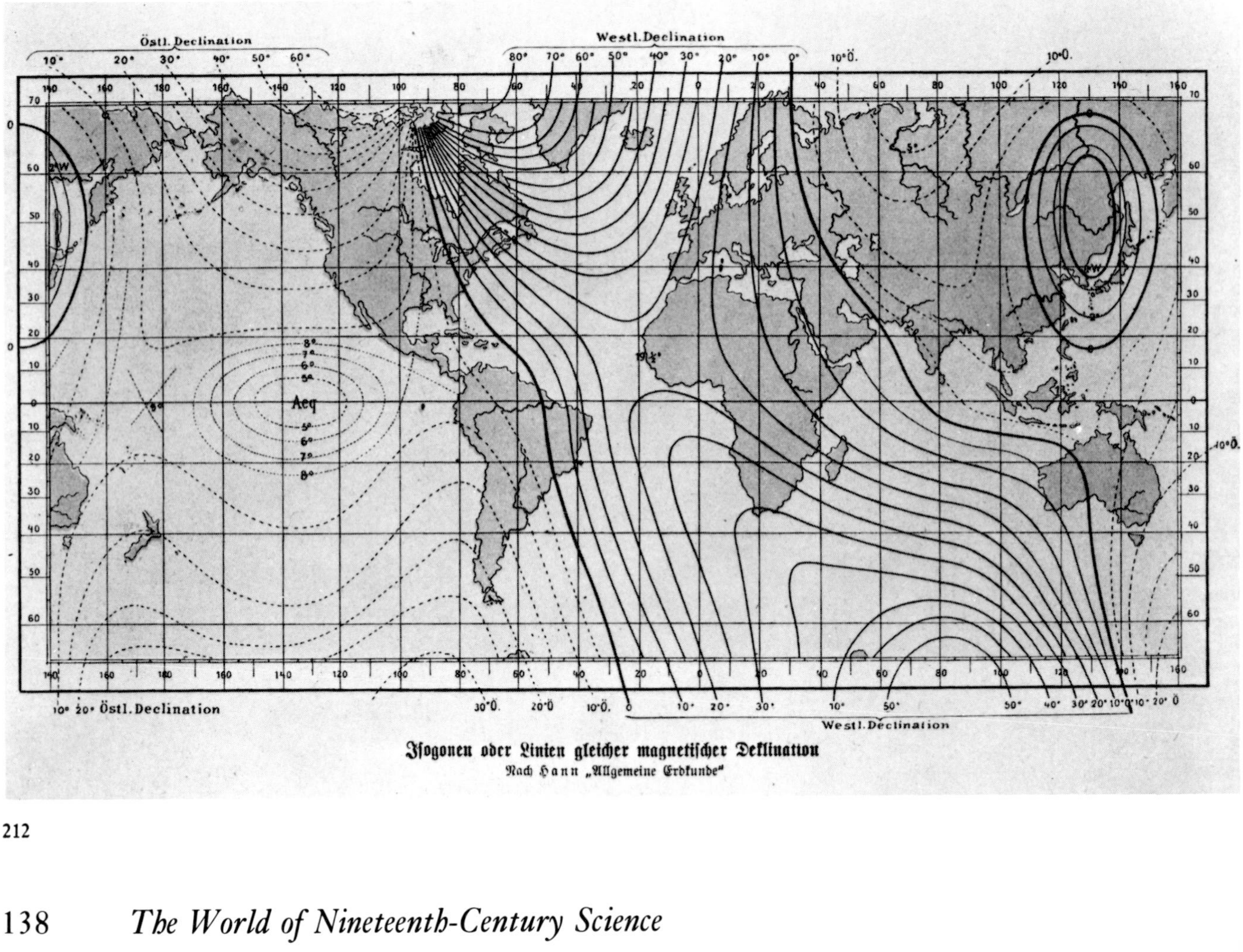
Östl. Declination
Westl. Declination
Aeq
Östl. Declination
Westl. Declination
Isogonen oder Linien gleicher magnetischer Deklination
Nach Hann „Allgemeine Erdkunde“

212

211. *The magnetograph at Kew.* Simple observations by amateurs were soon replaced by accurate and carefully regulated observations made by professionals. The magnetograph at Kew, near London, stood on four stone pillars to prevent casual vibrations. The octagonal case in the middle contained the recording apparatus, which automatically and continuously recorded the horizontal and vertical components of the earth's magnetic field, as well as the magnetic intensity and declination.

212. *The terrestrial magnetic web.* From thousands of isolated measurements it was possible to map the earth's magnetic field. This German map has lines of equal magnetic declination for the earth as a whole.

213. *A meteorological spectacular—the aurora borealis.* The northern lights are one of the most impressive displays in the earth's atmosphere. Not until the nineteenth century was it learned that there is a relationship between the earth's magnetic field and the aurora. Discovery of the exact relationship did not occur until the twentieth century.

213

214

214. *Glaciers and the earth's surface.* Glaciers had been known in Switzerland for centuries. It was only in the nineteenth century, however, that they were recognized as important agents in the shaping of the earth's surface. One impressive example, shown here, was the glacier on Mont Collon in Switzerland. The size of the trees in the foreground gives an indication of the size of the glacier.

215. *A well-dressed glaciologist.* William Buckland, a British clergyman, was one of the foremost amateur geologists in the first half of the nineteenth century. Here he is shown preparing to ascend a glacier for study. One wonders if the top hat was absolutely necessary.

215

216

216. *Vesuvius in 1843.* Volcanoes were obvious geological features that had long fascinated students of the earth. The eruption of Vesuvius in 1843 attracted attention throughout Europe. The artist who made this sketch was struck by the secondary cones erupting within the great crater.

217. OVERLEAF: *Turner's painting of Vesuvius.* The spectacle of a volcanic eruption stirred artists as well as geologists. The great British painter William Turner has left this record of Vesuvius in full activity.

218

219

218. *A volcanic eruption in the South Seas.* Not all volcanoes were alike. Vesuvius had a large crater and only occasionally blew up entirely. The periodic eruption of Mauna Loa in the Hawaiian Islands revealed the extraordinary power of the subterranean forces. Note the size of the people in the foreground.

219. *A volcanic sea in Hawaii.* The sense of almost demonic forces at work is graphically conveyed in this 1891 illustration showing the crater of Kilauea in Hawaii.

220. *Primitive volcanic theory.* The glowing mass of liquid lava spewed forth during eruptions could not help but suggest the presence of molten subterranean lava lakes. It was hypothesized that volcanoes were created when this lava broke through the earth's crust at a fault, as shown here at Vesuvius in 1843. This theory ignored the complex structure of the earth's crust and later was modified considerably.

220

Some Explorations

Man has been exploring the earth's surface since his ancestors first ventured out on the African veldt. The pace and scale of these exploratory voyages increased enormously in the eighteenth and nineteenth centuries. The eighteenth century sketched in the major continental masses of the earth; the nineteenth century ventured into their interior and opened them up to European and American conquest and exploitation. Because the nineteenth century was also the greatest century of Western imperialism, it has often been simply assumed that explorers were driven by imperialist urges or desires. In some cases, this was undoubtedly true, but in most of the great explorations it was curiosity and a burning desire to penetrate where no Westerner had ever been before that drove people on. There was, in the nineteenth century, a literal explosion of the Western world overseas. It would be impossible within a short space to chronicle all these assaults on geographical darkness, but we can and must notice some of the more outstanding achievements.

The search for a Northwest Passage across North America to the Pacific began soon after the discovery of North America. The economic implications here were obvious; if such a passage could be found it would cut the sea route to the east by more than half. By the nineteenth century, the economic dimension had shrunk for by then it was clearly realized that any Northwest Passage would have to pass through seas that were frozen for a good part of the year. The search continued because men were obsessed with it and convinced that there must be a way from the Atlantic to the Pacific without rounding Cape Horn or the Cape of Good Hope. Obsessions are dangerous things and the northwest Canadian Arctic was strewn with the bodies of explorers who had failed. But their explorations were not failures. They investigated and mapped the north, discovered the magnetic North Pole, and prepared for the final assault on the terrestrial North Pole in the twentieth century. The meteorological and geographical data that were brought back to Europe and America were invaluable in the compilation of Humboldt's cosmic vision of the earth as a whole.

Both South America and Africa were "dark" continents when the nineteenth century began. South America had a considerable European settlement along the coasts, but the interior was almost totally unknown. Africa had barely been touched by Europeans. Both continents were to be traversed and described during the nineteenth century. One of the great explorations of South America that had immense consequences for the development of science was the voyage of H.M.S. *Beagle* on which Charles Darwin shipped as naturalist. It was Darwin's observations of South American flora and fauna that provided him with the basis for his later researches on evolution through natural selection. Africa was explored by less scientific types. Richard Burton and his companion, John Speke suffered incredible hardships out of stubbornness and a love for exotica. Dr. David Livingstone was a medical missionary who wandered for years throughout the jungle fastnesses of Africa, treating and converting the natives and periodically reporting his geographical and other discoveries to an astonished public at home. Henry Morton Stanley was a journalistic buccaneer of incredible will and courage who used Africa to increase his own fame and fortune. From the efforts of these men, and a host of others like them, there filtered back to Europe and America precious geographical, ethnological, botanical, and zoological data. Most of it had to await the twenteth century before it could be fully assimilated, but by the end of the nineteenth century, Africa had lost its status (and innocence) as *terra incognita.*

The purely scientific oceanographic expedition had originated in the eighteenth century, the best example being the voyages of Captain James Cook in the Pacific. In the nineteenth century, such expeditions multiplied and, more importantly, set out with much improved apparatus and personnel for the investigation of the earth. The most famous was probably that of the *Beagle*, but the most extensive and the richest in terms of raw data was the voyage of H.M.S. *Challenger* in the years 1872-1876. The *Challenger* was fitted out as a research oceanographic vessel and its goal was nothing less than the complete mapping—geographical, botanical, and zoological—of the southern seas. The results of the expedition fill more than forty large, quarto volumes, richly illustrated by many competent scientist-artists, and accompanied by technical discussions of the objects under study.

By the end of the century, no one could complain about the paucity of data on the nature and products of the earth's surface and seas. There was more information available than could be easily assimilated by a single generation. Indeed, scientists are still at work trying to fit together all the facts that were displayed in the learned folios and quartos of the nineteenth century. The large-scale description of the globe was complete. Why and how the structures, animals, and plants that had been described had developed remained to be answered. But this was the task of geology and evolutionary biology.

221

221. *The end of the Franklin expedition.* Exploration of the northlands often ended in tragedy. In 1847 and 1848, a mission searching for the Northwest Passage under the command of Sir John Franklin suffered the fate pictured here. Franklin and his entire company froze to death.

222

224

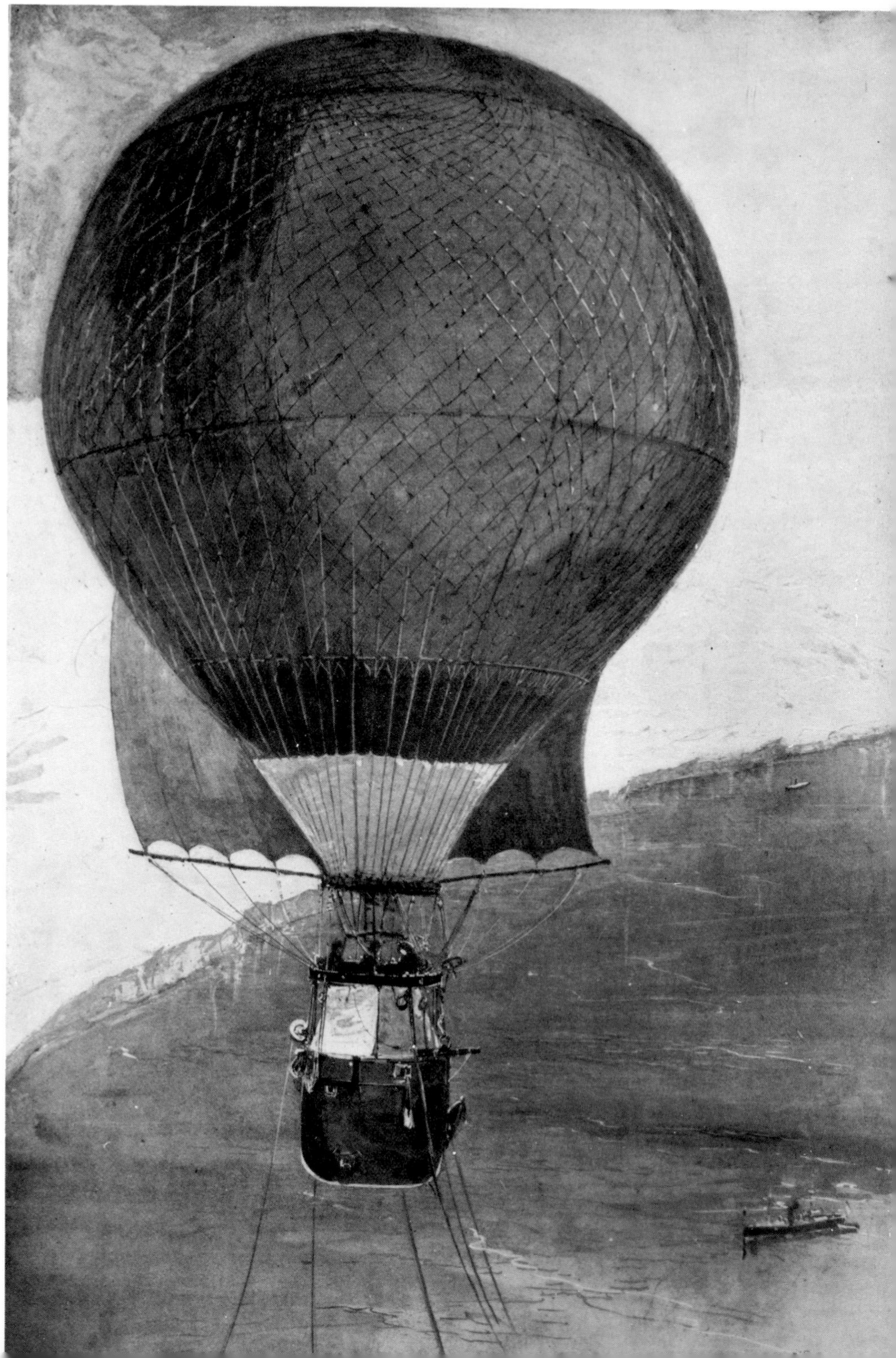

222. *The discovery of the North Magnetic Pole.* The beginning of the nineteenth century saw a flurry of intense exploration in the Canadian north, as expedition after expedition sought the Northwest Passage between the Atlantic and the Pacific. In the course of these explorations, James Ross discovered the North Magnetic Pole in 1831. An illustration of the discovery, dating from the end of the century, shows the declination needle pointing straight down. The position is 70 degrees, 5 minutes north latitude and 96 degrees, 46 minutes west longtitude.

223. *The Jeannette Arctic Expedition.* In 1880, an American expedition set out to explore the Arctic regions off the Siberian coast. Their ship, the *Jeannette*, was wrecked; and most of the crew and scientists aboard perished. Their bodies were found in 1881, testimony to the continued hazards of polar exploration.

224. *The Eagle departs for the North Pole (1897).* Journeys to the North Pole by ship or dogsled were both dangerous and uncomfortable. By the end of the century, however, it appeared that it might be possible to make the trip almost in luxury, by balloon. The *Eagle* was designed to sail over the pole while its crew remained warm and comfortable. On July 11, 1897, the *Eagle* set forth from Spitsbergen and vanished almost immediately. In 1930, the remains of the expedition were found on a remote island. There was to be no "royal road" to the pole.

225

FEZZAN
EGYPT
NILE
ARABIA
GREAT DESERT
AFRICA
NUBIA
ABYSSINIA

226

227

225. *The source of the Nile.* The Nile had been an object of curiosity since the time of Herodotus. In 1863, Captains John Speke and Robert Grant reported to the Royal Geographical Society that they had discovered the source of the Nile in the great body of water they christened Lake Victoria.

226. *Sir Samuel Baker's lecture on the anti-slavery expedition up the White Nile.* The time that elapsed between the exploration of a region and its political exploitation was short in the nineteenth century. Within twenty years of the discovery of the source of the Nile, the river's upper reaches were being used as a highway for European penetration into central Africa. Sir Samuel Baker's lecture on the expedition he led to stamp out the slave trade drew 1,500 people to the lecture hall. A talk combining the Dark Continent and slavery was irresistible to a Victorian audience.

227. *Sir Samuel Baker's expedition.* The expedition was made in some force. The steamboat, shown being hauled through the river grass, seems utterly out of place, but it was a powerful and awe-inspiring symbol of the British, who soon seized control of the entire length of the Nile.

228. OVERLEAF: *The Victoria Falls on the Zambezi.* In contrast with Sir Samuel Baker, who literally invaded Africa with an expeditionary force, the explorer David Livingstone penetrated to the heart of Africa almost alone. In his *Missionary Travels and Researches in South-Africa* (1857), he published this picture of the great falls on the Zambezi to which the name Victoria was given.

229

229. *HMS Challenger.* The motives behind exploration were many and complex, and often purely scientific concerns came low on the list. In 1872, however, HMS *Challenger* was fitted out for the specific purpose of undertaking the most extensive scientific expedition planned to date —to plumb the seas and bring back both specimens and information. The expedition lasted from 1873 to 1876 and greatly enriched knowledge of living forms and terrestrial formations.

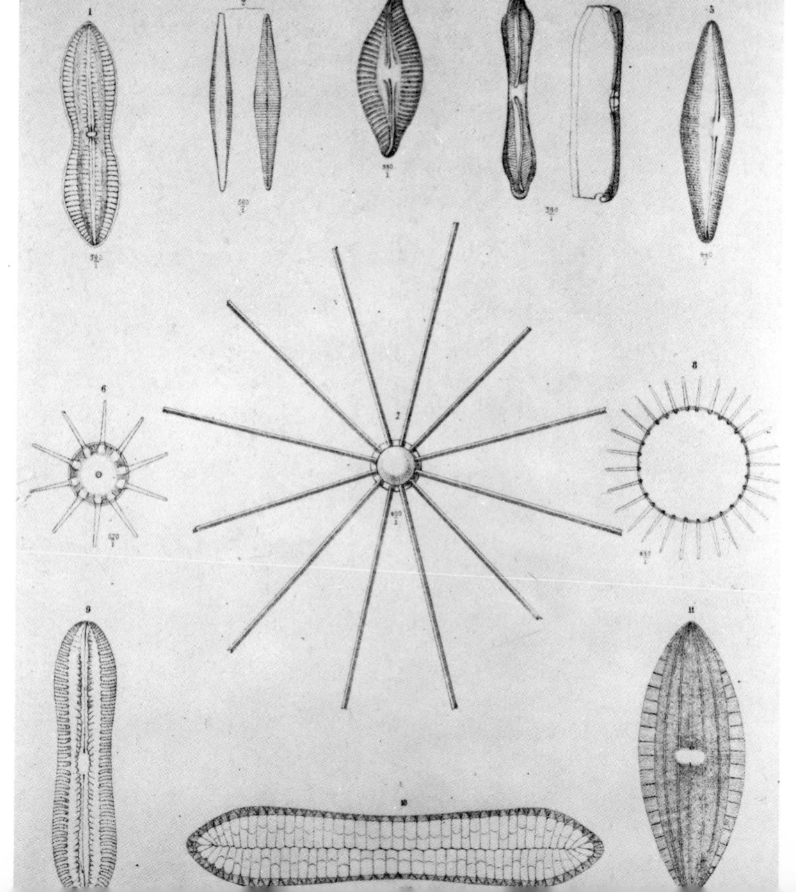

230

230. *Challenger expedition specimens: Diatomacea.* These minute shells were part of the class that had built up the white cliffs of Dover. Their forms are extremely delicate and beautiful.

231. *Challenger expedition specimens: shore fish.* Specimens were sought in every stratum of the oceans. Meticulous drawings or paintings were made of those found.

231

232. *Challenger expedition specimens: deep-sea medusae.*

233. *Challenger expedition specimens: Cirripedia.* No subject was too lowly to be studied. These barnacles may have come from the hull of the *Challenger* itself.

232

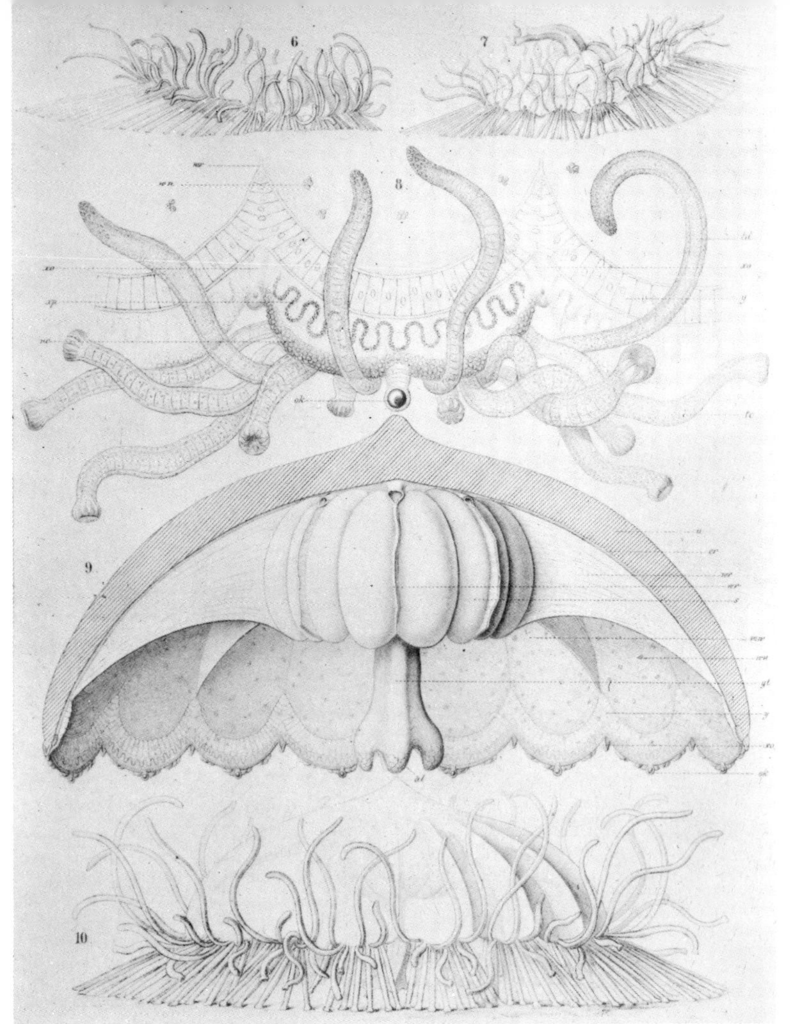

233

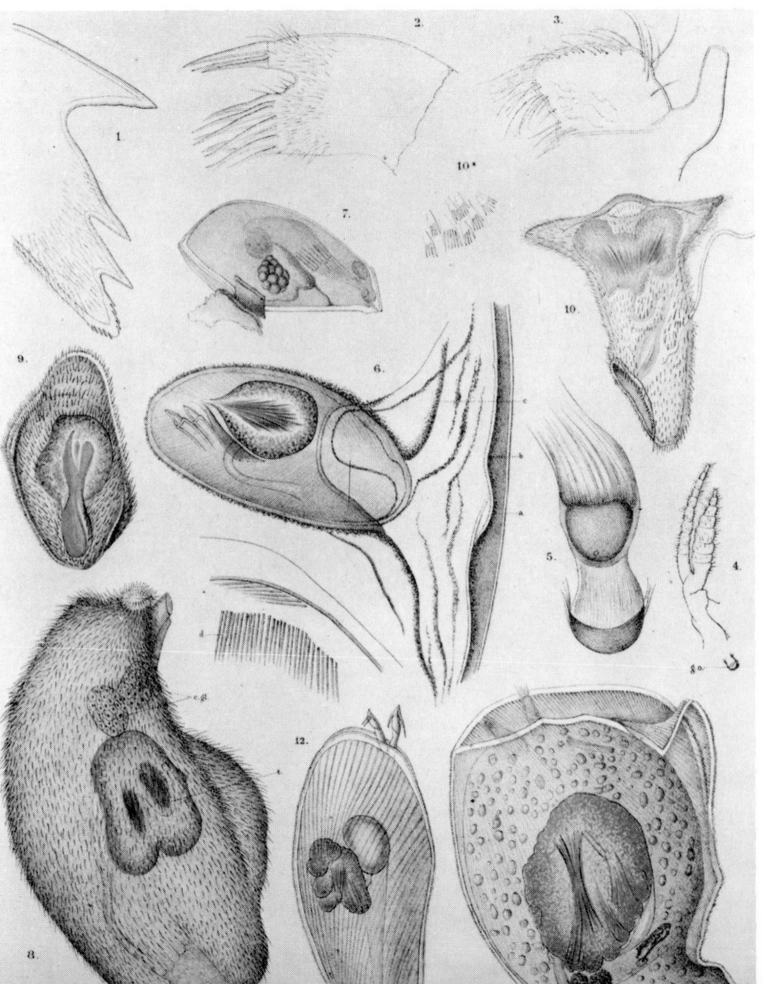

234

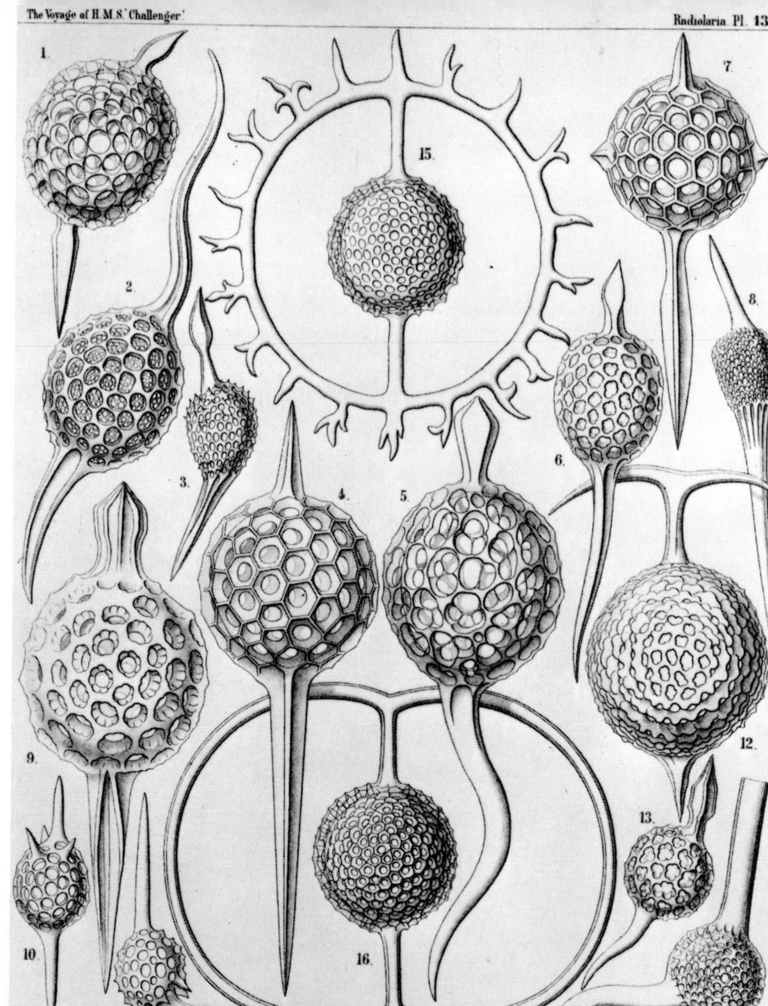

234. *Challenger expedition specimens: Radiolaria.* This figure and the two following were drawn by Ernst Haeckel, one of Germany's foremost biologists, who was fascinated by the forms of minute creatures such as these. Haeckel relied heavily on these researches in his later works on the formative forces of living creatures.

235

235. *Challenger expedition specimens: Radiolaria by Haeckel.*

236

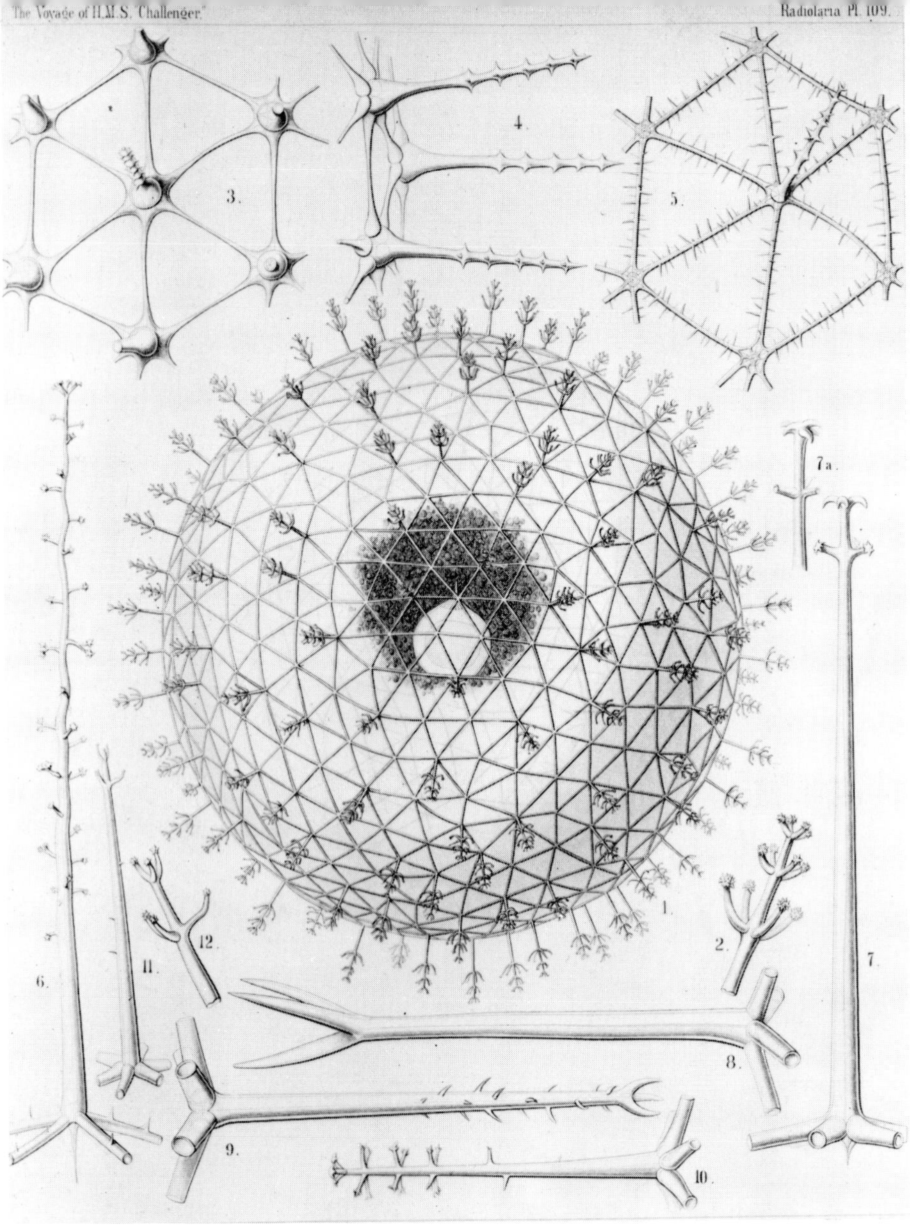

The Voyage of H.M.S. Challenger. Radiolaria Pl. 109.

236. *Challenger expedition specimens: Radiolaria by Haeckel.*

237. *Challenger expedition specimens: deep-sea fish.* The depths were searched for specimens, among which was this example. The artist-scientist who drew the illustration paid particular attention to the structure of the head.

238. *Challenger expedition specimens: Siphonophora.* Even the most delicate sea creatures were carefully collected and described. The beauty of living creatures can rarely have exceeded that of these siphonophores.

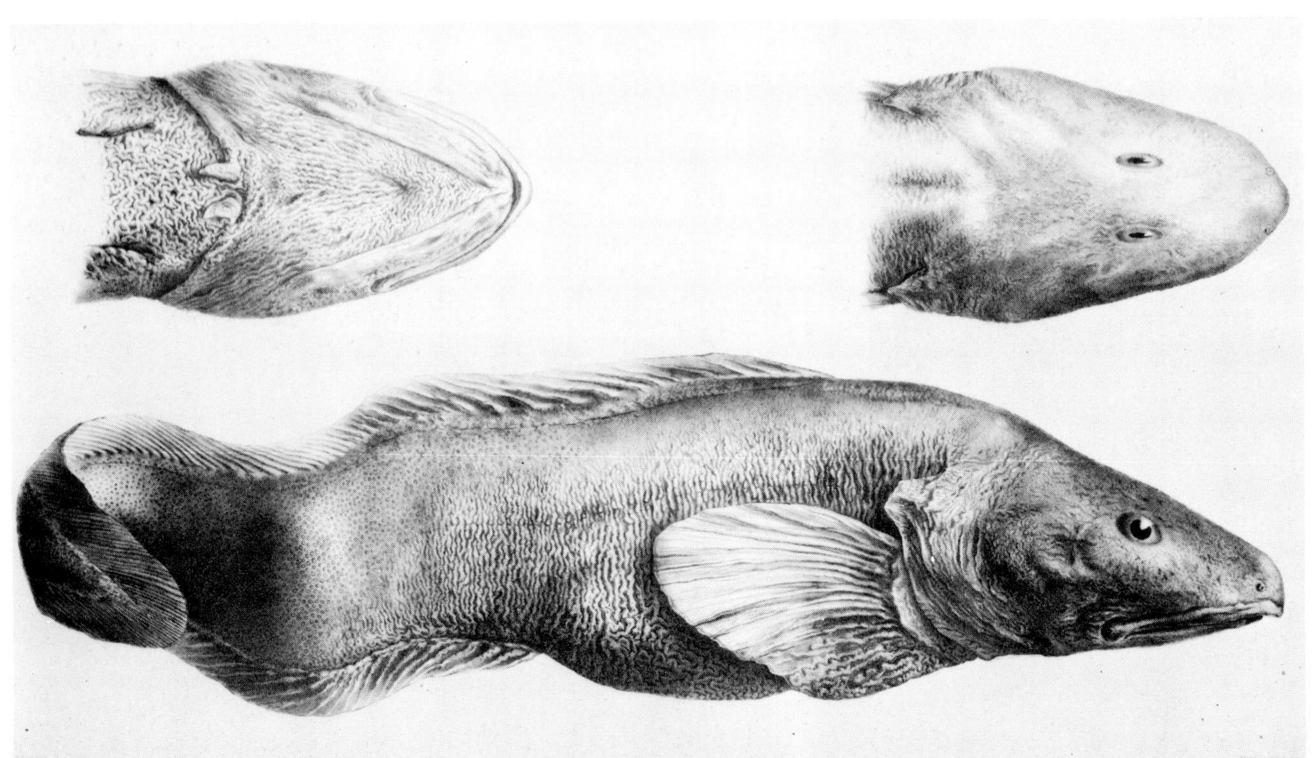

237

238

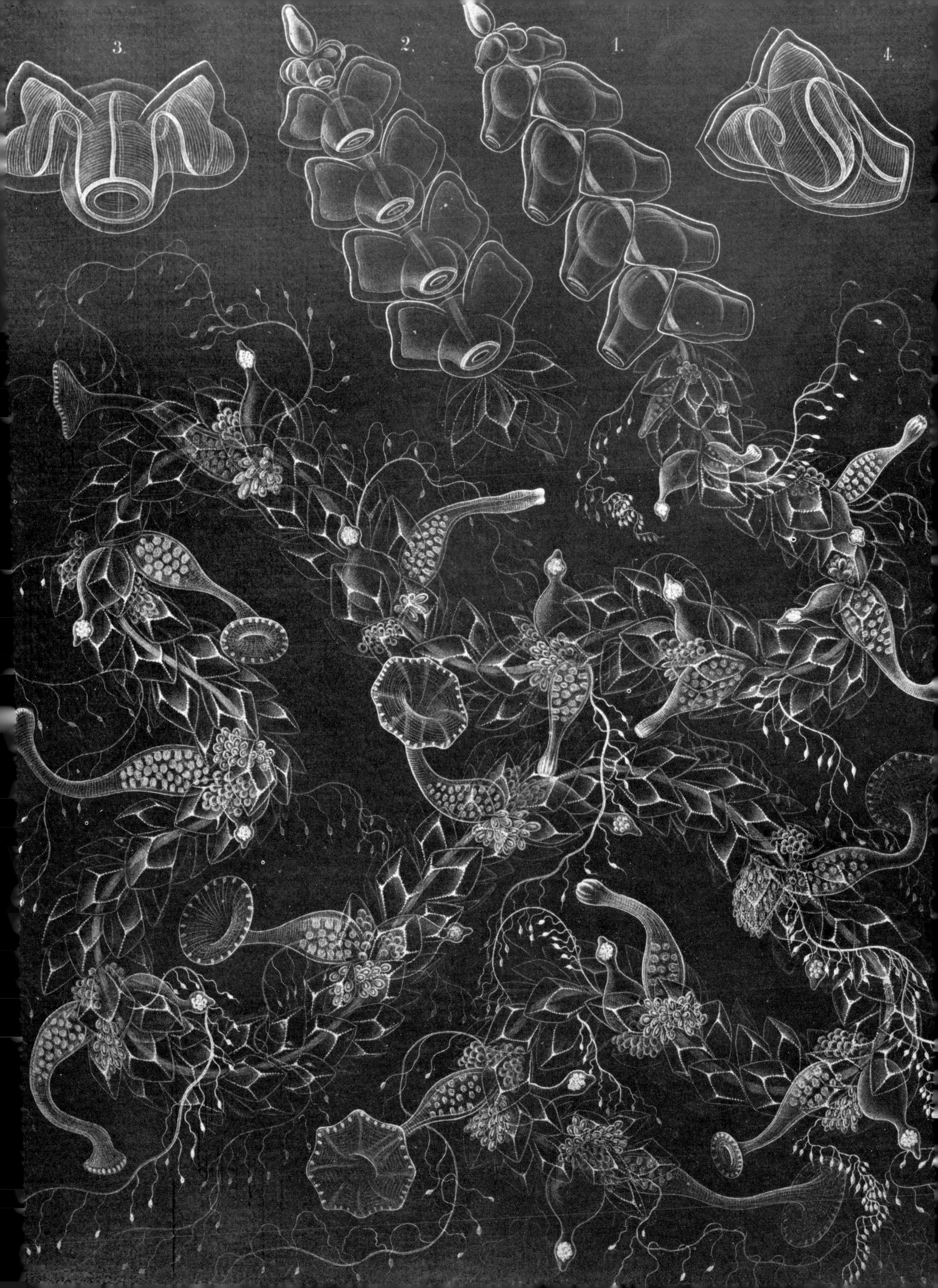
3.
2.
1.
4.

Geological Wonders

The Book of Genesis provided the Western world with a ready-made account of the formation of the earth. God had created the planet some 6,000 years before the nineteenth century began. He had populated it with plants and animals and man. In his wrath, he had seen fit to cover its surface with a great flood and, occasionally, to chastise sinful populations with fire and brimstone, as at Sodom and Gomorrah. By and large, however, the earth remained as he had made it; and for the faithful that was that. The problem was that it became increasingly difficult to reconcile the actual state of the earth with the biblical account. By the nineteenth century, something had to give and what gave was the Bible. It was, however, a fairly long and bitter battle. The first stage was the attempt to reconcile the biblical account with observations. In the eighteenth century in Germany, Abraham Gottlob Werner was able to give temporary scientific respectability to the Noachian flood as the primary geological force. His theory of Neptunism suggested that geological formations were the result of the flood, that most observable geological phenomena were the result of crystallization from the universal sea. The history of the earth's crust should, therefore, be read as one of precipitation or sedimentation from the waters that had covered the earth some six millennia before. Werner's theories found widespread acceptance, but they also ran into important pockets of opposition. Not the least of these was in Scotland, where James Hutton (1726-1797) took the then daring position that geology could say nothing of terrestrial origins or divine interventions. The history of the earth must be read from the knowledge of the forces at work at the present, and among these, Hutton singled out heat and fire as primary. Vulcanism (as Hutton's position was called) had the virtue of allowing geological change to continue through time but the defect of seemingly being too local to account for the whole structure of the earth. A third point of view was that put forward by Georges Cuvier in the first decades of the nineteenth century. Cuvier was a devout French Protestant who had no desire to overturn Scripture but who also realized that Scripture was no infallible guide to geology or paleontology. Cuvier was struck by the richness, diversity, and peculiarity of the fossil record. By his time it was becoming clear that many of the fossil animals no longer existed on the earth. God, apparently, took away as well as gave, in the zoological sense. Cuvier attempted to reconcile Genesis and geology by assuming that the earth had suffered a series of catastrophes in which whole animal and plant species had been wiped out and new ones created. Scripture described only the last such upheaval, leaving it to geologists to fill in the earlier examples of divine destruction and creation. This view still begged the geological question for it gave to the flood the primacy in the formation of certain features of the earth which were becoming increasingly difficult to account for in this fashion.

One of the more embarrassing geological facts for the pious was the existence of columnar basalt. Fingal's Cave in the Hebrides and the Devil's Causeway in North Ireland were prime examples of these huge, hexagonal columns of what appeared to be perfectly regular crystals of basalt. Humboldt reported another such formation in the New World. Since basalt was not water soluble, it was not likely that these great columns had "crystallized" from aqueous solution. Since, as well, there was other evidence near columnar basalt of intense and extensive volcanic activity, it seemed only logical to conclude that this formation, at least, was due to volcanic activity and not to the flood. It was a small victory, but an important one. So, too, was the argument that granite was extruded into rocks in a molten state, and had not "flowed" into them and then crystallized there from a water solution.

The existence of deep gorges also tended to

undercut the Neptunist and Catastrophist position. It was in these gorges that the various layers of the earth's crust could be seen in rather considerable detail and numbers. Could anyone seriously maintain that each separate stratum represented both a new catastrophe and another precipitation from the primeval ocean? The discovery and description of the Grand Canyon, after midcentury, laid to rest whatever lingering doubts there may have been about the inadequacies of the earlier theories and of Scripture as guides to an understanding of the earth's crust and structure.

239

239. *Columnar basalt in Fingal's Cave.* One of the most striking geological formations is columnar basalt. This formation figured prominently in the nineteenth-century dispute over the aqueous or volcanic origin of geological forms. Fingal's Cave in the Hebrides was a famous example of this peculiar natural phenomenon.

240. *Columnar basalt in Mexico.* Fingal's Cave was not geologically unique. The great naturalist and traveler Alexander von Humboldt reported this example from Mexico.

241. *The Grand Canyon.* No geological formation was as spectacular as the Grand Canyon. Its discovery and exploration were probably the most important geological contribution made by the United States in the nineteenth century.

240

241

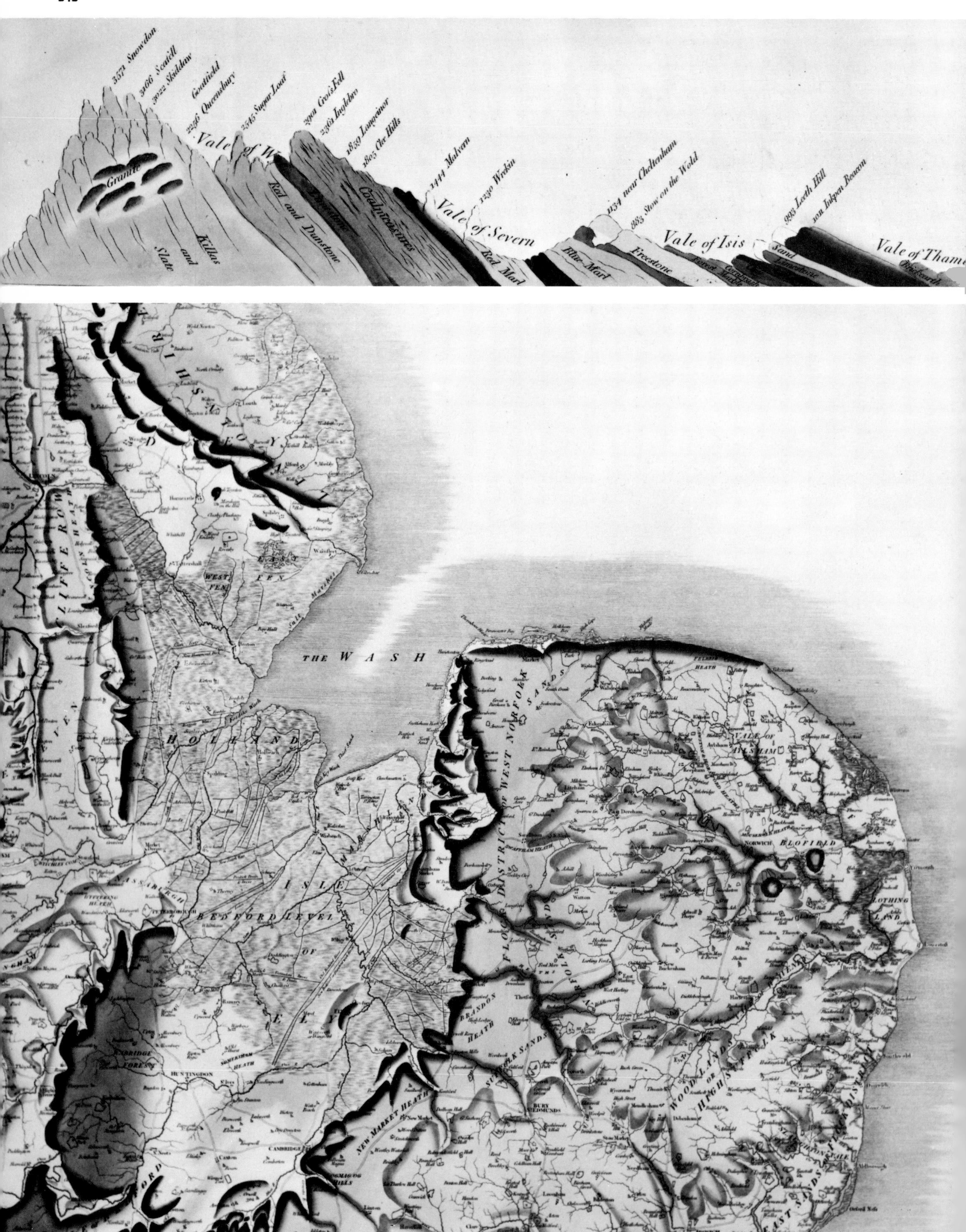
3571 Snowdon
3166 Scafell
3022 Skiddaw
Goatfield
2256 Queensbury
2545 Sugar Loaf
2901 Cross Fell
2361 Ingleboro
1859 Lanqainor
1803 Clee Hills
1444 Malvern
1230 Wrekin
1134 near Cheltenham
883 Stow on the Wold
993 Leeth Hill
1011 Inkpen Beacon
Granite
Killas and Slate
Vale of W
Red and Dunstone
Coalmeasures
Vale of Severn
Red Marl
Blue Marl
Freestone
Vale of Isis
Sand
Vale of Thame
Brickearth
THE WASH
HOLLAND
WEST FEN
BEDFORD LEVEL
ISLE OF ELY
DISTRICT OF WEST NORFOLK SANDS
FLOCK DISTRICT
BRANDON HEATH
SUFFOLK SANDS
NEW MARKET HEATH
GOGMAGOG HILLS
WOODLANDS OR HIGH SUFFOLK
SUFFOLK EAST SANDS
VALE OF AYLSHAM
BLOFIELD
LOTHING LAND
NORWICH
CAMBRIDGE
HUNTINGDON
BURY ST EDMUNDS
DOWNS

242. *William Smith's delineation of strata.* The key to gross geological formations was the observation of the strata, which could be seen when vertical cuts of some depth were made in the earth's surface. William Smith first systematically mapped these strata in his *A Delineation of the Strata of England and Wales* (1815), from which this illustration is taken.

243. *A geological map of England.* In William Smith's 1815 map of England, a detail of which is reproduced here, different colors were used to indicate geological features.

244. *Devonian and Carboniferous rocks in the gorge of the Chusovaya River (Russia).* The work of deciphering the geological record led geologists everywhere. The various layers that could be recognized in gorges such as this appeared to be of fundamental importance, although it was not easy to tell just what they meant.

245. *The Belaya Ravine in Russia.* R. I. Murchison, a retired veteran of the Napoleonic Wars, was an amateur geologist of considerable talent. Here he uses strata as a guide to their fossil contents. It was gradually realized that similar fossils are found in similar strata.

244

245

The lower beds *a*, *b*, *c*, *d*, *e*, *f*, consist of various sands and marls, in which ichthyolites are disseminated, but of which the bed *d* is a complete congeries of fish-bones surmounted by a copious mass of red, white, and green argillaceous marls. Then follow bituminous schists (*g*, *h*, *i*) with courses of bad coal, constituting the bottom beds of the carboniferous deposits, which, after other alternations of sands and marls (*k*, *l*), are followed by the carboniferous limestone *m*, *n*, with many characteristic fossils, including even species which are well known in Britain, such as the *Productus hemisphæricus*, *P. punctatus*, and *P. semireticulatus.*

The Structure of the Earth

The early secular theories of the earth in the nineteenth century were both crude and naive. They generally assumed the earth to consist of a warm or hot interior core, surrounded by a crust through which volcanic activity penetrated to cause massive, but local, upheavals. Some Neptunist origins to this crust could be conceded, as was the case with Amos Eaton, but the real relationships between the earth's crust and the forces shaping it were totally obscure. What had to be done, therefore, was painstakingly to analyze and scrutinize the local geological record in the hopes of discovering significant clues to geological action. Global insights, like Snider's, that the continents had originally all being joined together, did not really advance geology since no one then could even begin to suggest how the separation of the continents had taken place. So, a generation of geologists worked long and hard at the task of unraveling the history of the earth. And, gradually, this history became clearer. Given enough time (and this in itself was a revolution to secure the amount of time necessary), it could be shown that wind and rain and ocean waves and glaciers, with occasional catastrophic interruptions, could, in fact, account for most of the observable geological formations.

The understanding of these formations owed something to the manmade revolutions of the nineteenth century. Natural gorges revealed some details of the earth's crust, but manmade ditches called these details forcefully to the attention of those making them. Railroads had to be cut through hills and the Erie Canal cut through rich geological fields. Only the blind could ignore what was laid open before man's eyes. William Smith in England was the first to recognize the importance of the strata that were laid bare by the passage of the railroads. His great geological map of England was organized around the strata that he illustrated so well. The principle that similar fossils indicated identity of strata even though the strata were separated from one another by long distances provided the firm foundation for the scientific study of the earth's crust. The fossil record permitted geologists to trace similar formations over large distances and follow them through the various warping and folding processes to which they had been subjected. In this way, it became possible to gain some glimmering of how mountains had arisen, been eroded, thrust down into sea basins, and forced upward again over geological time. In short, the history of the earth became legible.

The final element needed for an understanding of strata was the realization that plant and animal forms had evolved over the course of time. Once it was possible to follow the process of speciation, the history of life provided the proper sequence for the history of the earth. Profiles of strata could be proposed that reflected the development of the earth. Geology had become a mature science and one, be it noted, that had passed from infancy to maturity entirely within the nineteenth century.

246. *Structure of the earth.* William Smith's work on strata came from close observation, but strata also could be deduced from theory. In 1830, at Albany, New York, Amos Eaton published *Geological Textbook.* Figure 1, at the top, illustrates the laying down of strata from the primitive ocean that covered the earth. The perfect serenity of this aqueous world was violently disturbed when, in Eaton's words, "in due time the combustible materials marked 1, 2, 3, 4, 5, 6 were ignited and produced the changes exhibited in Fig. 2."

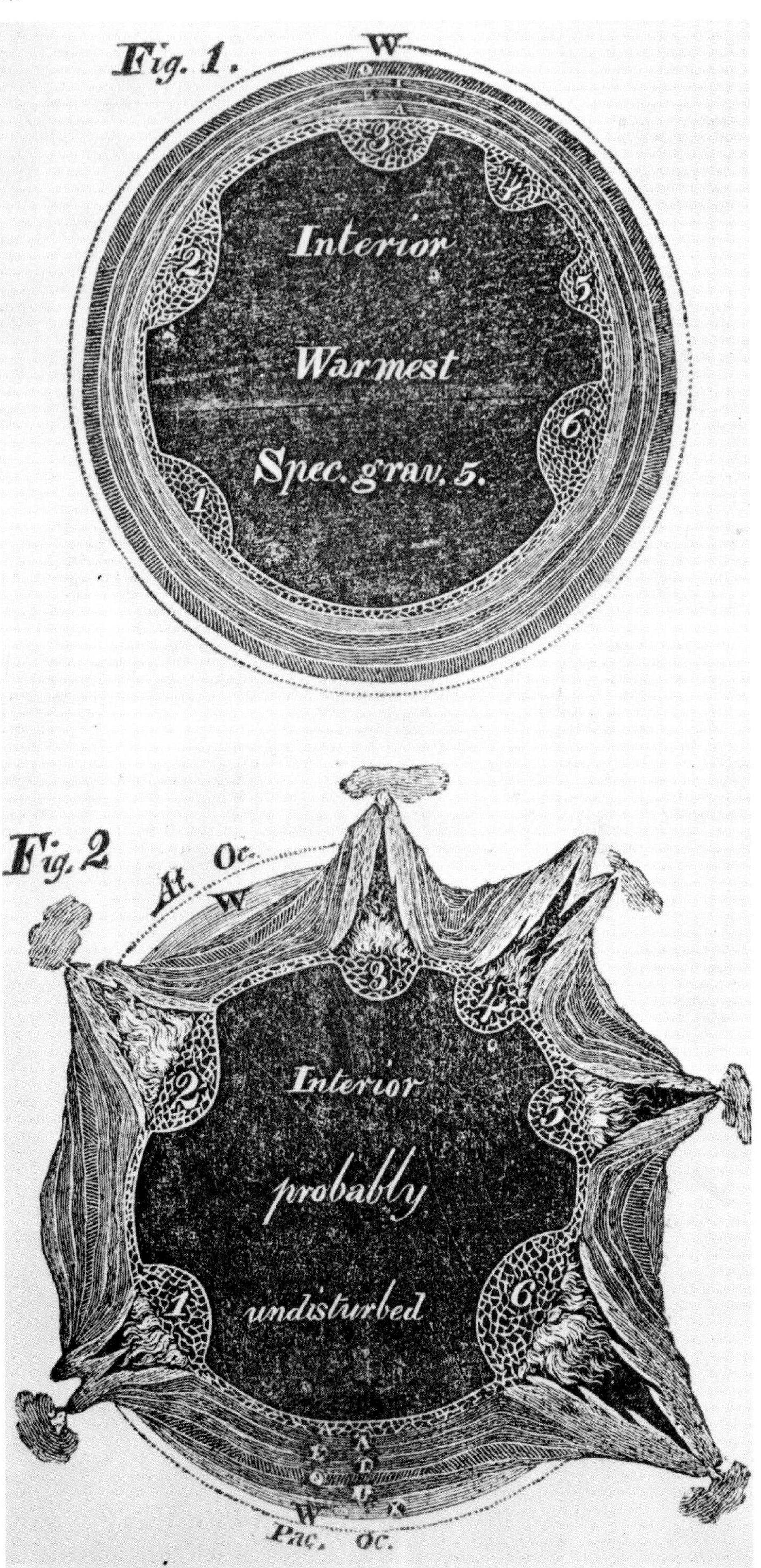

247

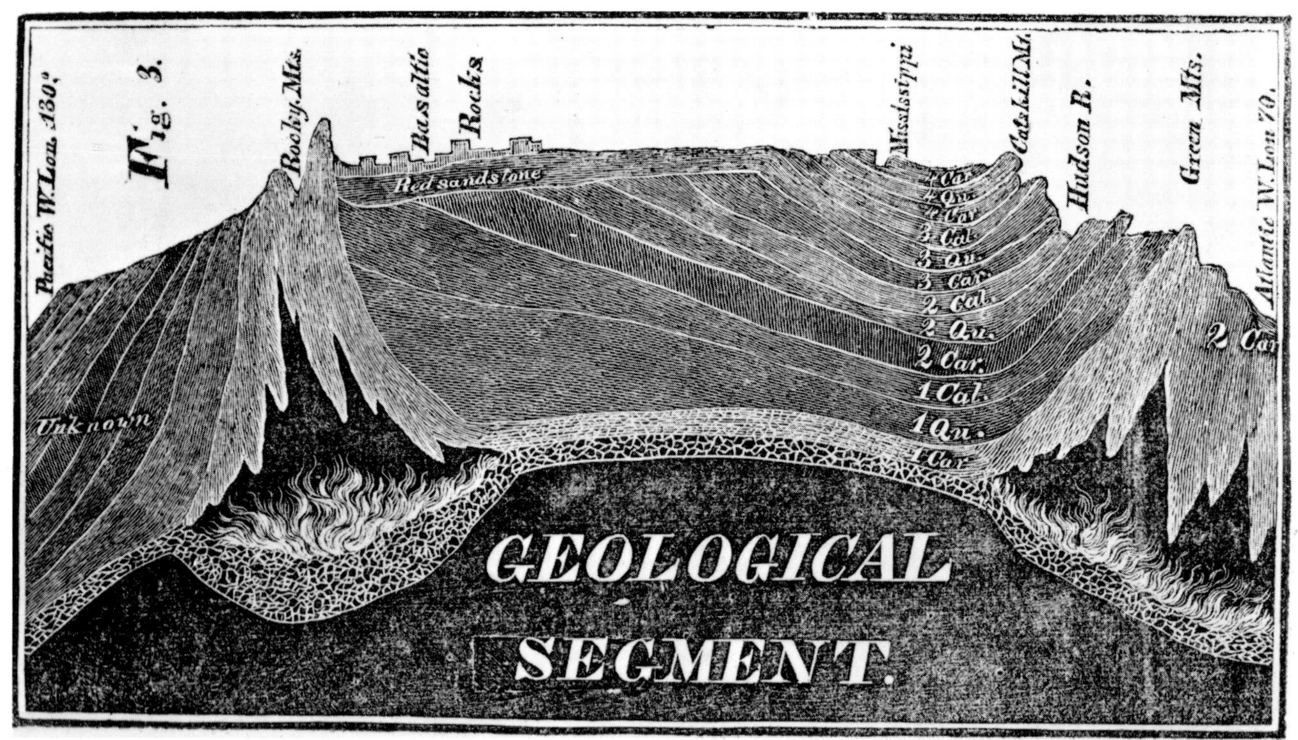

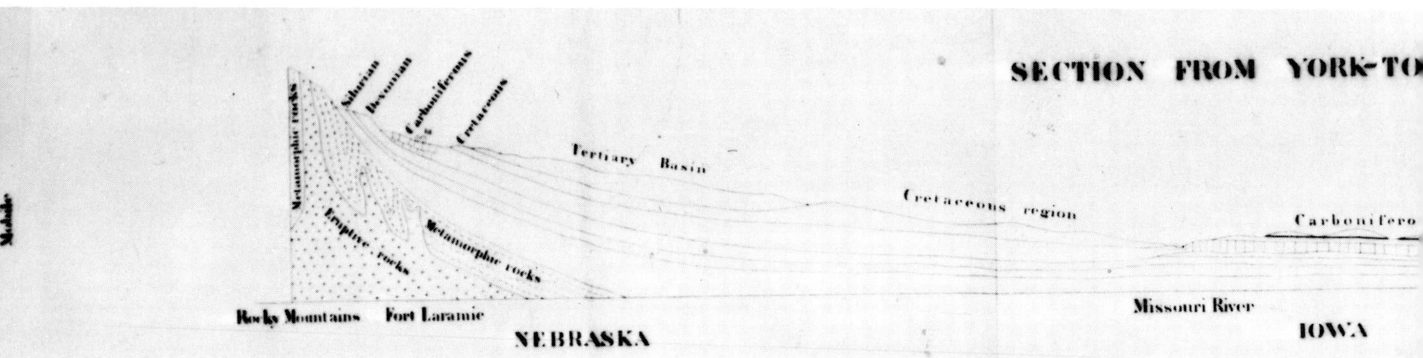

248

247. *A geological segment, being the whole of North America.* Eaton here lays out a cross section of North America from the Atlantic to the Pacific. The abbreviations Car., Qu., and Cal. stand for Carboniferous, Quartzose, and Calcareous formations, respectively. The second layer from the bottom is the "combustible materials, . . . now supposed to be too nearly exhausted" to produce active volcanoes but "still sufficient for ordinary earthquakes."

248. *The North American continent, 1853.* The progress of geology can be appreciated by comparing this section of North America to the Rocky Mountains with the previous figure. There are still large blanks to be filled in, but the true structure is becoming clear.

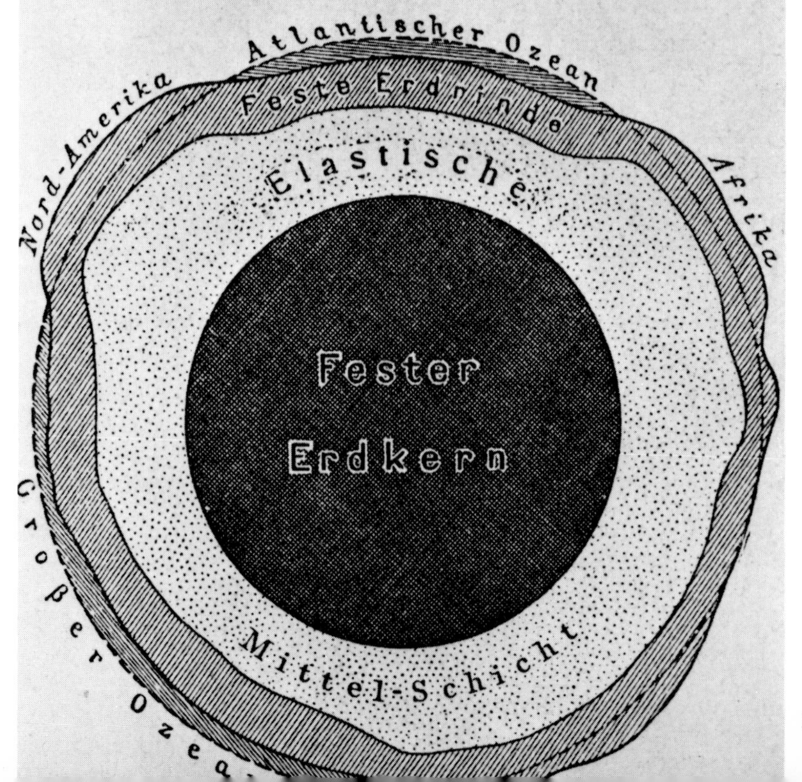

249

249. *The earth as a whole, at the end of the nineteenth century.* By the end of the nineteenth century, the basic structure of the earth seemed clear, as this German engraving indicates. At the center was the core (labeled *Fester Erdkern*), which gave the Earth its magnetic field. Above this rigid core was the elastic middle stratum (*Elastische Mittel-Schicht*), which was very hot and the source of volcanoes. Above this was the earth's crust (*Feste Erdrinde*), on which the oceans rested. The continents were merely thicker parts of this crust, and the whole crust was without break.

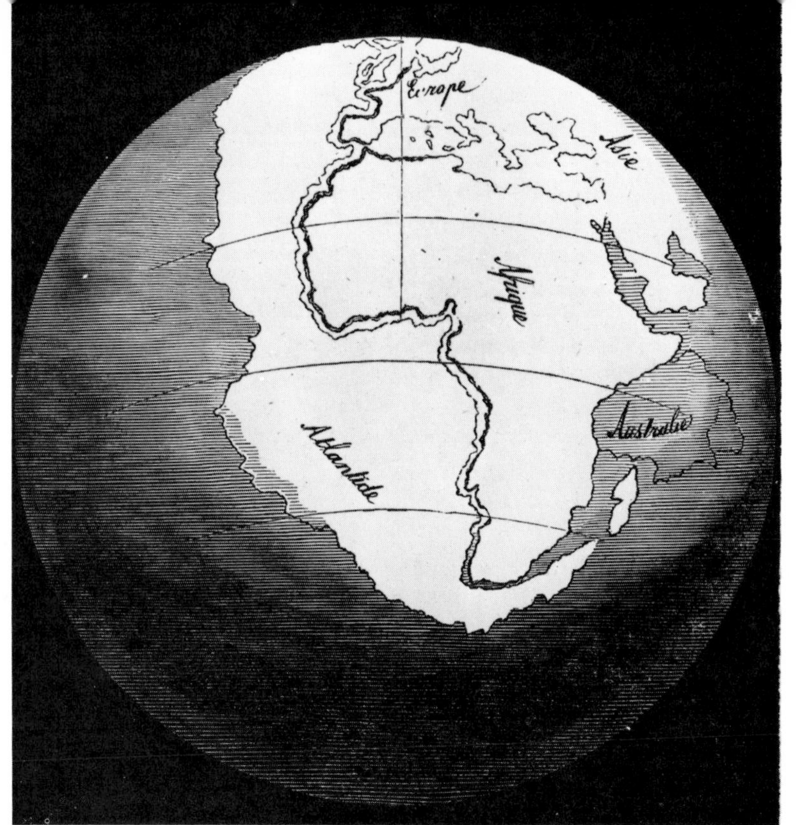

250a

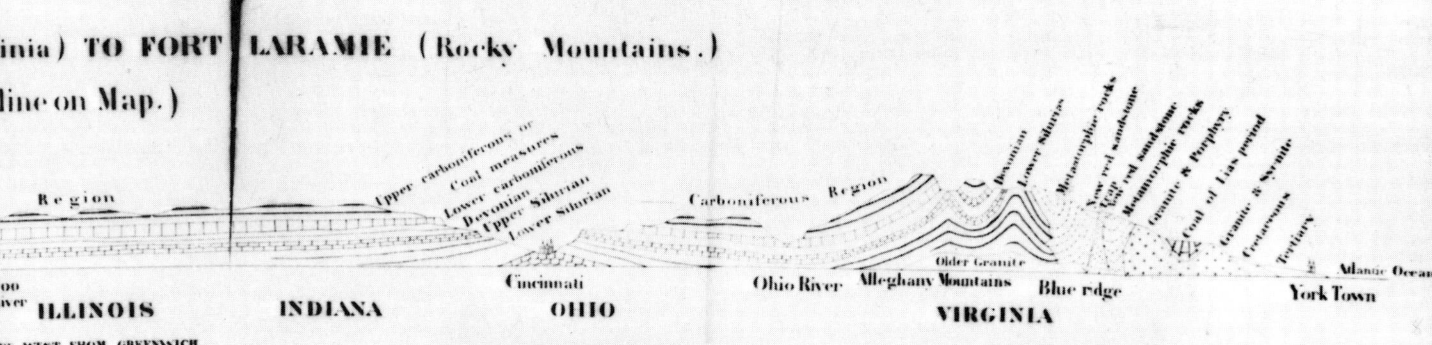

250 a, b. *The origin of the continents.* The dynamics of the earth's crust was almost a total mystery. The first person to notice the curious fact that the continents would fit together, and hence that they may have drifted apart, was not a geologist. Antoine Snider was a scientific dilettante who published a "sensational" book in 1859 with the title *Creation and Its Mysteries Unveiled.* These illustrations, showing separation of the continents, appeared in it. In the twentieth century, the idea became known as Wegener's hypothesis, and it is now basic to the theory of plate tectonics.

250b

13
12
11
10
9
8
16
16

251. *A geological cross section.* In the nineteenth century, the problem of geological dynamics had to be attacked on a scale smaller than the global. Sections of the earth could be isolated and an attempt made to reconstruct their geological history, as in this illustration, which shows the influence of igneous rock on the earth's crust. Molten magma from the earth's interior not only erupts, as at the left, in active volcanoes, but also forces itself through strata, giving rise to irregularities.

252. *A geological stress.* Strata revealed forces at work that were less catastrophic than volcanoes or the pressure of molten lava. Pictured here is a beautiful example of a simple "wrinkling" of the earth's crust as a result of relatively mild geological forces.

253. *An ideal cross section of the earth (1874).* By the end of the nineteenth century, the number of known strata had greatly increased; but the basic conception of the earth remained what it had been in Smith's time. It was still believed that the primary dynamic factor in the formation of the earth's crust was the action of heat from the earth's core. The strata shown, beginning at the top, are Alluvian, Diluvian, Pliocene, Miocene, Eocene, Cretaceous, Jurassic, Triassic, Permian, Carboniferous, Devonian, Silurian, Cambrian, Granitic, Glowing Magma.

253

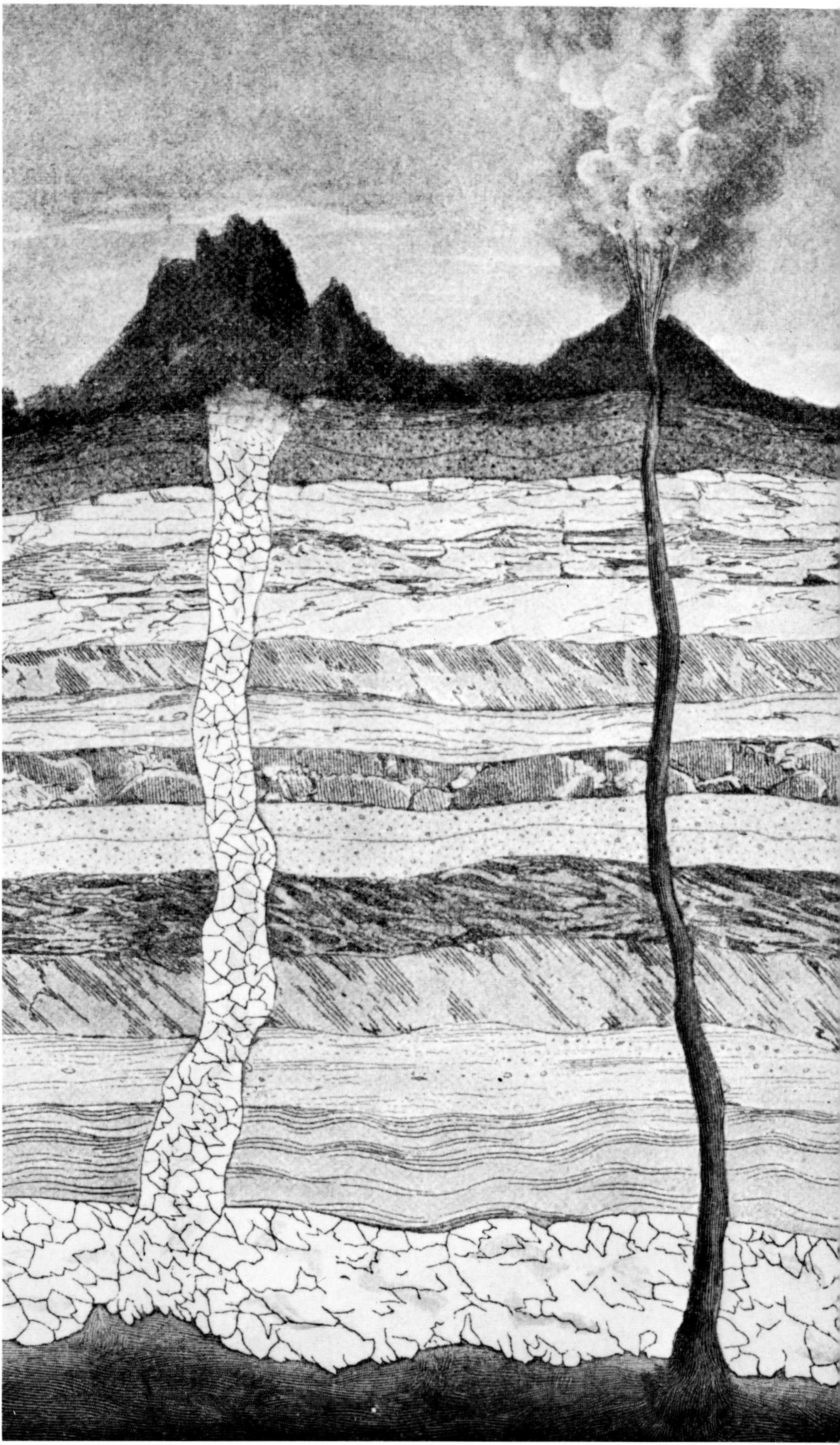

251

252

3
20
21
22
23
6
34
36
37
1
2
7
12
35
51
52
50
53
5
9
4
10
11
18
15
14
13
1
16
12
8
3

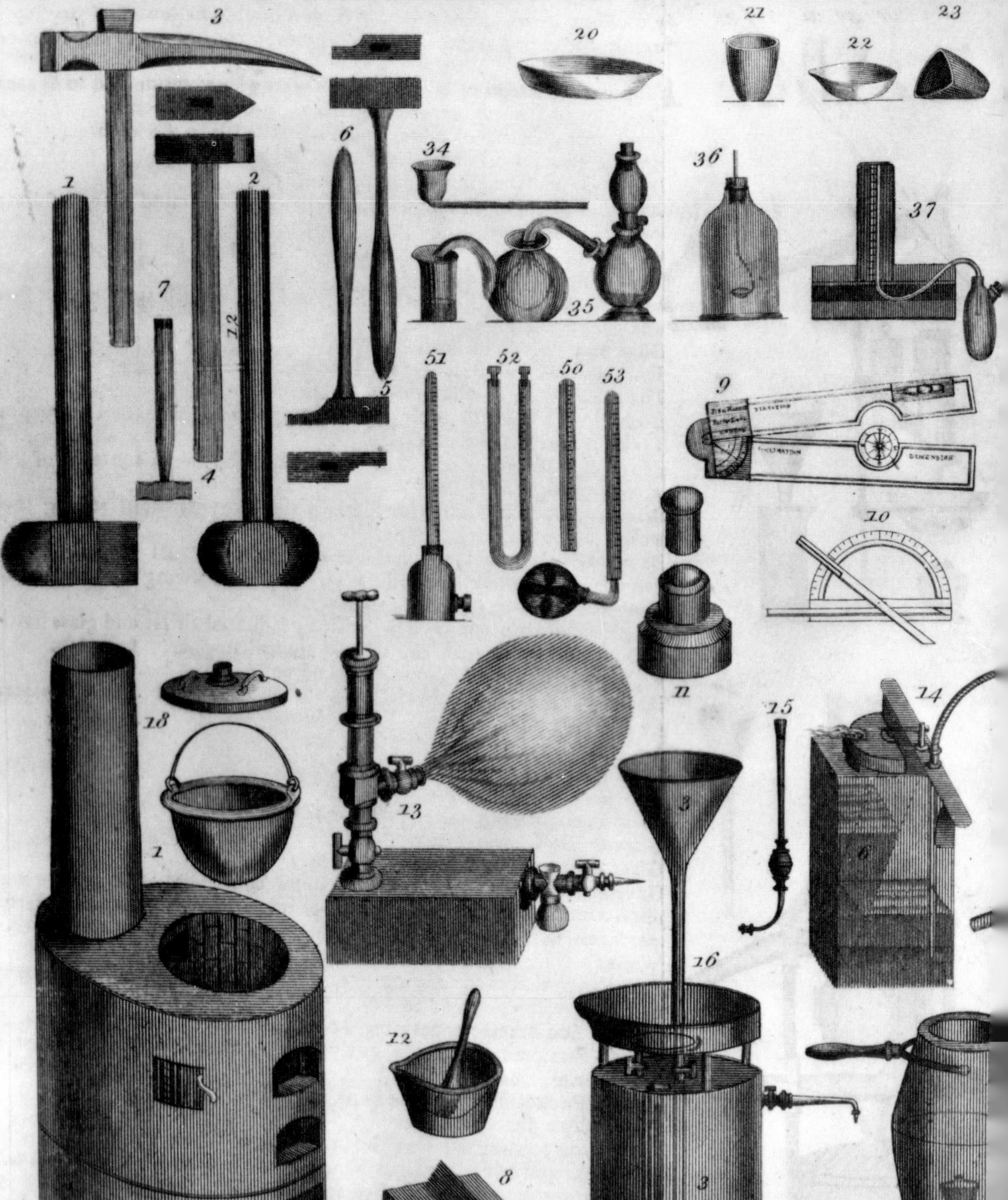

254

255

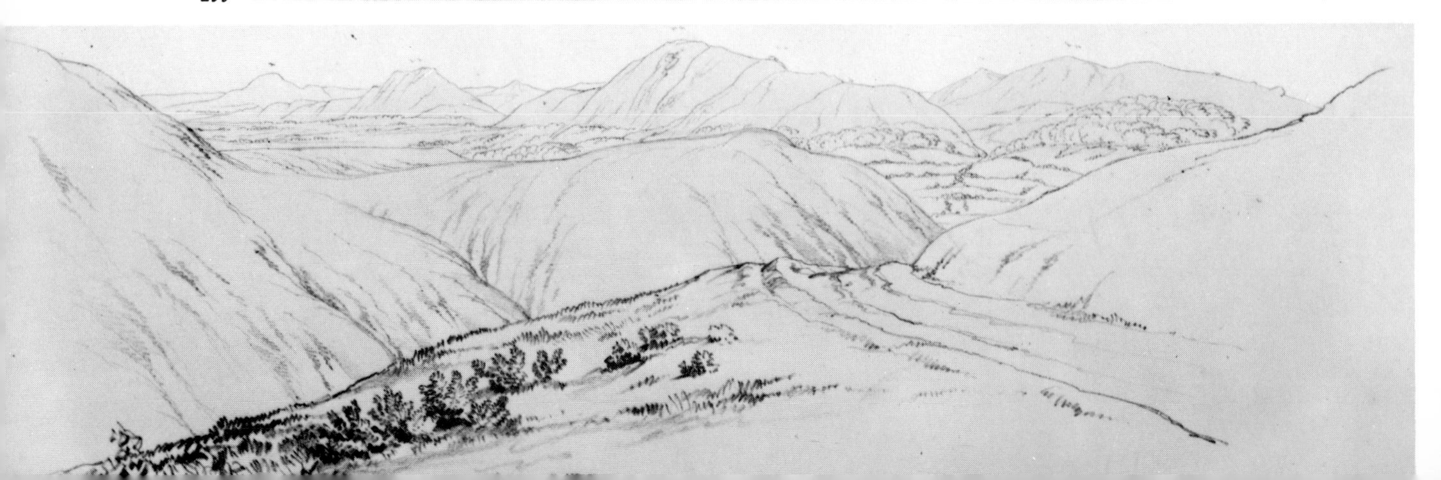

254. *Apparatus for the practical geologist.* Geology, like the other sciences, became highly professionalized in the nineteenth century and developed its own tools, most of which were borrowed from chemistry. The advertisement reproduced here appeared in the back of an early geological survey and illustrates the sorts of implements necessary for good geological work. Hammers of many different shapes obviously were necessary. So, too, were furnaces in which the assays of specimens could be made. Less familiar are objects such as the clinometer, for determining the direction, inclination, and dimensions of strata (9); a pocket goniometer (10); eudiometers for determining the composition of the air (50-53); and a mercurial pneumatic trough for collecting gases (37). The top row shows crucibles of various shapes.

255. *The Silurian system—the Caradoc Hills.* One of the first geological "systems" to be worked out was the Silurian. R. I. Murchison began his investigations with this lovely drawing of the Caradoc Hills in Wales.

256. *The geology of the Silurian system.* Beneath the "skin" of the Caradoc Hills were the strata that made up the Silurian system. Murchison's great achievement was to work them out.

256

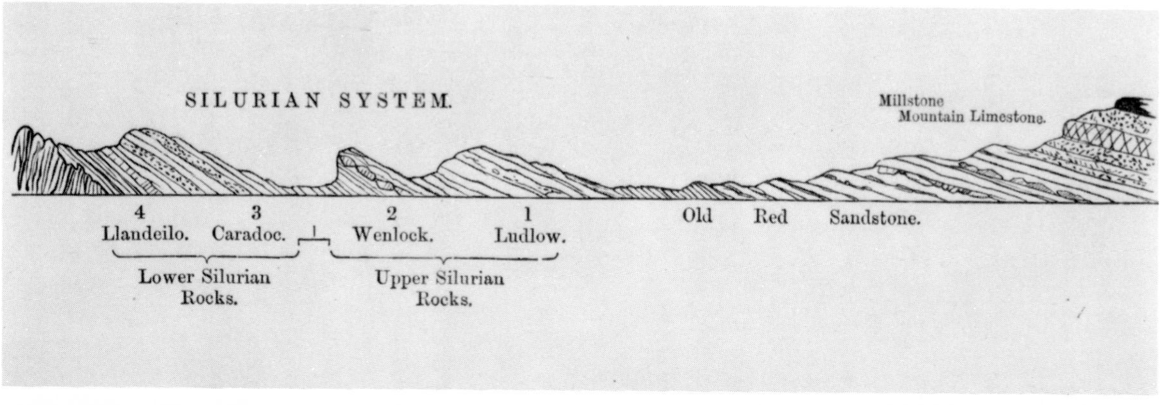

257a

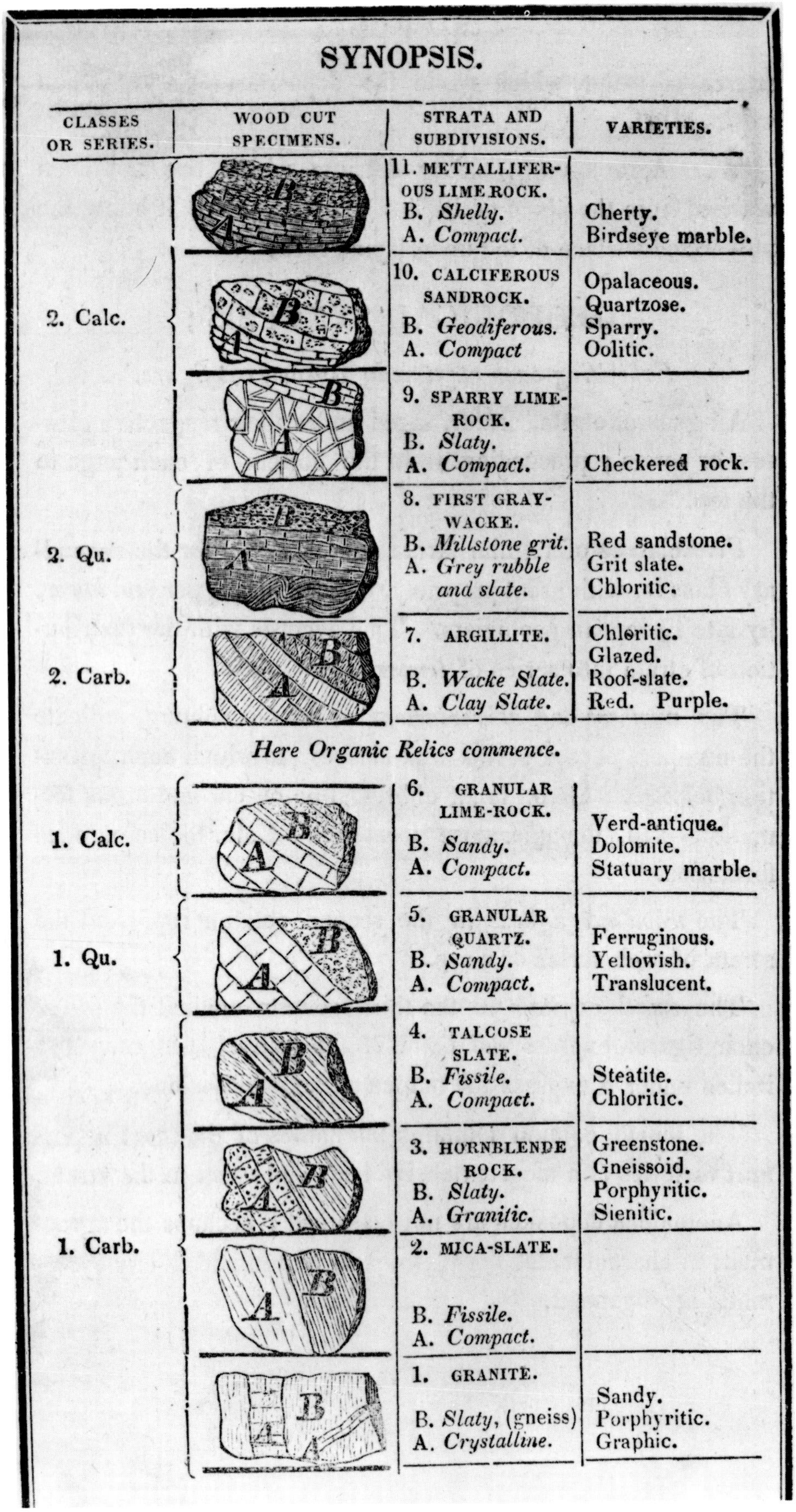

SYNOPSIS.

CLASSES OR SERIES.	WOOD CUT SPECIMENS.	STRATA AND SUBDIVISIONS.	VARIETIES.
2. Calc.		11. METTALLIFEROUS LIME ROCK. B. *Shelly.* A. *Compact.*	Cherty. Birdseye marble.
		10. CALCIFEROUS SANDROCK. B. *Geodiferous.* A. *Compact*	Opalaceous. Quartzose. Sparry. Oolitic.
		9. SPARRY LIME-ROCK. B. *Slaty.* A. *Compact.*	Checkered rock.
2. Qu.		8. FIRST GRAYWACKE. B. *Millstone grit.* A. *Grey rubble and slate.*	Red sandstone. Grit slate. Chloritic.
2. Carb.		7. ARGILLITE. B. *Wacke Slate.* A. *Clay Slate.*	Chloritic. Glazed. Roof-slate. Red. Purple.
		Here Organic Relics commence.	
1. Calc.		6. GRANULAR LIME-ROCK. B. *Sandy.* A. *Compact.*	Verd-antique. Dolomite. Statuary marble.
1. Qu.		5. GRANULAR QUARTZ. B. *Sandy.* A. *Compact.*	Ferruginous. Yellowish. Translucent.
1. Carb.		4. TALCOSE SLATE. B. *Fissile.* A. *Compact.*	Steatite. Chloritic.
		3. HORNBLENDE ROCK. B. *Slaty.* A. *Granitic.*	Greenstone. Gneissoid. Porphyritic. Sienitic.
		2. MICA-SLATE. B. *Fissile.* A. *Compact.*	
		1. GRANITE. B. *Slaty,* (gneiss) A. *Crystalline.*	Sandy. Porphyritic. Graphic.

257 a, b, c. *Strata and their geological contents.* The great importance attached to distinguishing strata came from the fact that each stratum, as these figures show, contains different geological entities. In America, Amos Eaton tried in 1830 to classify what he found in the strata that he had devised from the Wernerian theory of precipitations from aqueous suspensions. Eaton's depiction of strata is shown in Figure 247. Figure 257c appears on page 176.

257b

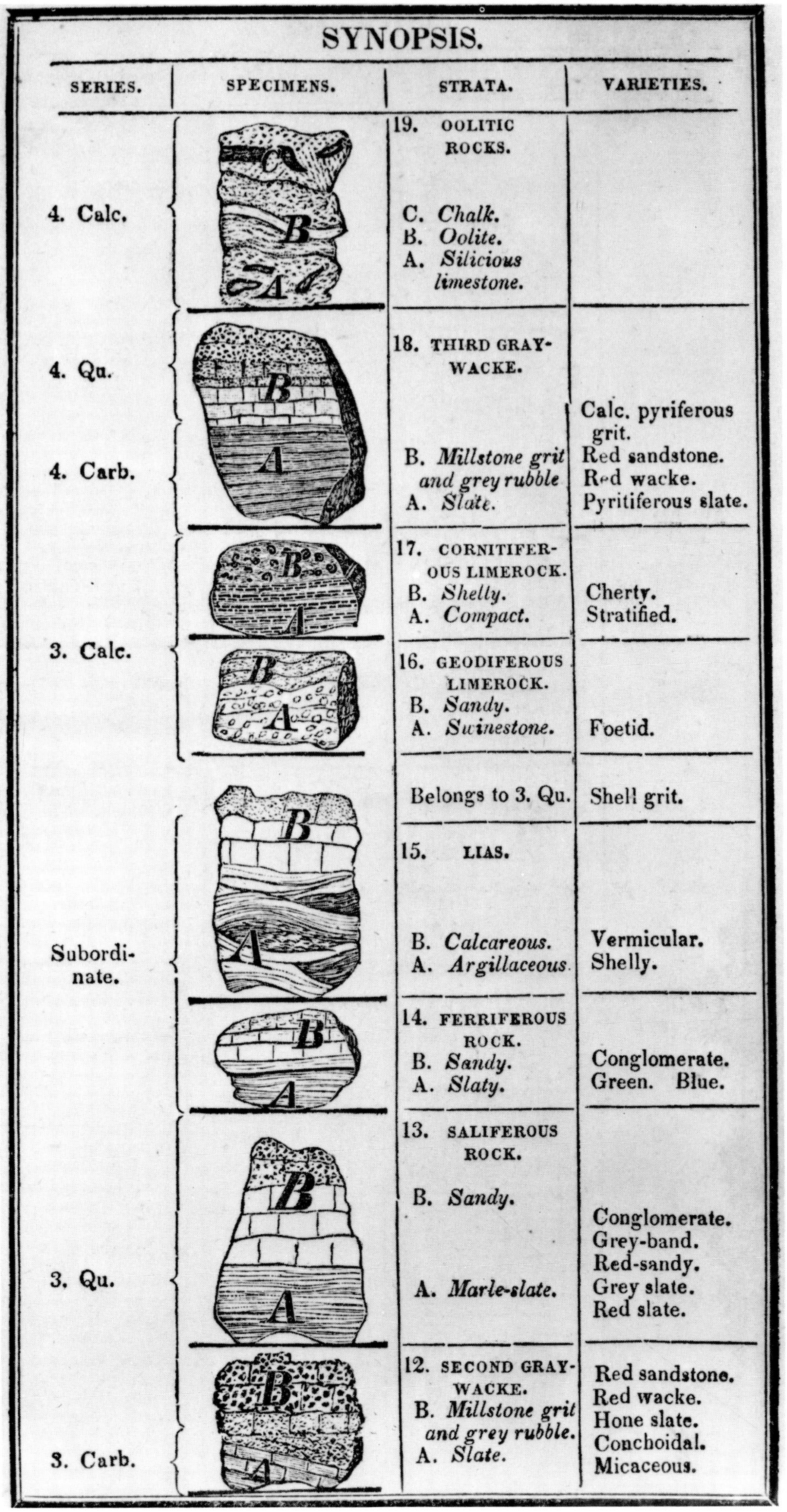

SYNOPSIS.

SERIES.	SPECIMENS.	STRATA.	VARIETIES.
4. Calc.		19. OOLITIC ROCKS. C. *Chalk.* B. *Oolite.* A. *Silicious limestone.*	
4. Qu. 4. Carb.		18. THIRD GRAY-WACKE. B. *Millstone grit and grey rubble* A. *Slate.*	Calc. pyriferous grit. Red sandstone. Red wacke. Pyritiferous slate.
3. Calc.		17. CORNITIFEROUS LIMEROCK. B. *Shelly.* A. *Compact.*	Cherty. Stratified.
		16. GEODIFEROUS LIMEROCK. B. *Sandy.* A. *Swinestone.*	Foetid.
		Belongs to 3. Qu.	Shell grit.
Subordinate.		15. LIAS. B. *Calcareous.* A. *Argillaceous*	Vermicular. Shelly.
		14. FERRIFEROUS ROCK. B. *Sandy.* A. *Slaty.*	Conglomerate. Green. Blue.
3. Qu.		13. SALIFEROUS ROCK. B. *Sandy.* A. *Marle-slate.*	Conglomerate. Grey-band. Red-sandy. Grey slate. Red slate.
3. Carb.		12. SECOND GRAY-WACKE. B. *Millstone grit and grey rubble.* A. *Slate.*	Red sandstone. Red wacke. Hone slate. Conchoidal. Micaceous.

257c

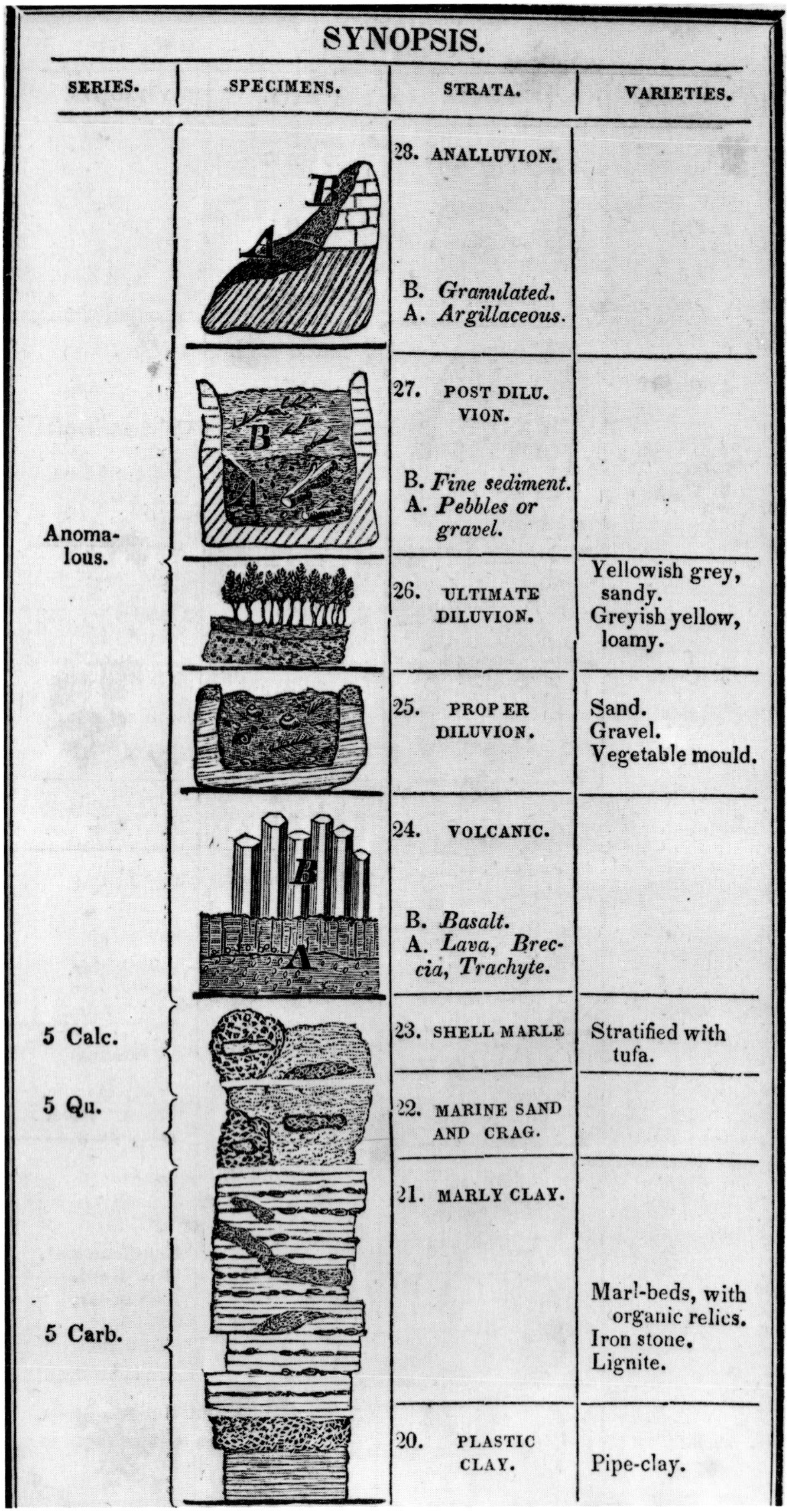

SYNOPSIS.

SERIES.	SPECIMENS.	STRATA.	VARIETIES.
Anomalous.		28. ANALLUVION. B. *Granulated.* A. *Argillaceous.*	
		27. POST DILUVION. B. *Fine sediment.* A. *Pebbles or gravel.*	
		26. ULTIMATE DILUVION.	Yellowish grey, sandy. Greyish yellow, loamy.
		25. PROPER DILUVION.	Sand. Gravel. Vegetable mould.
		24. VOLCANIC. B. *Basalt.* A. *Lava, Breccia, Trachyte.*	
5 Calc.		23. SHELL MARLE	Stratified with tufa.
5 Qu.		22. MARINE SAND AND CRAG.	
5 Carb.		21. MARLY CLAY.	Marl-beds, with organic relics. Iron stone. Lignite.
		20. PLASTIC CLAY.	Pipe-clay.

258

258. *The Landslip at Bindon, England.* The camera could record features of landscape, as in this photograph of the landslip at Bindon. There clearly has been some geological action involving the collapse of the ground in the center of the picture.

259. *The geologist's reconstruction of the landslip.* This section, when used in conjunction with the foregoing photograph, shows a geologist's work—in this case, accounting for the landslip in terms of known strata and forces.

Fig. 62.—*Section across the Bindon Landslip.*

Horizontal Scale, one inch to 75 yards. Vertical Scale, one inch to 600 feet.

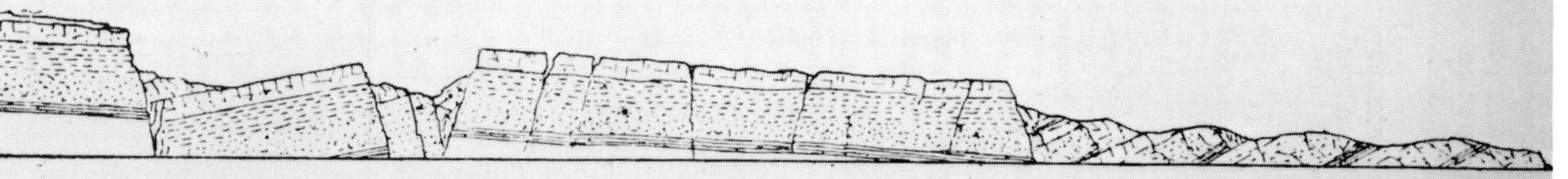

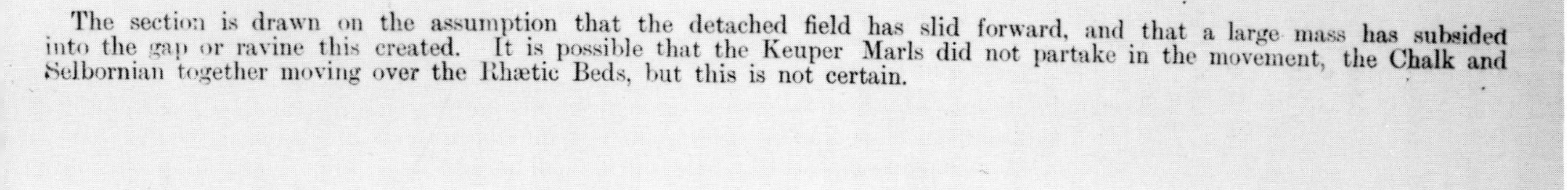

c. Lower Chalk.
b. Upper Greensand and Gault.
a. Rhætic Beds and Keuper Marls

The section is drawn on the assumption that the detached field has slid forward, and that a large mass has subsided into the gap or ravine this created. It is possible that the Keuper Marls did not partake in the movement, the Chalk and Selbornian together moving over the Rhætic Beds, but this is not certain.

59

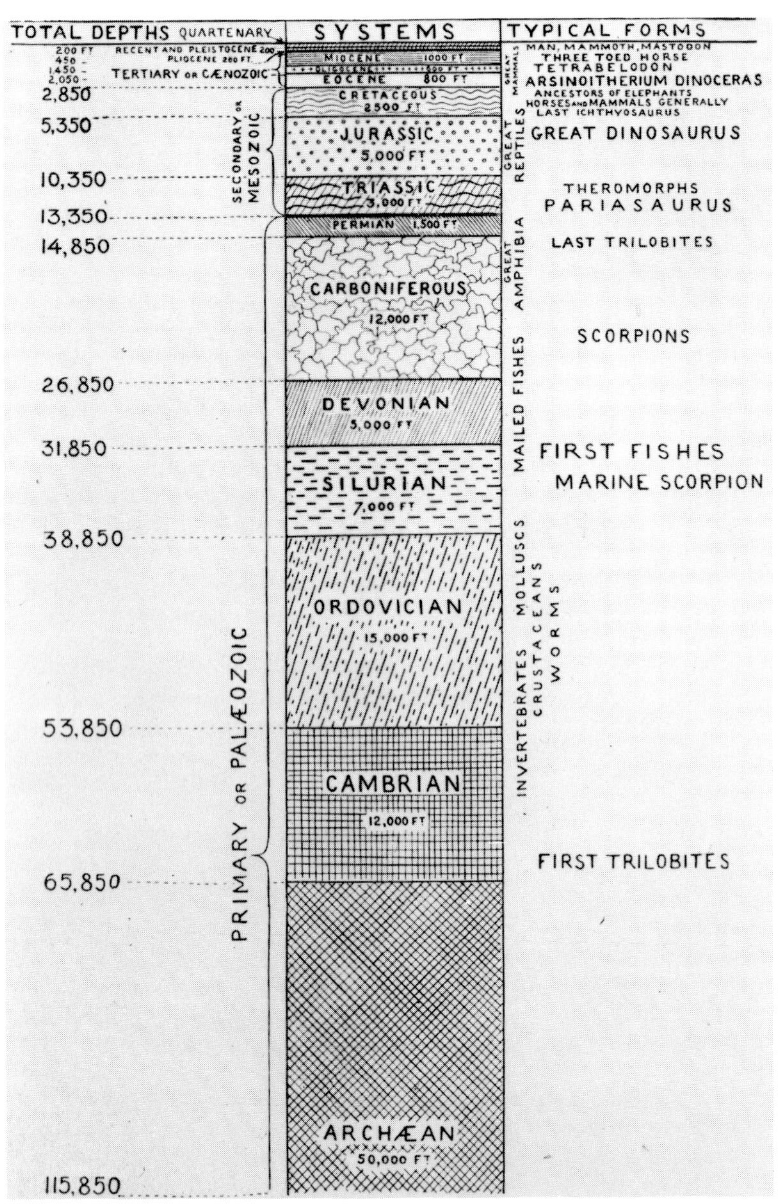

260. *Table of stratified rocks.* By the end of the nineteenth century, stratigraphy had become a much more refined and precise science. In this chart, dating from 1908, the geological history of the earth may be read at a glance.

261

Jurassic Rocks of Yorkshire

Sections.

To illustrate the Variations in Thickness and Mineral Characters of the Lower Oolites of Yorkshire.

South Yorkshire — Kirkby Underdale — Acklam — Kirkham — Brandsby — Hambleton Hills South — Hambleton Hills North — Rosedale — Whitby and Cleveland — Peak

261. *A stratigraphic profile of Yorkshire (1892).* Geologists ultimately were able to follow strata over some distances, despite disruptions and changes in thickness. This ability provided them with the opportunity to draw profiles, such as this one, that gave important clues to local geological history.

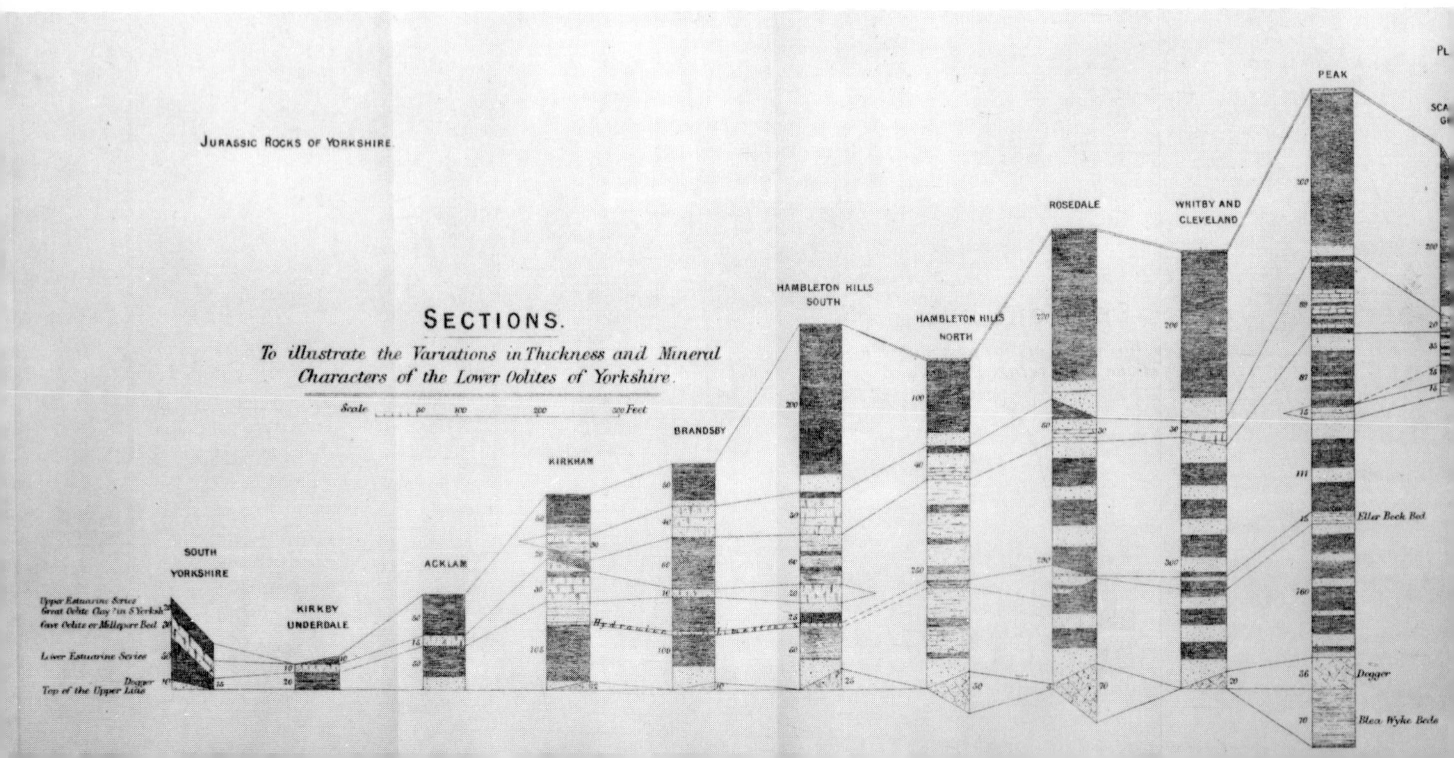

Creatures from the Past

The fossil record was both a key to the past and a fascination to the present. What, precisely, did these bones represent? Rationalists since at least the time of Leonardo da Vinci had insisted that the fossils were the remains of living creatures. Seashells found on mountaintops meant, for such thinkers as Leonardo, that mountaintops had once been sea bottoms. That simple logic was more than most people could bear and rather fantastic hypotheses were brought forward to get around it. It was seriously suggested that plant and animal fossils (the word originally simply meant anything dug up from a ditch or *fossa*) were planted in the earth by God either to test man's faith, or because he was trying out his plastic energies, or even because they were part of some divine message that men had not yet figured out. Anticlericals could be just as silly with fossils as the pious. Voltaire, in the eighteenth century, wrote that the oyster shells found in Alpine passes had been thrown there by medieval pilgrims on their way to Rome!

What nineteenth-century paleontologists achieved was to set the fossil record straight and use it to determine when the animals and plants preserved in stone had lived. The collection, reconstruction and classification, of fossil remains was an extraordinarily arduous task, but gradually some principles emerged that lightened the work. It was Cuvier who, through his intensive study of comparative anatomy, created the firm foundation for paleontology. His analysis of living forms led him to an intimate knowledge of the connections between animal form and function so that he could correlate anatomical structures within a single organism. He once made the boast that if he were given the tooth of a mastodon, he could reconstruct the whole animal from it. The tooth would offer some clue to the size of the mature animal. By its form and pattern of wear, Cuvier could tell that the animal was an herbivore, and this would permit him to reconstruct the necessary jaw structure to contain the tooth. The jaw, in turn, determined the size and rough shape of the head, and the total size of the beast would require certain structures for legs, tail, and so on. Since paleontologists often had little more than a few bones with which to work, Cuvier's insights were invaluable to the science.

The reconstructions of ancient life forms fascinated the general public. In the early years of the century, skeletons of animals such as the megatherium or the hydrarchos or sea serpent drew crowds who could only wonder at God's creative power. When the Darwinian theory of evolution was put forward, fossils took on a new significance. These were no longer God's whimsies, but the actual ancestors or enemies of prehuman life forms. There was a lesson to be learned from these bones. Animals could become extinct and it behooved man to understand how and why. More importantly, animal forms could change through time and species could evolve. The human species was clearly implicated in this vast scheme of things and man had better begin to look at his evolutionary past if he wished to understand his present. It was the realization of this fact that worked the greatest revolution in thought in the nineteenth century. The theory of evolution firmly anchored humanity in animality. The vision of man that emerged from this was vastly different from the one that Western man had cherished since the time of the Greeks.

262. *The Asiatic deluge.* The presence of animal and plant remains in geological strata could be explained in orthodox terms by calling upon the great Deluge described in Genesis. Here the popularizer of science Louis Figuier gives his impression of what the Flood looked like in Asia.

262

263

263. *The Deluge in Europe.* Here the waters sweep over the European landscape. From watery graves, the remains of animals and plants passed into the fossil record, to be discovered by the geologists of the nineteenth century.

264

265

264. *The remains of the Flood.* In 1823, Rev. William Buckland published his account of the fossil remains of the Flood, *Reliquiae diluvianae.* In the Dream lead mine in Derbyshire, he discovered a cave filled with the bones of animals he believed had been drowned in the Noachian deluge. It was a find that captured the imagination of geologists everywhere and spurred similar geological investigations.

265. *Georges Cuvier and the meaning of fossils.* Most fossils were of species unknown to nineteenth-century zoologists. Georges Cuvier, in France, suggested that they belonged to groups of previous creations that had been wiped out by the Deity. More important, he used his unparalleled knowledge of comparative anatomy to show others how whole creatures could be reconstructed from a few bones and pieces. This illustration, published around 1900, shows Cuvier examining a fossil.

266. *Fossils and strata.* One of the main clues to the history of the earth was discovered in the nineteenth century: that everywhere similar fossils appeared in the same stratum. Thus the appearance of a particular fossil species in a stratum in England and in France meant that the stratum was the same in the two countries. Moreover, it soon became clear that the deeper the stratum, the more "primitive" the fossils, as this table illustrates.

26

FIG. 4.—*TABLE OF STRATIFIED ROCKS.*

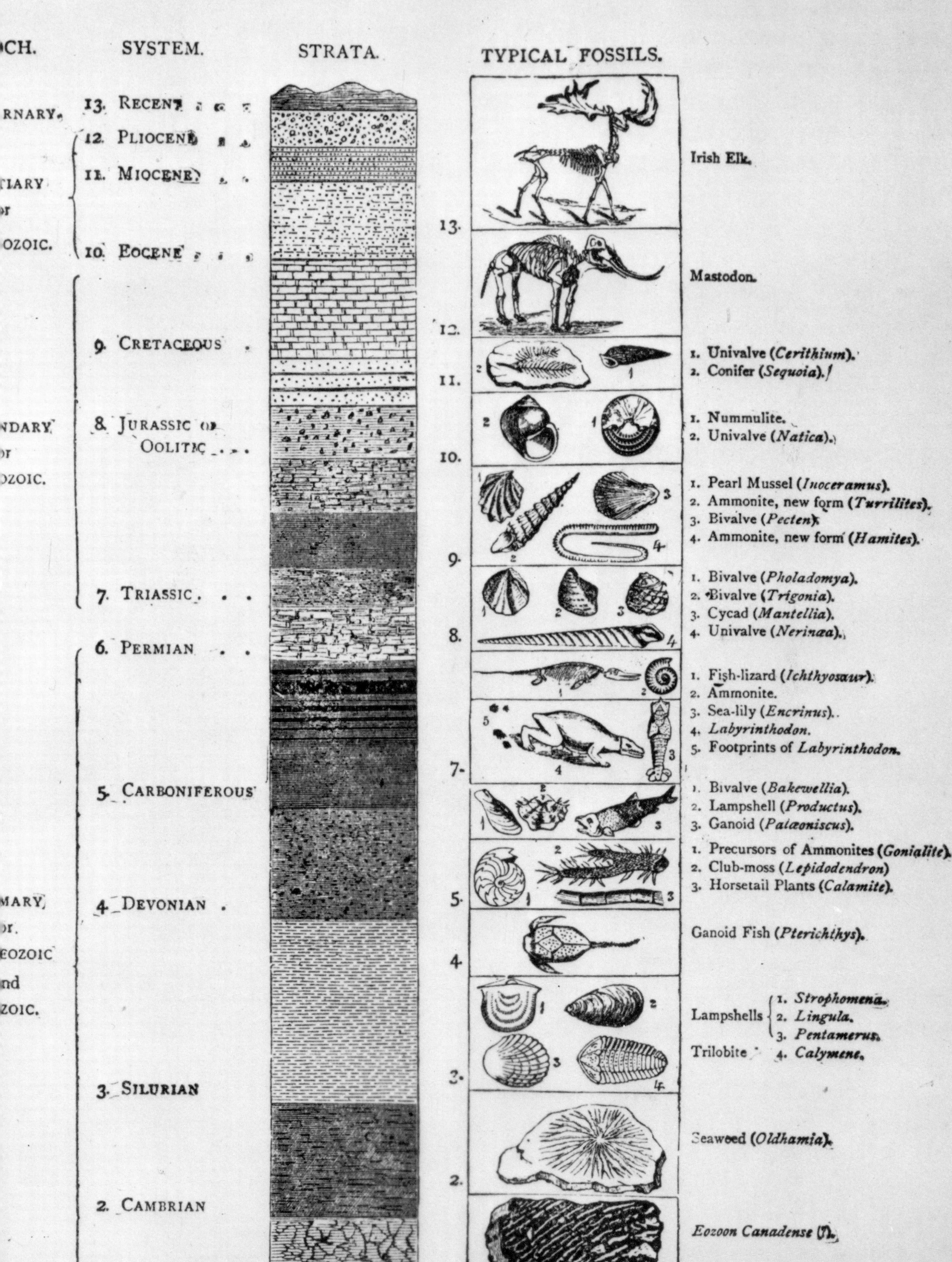

CH.
SYSTEM.
STRATA.
TYPICAL FOSSILS.
RNARY.
13. RECENT
12. PLIOCENE
11. MIOCENE
TIARY
or
OZOIC.
10. EOCENE
9. CRETACEOUS
NDARY
or
OZOIC.
8. JURASSIC or OOLITIC
7. TRIASSIC
6. PERMIAN
5. CARBONIFEROUS
MARY
or.
EOZOIC
nd
ZOIC.
4. DEVONIAN
3. SILURIAN
2. CAMBRIAN
1. LAURENTIAN
13.
12.
11.
10.
9.
8.
7.
5.
4.
3.
2.
1.
Irish Elk.
Mastodon.
1. Univalve (Cerithium).
2. Conifer (Sequoia).
1. Nummulite.
2. Univalve (Natica).
1. Pearl Mussel (Inoceramus).
2. Ammonite, new form (Turrilites).
3. Bivalve (Pecten).
4. Ammonite, new form (Hamites).
1. Bivalve (Pholadomya).
2. Bivalve (Trigonia).
3. Cycad (Mantellia).
4. Univalve (Nerinæa).
1. Fish-lizard (Ichthyosaur).
2. Ammonite.
3. Sea-lily (Encrinus).
4. Labyrinthodon.
5. Footprints of Labyrinthodon.
1. Bivalve (Bakewellia).
2. Lampshell (Productus).
3. Ganoid (Palæoniscus).
1. Precursors of Ammonites (Gonialite).
2. Club-moss (Lepidodendron)
3. Horsetail Plants (Calamite).
Ganoid Fish (Pterichthys).
Lampshells 1. Strophomena. 2. Lingula. 3. Pentamerus.
Trilobite 4. Calymene.
Seaweed (Oldhamia).
Eozoon Canadense (?).

267

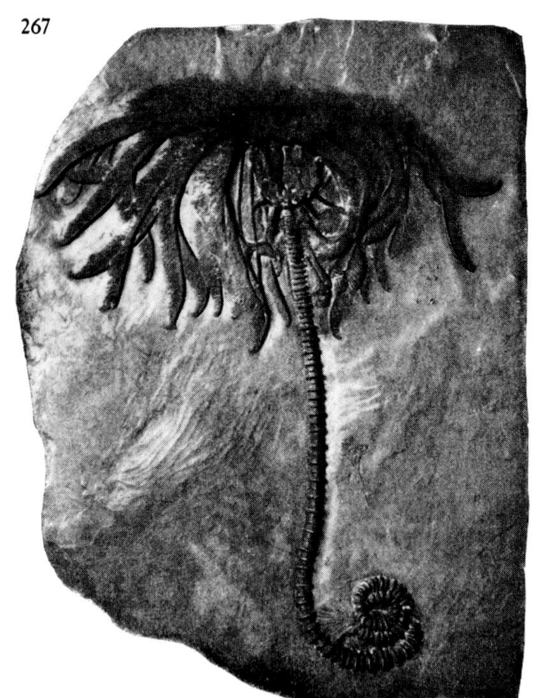

268

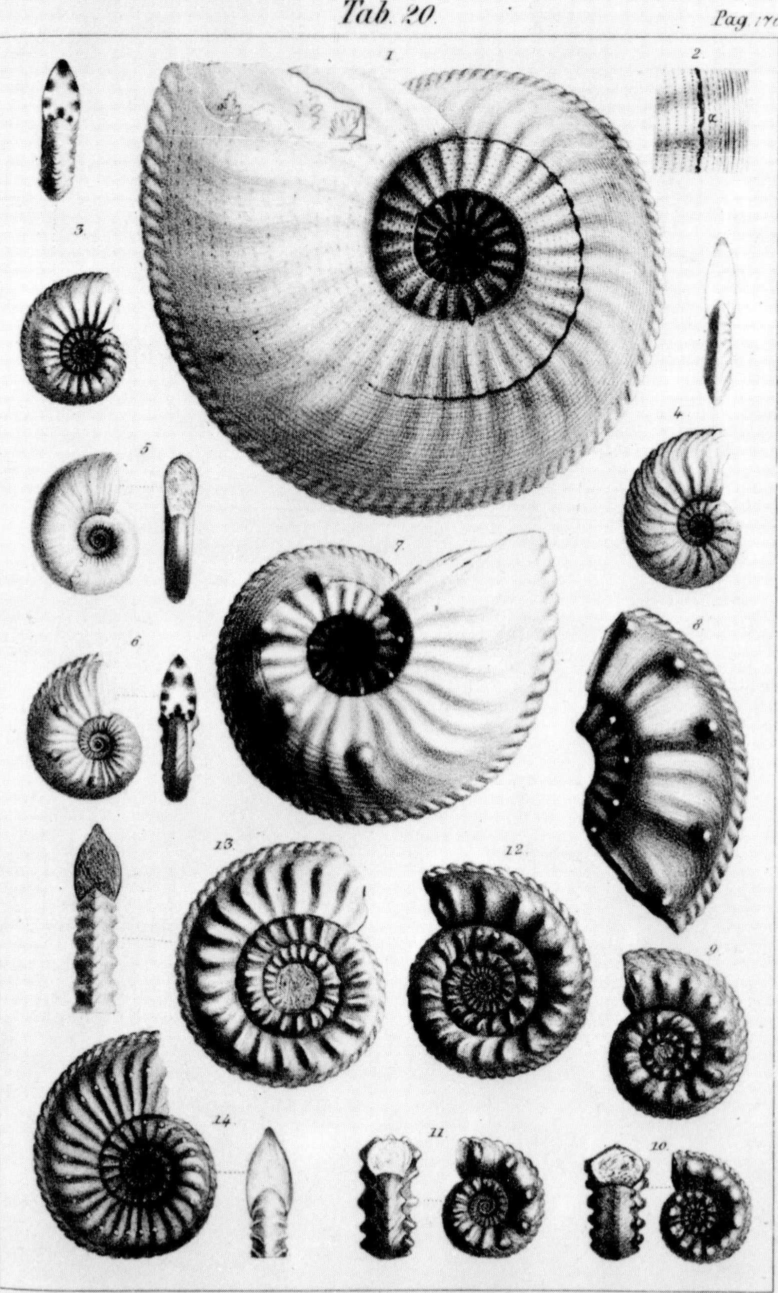

267. *An ancient sea lily.* Fossil forms were often of the utmost delicacy and beauty, as this example demonstrates.

268. *Fossil shells.* Shells were particularly plentiful in fossil beds. Their presence in modern dry-land locations led to the inescapable conclusion (which had already been reached by Leonardo da Vinci) that the dry land had once been the bottom of the sea.

269. *Trilobites.* The existence of primitive forms such as trilobites stimulated a good deal of speculation. Had trilobites been wiped out by the Flood, or were they merely steps in the early development of more complex creatures? Extinct primitive species almost drove one to think in evolutionary terms.

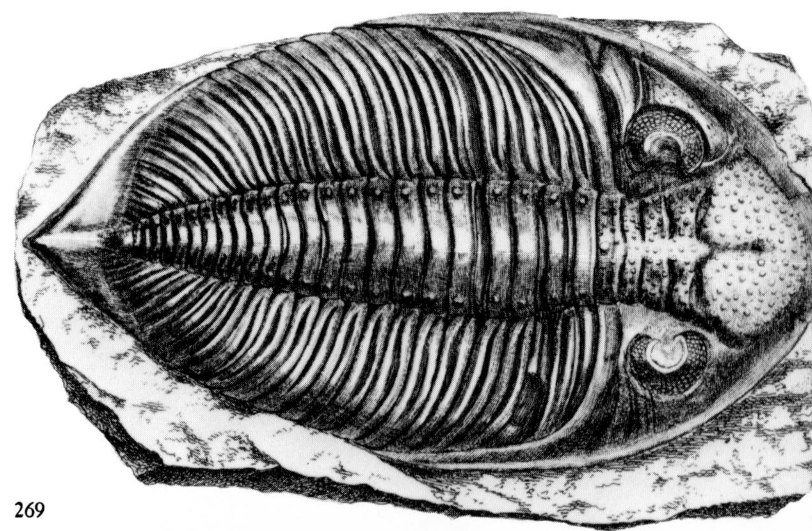

269

270. *Fossils in ancient rocks.* Increasingly, the question obtruded itself as to how fossils had gotten where they were. Clearly, ancient rocks had not been laid down within the last few millennia. Since the Silurian system was old, fossils embedded in it also were old—far older, in fact, than current theories of the earth allowed. A specimen such as this horseshoe crab embedded in Silurian rocks presented a real problem. Had it been created "in the beginning," as Genesis suggested? How had it been buried so deeply if the Flood had lasted only 40 days and nights? And why had a marine creature perished in the Flood?

271. *A fossil fish.* Ancient creatures were not imperfect foreshadowings of modern species but, as this fossil shows, highly developed organisms. What had happened to the species that had become extinct? Why had they died out? Questions such as these troubled paleontologists in the first half of the nineteenth century.

270

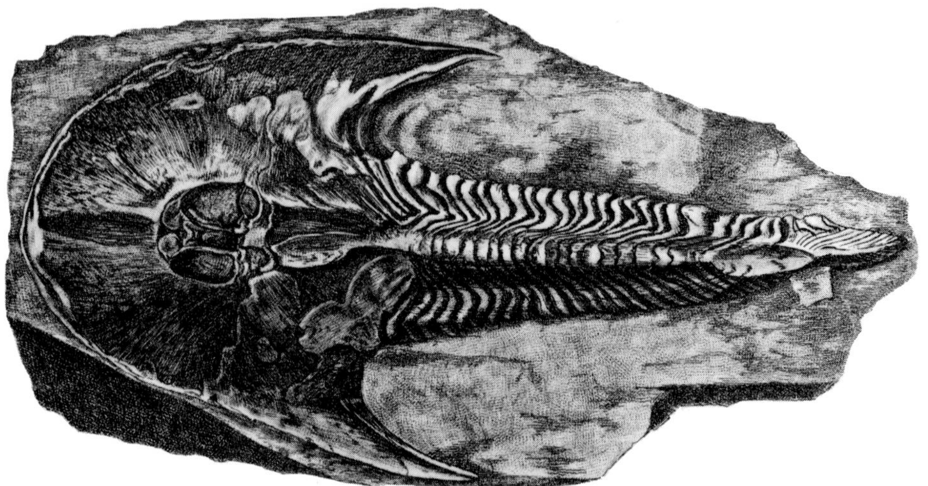

271

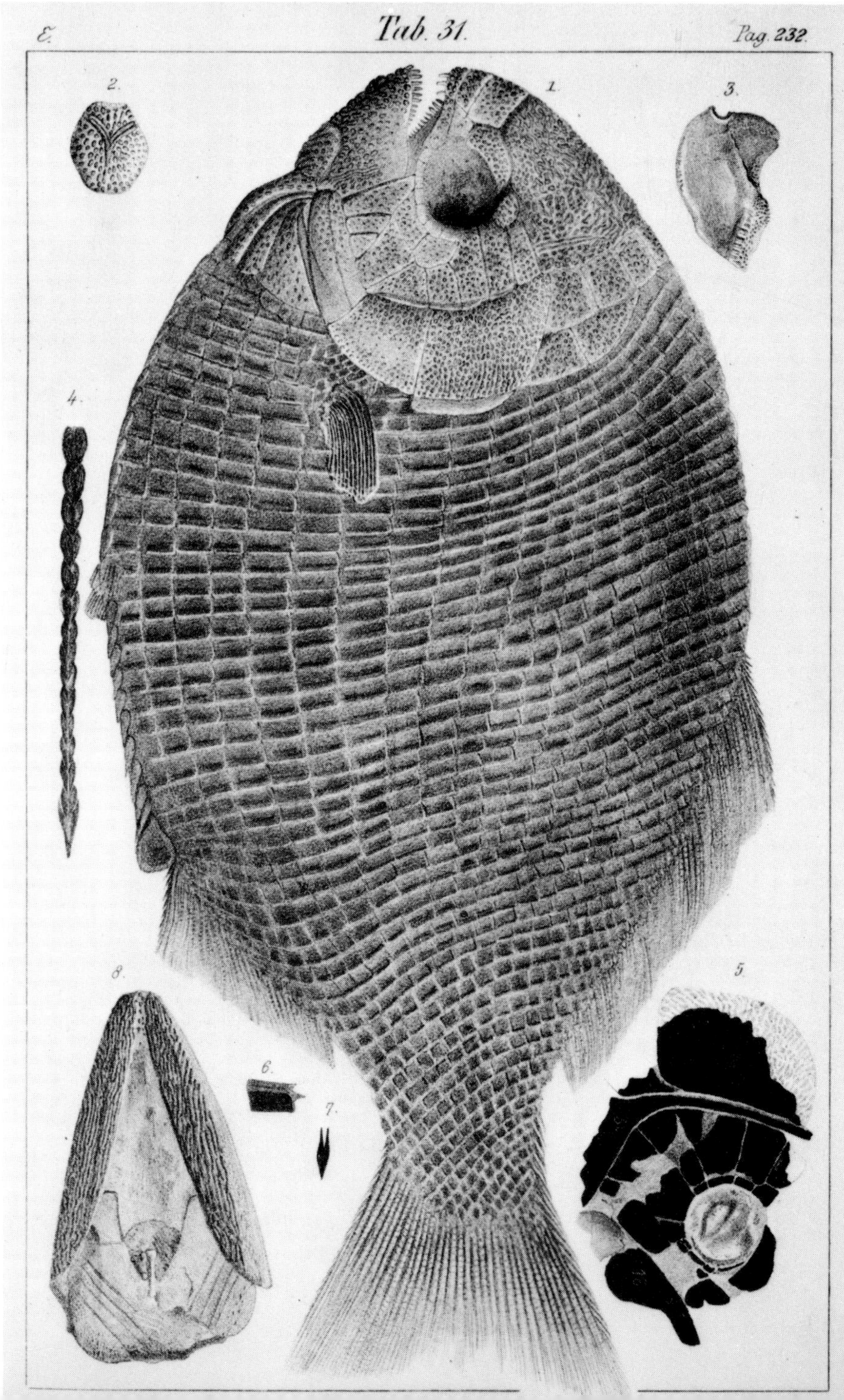

E. Tab. 31. Pag. 232.

272. *A fossil monster*. The paleontological world was a fascinating one, for it contained creatures that almost defied reason. This armored monster, *Glyptodon daedicurus*, seemed unbelievable. Why had nature designed such an animal? And how or why did it survive?

273. *From the age of reptiles*. It gradually became clear that there had been different geological ages during which different life forms had been dominant. This *Pelosaurus*, a beautiful example from the age of reptiles, was of only modest size.

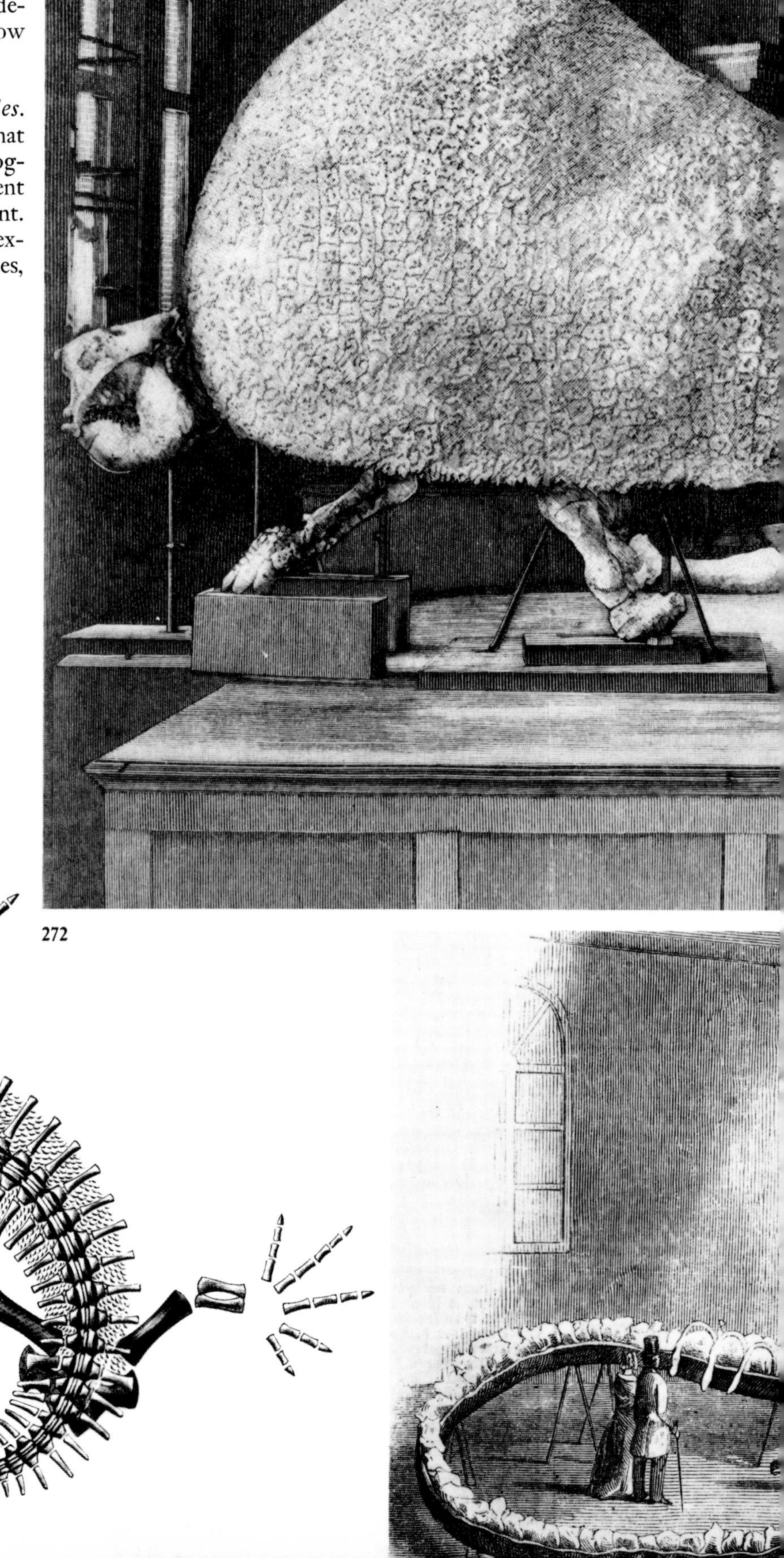

272

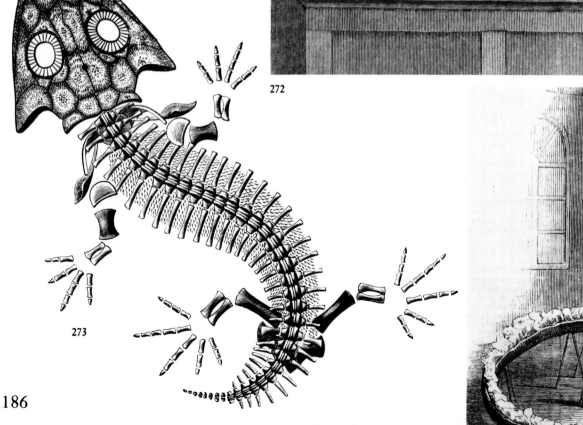

273

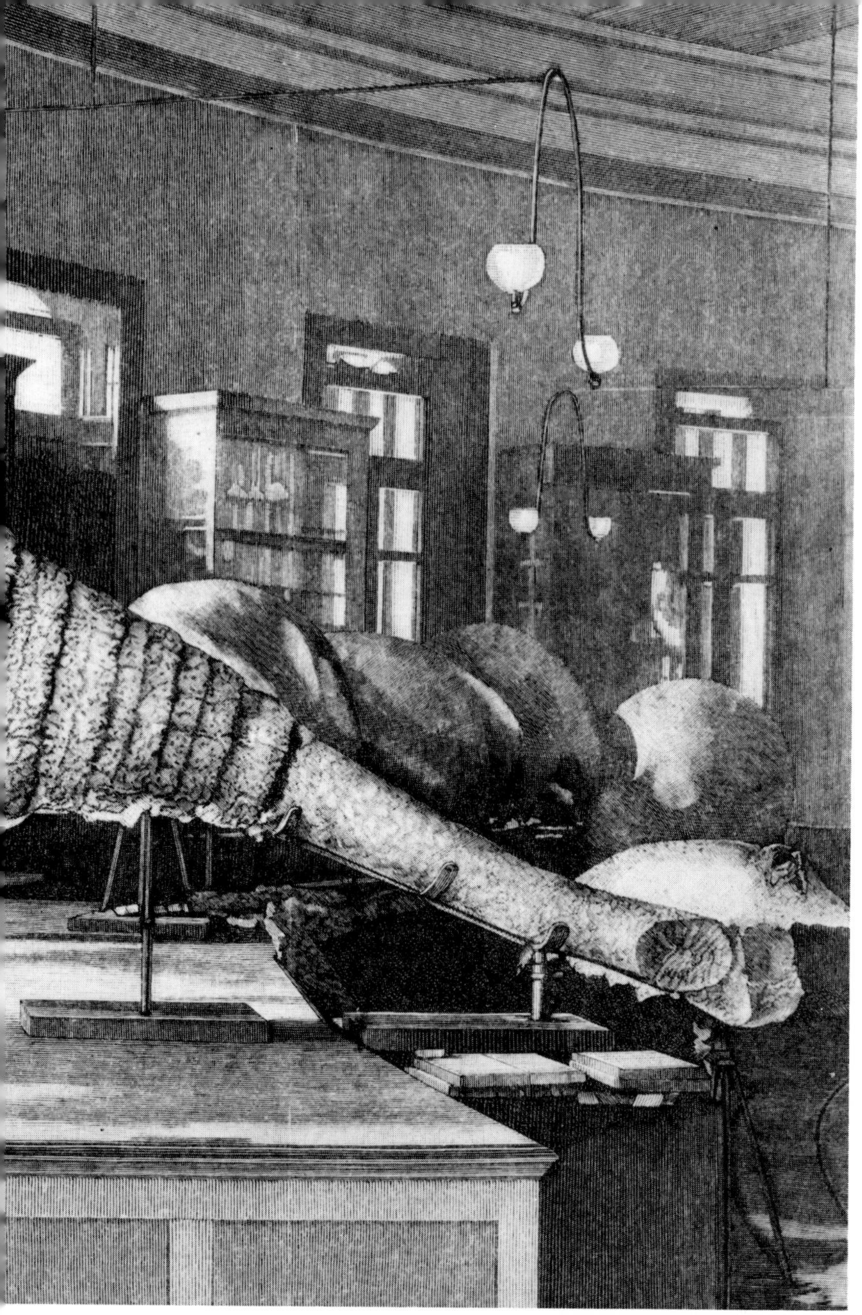

274

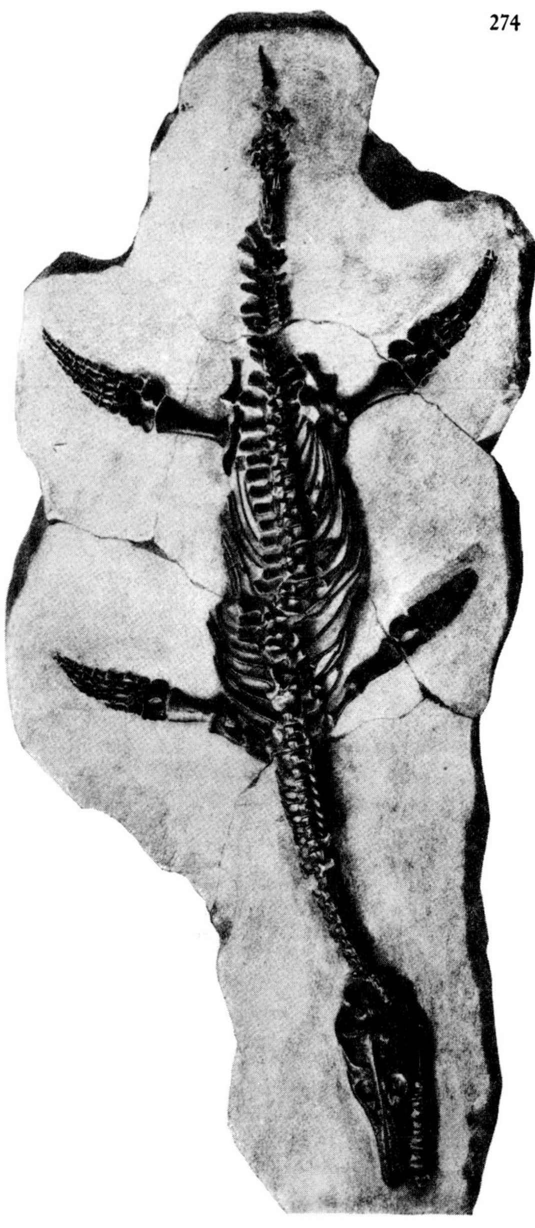

274. *A giant reptile.* The discovery of dinosaurs dates back to the eighteenth century, but serious study of them came in the nineteenth. This *Ptesiosaurus* was discovered at midcentury. Its great size and obvious ferocity were awe-inspiring. The antediluvian world was becoming less and less Edenlike and more and more fearsome.

275. *The "Hydrarchos" on exhibit.* Some ancient reptiles could be accepted by nineteenth-century man with fair ease. Sea serpents had long been credible subjects for sea stories; this fossil was considered to be such a creature.

275

277

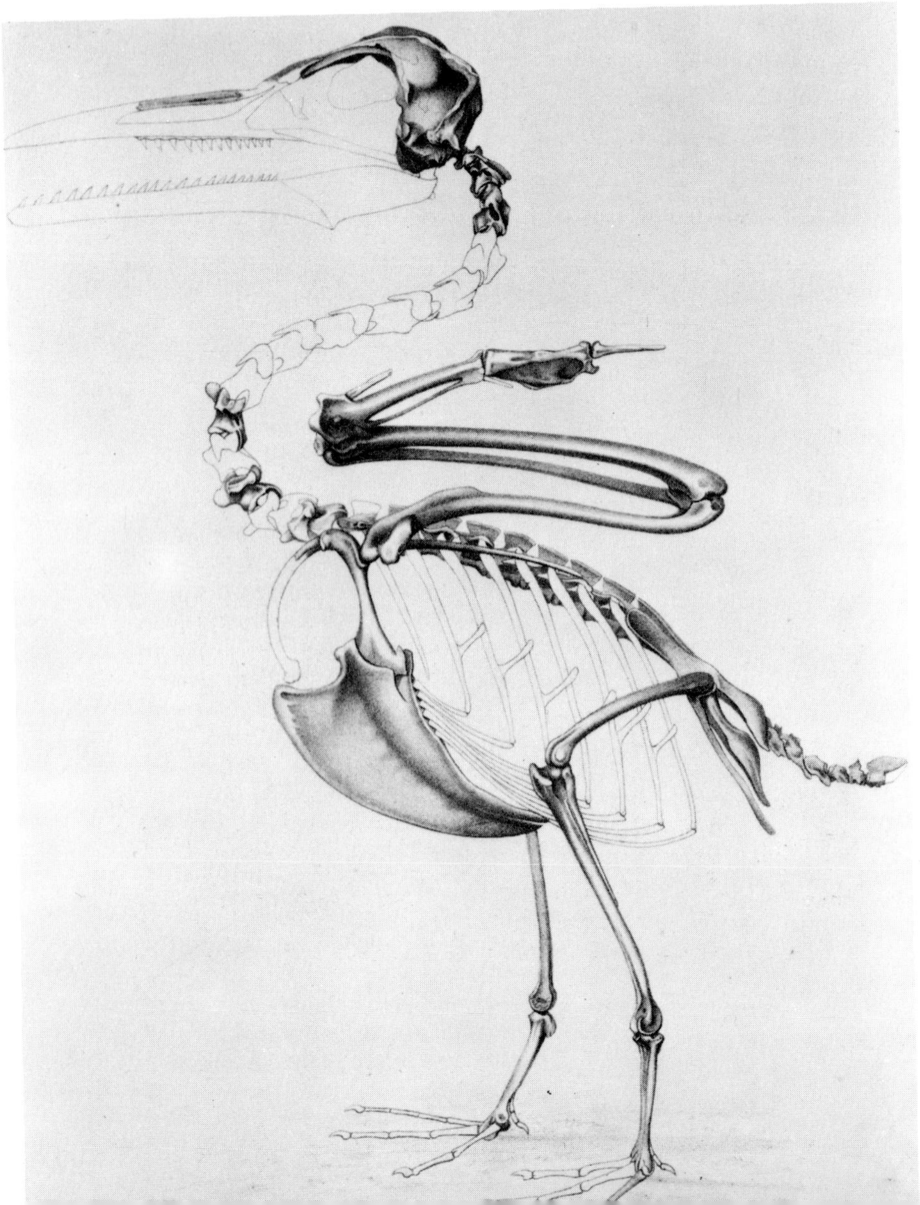

276. *The bird-reptile: Archaeopteryx lithographica.* The publication of Charles Darwin's *Origin of Species* in 1859 provided a new perspective for paleontology, suggesting that species evolved through the accumulation of small adaptive variations. Paleontologists therefore expected to find transition forms that exhibited the characteristics of two different species. In 1861, the fossil shown here was discovered. Having many of the characteristics of a reptile as well as the feathers of a bird, it seemed to provide dramatic proof of Darwin's theory.

277. *The toothed bird. Archaeopteryx* was not unique. Other "transition" bird fossils were soon found. The specimen shown here was subjected to intensive study by Othniel Charles Marsh, the father of American paleontology. It was an enormous bird, equipped with a fearsome set of teeth that must have made it an excellent predator. The teeth were, of course, part of its reptilian inheritance.

278. *The Megatherium.* Next to dinosaurs, the most astonishing and popular fossil for the nineteenth-century public was the *Megatherium*, the first skeleton of which had been found in 1789 in South America. The fascination of the *Megatherium* lay in its enormous size—some 14 feet long and 8 feet high—and in its resemblance to modern bears. The claws on its forepaws indicated that, like bears, it dug for roots and insects. But before Darwin, it was difficult to understand what the relationship might be between the *Megatherium* and the modern bear.

279. *The Megatherium in Spain.* This is a woodcut of the first *Megatherium* skeleton to be found. Crude in both drawing and conception, it appeared in a Spanish illustrated weekly in 1836.

278

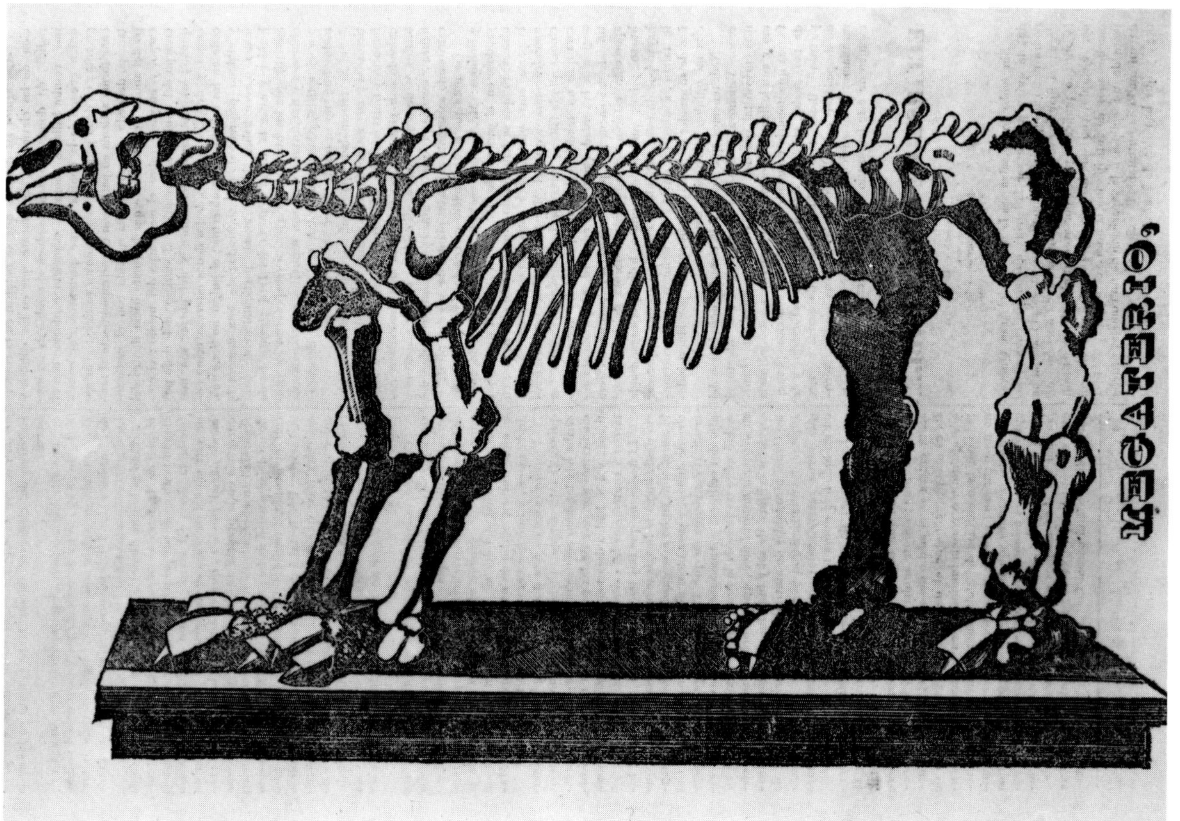

279

280

281

280. *A Spanish reconstruction of the Megatherium.* Though Cuvier's lessons in comparative anatomy were known in Spain, they hardly were fully assimilated. This imaginary reconstruction of a *Megatherium* does not even correspond with the skeletal remains, and demonstrates the level of scientific culture in Spain during the midnineteenth century.

281. *Dinocerate.* Enormous mammals became rather common in paleontology as the nineteenth century progressed. This dinocerate was another object of study by Othniel Marsh.

282. *A paleontological nightmare.* It was natural for popularizers of science to exploit the fossil world. Louis Figuier was a *vulgarisateur* of enormous productivity whose works inspired and, as this cartoon suggests, perhaps terrified a whole generation of European youth.

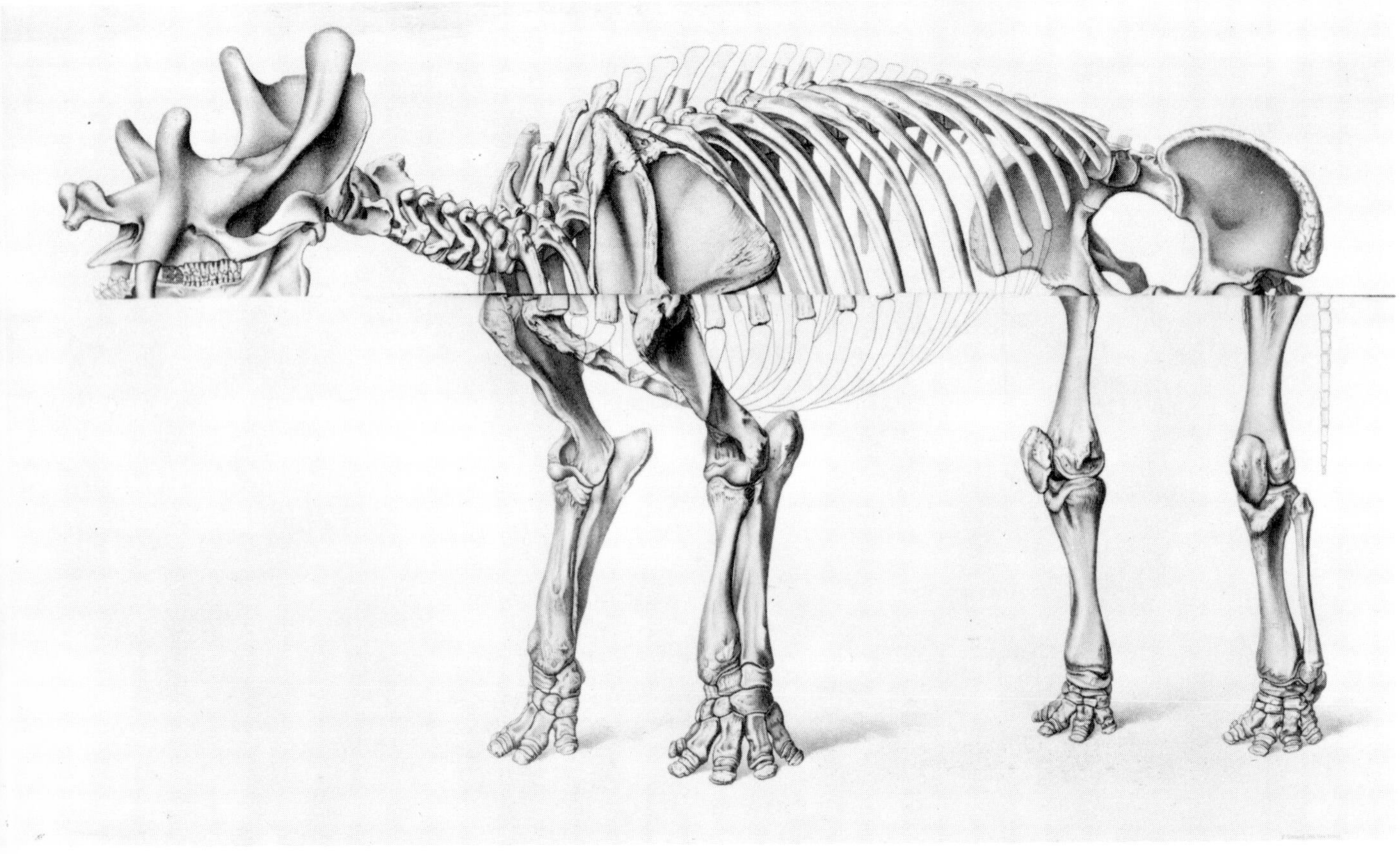

A LITTLE CHRISTMAS DREAM.

Mr. L. Figuier, in the Thesis which precedes his interesting Work on the World before the Flood, condemns the Practice of awakening the Youthful Mind to Admiration by means of Fables and Fairy Tales, and recommends, in lieu thereof, the Study of the Natural History of the World in which we live. Fired by this Advice, we have tried the Experiment on our Eldest, an imaginative Boy of Six. We have cut off his "Cinderella" and his "Puss in Boots," and introduced him to some of the more peaceful Fauna of the Preadamite World, as they appear Restored in Mr. Figuier's Book.

The poor Boy has not had a decent Night's Rest ever since!

III

Man

The Evolution of Man

Man's place in nature has always been a matter for controversy in the Western world. Most early societies recognized his divine nature, but every society also was aware of his beastliness. Furthermore, as knowledge of living forms expanded, the gulf between man and the other members of the animal creation narrowed. As early as 1699, man's affinity with the orangutan was noticed with the publication of Edward Tyson's *Orang-Outang, Sive Homo Sylvestris: or, the Anatomy of a Pygmie Compared With that of a Monkey, an Ape, and a Man.* The orangutan was more or less considered part of the human species, only on a lower level. In 1737 Linnaeus put man in with the rest of the animal creation in the first edition of his *Systema naturae.* In the tenth edition, the races of man included *Homo ferus* (Wild Man) and *Homo monstrosus* to include those "border" cases between man and the animals. Thus, the continuity between the animal creation and man was stressed although no one had yet explicitly insisted upon the evolution of man from the lower animals, making the connection genetic. It was still possible to argue in the eighteenth century that the affinities of man to the beasts merely reflected the divine desire to create each link in the Great Chain of Being. One would, therefore, expect neighboring links to look very much alike, but each link was a separate species clearly differentiated from its neighbors. Man stood eternally apart from the rest of the creation by reason of the existence of his immortal soul and his reason.

This doctrine remained orthodox throughout the first half of the nineteenth century in spite of the emergence of evidence that challenged it. When the bones of Neanderthal man were discovered in 1857, Rudolf Virchow, one of the leading pathologists of the day, sneeringly dismissed the idea that they belonged to a more primitive race than *Homo sapiens.* They were, said Virchow, examples of pathological causes and the cranium with its beetling brows was that of an idiot. Man had been created six thousand years before, had almost immediately passed into the state of civilization, and his past was recorded by history. Bones and stone implements and evidence for a long prehistorical existence of man were all nonsense and best ignored.

The publication of Darwin's *On the Origin of Species* (1859) and *The Descent of Man* (1871) virtually clinched the case for evolution of species against the special creation of species espoused by religious fundamentalists. In the course of the preceding century, other naturalists, including Darwin's grandfather, Erasmus Darwin, had proposed theories of evolution, and in fact the concept had ancient roots. But Darwin provided what has survived as the most satisfactory explanation of how such an evolutionary process takes place. He also compiled a mountain of evidence in support of his theory. In the *Origin,* Darwin showed how artificial selection by man of various traits could lead to wide variations in animal and plant species. The ordinary rock pigeon, for example, could be turned into rather bizarre types such as the tumbler pigeon which tumbled while flying, or the puffer pigeon that could puff out its crop to enormous proportions. Species were not so constant as scientists had claimed and could be changed. Nature effected changes through natural selection. At the beginning of the nineteenth century, the Reverend Thomas Malthus had shown in his *Essay on Population* that there was always a competition for food among human populations that prevented most human beings from surviving. Darwin seized upon this idea as a basis for the process of natural selection; that is, a mechanism for weeding out those variations within a species which were not advantageous to survival. Thus, variations that aided the organism in adapting to its environment were passed along at the expense of less successful ones. In order for speciation to take place under these conditions, great amounts of time were necessary—far more than were allowed by the Bible. By 1859 geology had also stretched the biblical time-span beyond recognition and, in a few short years, the scientific time-scale simply exploded. The earth was millions of years old; man and his intelligence had evolved from the lower animals and had spent untold eons in a state of savagery. It was the task of science to find the evidence for these statements and fill in the evolutionary history of mankind. It was a task eagerly accepted by Darwin's younger contemporaries. From embryology, paleontology, geology, anthropology, and ethnology the evidence flowed in. Evolutionary principles were even used to interpret the evolution of the rifle bullet! What was lost was the conviction of man's special relationship with the divinity. What was gained was a new sense of the possibilities of human development. Man had emerged from darkness and savagery in an evolutionary process in which chance played a fundamental role; perhaps, through science, man could rise even higher. To the question of whether man was ape or angel, posed by the Victorians, evolutionists could answer that he was both. He had been ape, he would be angel.

283

283. *Man's place in nature as the nineteenth century began.* From the onset of the Scientific Revolution in the sixteenth century, the ultimate question to be asked was "Where does man fit into the scheme of nature?" At the beginning of the nineteenth century, as this frontispiece to the first volume of a journal of natural history indicates, man was still considered the center of cosmic concern. The stars and their locations in the houses of the zodiac were assumed to influence man's organic fate.

284. *Woman's place.* Woman, like man, was a focus of celestial influences; but there was a clear separation of the sexes' natural roles. Man was formed for contemplation and action; woman was, as Scripture stated, created for man.

285. *The wild man of the woods —the orangutan.* The gulf between man and the animals was not considered particularly wide in the early nineteenth century. The discovery of the orangutan was easily assimilated to the doctrine known as the Great Chain of Being, which went back to Aristotle and was based on the gradation of natural forms. Thus the orangutan could be considered quasi-human and was depicted that way.

284

285

286

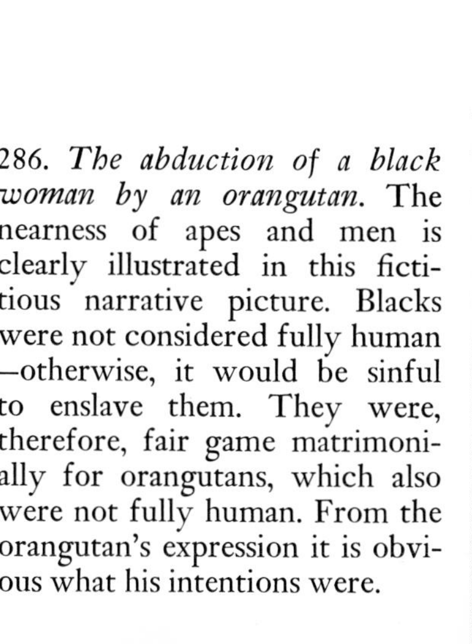

286. *The abduction of a black woman by an orangutan.* The nearness of apes and men is clearly illustrated in this fictitious narrative picture. Blacks were not considered fully human —otherwise, it would be sinful to enslave them. They were, therefore, fair game matrimonially for orangutans, which also were not fully human. From the orangutan's expression it is obvious what his intentions were.

287

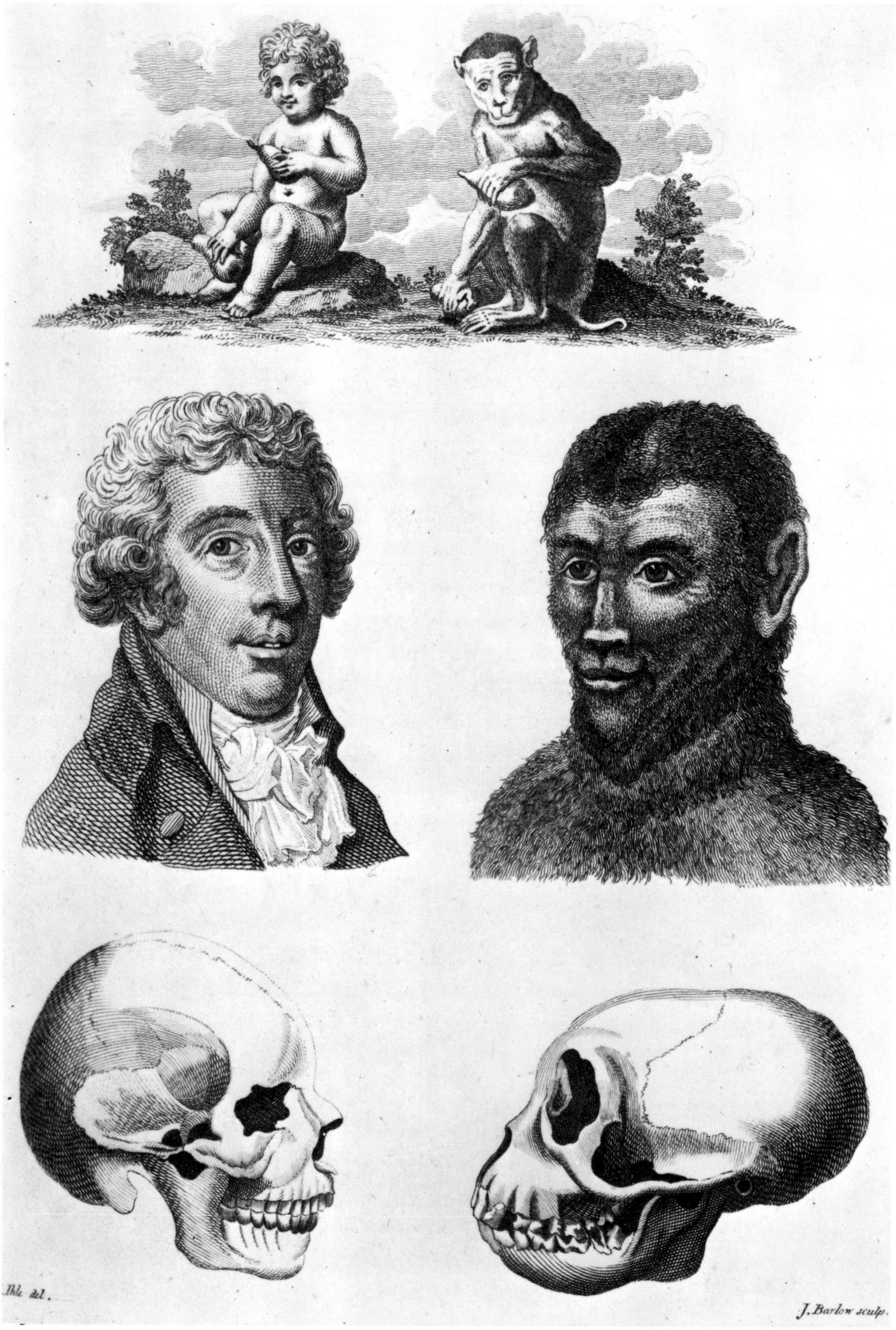

287. *The comparative anatomy of man and orangutan.* The affinity between man and the higher apes was clearly recognized at the beginning of the nineteenth century. Merely an "affinity" was recognized, however —not a direct relationship.

288. *Darwin the naturalist.* The doctrine of the Great Chain of Being was to be upset by Charles Darwin, who radically altered man's view of his place in nature. When Darwin joined the ship's company of HMS *Beagle* in December 1831, however, no one could have foreseen what the voyage would lead to, in terms of a revolution in biology. Darwin shipped out as a naturalist whose duty it was to collect specimens and make geological observations of the lands visited by the *Beagle* on her oceanographic voyage to South America. He is shown here at the age of 33 with his son.

289. *Darwin's home and study at Down, Kent.* It was in this house and study that the ideas, born during the voyage of the *Beagle* and shortly thereafter, began to take firm shape in Darwin's mind. He collected evidence for his theories for more than 20 years after the idea of natural selection first came to him in 1837 but hesitated to publish them, fearful of the storm he knew would be raised.

288

289

290

THE LION OF THE SEASON.
Alarmed Flunkey. "MR. G G-G-O-O-O-RILLA!"

291

290. *Punch: a view of natural selection (1877).* Probably no great scientific theory has been so misunderstood and misrepresented as Darwin's theory of natural selection. This cartoon purports to caricature the "Darwinian" idea that steady use of an organ literally would lead to its increase in size. The idea was Lamarck's, not Darwin's.

291. *Man's place in post-Darwinian society.* In his *Origin of Species,* Darwin had given only the faintest hint that man, like other animals, had evolved from lower forms. The opening up of Africa to European explorers led to the discovery of the gorilla, an animal whose almost human attributes made scenes like this seem possible to those who misinterpreted Darwin's theories.

292. *The descent of man.* In 1871, in *The Descent of Man,* Darwin published his theory of the evolution of man from lower forms. The doctrine was eagerly espoused in Germany by Ernst Haeckel, who prepared this schematic history of man's slow progress to the present. Note that no. 23, clearly lower on the evolutionary scale than 24, is obviously human but black.

1a
2
3
4
5
6
7
8a
8
9
10
11
12
13
14
17
15
16
18
19
20
21
23
22
24

293

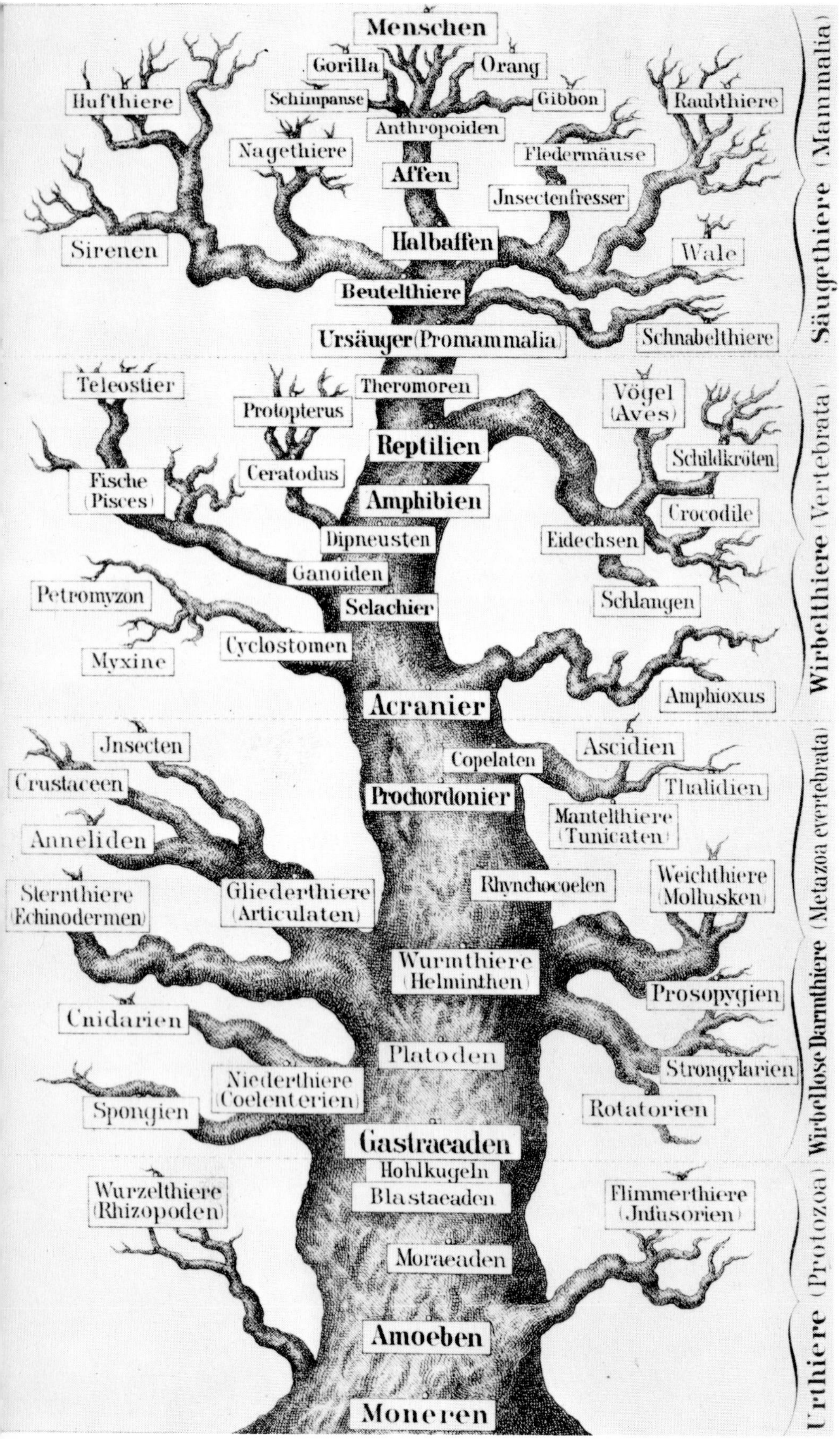

294a

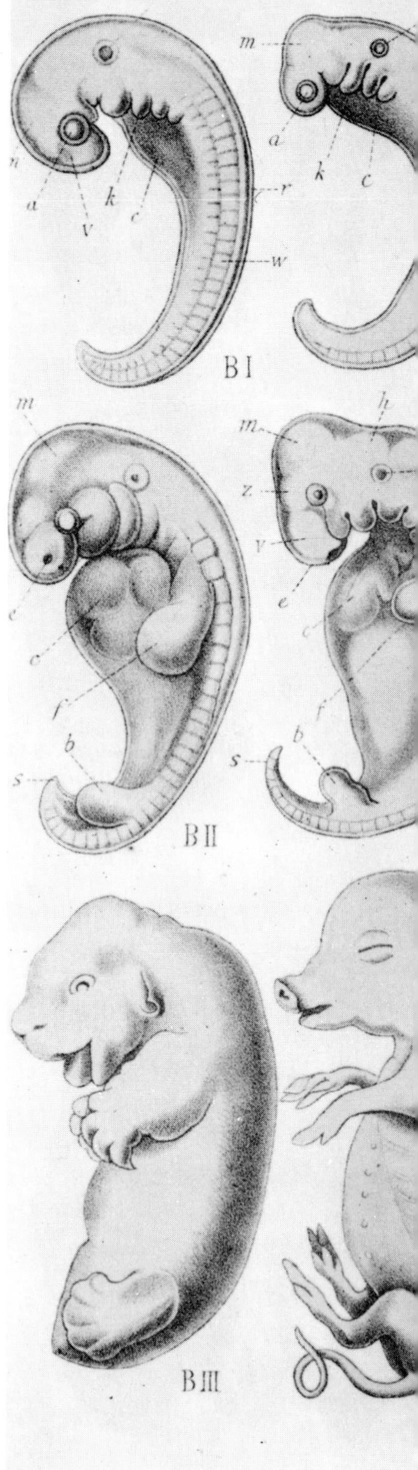

293. *Man's place in living nature.* In somewhat greater detail, Haeckel here pictured the tree of life with its special branches. At the bottom were the Protozoa or unicellular organisms. At the top was man. An important point Haeckel makes in this representation is that man is *not* descended from contemporary apes, but that both man and the apes have a common ancestor.

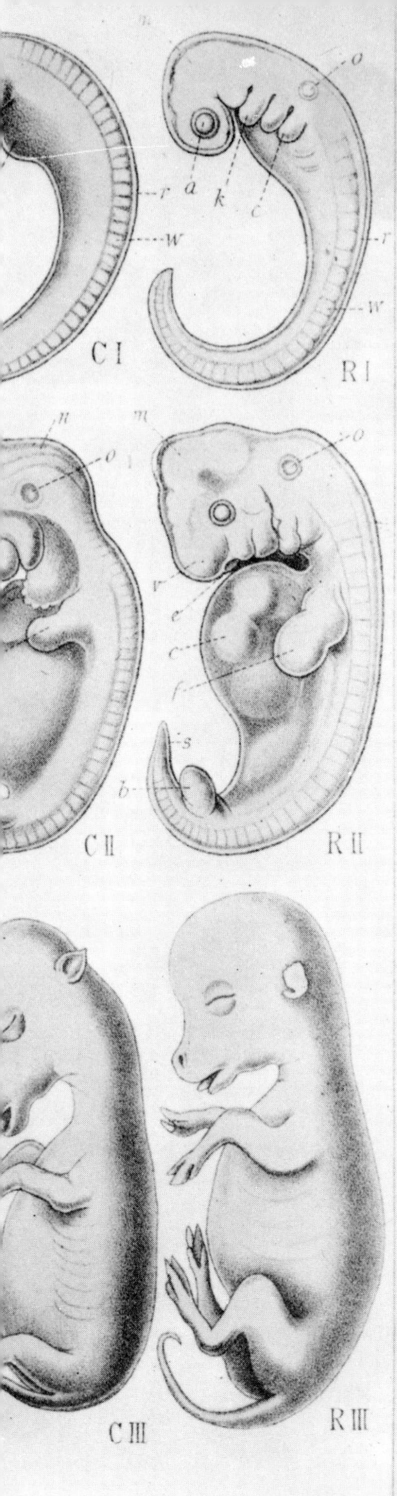

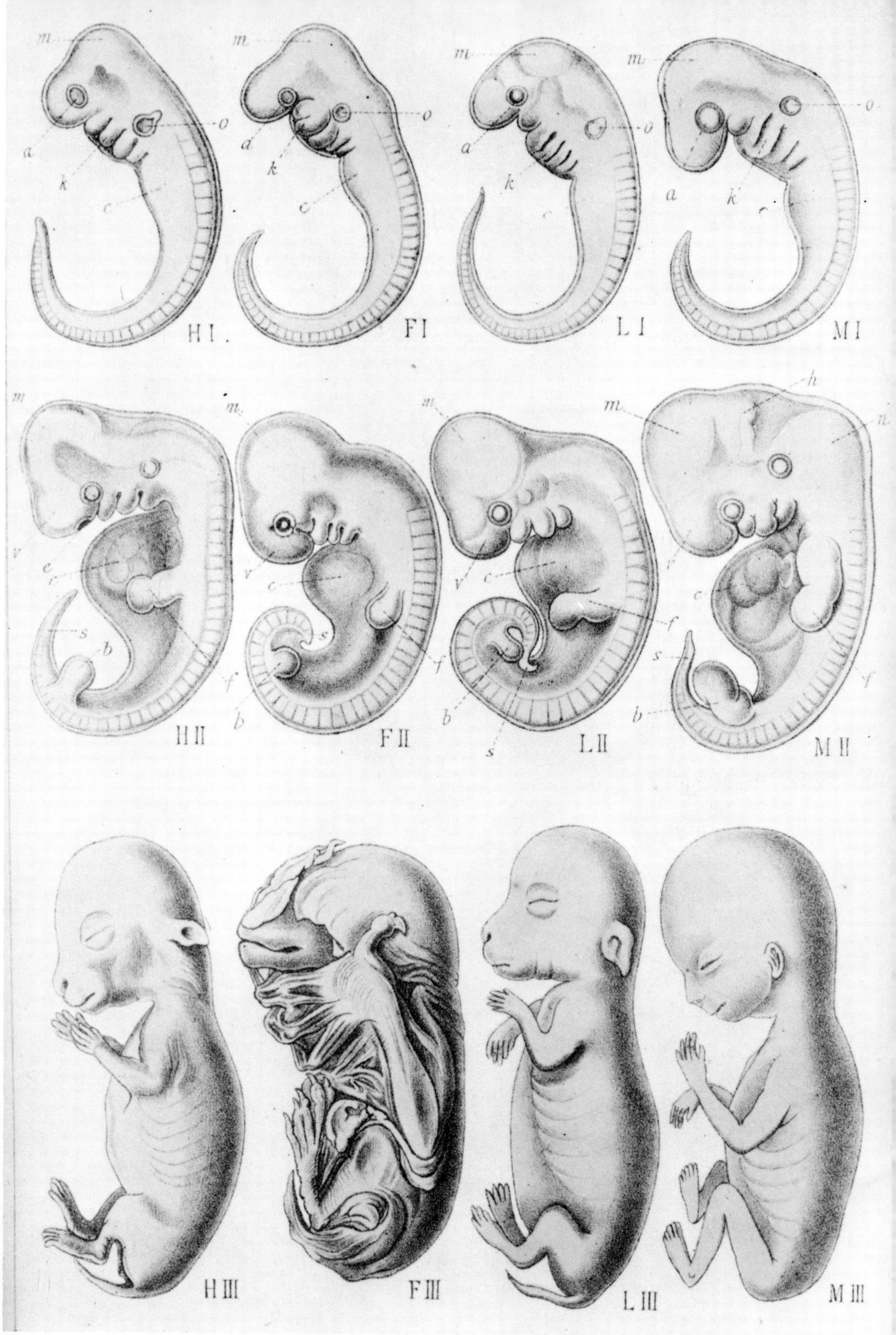

294 a, b. *The evolutionary ancestry of man.* Darwin's theory, drawn from observations of mature creatures, shed great light on the field of embryology. Haeckel used embryology to show that the human fetus is indistinguishable from the fetuses of other animals until almost the last stages of its development. This, he felt, reflected the common evolutionary history of the class Mammalia, to which man belongs. It was Haeckel who coined the phrase "Ontogeny recapitulates phylogeny," meaning that the mammalian embryo passes through all the evolutionary stages of all the species, becoming recognizable only at the end of the process. The embryos shown, at three stages, are, from left to right: Marsupial, Pig, Roe Deer, Cow, Dog, Bat, Rabbit, Man.

294b

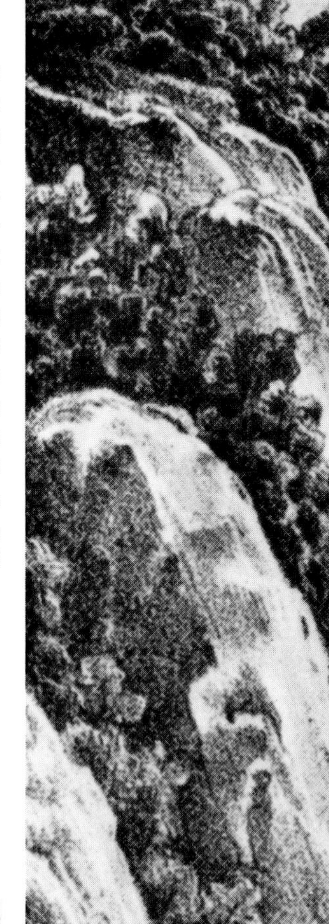

295. *Evolution and bourgeois values.* Haeckel was not only a naturalist of some distinction; he was also the founder of a new religion, Monism, which denied traditional values and morality. His aim was to shock the staid, conservative German burghers of his day, and this frontispiece to one of his books seems specifically designed to do just that. Humanity's place among the apes could be shown no more graphically than in this picture of a German *fräulein,* of Brunhilde-like proportions, combing her hair in the midst of her simian relations.

295 296

296. *A public reaction to Darwin.* Darwin threatened the whole moral and religious structure, which had been erected on the assumption that man had been created by God for a special divine purpose. To say that man evolved was to deny man's divinity and place him with the animals. Such a position had obvious political overtones, and politicians were forced to align themselves for or against it. Benjamin Disraeli, sometime Prime Minister and leader of the Conservative Party, took a conservative position, as this caricature shows.

297. *The Neander Valley.* Darwin's theory was only that until fossil men were discovered. It was in the Neander Valley of Germany that the first important human fossil was discovered in 1856.

298. *The Neanderthal cave.* It was in this cave that the bones of Neanderthal man were found. A cross section of the site shows: the cave (*a*); the terrace of the cave (*b*); the crevasse running through the cave (*b-c*); a loam deposit 10 to 12 feet thick (*d-e*); the Düssel River (*f*).

297

298

299

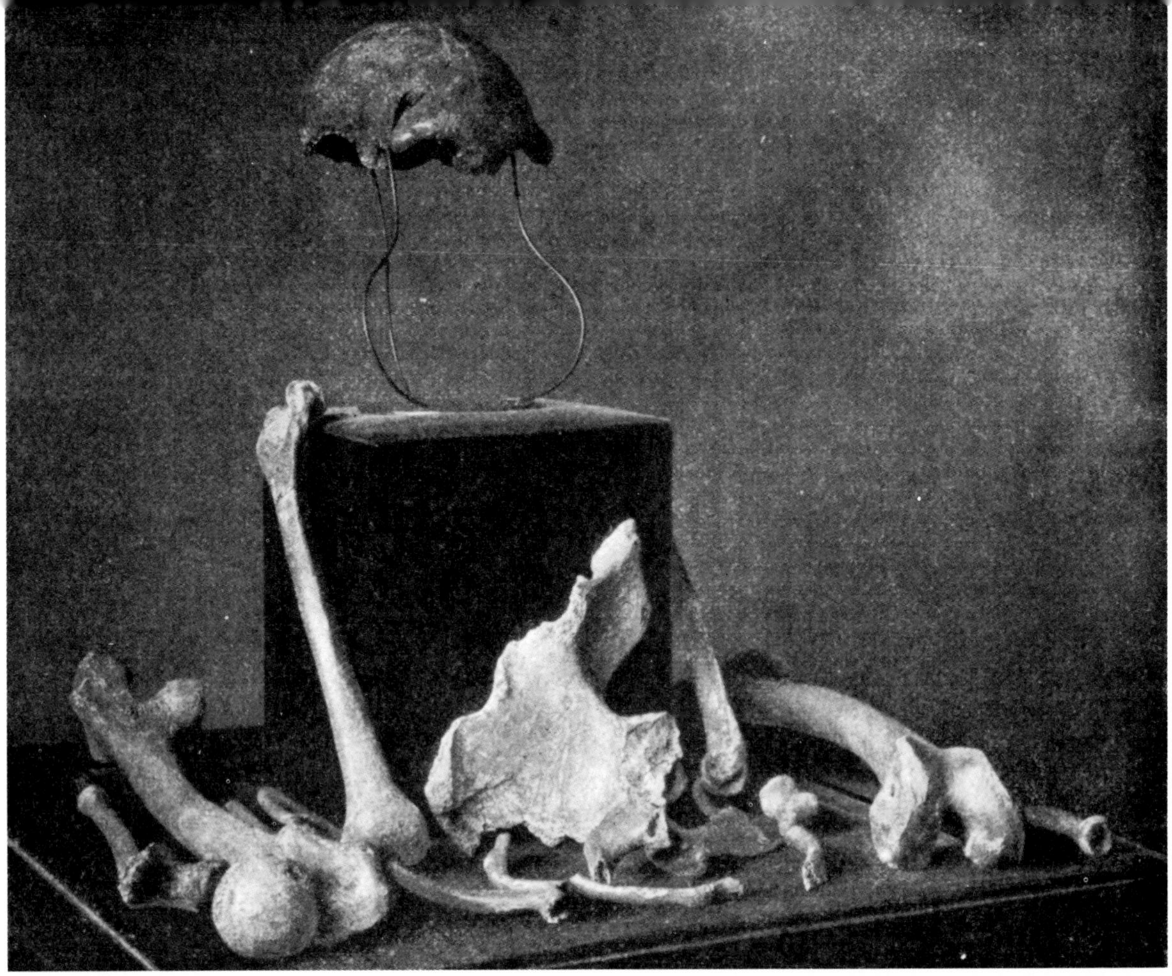

300

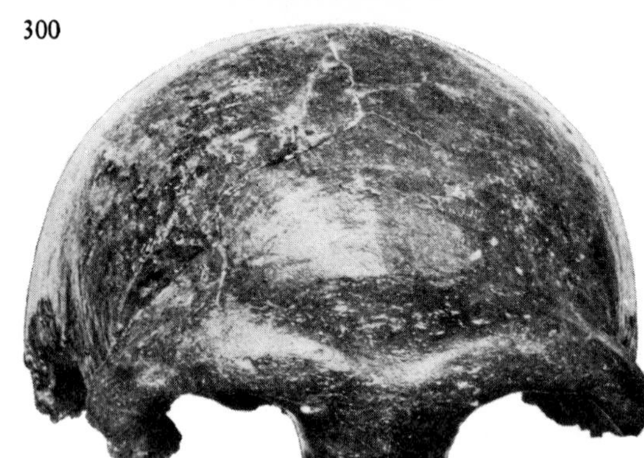

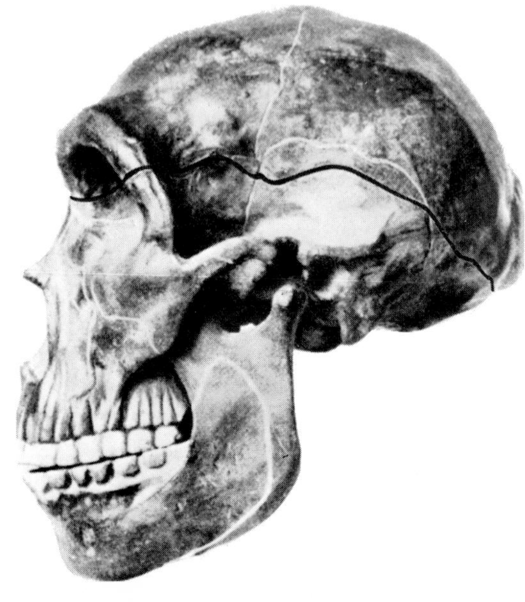

301

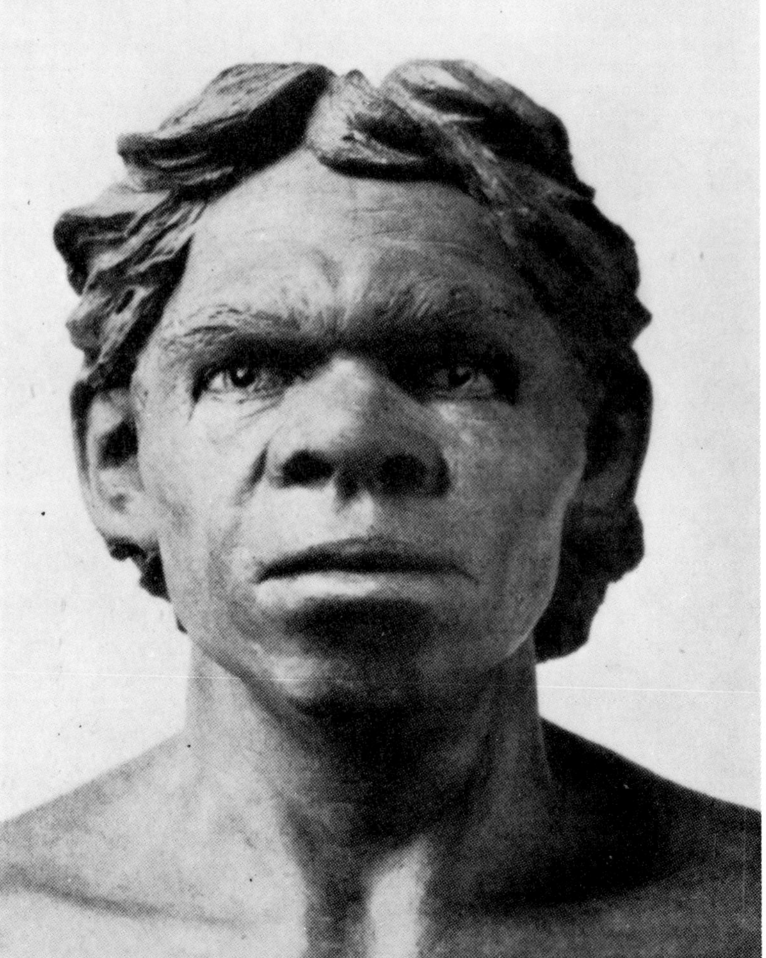

302a

302b

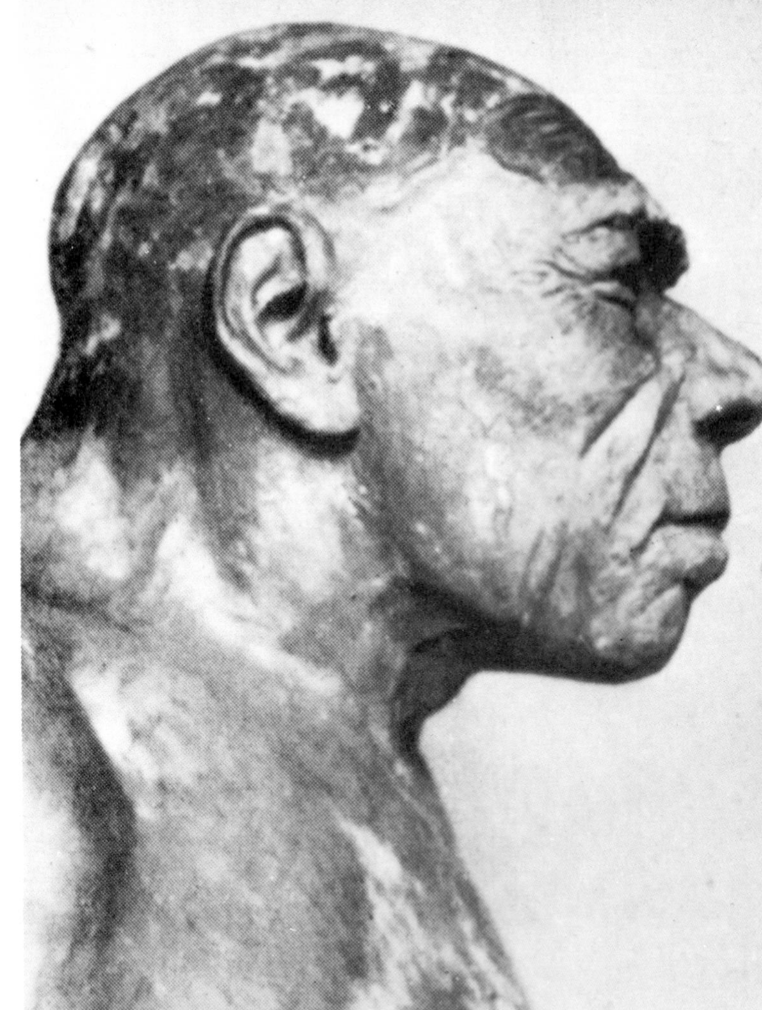

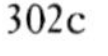
302c

299. *The fossil remains of Neanderthal man.* These bones were obviously human. They equally obviously were not the bones of any human race known in the nineteenth century.

300. *The skullcap of Neanderthal man.* This skull fragment showed that Neanderthal man had a fairly large brain capacity. Its shape indicated that the brain might be distributed somewhat differently from modern man's. Note the heavy brow ridge and the sloping forehead.

301. *Reconstruction of Neanderthal man's skull.* Using the cap (marked by the dark line) and the principles of anatomy, it was possible to reconstruct the whole skull. The result was undoubtedly human, but bizarre.

302 a, b, c. *A reconstruction of Neanderthal man.* From the bones, these examples were sculpted to show Neanderthal man as a youth, a mature man, and an old man. The evidence indicated that Neanderthal's intelligence was comparable with that of *Homo sapiens*, despite his sloping forehead and heavy brows.

303

303. *An imaginary fossil man.* Before the discovery of Neanderthal man, there had been lively speculation about what antediluvian man had looked like. This picture dates from 1840—long before Darwin's theory was published. This "fossil" man was only a short step from the apes. His long muzzle and peculiar feet were nonhuman. It is an interesting speculation, coming, as it did, from Spain, where such ideas generally seemed out of place.

304. *The life of prehistoric man.* The discovery of fossil man stimulated the artistic imagination. How had prehistoric man lived? What were his customs? Given the fossil record, the enterprising artist could let his imagination fill in the details and provide a scene like the one reproduced here, showing man the hunter after the last ice age.

305. *A Russian mammoth hunt.* The life of prehistoric man attracted even academic artists. The drama of the scene guaranteed public interest, and the artist could use his academic training to good effect. The prostrate figure in the foreground of this painting could have come from almost any standard study of the male nude in the nineteenth century. It was the mammoth that provided the drama.

304

305

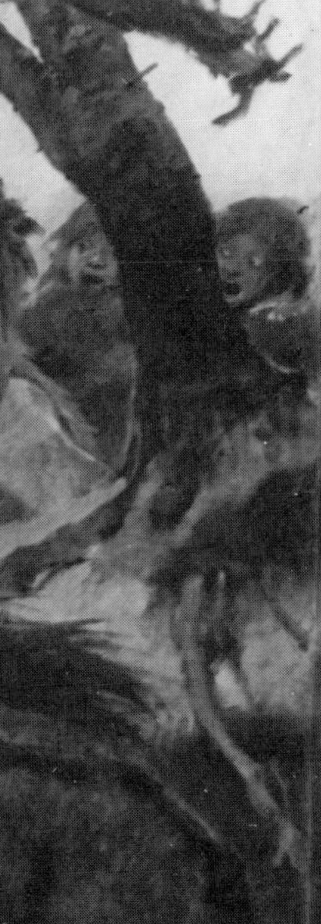

The Science of Man

Ape or angel, man was subject to scientific laws and there was ample stimulus in the nineteenth century to seek out these laws and discover a science of man. At the beginning of the century, enthusiasm for the search and for the subject tended to carry people away. In Germany there grew up a group of nature philosophers who felt that the secrets of the universe, including that of man's nature, were to be found through speculative reason. Man's reason, they argued, was a dim but exact reflection of divine reason. Therefore, if the reason were let loose and permitted to speculate freely it would discover all the secrets of God's creation. It was a noble and high-flying dream that did produce some interesting ideas but that failed to unlock the mysteries of the universe. Man was more than a sexual, rational, emotional, creature whose nature could be deduced from a few simple axioms. The main road to anthropological enlightenment still led through hard work, experiment, and careful observation and could not be passed over by speculation.

The main lines of gross human anatomy had been made clear before the nineteenth century began. It was possible to *represent* the human body without understanding it. Most of the gross organs were well known and well described, although many of their functionings were still obscure. There were still frontiers. Human generation was imperfectly understood, as was generation in general. Von Baer did not isolate and observe the mammalian ovum until the 1830's and it was not until some time later that the human ovum was discovered. It had, of course, long been suspected that embryos arose from ova, but suspicions cannot be substituted for facts in science. Gradually, throughout the century, the science of embryology advanced until it was possible finally to comprehend every stage of infantile human development from conception to birth. This work added a significant new stage to the ancient representation of the ages of man.

One of the more fascinating frontiers in anatomy and physiology was that involving the brain and behavior. Some primitive work had been done in the eighteenth century, and Swedenborg had been able to show from clinical evidence that brain function was localized. His studies of injuries to the head enabled him to illustrate how various motor functions were destroyed when specific parts of the brain were injured. This work was vastly extended in the nineteenth century until it was possible to draw fairly accurate maps of the brain showing precisely what area controlled what function or activity.

The simple fact of brain localization provided the scientific basis for one of the more popular pseudosciences of the time. If brain function was localized, then it was no large leap of the imagination to suspect that the surface of the brain reflects, in form, its function. This reflection, in turn, ought to be passed on to the skull so that a careful "reading" of the skull ought to enable a skilled practitioner to discover peculiar talents and abilities. So phrenology was born and swept over Europe and America. Fantastic tables of characteristics and talents were drawn up and the skull was divided into minute segments supposedly localizing these psychological qualities. People paid their money and probably felt they had received good value. The smart phrenologist, after all, like counsellors since time immemorial, has considerable insight and tells a patient or a subject some things of value and importance.

Psychological qualities were not so easily analyzed as phrenologists hoped. They came from deeper within the human being than the bumps on his skull. Some of their external effects could be produced artificially by external stimulation such as electricity, but this only proved that their real causes in the individual were more subtle than mere gross anatomy. Darwin, indeed, sought their origins in the evolutionary history of the race. In his *Expression of the Emotions in Man*, he tried to show how fear, anger, joy, and terror were all expressed in ways that had adaptive value for the human organism. In this basic sense, a man's character was as much the product of his evolutionary history as of his own personal experiences.

The problem of character had concerned natural philosophers since the time of the Greeks. An apparently fruitful attack on it was made in the nineteenth century by Blumenbach and his followers. Blumenbach was a famous anatomist and one of the fathers of anthropology. He believed in precise measurement rather than speculation. His science of physiognomy was built upon exact observations, but proved as illusory as the fantasies spun by the nature philosophers. Yet his approach did contain the kernel of valuable anthropological practice. Blumenbach was convinced that human character literally shines from the face or physiognomy of the individual. This "illusion," if such it is, lies behind some of the greatest painting in the world and is commonly held by most people. We instinctively mistrust the shifty-eyed individual, consider the person with a low, sloping forehead slightly subnormal in intelligence, and think gross features generally to be indications of gross character. What Blumenbach wished to do was to take these rather vague terms and make them mathematically exact. What was, in fact, an abnormally sloping forehead? To answer that question one had to have two kinds of data: statistical figures on the average slope of foreheads and mathematically precise measurements of the slope of foreheads. When all that could be measured was measured, it ought then to be possible to correlate a measured feature with observed character. Physiognomy was to be as precise and accurate as science could make it.

Although Blumenbach's dream of creating a new science faded by the middle of the century, his methods continued to fascinate anthropologists. The physical sciences, after all, were accomplishing prodigies through accurate measurement so it seemed only natural for proponents of the new science of physical anthropology to embrace measurement eagerly and embark upon a course of measuring every aspect of the physical nature of man susceptible to measurement. In 1840 Anders Retzius invented the cephalic index, the proportion of the breadth to the length of the head. Dolichocephalic crania are those whose ratio of breadth to width is less than four-fifths; people with cephalic indices of less than 0.8, then, are long-headed. Brachycephalic crania are those whose ratio of breadth to width is greater than four-fifths, and a cephalic index greater than 0.8 makes one round-headed. This index appeared to offer valuable clues about the races of man and their migrations during historic and prehistoric times.

Although the early work focused on skulls, it soon became clear that if anthropometry were to be effective, the whole man must be subjected to measurement. Instruments to accomplish this were devised and large-scale measurements were made on considerable numbers of subjects. The statistical mathematics to handle these measurements were devised by men such as Adolph Quetelet in Belgium and it became possible to graph or otherwise represent the results of these investigations. It must be confessed that the ultimate product was disappointing. Until the laws of human genetics were discovered—and this did not occur until after 1900—there were few really meaningful conclusions that could be drawn from this orgy of measuring. To what end does one tabulate the number of adult males with fair eyes and dark hair in the British Isles?

306 a, b. *The end of the world.* Knowledge of man's origins inevitably led to speculation about man's fate. The discovery of the second law of thermodynamics seemed to imply extinction through energy starvation. Here are a pair of fin-de-siècle depictions of man's last days on Earth, just as the lights go out on a cosmic scale.

307. *The nature of man.* Finding out where man came from and where he is going does not answer the question of what man is. This was the domain of anatomy and physiology. Both sciences made great strides in the nineteenth century, but neither could escape the prevailing intellectual currents. In early nineteenth-century Germany, under the influence of speculative philosophy, natural phenomena were viewed as resulting from conflicts between positive and negative forces. This chart depicts human physiology in these terms. On the left are the negative forces—femaleness, necessity, heaviness, stolidity, and carbon dioxide; on the right are the positive forces—maleness, liberty, light, expansivity, heat, and oxygen. In between is a schematic representation of human anatomy and of the functions that arise from this conflict of forces. The whole represents a microcosm that reflects similar conflicts in the macrocosm of the universe.

308. *What is woman, from an embryological point of view?* Even something so familiar as a woman's face was not constant. Haeckel here provides a modern scientific version of the classical theme of the ages of man.

306a, b

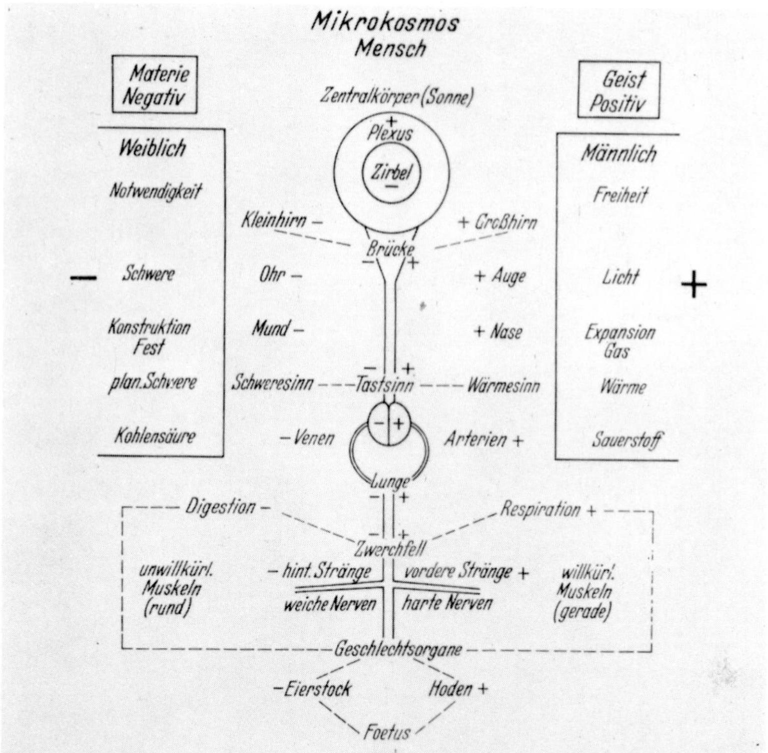

307

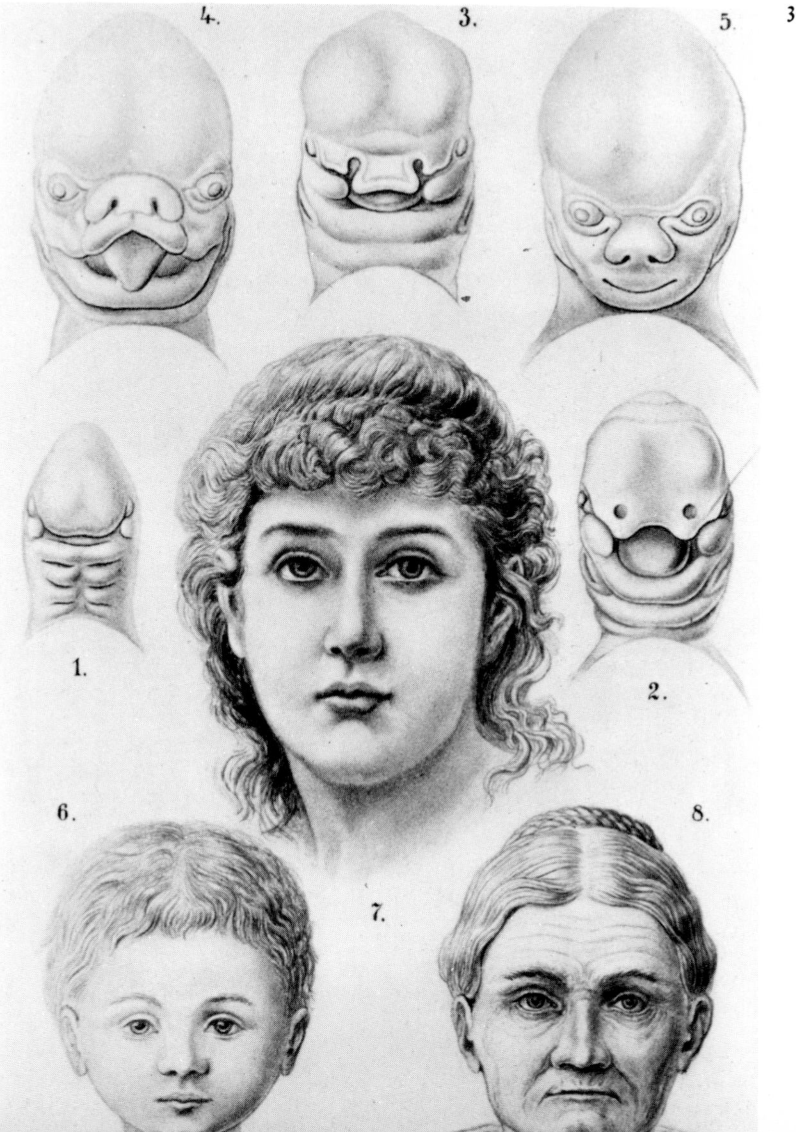

308

1
2
3
4
5
6
7
8
9
10
11
12
a
b
c
d
e
f
g
h
i
k
l
m
n
o
p
q

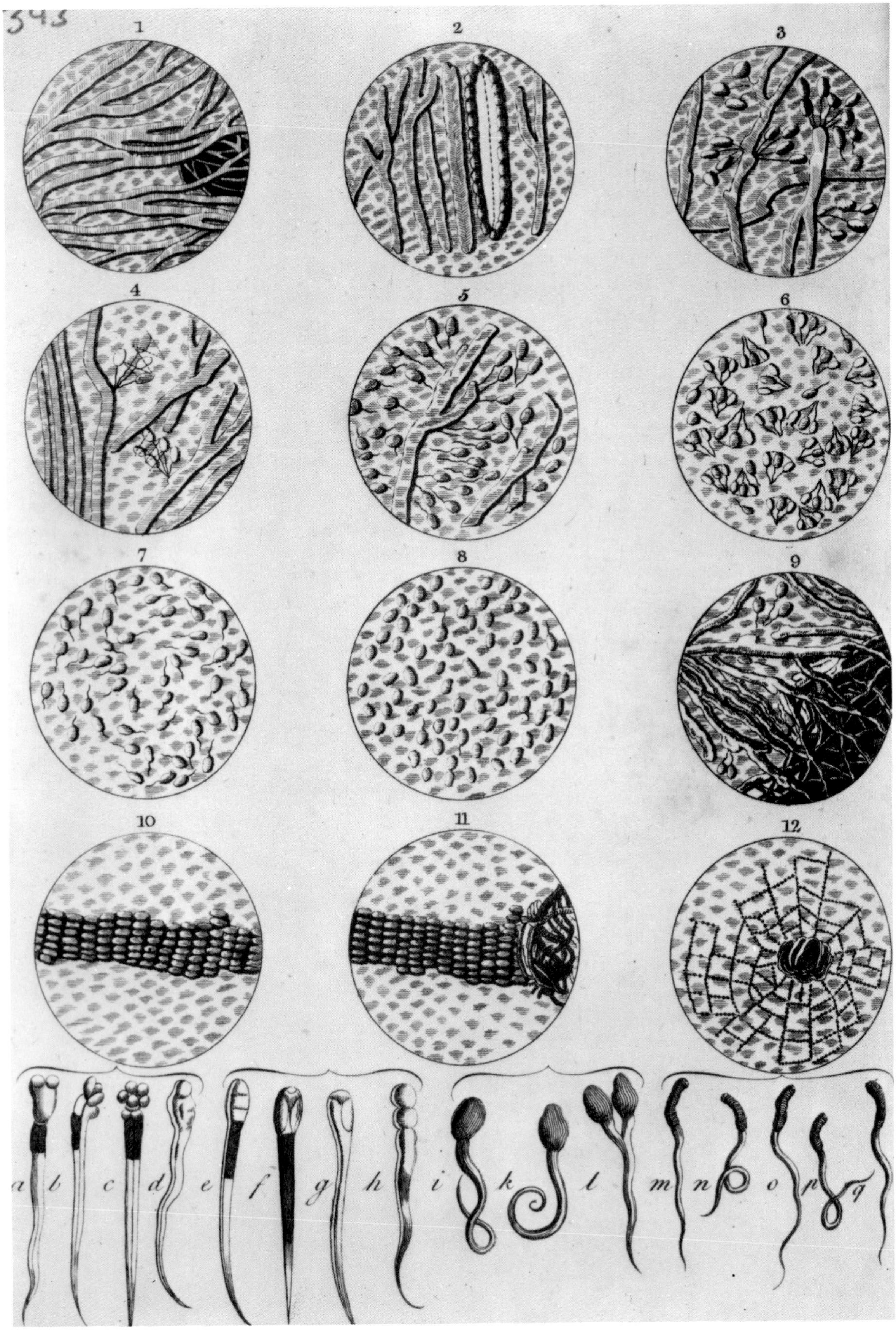

309. *Generation and spermatozoa: Buffon's theory*. Probably the most careful microscopic study of seminal fluid in the eighteenth century was made by George-Louis Leclerc, Comte de Buffon (1707–1788). Views 1–12 are drawings of what he observed. He considered the fibers essential and insisted that the small particles were not "animals"; their "tails," he thought, were fibrous remnants that impeded their motion. According to Buffon, these particles were organic molecules that carried the vital principle with them and helped to begin the organization of the embryo; their motion was evidence of their vitality. On the bottom are drawings of the spermatozoa of rabbits (*a,b,c,d*), dogs (*e,f,g,h*), rams (*i,k,l*), and roosters (*m,n,o,p,q*). Significantly, these were taken from Antoni van Leeuwenhoek's descriptions, which appeared in the seventeenth century. When we compare Leeuwenhoek's work with Buffon's, it is apparent that microscopy did not progress far in the eighteenth century.

310 a, b. *Formation of the human fetus (ca. 1800)*. The fetal history of man was obscure. The moment of conception is shown here in almost spiritual terms, and even a month-old embryo is crudely and inaccurately depicted. Only in its later stages did scientists have a clear idea of what a human embryo looked like.

311. *The development of the embryo (1908)*. The great strides made in human embryology during the nineteenth century may be appreciated by comparing this figure with the preceding one. Every stage of fetal development can now be shown in detail.

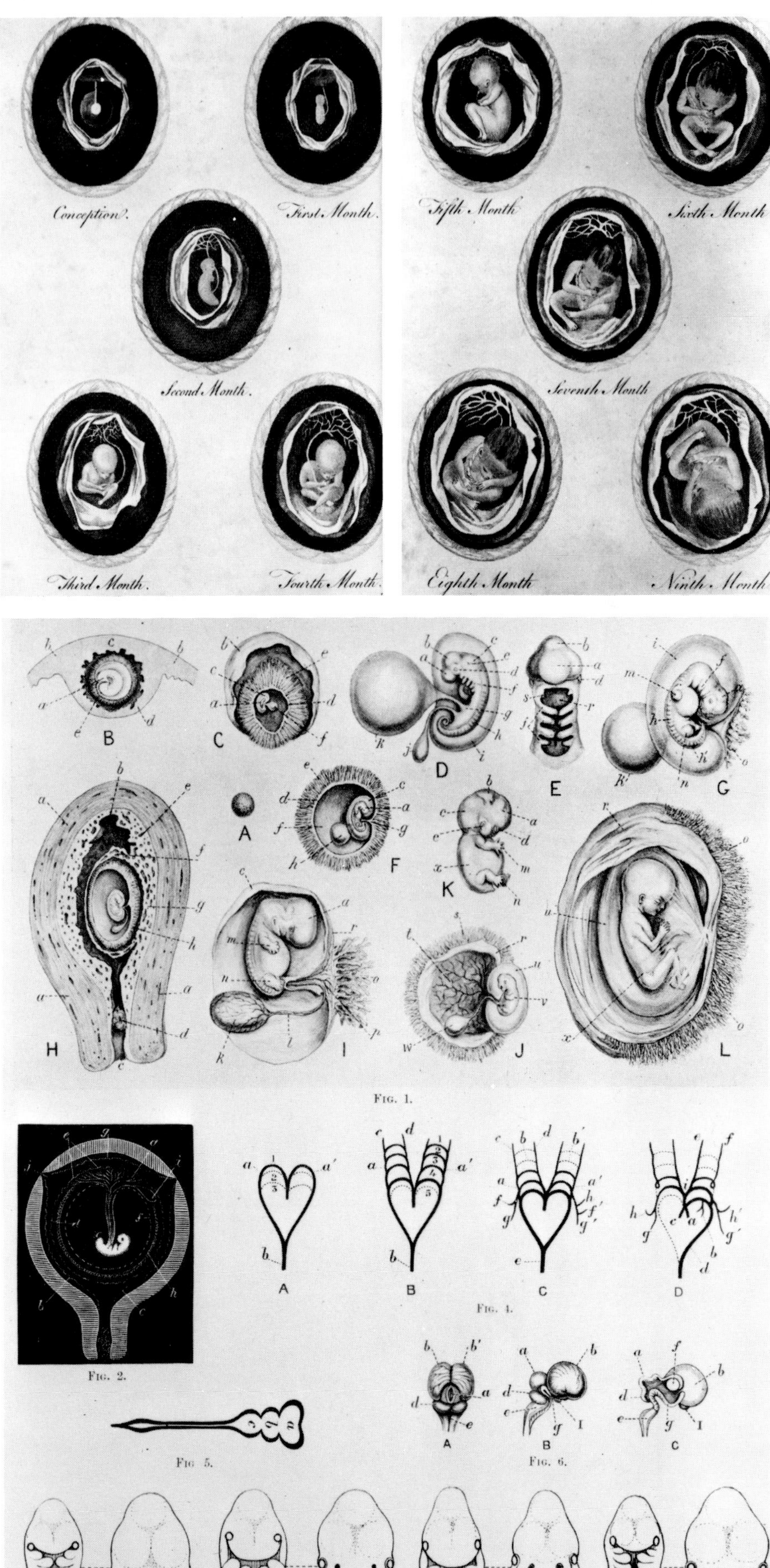

312

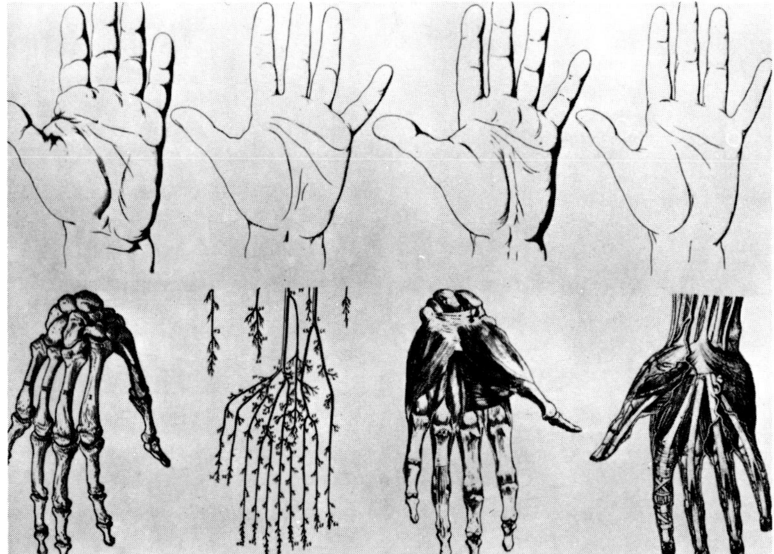

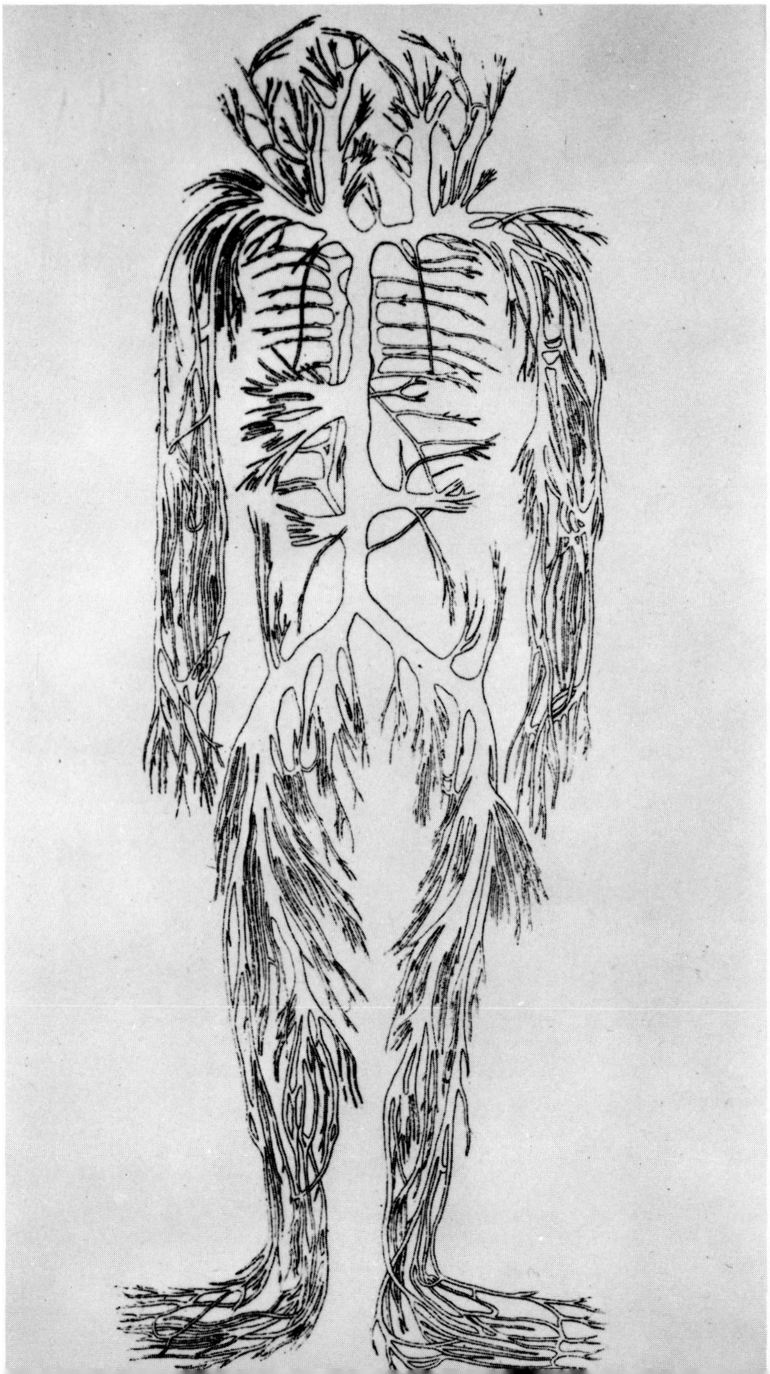

312. *The anatomy of the hand (1833)*. Charles Bell was one of the great gross anatomists of the early nineteenth century. His anatomy of the hand fits into a tradition that dates back to Vesalius, in which the gross features are described but no attempt is made to penetrate to the microscopic level.

313. *The circulatory system (1826)*. This purports to be the gross anatomy of the vascular system, drawn by a medical doctor, in the early nineteenth century. It is, at best, a schematic treatment and, at worst, simply wrong.

314. *Gross anatomy of the human body (1826)*. This view of human anatomy suffers by comparison with the works of professional anatomists of the eighteenth, or even the seventeenth, century. To be sure, the figure came from a "popular" work; but it was from such works that most people of the time drew their knowledge of the human body.

313

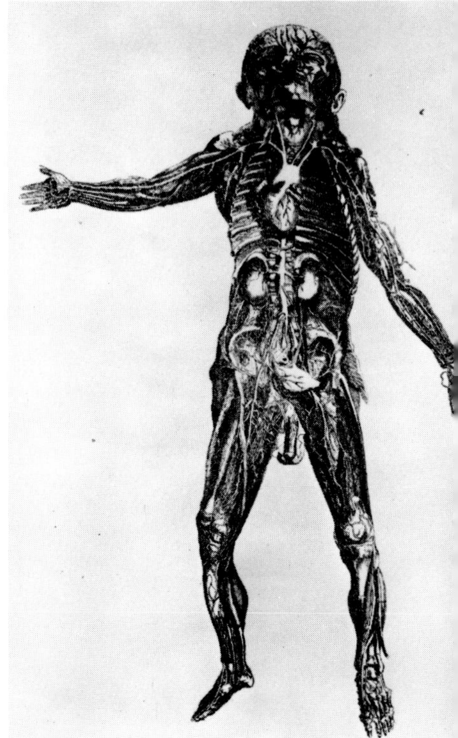

314

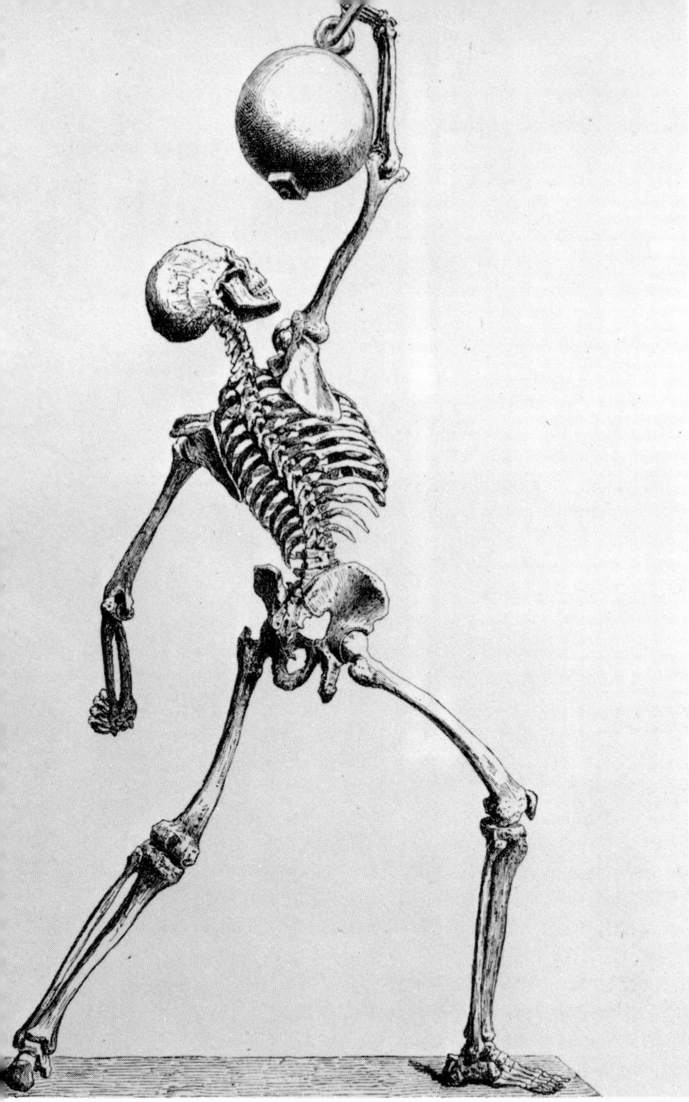

315 316

315. *The human skeleton (1894).* The anatomical tradition founded by Vesalius remained strong throughout the nineteenth century. The combination of anatomical accuracy and dramatic pose served to enliven a somewhat grim subject.

316. *Man's muscular anatomy (1894).* The Vesalian tradition lent itself well to such popular illustrations as this. The play of muscles is clear, and the subject has a certain amount of dramatic and artistic tension to capture the viewer's interest.

317. *The blood vessels of the heart.* The real advances in the nineteenth century were made in the investigation of separate organs and tissues. In this 1894 German view, the blood vessels in the heart that are visible to the naked eye are well laid out.

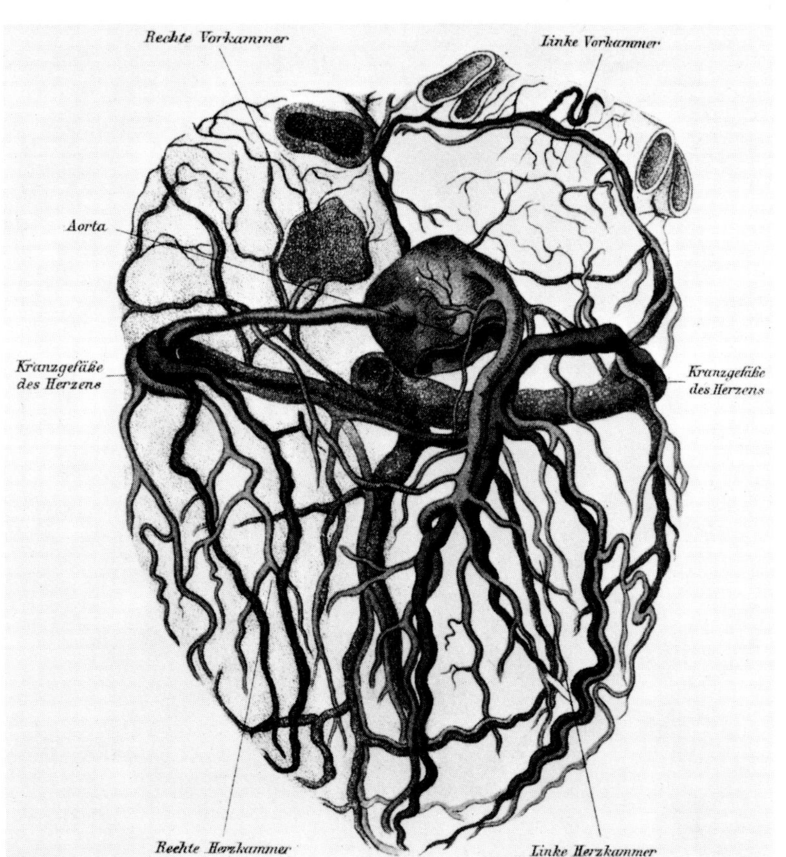

317

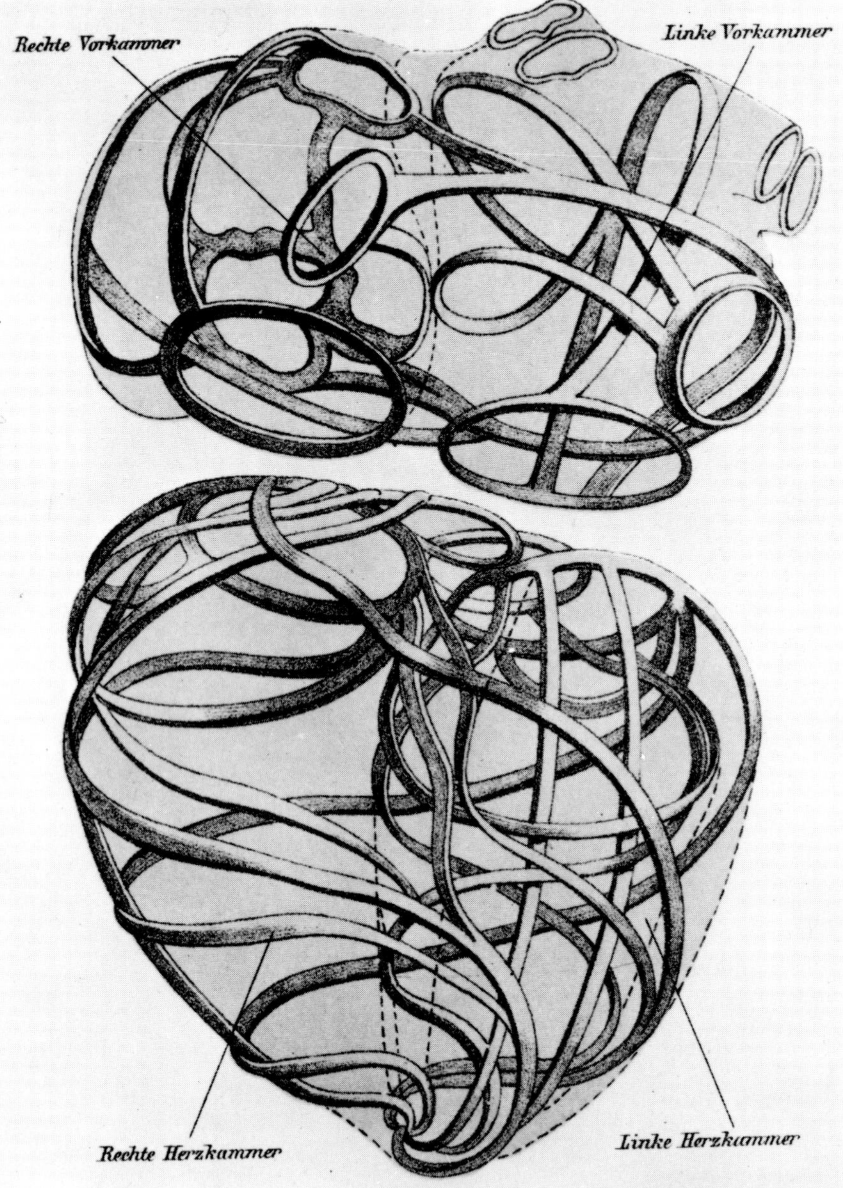

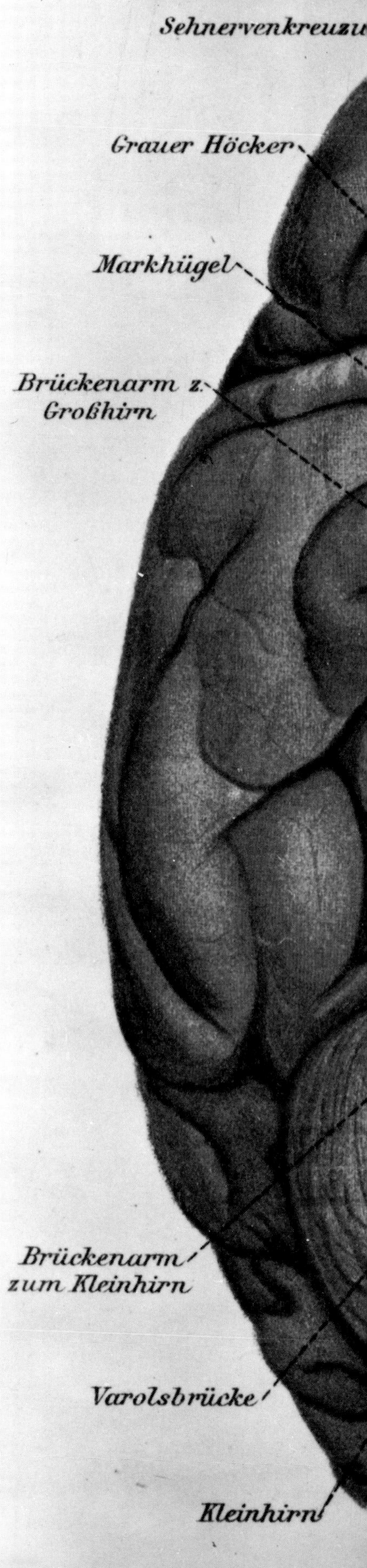

318. *A model of the heart as a muscular machine.* Since the time of William Harvey, it had been known that the heart is a pump. Here, in the late nineteenth century, was a "layout" of the musculature of the cardiac pump in abstract detail, illustrating how the machine worked.

319. *The gross anatomy of the brain.* The anatomy and physiology of the brain and the nervous system were objects of special attention in the nineteenth century. The gross anatomy of the brain had been fairly well worked out in the eighteenth and early nineteenth centuries; this view of the brain's underside was standard in medical texts of the nineteenth century.

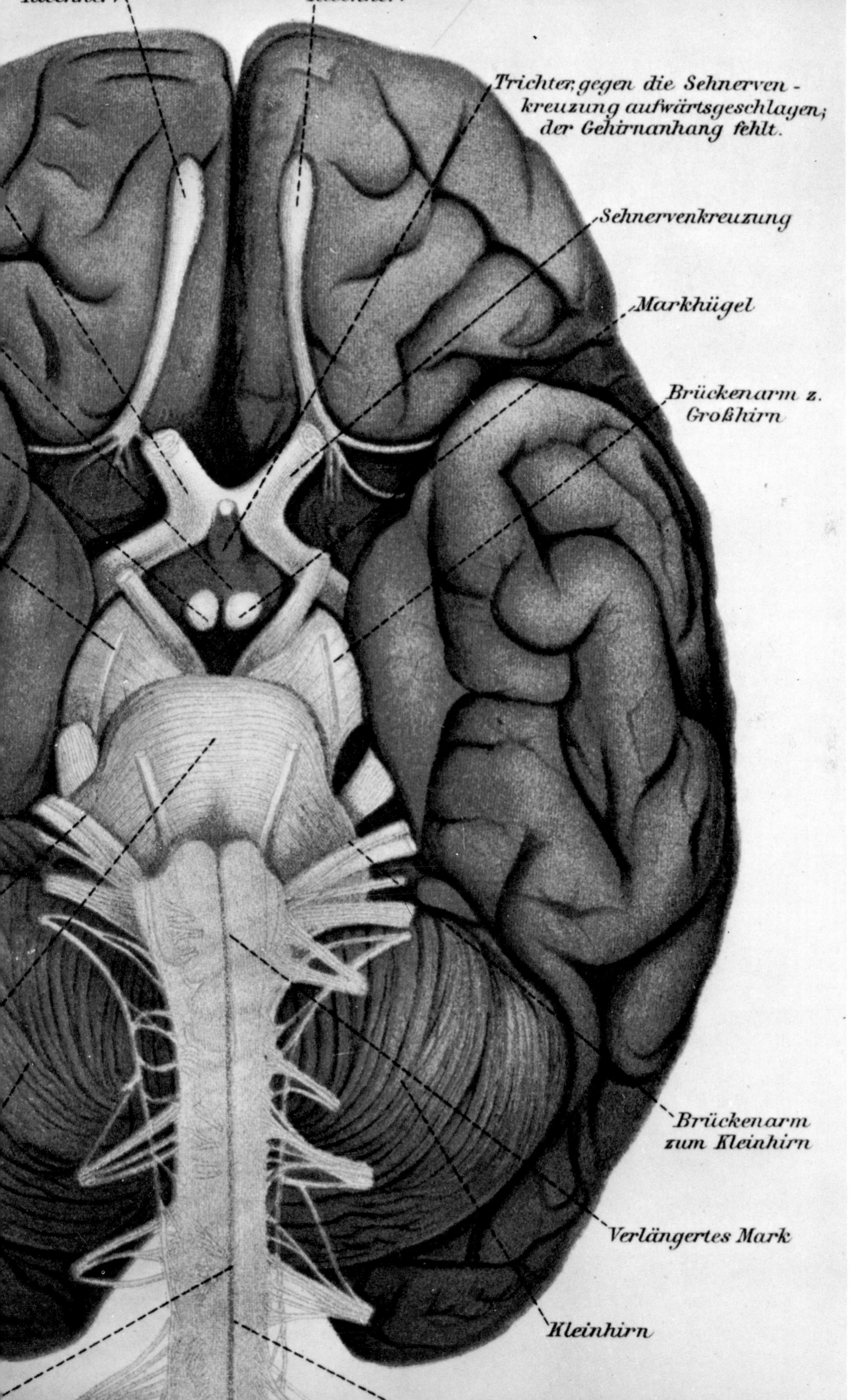
Riechnerv
Riechnerv
Trichter, gegen die Sehnerven-kreuzung aufwärtsgeschlagen; der Gehirnanhang fehlt.
Sehnervenkreuzung
Markhügel
Brückenarm z. Großhirn
Brückenarm zum Kleinhirn
Verlängertes Mark
Kleinhirn

320

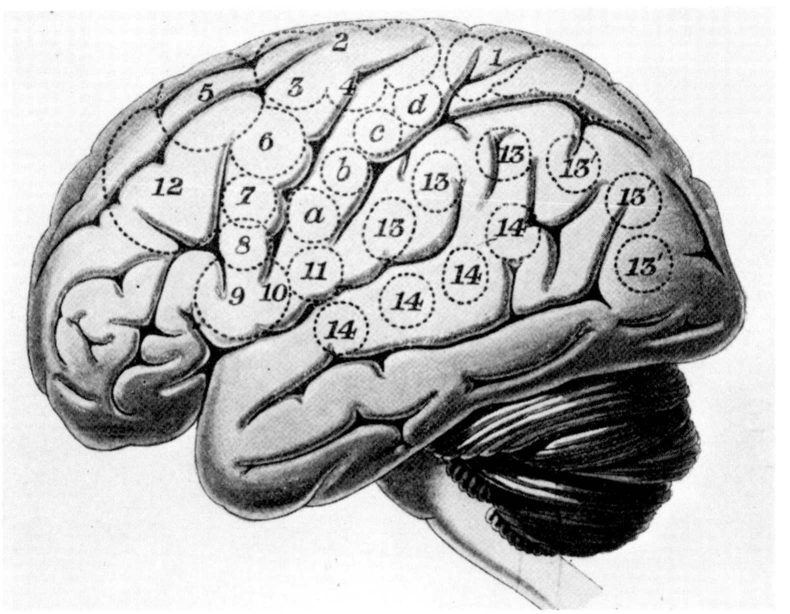

320. *Localization of brain function.* It was in the nineteenth century that the correlation of brain anatomy and bodily function was made in some detail. The areas marked had the following relation to function: 1. center for movement of the opposite leg and foot (the left side of the brain controls motions on the right side of the body and vice versa); 2, 3, 4. centers for complex movements of the arms and legs; 5. forward extension of arm and hand; 6. turning (supination) of hand and flexion of the forearm; 7, 8. elevation and depression of the angle of the mouth; 9, 10. movement of the lips and tongue; 11. retraction of the angle of the mouth; 12. movement of the eyes; 13, 13′. vision; 14, hearing; *a, b, c, d.* movements of the wrists and fingers.

321. *Brain function and skull formation.* The idea that certain areas of the brain are responsible for certain physiological functions dates back to the eighteenth century. The belief that these areas caused skull deformations, and that these deformations could be used as clues to character, intelligence, and aptitudes, was the basis of the nineteenth-century "science" of phrenology. Here the various areas of the skull are delineated and then correlated with such attributes as sexual desire (1), color vision (6), philosophical insight (21), and even theosophy (25).

322. *The development of the skull.* By focusing attention on the skull, phrenology had some beneficial side effects. This careful examination of the development of the human skull from infancy to maturity and old age, by the accomplished anatomist Robert Froriep, provides an interesting variation on the theme of the ages of man.

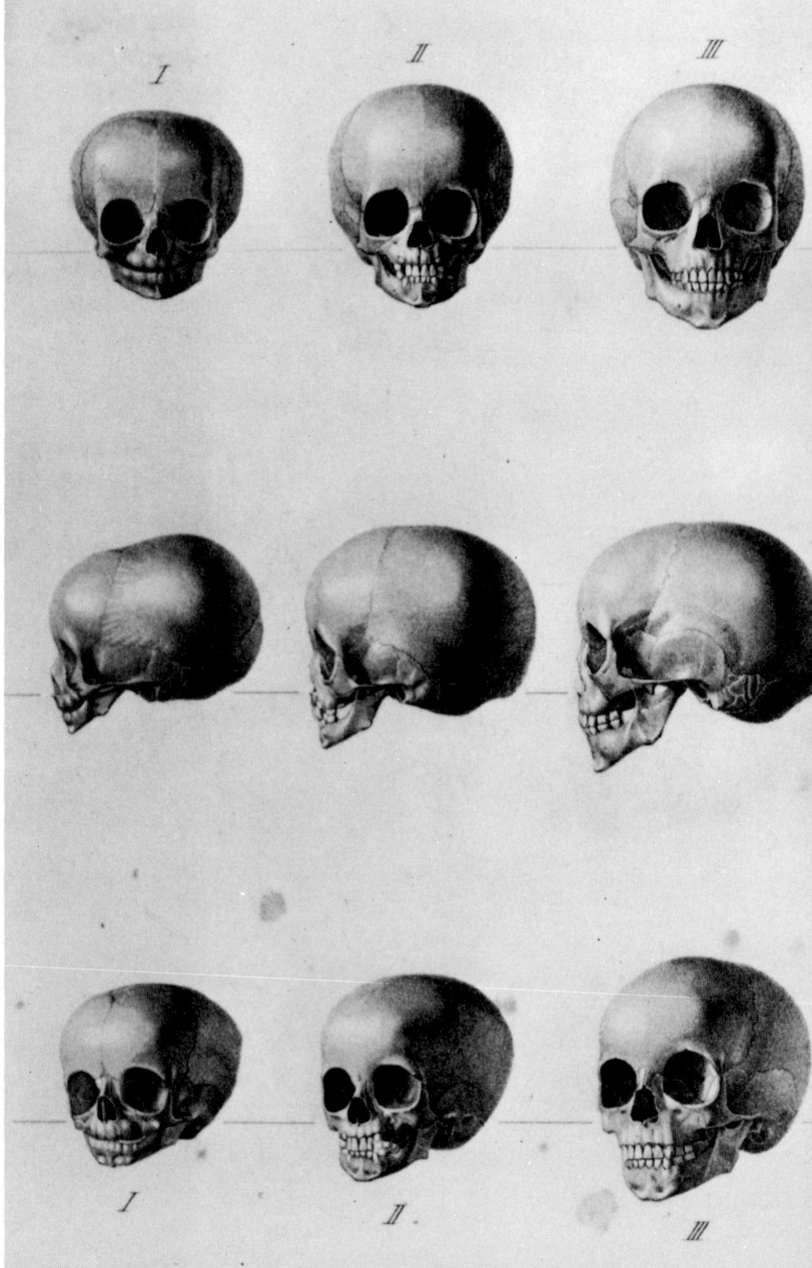

321

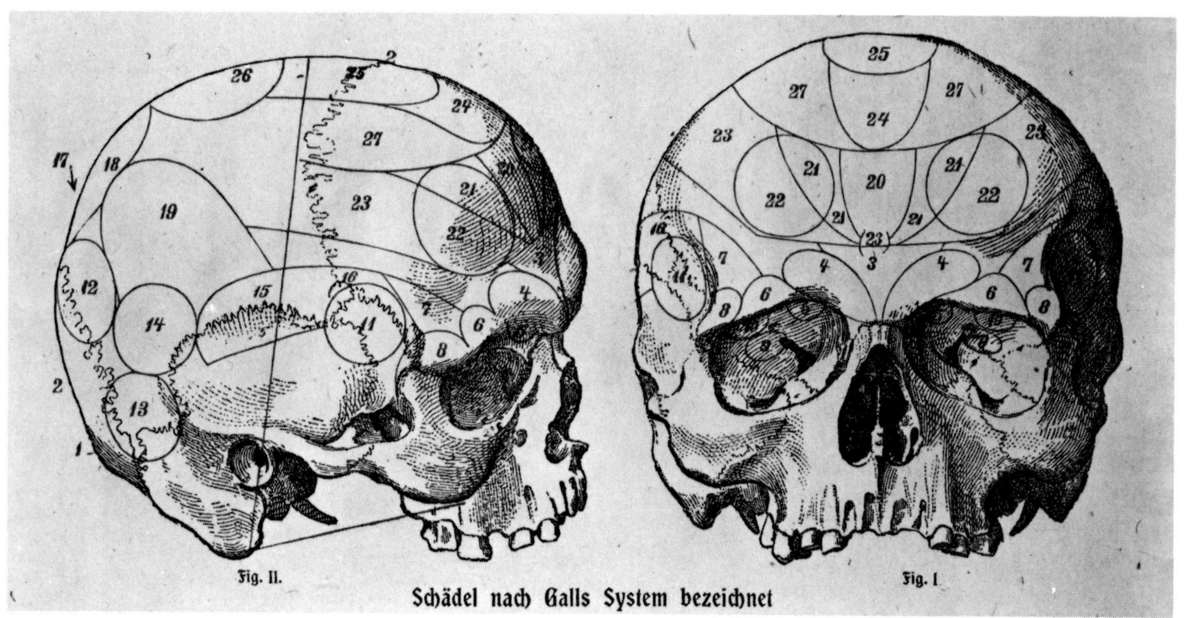

Schädel nach Galls System bezeichnet

322

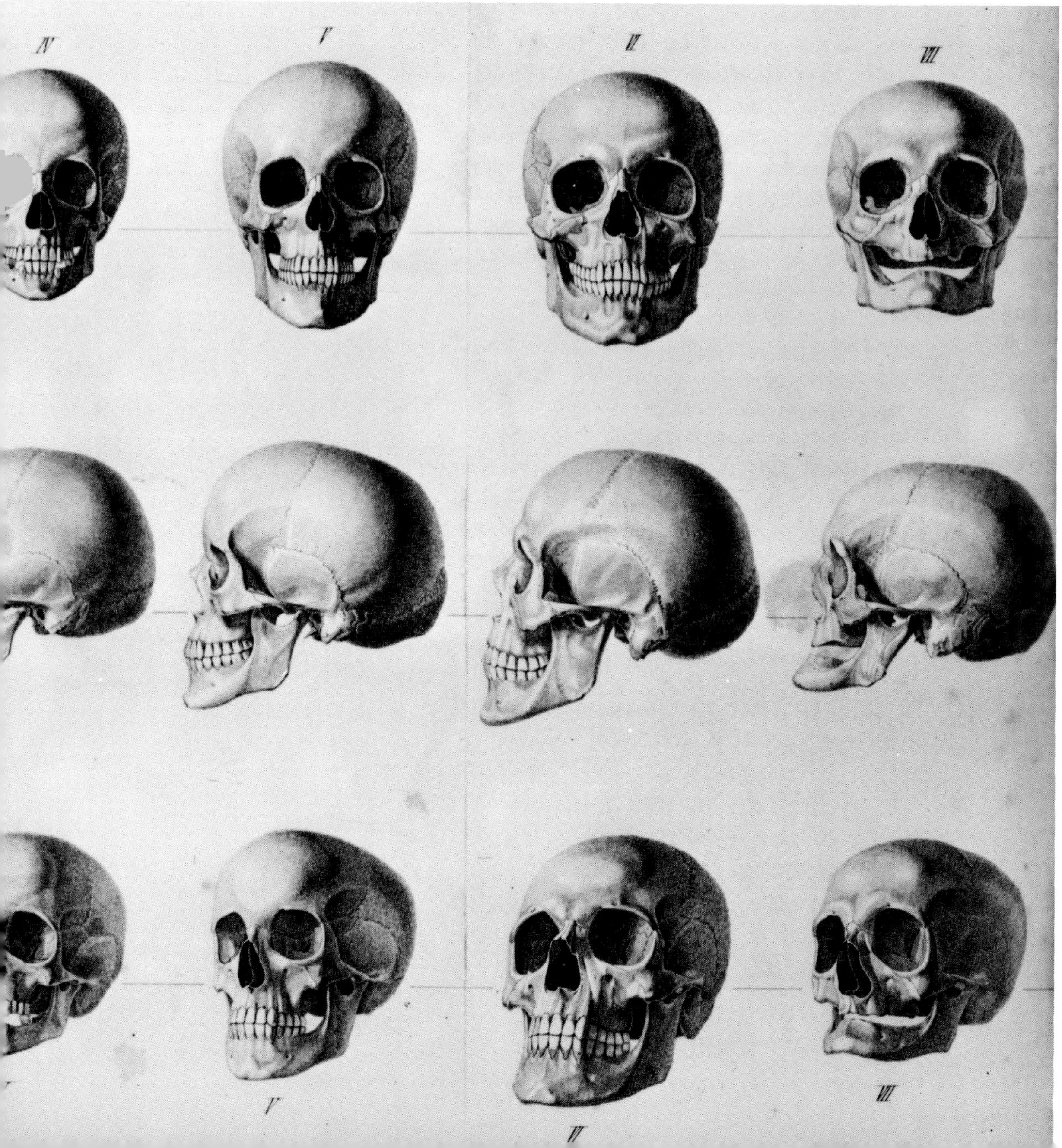

323

324

323. *From frog to man.* The phrenological movement called attention to minor variations in skull conformation and to the gradations that exist between different types of skulls. This illustration, drawn long before the time of Darwin, is interesting in that it illustrates how few steps are necessary to transform a frog into Apollo.

324. *The phrenologist.* Phrenology served the same purpose in the nineteenth century that psychological testing does today (and with about the same success). Ambitious parents took their children to the phrenologist to learn their aptitudes, so that their futures could be carefully and rationally planned.

325. *A phrenological map.* The science of phrenology was taken very seriously, as this German illustration indicates. Every aspect of human character could be read from the examination of the skull, and such charts as this were essential for the practitioner of the phrenological art.

326. *An old maid's skull phrenologized.* Phrenology lent itself marvelously to caricatures, some of which, like this example, were rather cruel.

325

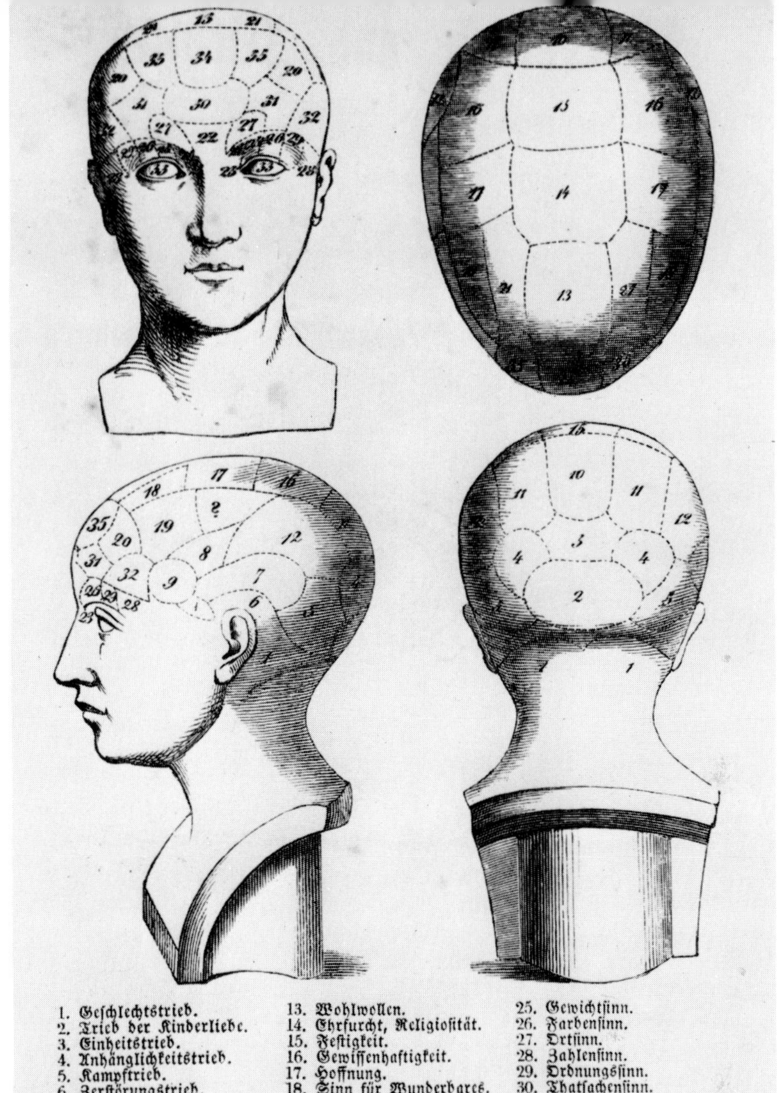

326

327

328

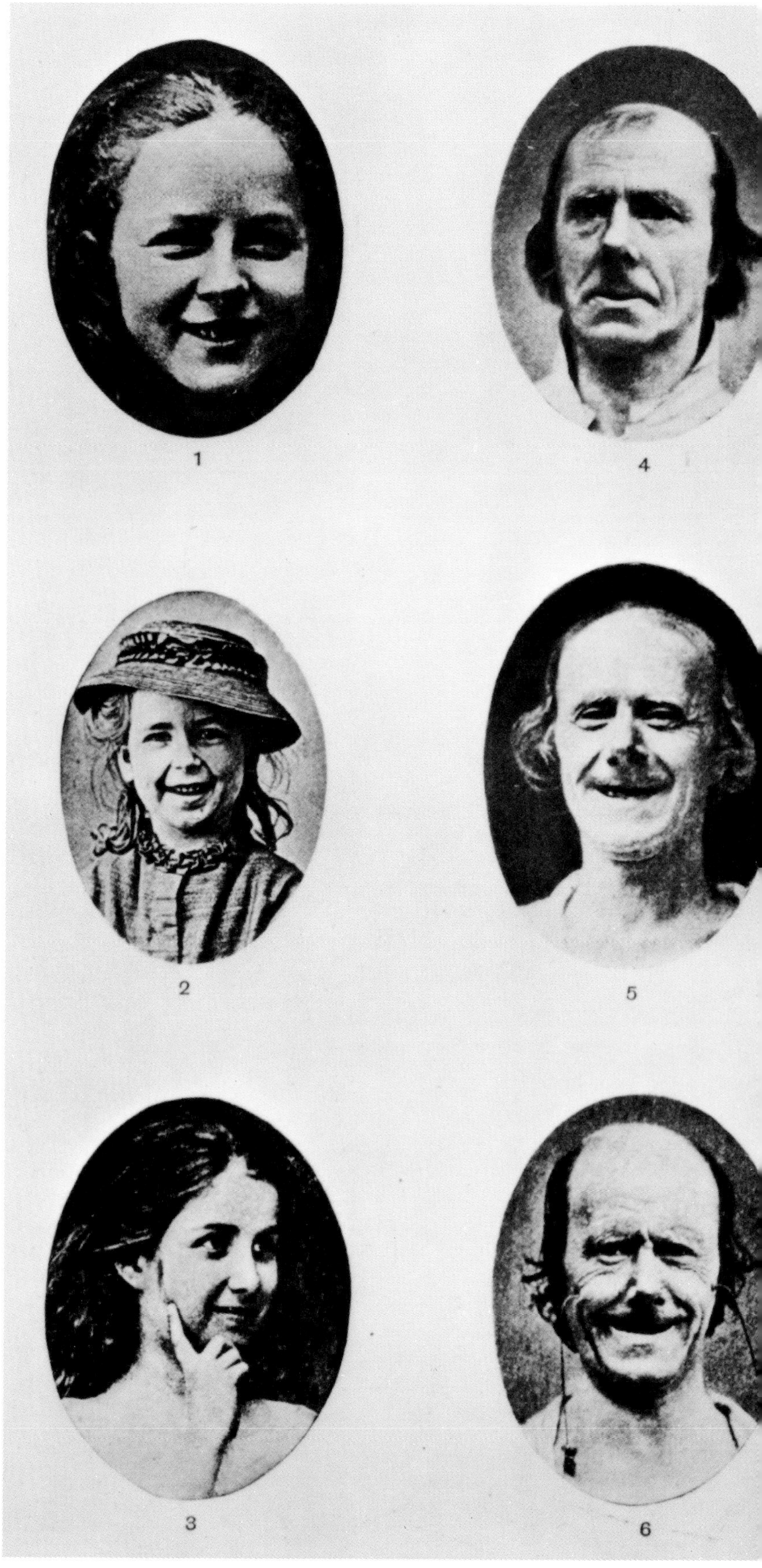

1 4

2 5

3 6

327. *The expression of mental states.* Phrenology was merely one attempt to get inside the human psyche in the nineteenth century. It was finally dismissed as a pseudo science, but other avenues remained to be explored. Charles Darwin's *The Expression of the Emotions in Man and Animals* endeavored to show how, under emotional stress, external physiognomy reflected the evolutionary history of species. Darwin used this picture of stark terror in a human being to illustrate physiological reactions to danger and their expression on the face.

328. *Pleasure and physiognomy.* The smile, Darwin showed, was a rather complex neuromuscular process that had been adapted from other activities in order to show pleasure. A smile could be artificially induced by the application of electricity to the proper facial muscles, as in illustration no. 6 here.

329. *Physiognomy and psychology.* The idea that a person's appearance reveals his or her character is as old as realistic art in Western civilization. The nineteenth century, characteristically, tried to build a science around it. The founding father of the science of physiognomy was the eighteenth-century anatomist Johann Kaspar Lavater. In his *Essays on Physiognomy*, he illustrated basic "types." Figure 1 at top shows "mildness of mind without powerful or enterprising strength"; Figures 3 and 4 represent "power and goodness, fortitude and condescension." At bottom, the attempt is made to reduce character to geometry.

330. *Head form and science.* The shapes of skulls continued to fascinate nineteenth-century scientists even after phrenology and physiognomy were shown to be scientifically invalid. Surely there was *some* meaning to be found in skulls. Adolphe Quetelet, one of the first biometricians, here laid out the basic mathematical grid for taking measurements of heads, with the clear hope that these measurements would prove to be correlated to other factors.

329

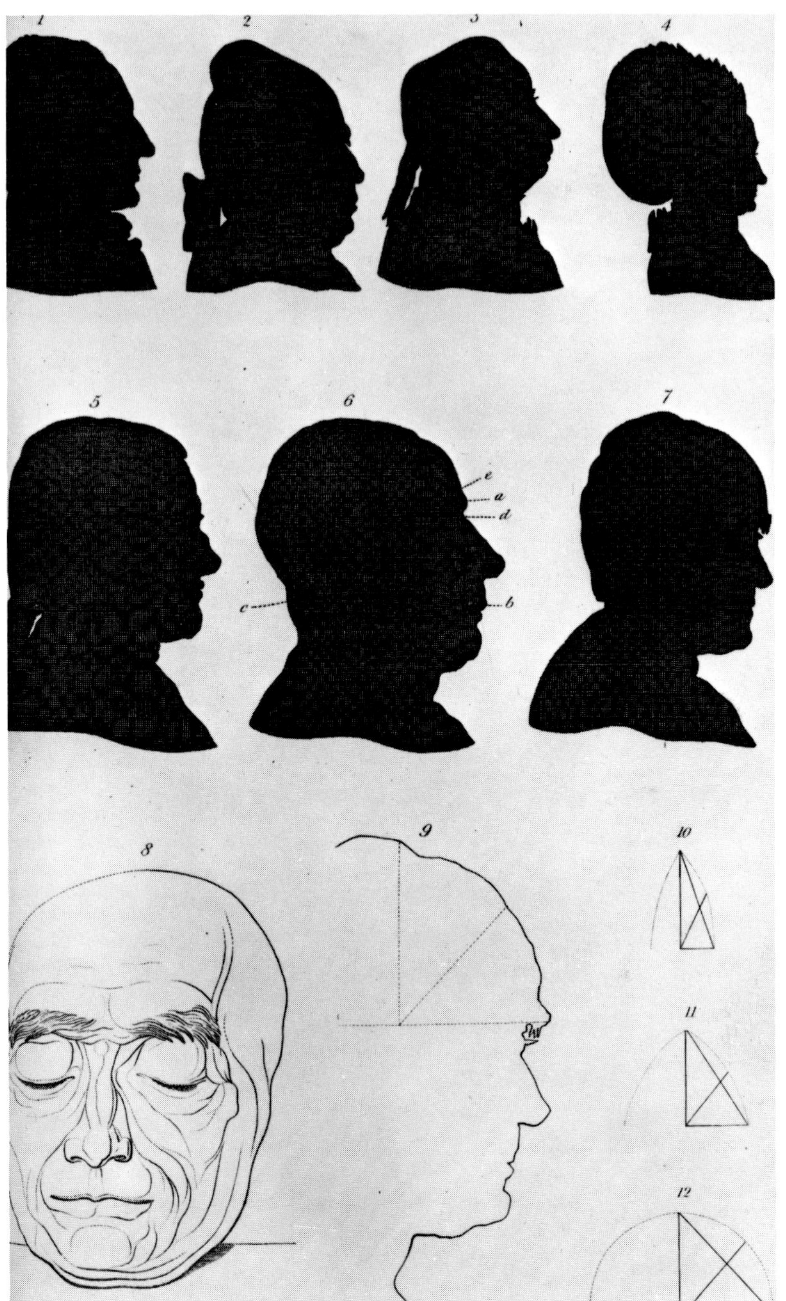

330

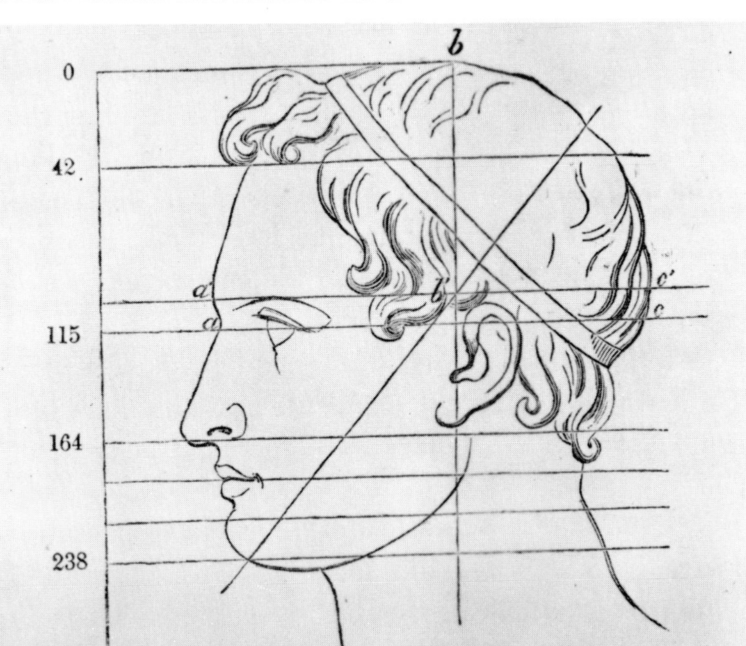

331

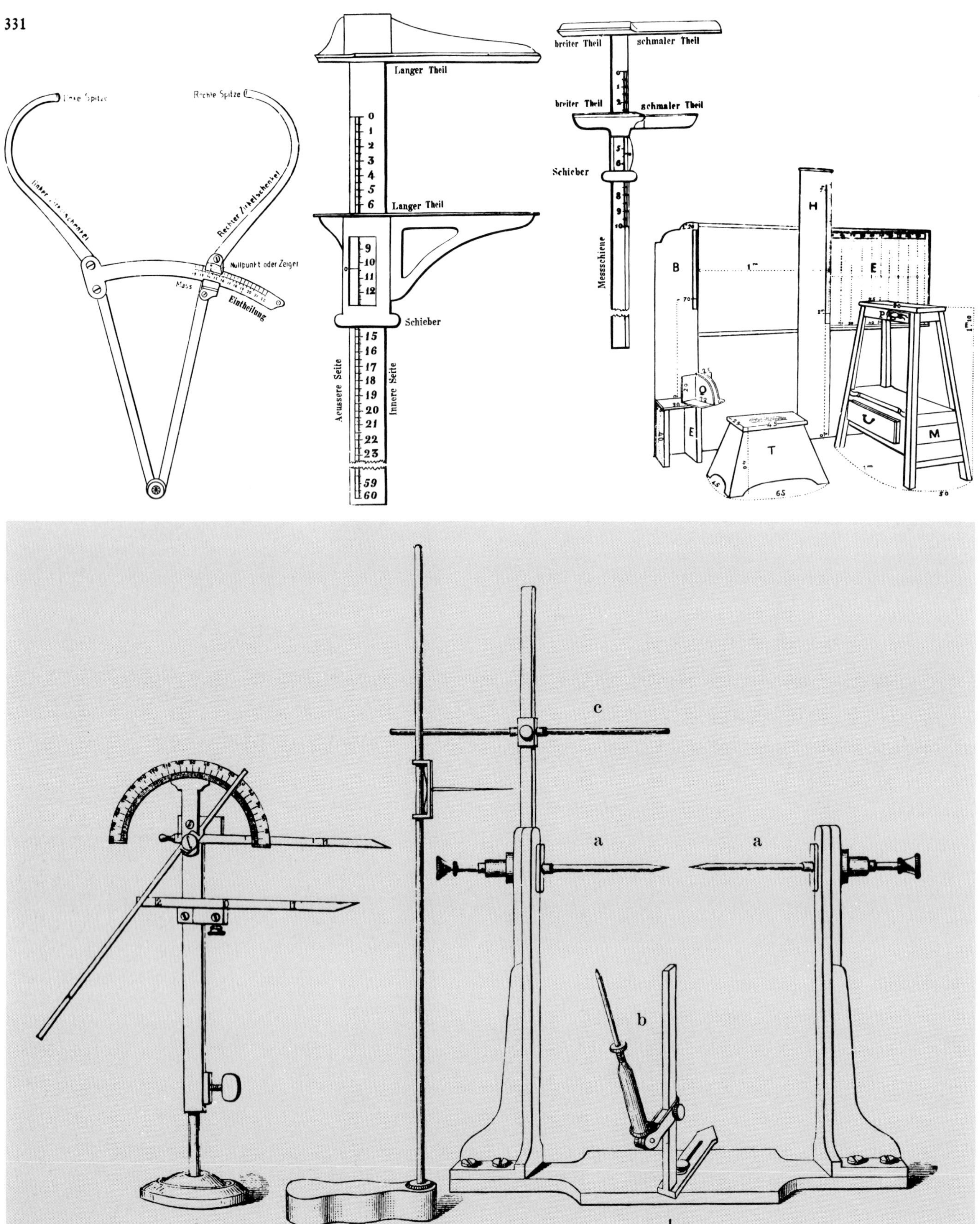
Rechte Spitze
Rechter Zirkelschenkel
Nullpunkt oder Zeiger
Mass
Eintheilung
Langer Theil
Langer Theil
Schieber
Aeussere Seite
Innere Seite
breiter Theil
schmaler Theil
breiter Theil
schmaler Theil
Schieber
Messschiene
H
B
E
P
Q
E
T
M
c
a
a
b
3
2
1

332

331. *Anthropometrical instruments.* Concluding that anthropological science was measurement, anthropologists developed instruments for measuring crucial human dimensions. The most important part of the body was the head, and the various calipers shown here in a German illustration were devised to measure its width, breadth, and height. At right is an array of standards for measuring other bodily features.

332. *The measurement of cranial angles.* As Quetelet and Lavater had suggested, the various angles in the skull, such as slope of forehead and jaw angle, seemed fundamentally important. This apparatus was used to determine these angles. The skull was fixed on the points of the instrument marked *a,a* and then the goniometer (3) was applied to the various points. The instrument at center (2) was used for measuring vertical distances.

333. *A cranial measurement.*

334. *Round heads and long heads.* From the data of anthropometry there emerged what appeared to be a basic fact. The "races" of man differed in skull conformation. On the left is the long-headed skull of an African Negro; on the right, the round-headed skull of a Finn. Long heads were christened dolichocephalic; round heads, brachycephalic. This division was considered vital to an understanding of the classification of human beings.

333

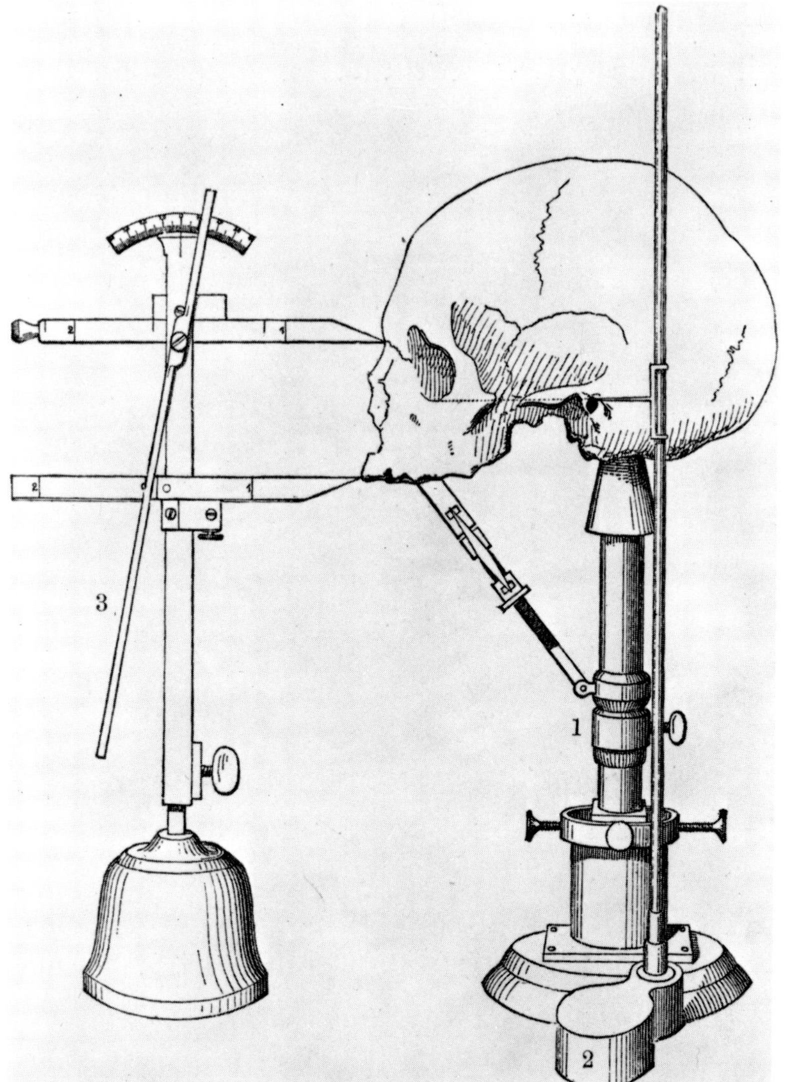

334

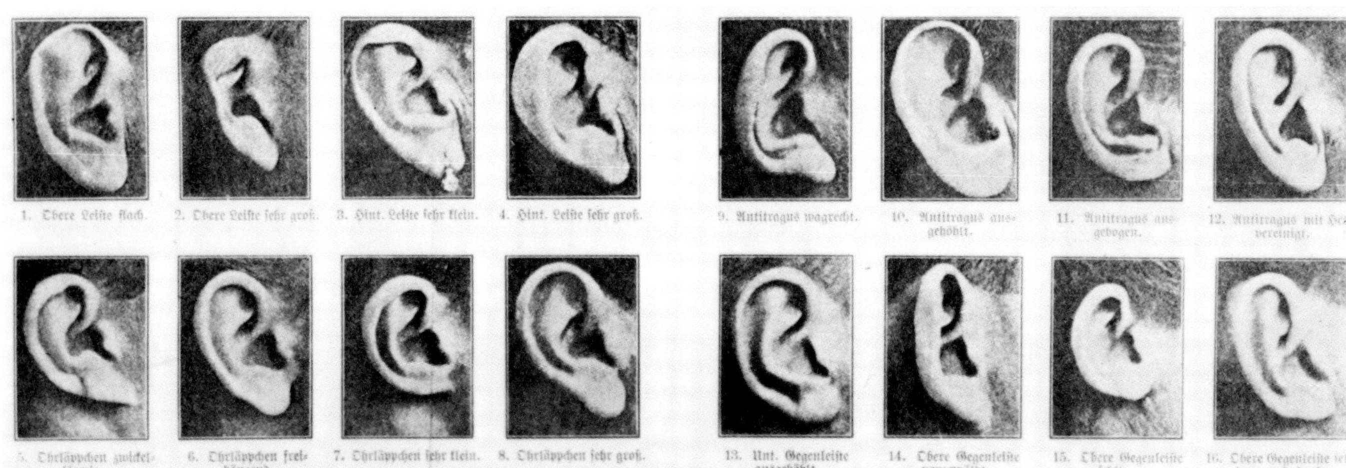

335. *Ears and individuals.* The search for individual characteristics was as keen as the search for racial traits. Ears, it was found by anthropometricians, are never exactly alike from one person to another. The attempt was made, therefore, to classify the different forms that occurred.

336. *Anthropometry and crime.* One of the more grisly crimes of the late nineteenth century involved the young lady shown above, who was suspected of having committed a murder while hypnotized. Here her "ear prints" are being taken by the "Anthropometrical Service" of the Paris police. This measurement and others served the same purpose that fingerprints do today.

337. *The measurement of man.* Emphasis on skulls gradually gave way to interest in all the measures of man. The new approach was designed to build up a large quantity of data that encompassed the totality of human physical and environmental characteristics. Here, note particularly the third column in this regard: rank and occupation are given as much attention as age or hair color. The absence of genitalia in the figure is an interesting commentary on Victorian taboos, even in anthropology.

336

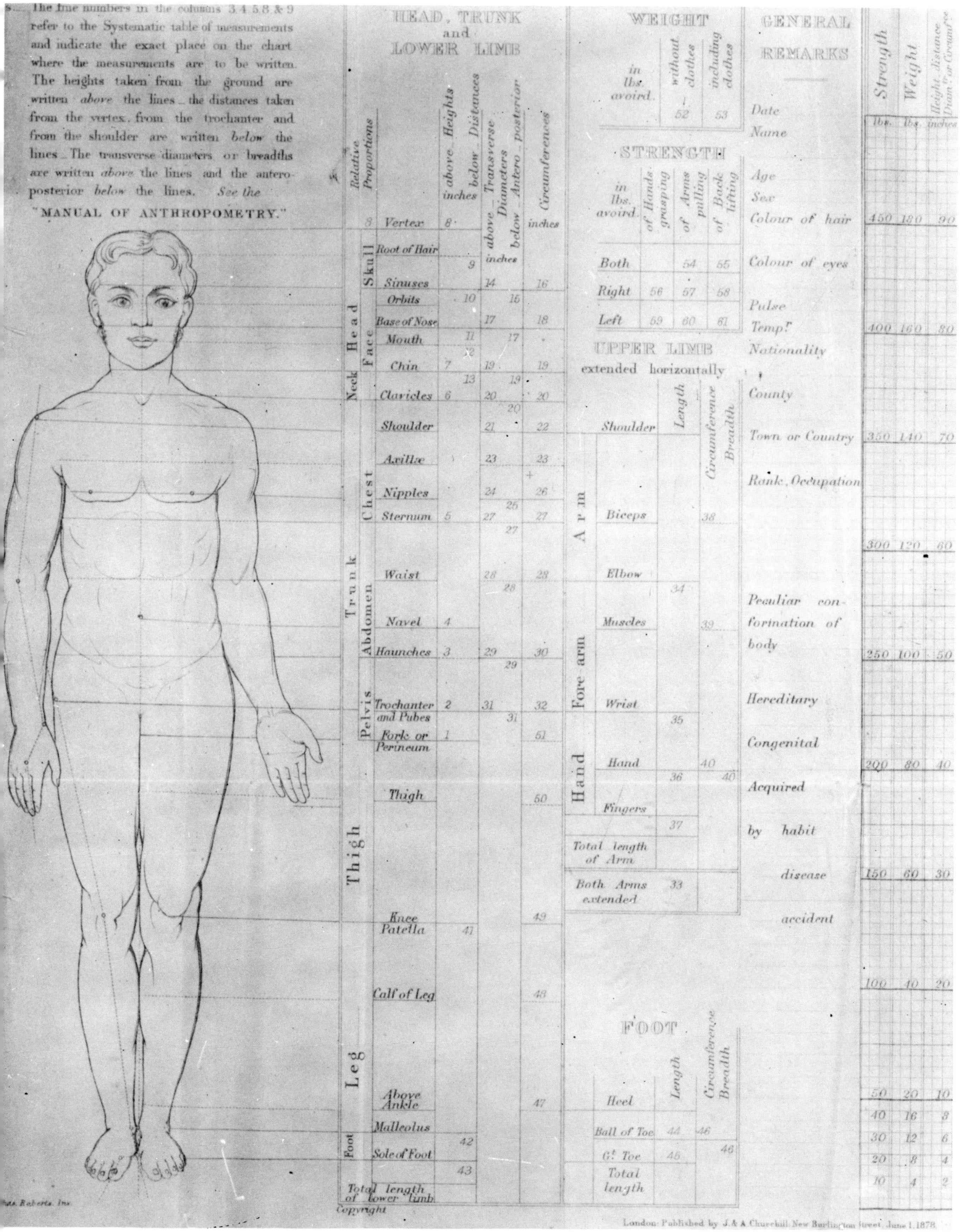
The fine numbers in the columns 3.4.5.8.&9 refer to the Systematic table of measurements and indicate the exact place on the chart where the measurements are to be written. The heights taken from the ground are written above the lines _ the distances taken from the vertex, from the trochanter and from the shoulder are written below the lines _ The transverse diameters or breadths are written above the lines and the antero-posterior below the lines. See the "MANUAL OF ANTHROPOMETRY."
HEAD, TRUNK and LOWER LIMB
Relative Proportions
above _ Heights
below _ Distances
inches
above _ Transverse Diameters
below _ Antero _ posterior
inches
Circumferences
inches
Vertex
Root of Hair
Sinuses
Orbits
Base of Nose
Mouth
Chin
Clavicles
Shoulder
Axillæ
Nipples
Sternum
Waist
Navel
Haunches
Trochanter and Pubes
Fork or Perineum
Thigh
Knee Patella
Calf of Leg
Above Ankle
Malleolus
Sole of Foot
Total length of lower limb
Skull
Face
Head
Neck
Chest
Abdomen
Trunk
Pelvis
Thigh
Leg
Foot
WEIGHT
in lbs. avoird.
without clothes
including clothes
STRENGTH
in lbs. avoird.
of Hands grasping
of Arms pulling
of Back lifting
Both
Right
Left
UPPER LIMB
extended horizontally
Length
Circumference
Breadth
Shoulder
Biceps
Elbow
Muscles
Wrist
Hand
Fingers
Total length of Arm
Both Arms extended
Arm
Fore arm
Hand
FOOT
Length
Circumference
Breadth
Heel
Ball of Toe
Gt. Toe
Total length
GENERAL REMARKS
Date
Name
Age
Sex
Colour of hair
Colour of eyes
Pulse
Tempr.
Nationality
County
Town or Country
Rank, Occupation
Peculiar conformation of body
Hereditary
Congenital
Acquired
by habit
disease
accident
Strength
Weight
Height, distance Diam. or Circumference
lbs. lbs. inches
Copyright
London: Published by J. & A. Churchill New Burlington Street, June 1.1878.

338

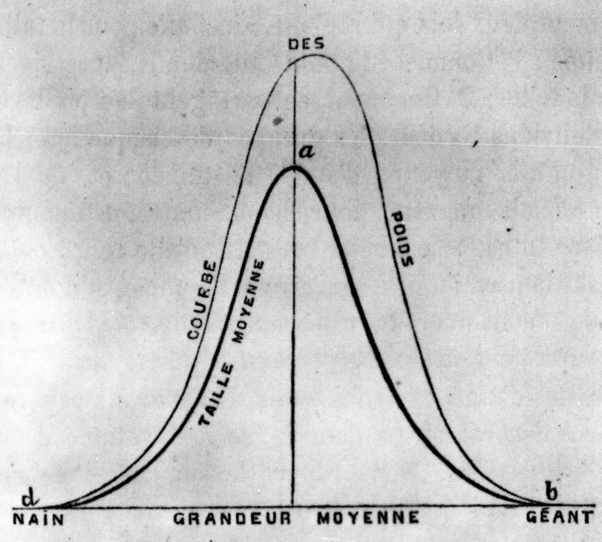

339

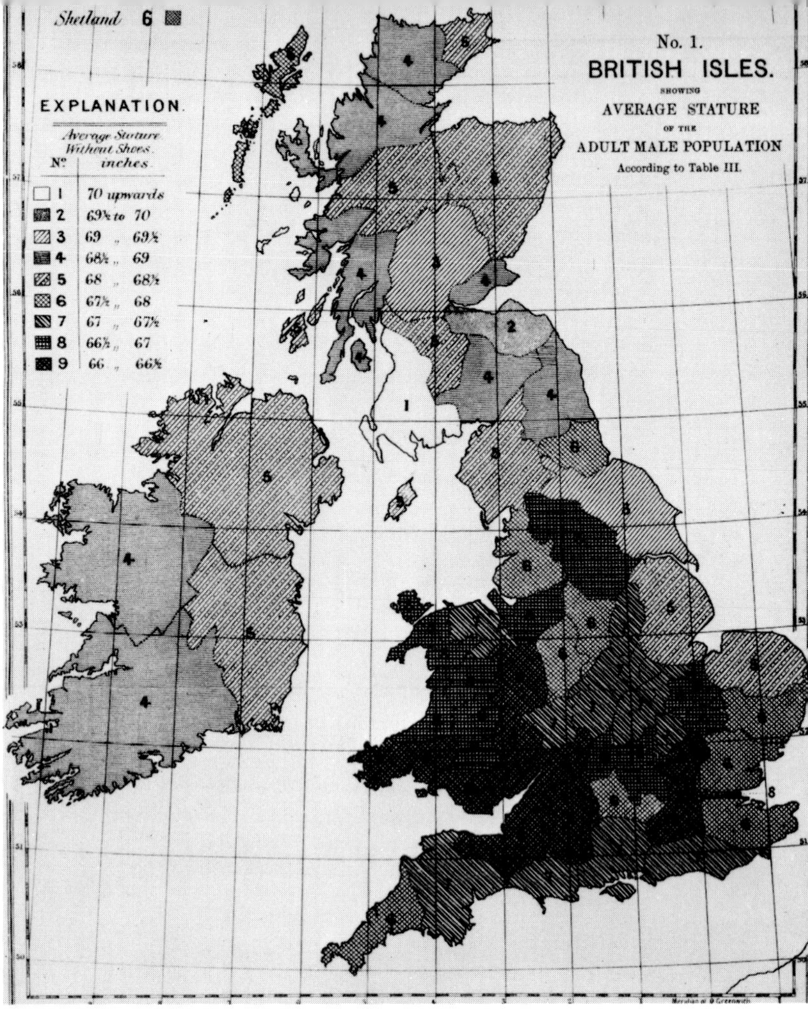

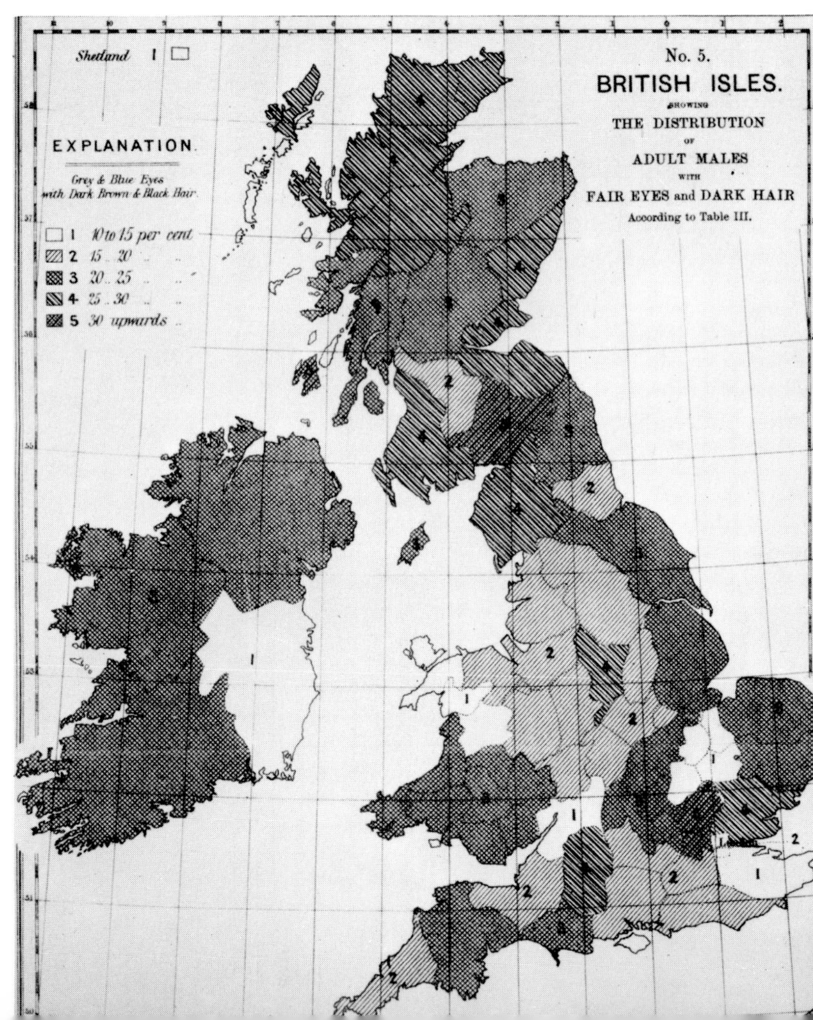

338. *Distribution of height and weight.* The famous "bell curve" was made familiar through nineteenth-century anthropology. As we would expect today, height and weight are distributed throughout the general population according to the laws of probability, from dwarfs on the left, in this chart by Quetelet, to giants on the right. This showed that some human characteristics could be quantified.

339. *The geographical distribution of height.* Height is hereditary. Height also varies geographically, as any British visitor to France recognizes immediately. The geographical distribution of heights, therefore, might be expected to shed some light on racial and migration patterns of mankind. Hence, such maps as this example. Note how generally small in stature the population of the British Isles was in 1878. South of the Firth of Clyde, however, the average height was 70 inches or more.

340. *The distribution of adult males with fair eyes and dark hair.* Other characteristics, or combinations of characteristics, besides height and weight could be mapped. Shown here is the distribution of blue or gray eyes with brown or black hair in Great Britain and Ireland. No firm or really useful conclusions can be drawn from these data, but they represent a tribute to anthropological persistence.

The Races of Man

Much of anthropometry was inspired by the hope that precise measurements of large numbers of individuals would permit the races of man to be scientifically described and separated. The problem of races was particularly acute throughout the entire nineteenth century. The greatest expansion of slavery in the United States took place after Eli Whitney's invention of the cotton gin at the beginning of the century. Morally sensitive slaveholders could salve their consciences if they could prove to their own satisfaction that Negroes were not members of the same human race as they. Abolitionists, of course, were intent on proving exactly the opposite. Biblical scholars could enter this increasingly acerbic debate by citing Adam and Eve as evidence for the unity of the races of man, only to have Ham, Shem, and Japeth hurled into their teeth as the God-given progenitors of separate peoples with one-third of the resultant races condemned forever to be hewers of wood and drawers of water. Later in the century, political developments in Europe stimulated political racism to explain the collapse of cherished institutions and the political decline of privileged segments of society. Gobineau propounded a theory of the rise and fall of states based on the notion that civilizations arose when two or more races mingled and stimulated each other's special talents. But the process of mixing ended in decline, because the talented were swamped by the inferior elements in the mixture. Modern France, he thought, was doomed to lose its preeminence under the weight of the Latin or "Gallo-Roman" population visibly decadent in Italy and elsewhere along the Mediterranean. The Slavs were not merely a separate culture from Westerners desiring political independence, but a separate race from Westerners whose peculiar qualities had their origin in racial roots. Finally, as the Western world expanded into the rest of the world, Western man made intimate contact with strange and new peoples with strange, new, and bizarre ways. Could the bewildering variety of languages, or even of sounds that men could make, be the attribute of a single race or was it not necessary to insist upon essentially different racial origins of the various peoples of the world just to understand their linguistic, cultural, and political variety? And, of course, if it could be shown that some of the new races were "inferior," then it was not so difficult to preach their annihilation if they stood in the way of Western expansion.

The scientific study of race, then, was not conducted in the calm atmosphere of academe but in the maelstrom of political, economic, and cultural struggle. If the study of human races made progress in spite of much acrimony, it was because the embattled ethnologists gathered abundant data about the varieties of mankind and did so with considerable accuracy. But it was not until the turn of the century that a new generation of scientists, including Franz Boas, revolutionized the study of anthropology by developing the notion of culture and abandoning the search for "proofs" (of inferiority or superiority) in physical traits such as the color of hair and eyes or the measurements of skulls (cranial index).

One of the more striking aspects of the history of the human species in the nineteenth century was the enormous expansion of the number of facts available. At the beginning of the century, contact with other peoples was not at all widespread and much of what was "known" in Europe was based upon hearsay rather than close observation. The illustrations reproduced here from Silby's *Magazine of Natural History* are extremely informative in this regard. Without exception, they are highly idealized portraits of various types of humanity from different parts of the world. The women are uniformly beautiful and the men handsome. Only their strange color, or clothing, or weapons define them as different from Europeans. This is library anthropology, without firsthand contact, based upon the reports of travellers. As the explorers of the nineteenth century opened up the various territories of the world, more accurate depictions of their inhabitants could be made and spread through the West. Captain Fitzroy's Fuegian is ugly and probably accurately drawn. Through photography the crude or idealized portrayal of the unfamiliar could be eliminated to a great extent and one could actually tell what a native of a strange land looked like.

As the number of native "races" multiplied, the whole concept of race began to lose its sharp outlines. This, together with Darwin's work that stressed the wide variations that could exist within the same species, tended to discredit the notion of separate races of mankind. Although the idea of separate races of man was to survive the nineteenth century, it was during this century that sufficient anthropological data was collected to provide its opponents with some powerful arguments.

341

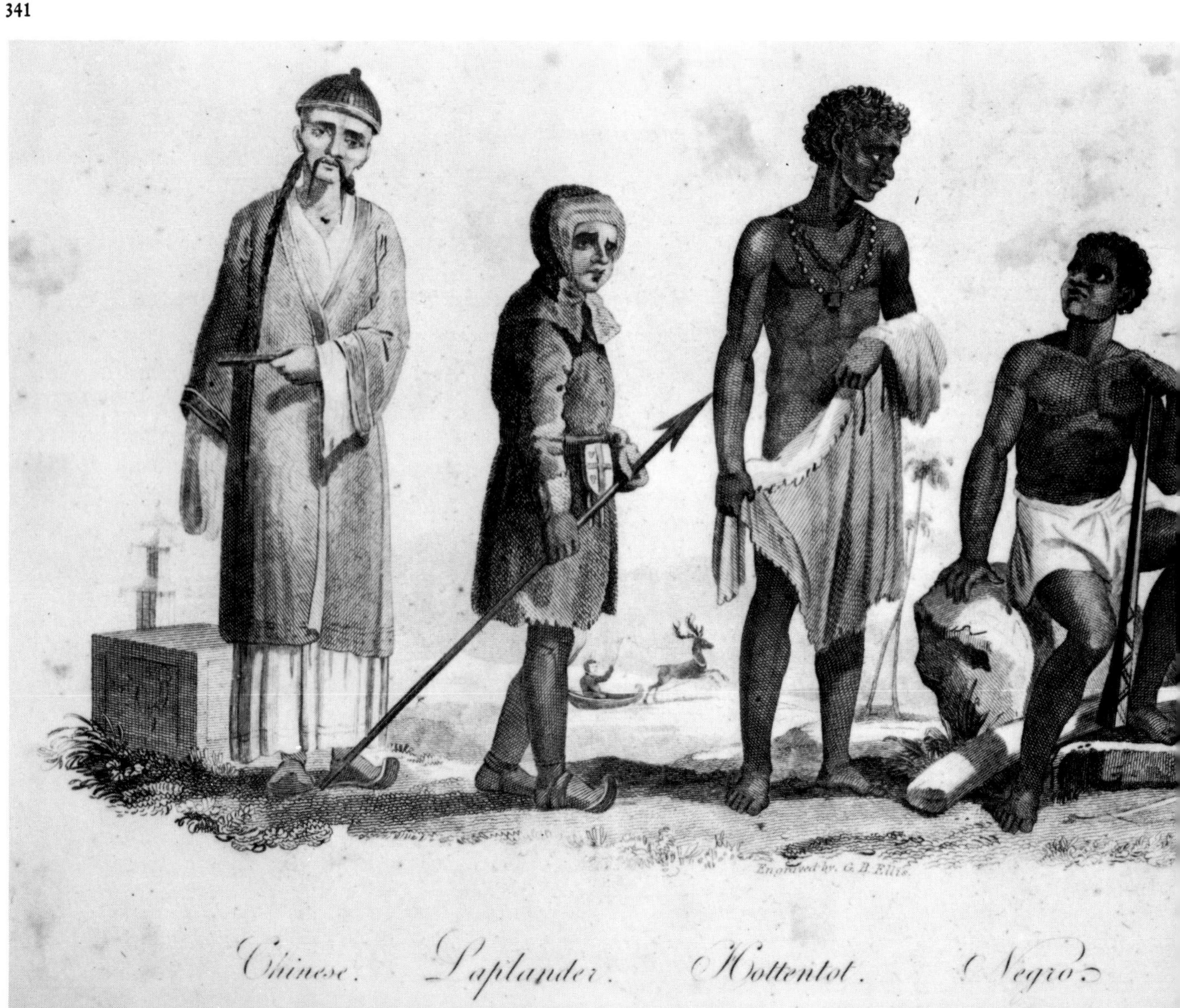

341. *The races of man.* Behind the anthropometrical movement was concern for distinguishing the races of man. It was hoped that racial differences would show up plainly in measurements. Differences had fascinated Europeans since they first set out to explore the world. In the eighteenth century, *History of the Earth* by Oliver Goldsmith, poet and polymath, provided the general European public with five examples of alien races.

342. *Five races later in the nineteenth century.* Armchair anthropology gave way to observation, and even the popular press reflected the difference. These examples from a popular work published in 1875 have considerably greater "reality" than the pictures published as late as 1830 to accompany Goldsmith's book.

342

1. MONGOLIAN. 2. MALAY.
3. CAUCASIAN. 4. NEGRO.
5. AMERICAN INDIAN.

343 344

345 346

347 348

343. *An early survey of the races of man.* When the nineteenth century began, Europeans had made contact with most of the major cultures and societies of the world. There was intense curiosity about these peoples, and a large literature grew up to satisfy it. The level of accuracy was not always the highest. Some of the portraits reproduced here and in Figures 344–348 were made from life; others were highly romanticized versions based on hearsay. This picture of an Australian aborigine, said to be "An Exact Portrait," does appear to have been drawn from life.

344. *A Circassian and a Georgian woman.* This probably is not drawn from life, but from reports. Circassian women were famous for their beauty and for their voluptuous figures, which may account for the Circassian woman's pose.

345. *A Senegal woman.* This is an idealized and romanticized portrait. Only color, hair, and costume distinguish this from a standard, academic European nude.

346. *A Senegal male.* Another romanticized picture that obviously was not drawn from life.

347. *A Hottentot woman.* The European perception of an African woman in 1800 was exotic but familiar. Except for her dark skin, the woman looks like a neighbor.

348. *An American Indian.* This almost Homeric figure is impossible to identify as a member of any known American tribe. It seems to have been a creation of the artist's imagination.

349. *A native of Tierra del Fuego.* Natives were observed and depicted by the scientific expeditions that went out during the nineteenth century to explore the globe. There is little doubt that this is a portrait of an actual inhabitant of Tierra del Fuego, drawn by the artist aboard HMS *Beagle.*

350. *A Sioux chief.* The photograph was of fundamental importance in providing accurate representations of what had been observed. This Sioux chief should be compared with the Indians shown in Figures 341 and 348.

351. *A young girl in Tierra del Fuego.* Compare this with Figure 349.

352. *A Welsh type of Montgomeryshire.* The camera offered the possibility of recording supposed racial types as "standards" against which other races could be compared, for the nineteenth century firmly believed in the existence of races. This is a nineteenth-century version of what a "typical" Welshman looked like.

353. *Racial types: an Algerian Jew.*

349

350

351

352

353

354

355

356

357

354. *Racial types: Swazi-Bantu women.* Compare this photograph with Figure 347 for some appreciation of how photography influenced the ability of Europeans to see and understand other peoples in the world.

355. *Racial types: Kurumba of the Nilgiri Hills in southern India.*

356. *Racial types: Agni Negro of Krinjabu in West Africa.*

357. *Racial types: Mentawai Islanders of the East Indies.*

358. *Racial types: Nias Islanders of the East Indies.*

359. *The races of man at the end of the nineteenth century.* As the preceding figures have graphically illustrated, the "types" of mankind were multiplied throughout the nineteenth century as knowledge of man increased. By the end of the century, the "races" of man had been expanded almost fivefold. The concept of race was still strongly held, but its anthropological basis was being corroded by the very number of races that had to be accounted for as separate entities.

358

359

IV

The Living World

The Variety of Life

The discovery of new living forms followed the course of the discovery of new geographical areas in the nineteenth century. There was a large public interested in the new lands and their new inhabitants and the number of works devoted to both geographical and biological descriptions mounted rapidly. The standards were not always of the highest, nor were the methods of presentation. One of the most popular works describing the surface of the globe and its flora and fauna was Oliver Goldsmith's *History of the Earth* which, although written in the eighteenth century, continued to sell well up to the last quarter of the next century. There was little method in Goldsmith's presentation except for such broad categories as mammals, insects, and so on. What fascinated his readers was obviously just a glimpse at the creatures themselves. Beetles and spiders and bears and baby alligators were all increasingly removed from the urban dwellers who made up Goldsmith's audience. More exotic creatures were guaranteed to capture the public fancy as well. The duckbill platypus was featured, for example, in a number of popular journals to illustrate some of the bizarre creatures that could be found in equally bizarre places. Deep-sea fishes also seem to have exercised a strong fascination for the European reading public, for the great illustrated weeklies periodically featured pictures of the peculiar things that inhabited the deepest parts of the ocean.

The expansion of knowledge of species had an obvious effect upon the attempts to sort out plants and animals and put them into some kind of order. It put taxonomy under enormous pressure, relieved only by Darwin's evolutionary theories. Armed with the idea of natural selection, biologists could then make sense of the many and various adaptations to be found in the strange new specimens brought back by travellers or scientists.

360

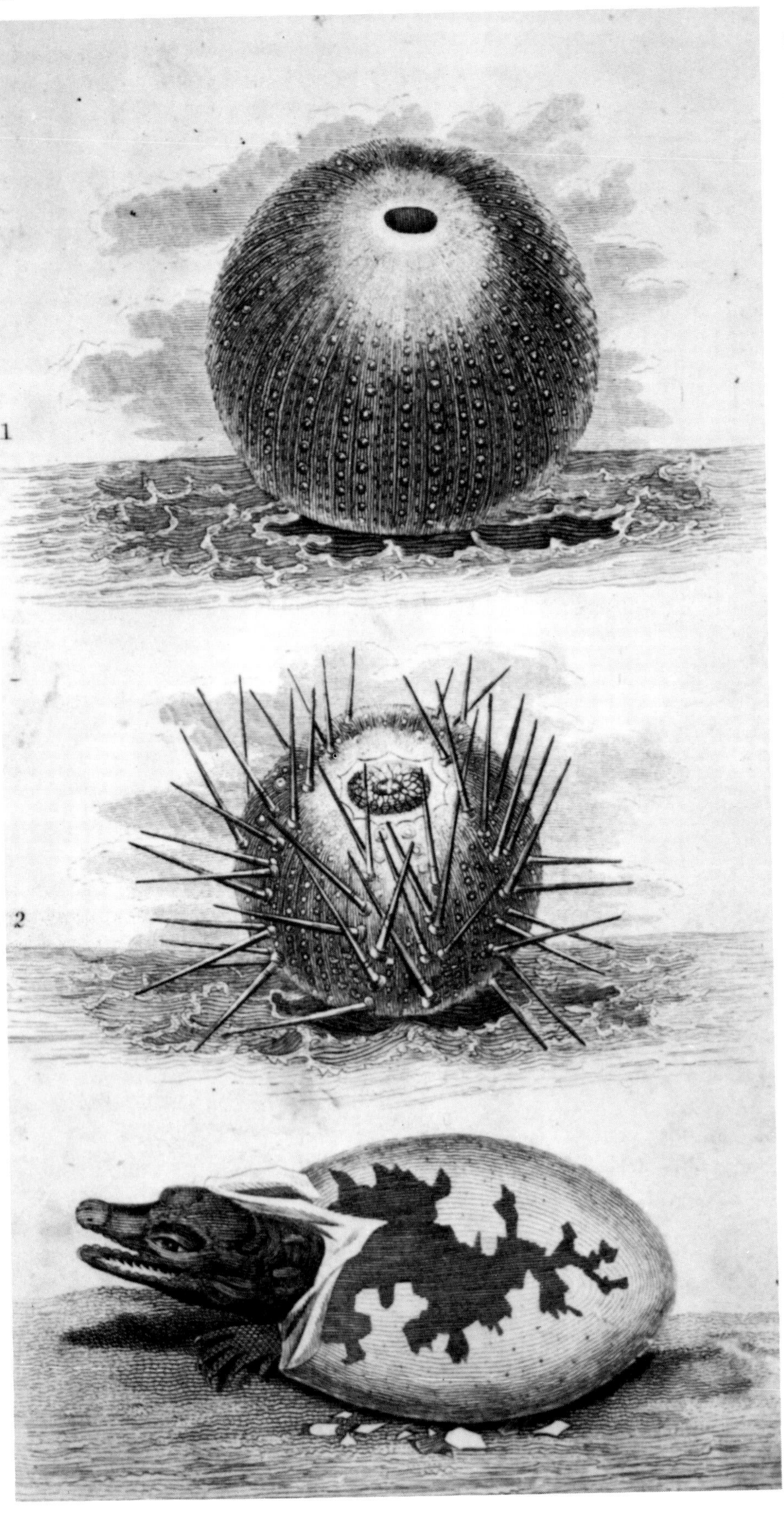

360. *The creatures of the earth: the view from the late eighteenth century.* There existed a large popular audience throughout the eighteenth and the nineteenth centuries for descriptions of living creatures on the earth and beneath the seas. One of the most popular works to reach this audience was *History of the Earth* by the British poet Oliver Goldsmith. This work went through 17 editions in Great Britain alone between its first publication in 1774 and 1876. The verbal descriptions were simple and anecdotal; the plates were clear, though often inaccurate. Still, they did give some idea of the enormous variety of living forms. This plate, showing sea urchins and a hatching alligator, illustrates Goldsmith's tendency to put together things that actually have little in common. Yet, there is a point: sea urchins are not very interesting, but a hatching alligator does catch the eye.

361. *A mixed bag of creatures.* The relationships among the creeping and crawling things of the earth were not clear when Goldsmith wrote. Hence, various creeping and crawling animals were put together just because they did creep and crawl.

362

362. *Some mammals.* Goldsmith's uncertainty about relationships among species is nicely shown here. With a rat, a dormouse, and a marmot, he places a rather odd beast labeled vaguely as "Animal of the Bear Species."

363. *Larger mammals.* The hippopotamus particularly fascinated Europeans of the late eighteenth century. Goldsmith's example looks as if it had been drawn from descriptions. Note the odd feet and toes, and the proportion of the head to the body.

364

364. *The hippopotamus.* Another popular view was this one, which appeared in 1800. The gentle "river horse" appears as a ferocious beast with dentition worthy of the saber-toothed tiger.

365. *The hippopotamus as described by a scientist.* Georges Cuvier provided the scientific world and the public with a careful depiction of the hippopotamus. The peculiar dentition is accurately drawn, as is the animal as a whole.

365

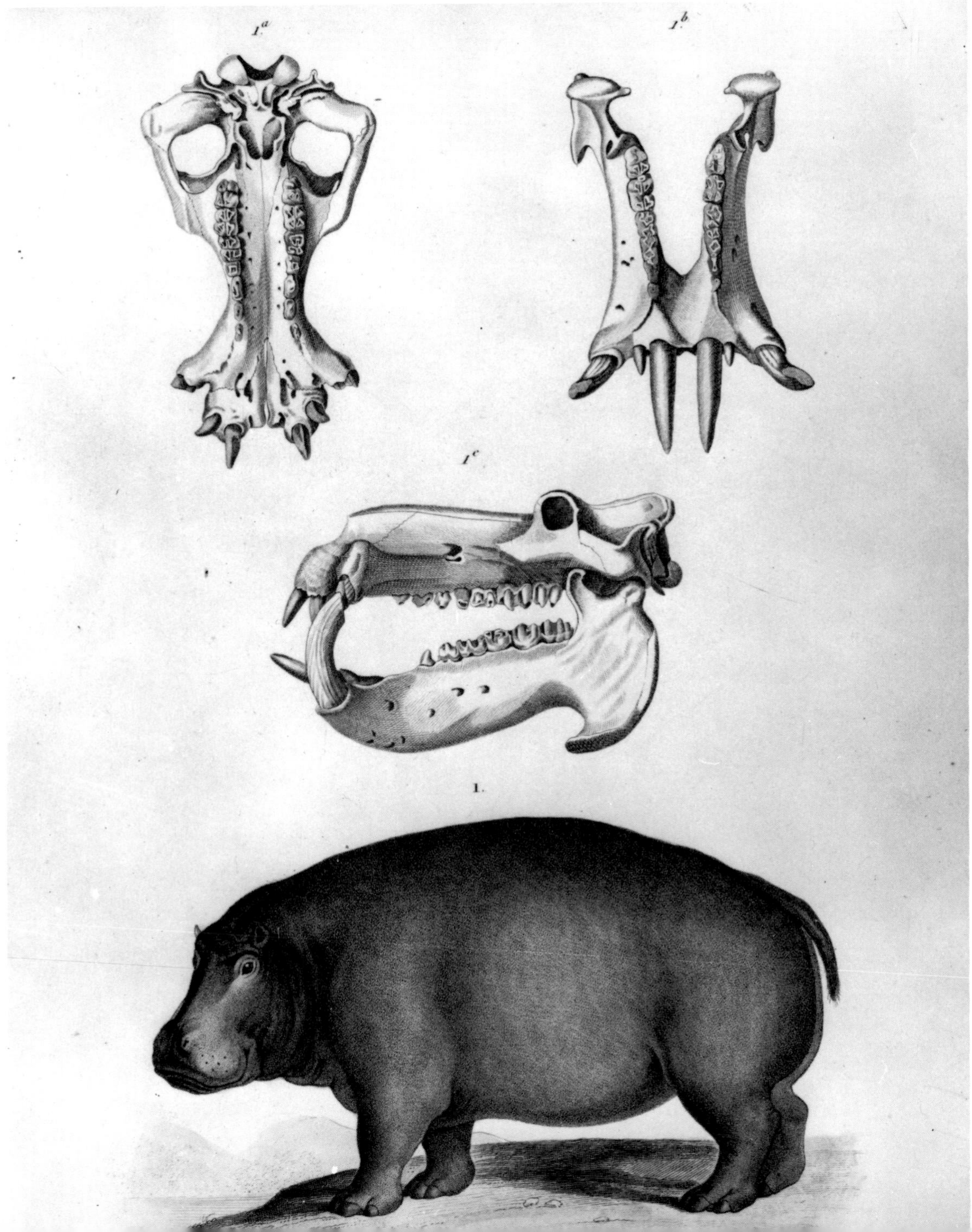

366. *The duck-billed platypus.* The hippopotamus was strange because its habitat and form were far removed from the European experience. The duck-billed platypus was strange in and of itself. It fascinated early nineteenth-century zoologists because it appeared to cut across standard taxonomic lines. What was one to make of an animal with a bird's bill that laid eggs and suckled its young? No one could answer until Darwin's theory of evolution provided the essential theoretical foundations of taxonomy.

367. *Peculiar fish with internal lights.* The exploration of the ocean depths brought many unusual specimens to light. This *Stomias boa*, shown in both lateral and frontal views, came equipped with lit "portholes" on its sides.

368. *Another denizen of the deep. Malacosteus niger* had two glowing spots under each eye. Such lights attracted prey, and had evolved to suit the unique ecological niche occupied by dwellers in total darkness.

369. *Deep-sea anglers.* The bizarre forms that evolutionary development could produce were well illustrated by these deep-sea fish. Lights at the ends of their filaments lured other fish to their death. One had to admire the cleverness of the design and wonder if it could have been produced solely by chance and natural selection.

366

367

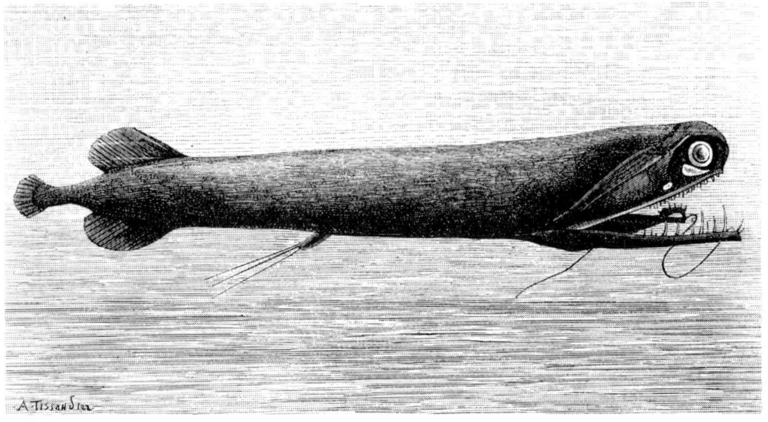
368

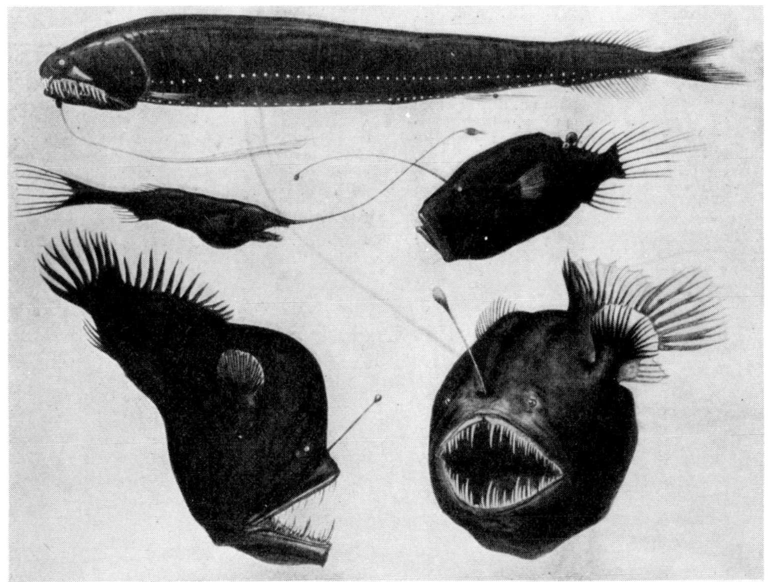
369

The Distribution of Life

Exotic specimens, of course, interested serious botanists and zoologists in and of themselves but of greater concern was the relationship between plant and animal forms and the environment in which they were to be found. Long before Darwin explicitly called attention to the adaptation of forms to environment, natural historians had realized that there was a correlation here that was important for understanding the nature of the forms to be found in a given geographical locality. So, at the beginning of the nineteenth century, the attempt was made simply to map the geographical distribution of life forms without any underlying principle being available to account for this distribution. As the century moved on, more and more specimens were brought to light and increasing numbers of plants and animals could be viewed against the background of their environment. These environments were now intensively explored. Swamps and mountains yielded up their inhabitants, as did jungles and deserts and arctic tundras. The importance of the relationship between an animal or plant form and its environment was at the foundations of the theory of evolution. It was while visiting the Galapagos Archipelago off the western coast of South America that Darwin was struck by the variations in similar inhabitants of the various islands. The finches, for example, on the different islands were all obviously finches, but their beaks differed significantly from island to island, reflecting the different conditions that obtained on each of the islands. With this clue, and the later idea of natural selection, Darwin was able to enunciate his theory of the origin of species through natural selection.

With the theory of natural selection, the geographical distribution of living forms took on new importance. Here was the evolutionary naturalist's laboratory. Species wandered about geographically and, as they wandered, they encountered new environmental conditions. Those that could adapt to these conditions survived and propagated; those that could not, died out and became extinct. What the naturalist now had to be equally conscious of was both the living specimen and its environment and the connections between the two. It was no longer enough simply to picture an animal or a plant as an individual in a colored plate to convey its essence. Its environment also had to be given, so that the living form could be seen in its ecological framework. That realization was one of the more important advances of nineteenth-century biology.

370. *Variation of vegetation with altitude.* The exploration of the globe forced nineteenth-century scientists to contemplate both the incredible variety of living forms and the laws that controlled their geographical variation. The geographical distribution of both plants and animals was to provide many valuable clues to the history of evolution. Here, in a schematic representation made around 1840, types of vegetation are correlated with altitude.

5000
8
8
4500
7
7
3800
6
6
3270
5
5
2330
4
4
1900
3
3
1270
2
2
630 m
1
1

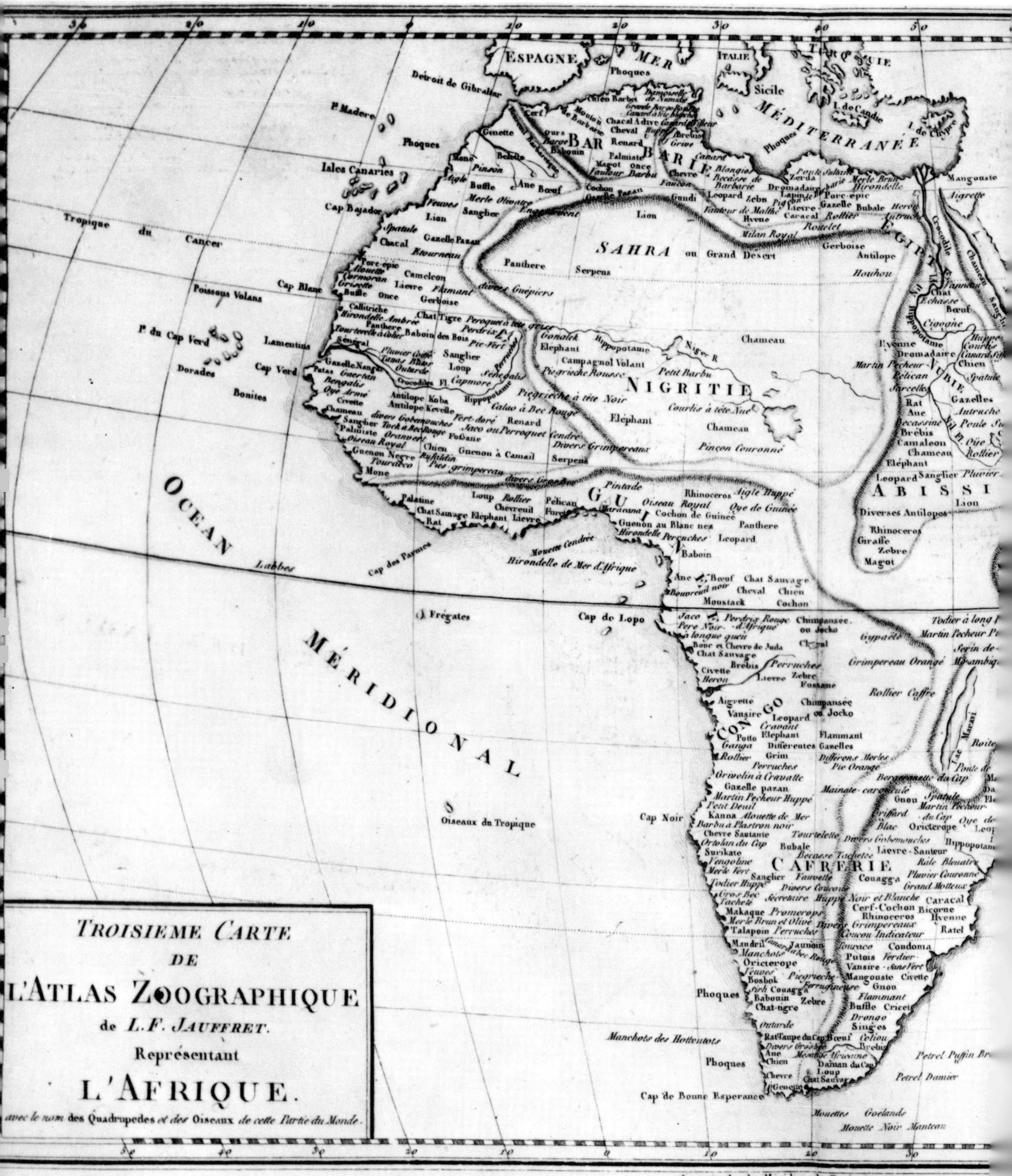

TROISIEME CARTE
DE
L'ATLAS ZOOGRAPHIQUE
de L.F. JAUFFRET.
Représentant
L'AFRIQUE.
avec le nom des Quadrupedes et des Oiseaux de cette Partie du Monde.
ESPAGNE
MER
MÉDITERRANÉE
ITALIE
Sicile
I. de Candie
I. de Chypre
Phoques
Détroit de Gibraltar
Pe Madere
Isles Canaries
Cap Bojador
Tropique du Cancer
Cap Blanc
Poissons Volans
Pe du Cap Verd
Lamentins
Cap Verd
Dorades
Bonites
BARBARIE
SAHRA ou Grand Desert
Panthere
Serpens
Lion
Chameau
NIGRITIE
Elephant
GUINÉE
Cap des Palmes
Labbes
OCEAN MÉRIDIONAL
Frégates
Cap de Lopo
CONGO
Oiseaux du Tropique
Cap Noir
CAFRERIE
Manchots des Hottentots
Phoques
Cap de Bonne Esperance
Mouettes
Goelands
Mouette Noir Manteau
Petrel Damier
EGIPTE
NUBIE
ABISSI
Mangouste
Aigrette
Gerboise
Antilope
Houhou
Hippopotame
Diverses Antilopes
Rhinoceros
Giraffe
Zebre
Magot
Chimpansee ou Jocko
Grimpereau Orangé
Rollier Caffre
Pintade
Rhinoceros
Oiseau Royal
Cochon de Guinée
Panthere
Leopard
Baboin
Hirondelle de Mer d'Afrique
Mouette Cendrée
Moustack
Cochon
Chat Sauvage
Zebre
Fossane
Flammant
Gazelles
Bubale
Couagga
Caracal
Hyenne
Ratel
Gnou
Longitude du Méridien de Paris.

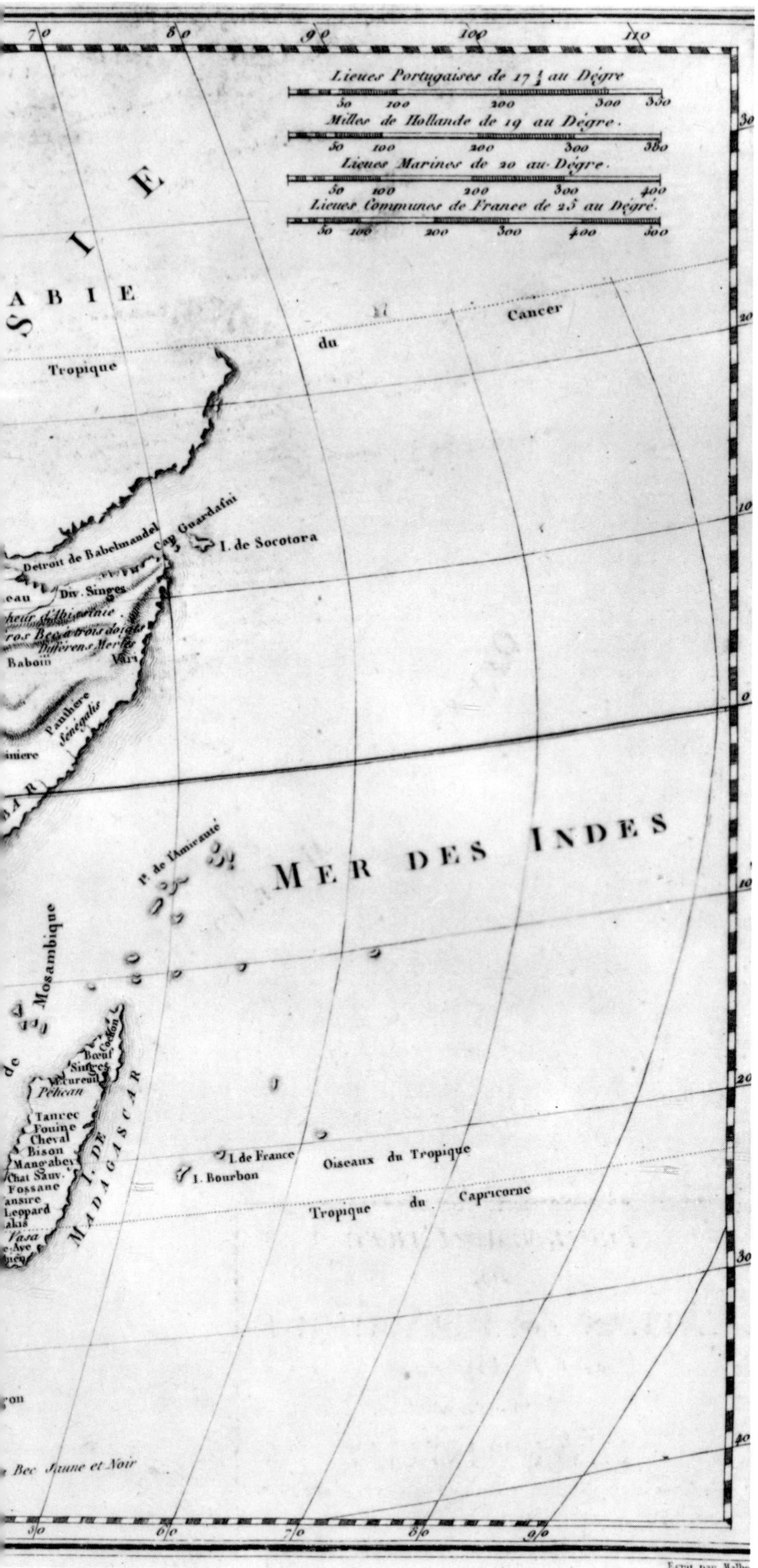

371. *The geographic distribution of animals.* It early became obvious that climate, which varied with geography, had an important effect upon the general distribution of plants and animals. Here, from a work published in 1800, is an attempt to locate species geographically in Africa. The blank spaces indicate lack of information, not a paucity of species.

372. OVERLEAF: *A closer look at the geographical distribution of animals (1800).* This map of the distribution of fauna in North Africa shows just how well the Barbary Coast was known to zoologists by the beginning of the nineteenth century.

I. Minorque
I. Majorque
PORTUGAL
ESPAGNE
MER
Détroit de Gibraltar
SAHRA ou
DÉSER

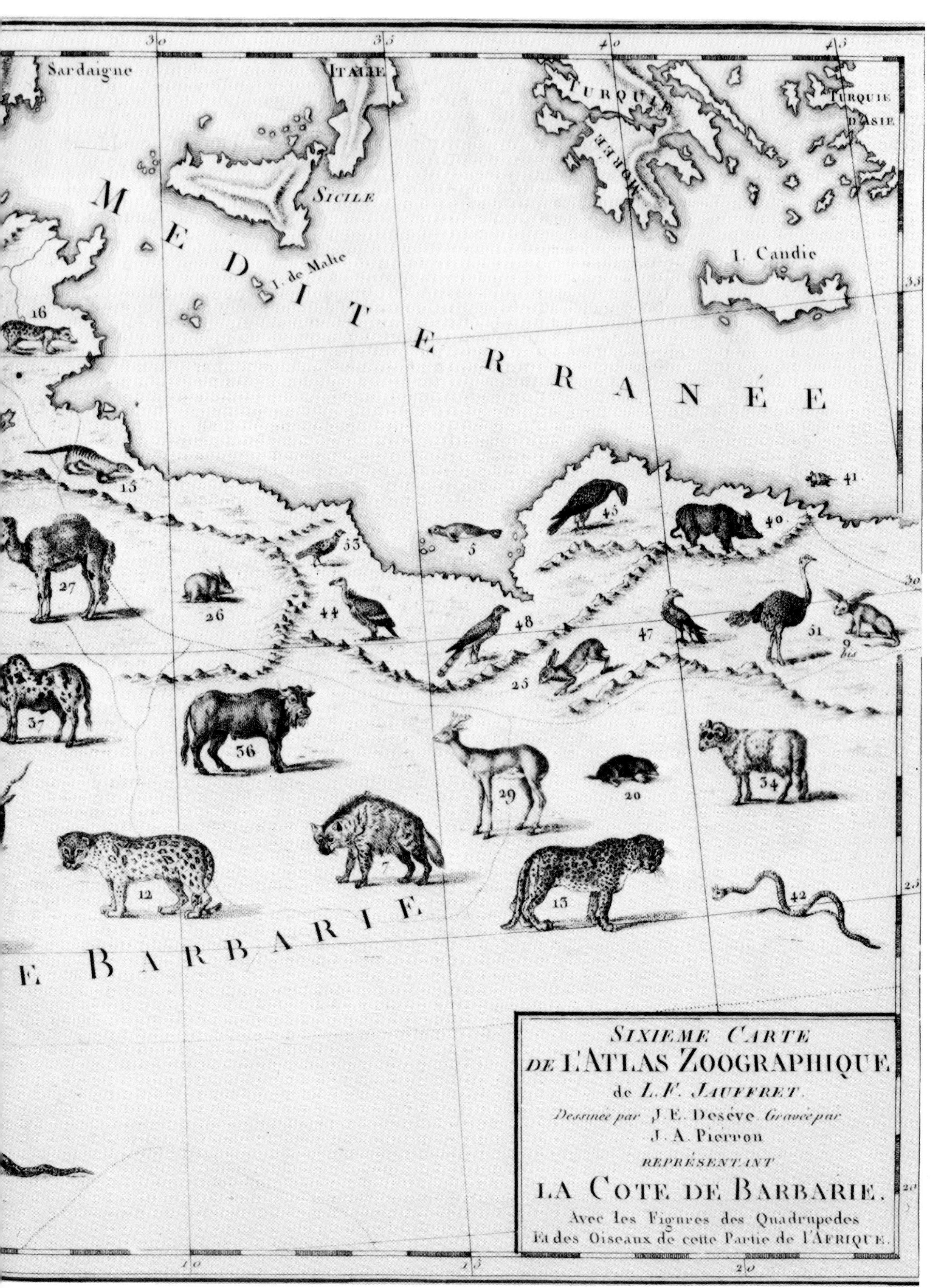

Sardaigne
ITALIE
TURQUIE
TURQUIE D'ASIE
MOREE
SICILE
I. de Malte
I. Candie
MEDITERRANÉE
E BARBARIE
SIXIEME CARTE
DE L'ATLAS ZOOGRAPHIQUE
de L.F. JAUFFRET.
Dessinée par J.E. Deséve. Gravée par
J.A. Pierron
REPRÉSENTANT
LA COTE DE BARBARIE.
Avec les Figures des Quadrupedes
Et des Oiseaux de cette Partie de l'AFRIQUE.

373

374

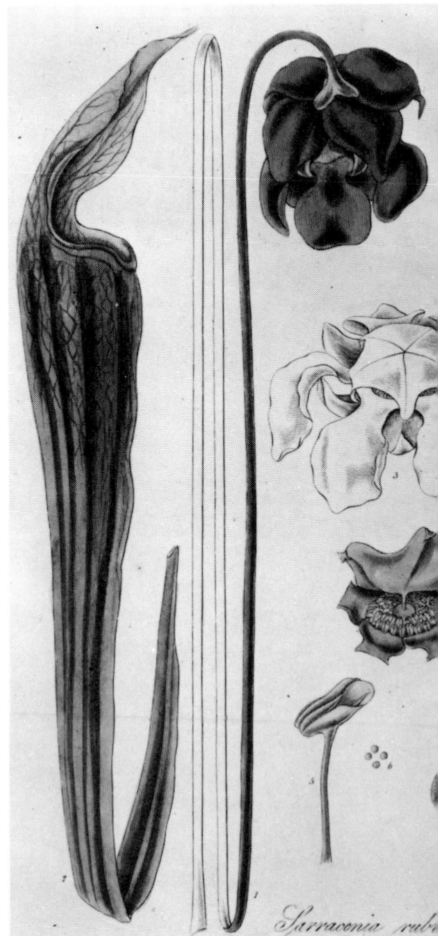

375

376

377

373. *A British botanist in the Himalayas.* No place was too high or too low for the eager scientists of the nineteenth century. Here the great British botanist J. D. Hooker, following the flag of empire in India, examines the flora of the roof of the world.

374. *Sarracenia rubra.* Exotic places yielded exotic plants. But, of course, what was exotic in Europe could be common in its own habitat. This red sidesaddle flower was an ordinary swamp plant in Georgia and the Carolinas.

375. *Eucrosia bicolor.* This specimen from Chile was both exotic and beautiful. The illustration points up the close connection between art and science in the nineteenth century.

376. *Edward Lear's drawing of Manypeeplia Upsidownia.* A bit of whimsy took advantage of the popularity of botany and science in the midnineteenth century.

377. *Algae hunting in Siberia.* The collection of exotic flora and fauna was open to everyone. The Russian government sponsored a Siberian expedition in the 1820's. One of its aims was to collect and depict the species of algae and seaweed that grew in eastern Russian waters.

378. *Selected Russian seaweeds.* Even seaweed could be made to look beautiful if enough care were taken. These are some of the specimens collected by the Russians in the 1820's.

378

379

379. *Brazilian rain forest.* Penetration of the tropical rain forest by scientists occurred for the first time in the nineteenth century. The result was such lush pictures as this. Note the jaguar at lower right.

380. *The quinquina tree.* When nineteenth-century men penetrated the tropics, they fell victim to regional ills such as malaria. The tropics themselves, however, provided medicines such as quinine, from the bark of the quinquina tree, that permitted Europeans to continue their explorations.

381. *The botanical landscape: the frozen North.* Armed with a camera, the botanist of the late nineteenth century could bring back more than specimens and sketches: he could return with photographs of the total botanical system. This view is of Russian Lapland in winter.

380

381

382

383

382. *The botanical landscape: a wooded river bank in the American Midwest.*

383. *The botanical landscape: a monsoon forest in Burma.* This photograph of a tropical jungle should be compared with the depiction of Figure 379. The photograph is both less romantic and more accurate.

384. *The botanical landscape: a tropical savanna in Africa.* The "hills" in the foreground are termite mounds. The rest of the landscape is rather severe.

385. *The botanical landscape: the American desert.* The American desert was not as hot as a tropical savanna, but it was drier. Giant cacti replaced the trees and shrubs of the more humid savanna. They and a few shrubs and grasses were about all that could grow.

384

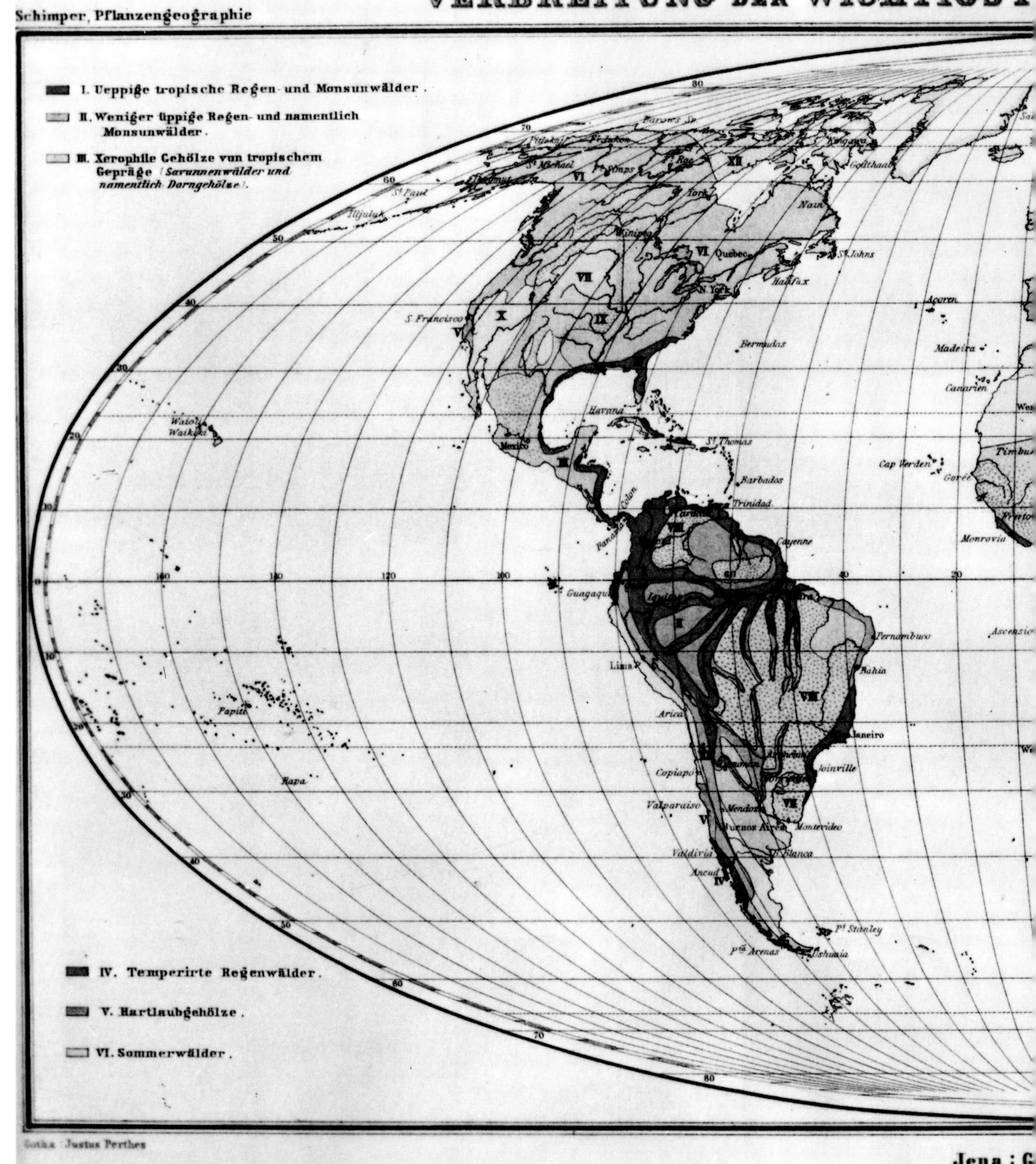

386. *A botanical world view.* From a consideration of all the individual botanical environments known by the end of the nineteenth century, it was possible to put together this map, published in Germany in 1898, which illustrated the distribution of the world's major botanical groups.

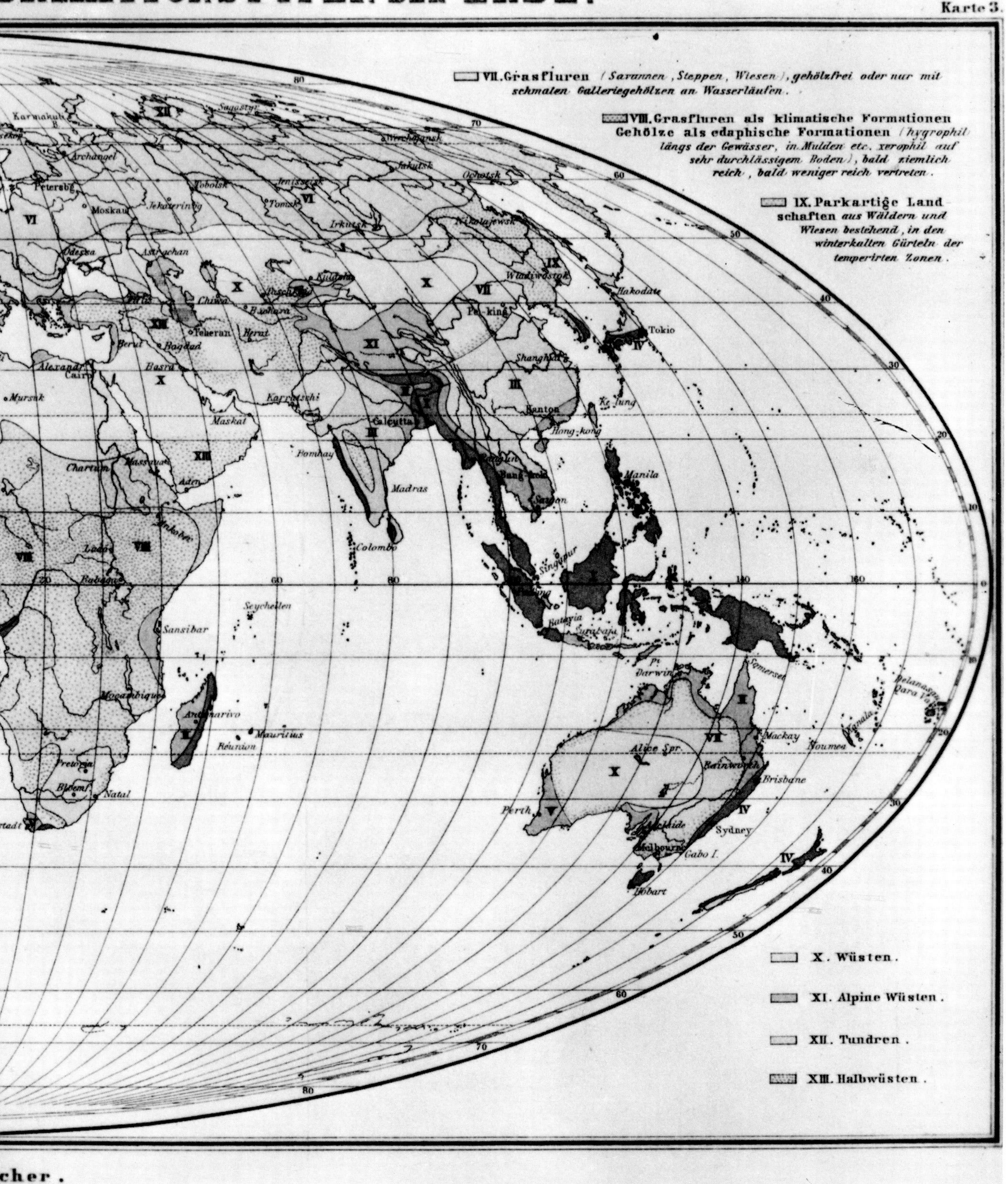
ORMATIONSTYPEN DER ERDE.
Karte 3.
VII. Grasfluren (Savannen, Steppen, Wiesen), gehölzfrei oder nur mit schmalen Galleriegehölzen an Wasserläufen.
VIII. Grasfluren als klimatische Formationen Gehölze als edaphische Formationen (hygrophil längs der Gewässer, in Mulden etc., xerophil auf sehr durchlässigem Boden), bald ziemlich reich, bald weniger reich vertreten.
IX. Parkartige Landschaften aus Wäldern und Wiesen bestehend, in den winterkalten Gürteln der temperirten Zonen.
X. Wüsten.
XI. Alpine Wüsten.
XII. Tundren.
XIII. Halbwüsten.
Archangel
Moskau
Tobolsk
Tomsk
Irkutsk
Jakutsk
Ochotsk
Nikolajewsk
Odessa
Astrachan
Chiwa
Buchara
Teheran
Bagdad
Basra
Alexandria
Kairo
Mursuk
Maskat
Aden
Chartum
Massaua
Kurratschi
Calcutta
Bombay
Madras
Colombo
Peking
Shanghai
Kanton
Hong-kong
Wladiwostok
Hakodate
Tokio
Manila
Saigon
Singapur
Batavia
Surabaja
Seychellen
Sansibar
Mocambique
Antananarivo
Mauritius
Réunion
Pretoria
Natal
Pt. Darwin
Somerset
Mackay
Brisbane
Alice Spr.
Perth
Adelaide
Melbourne
Sydney
Gabo I.
Hobart
Noumea
cher.

387

388

387. *Securing zoological specimens.* Animals, as well as plants, were eagerly sought as Europeans discovered new and exotic lands. A thriving trade arose, as here illustrated, in all kinds of beasts and birds for which Europeans were willing to pay good money. Alfred Russel Wallace, who developed a concept of evolution similar to Darwin's, made his living as a young man by capturing animals for European zoos.

388. *The zoological landscape: East Africa.* It gradually became clear to Europeans that animals lived in specific environments that had to be comprehended in order to understand the animals' origins and habits. One of the pioneers in the study of animal ecology was Alfred Russel Wallace, who, as a trapper of wild animals, was perhaps more conscious of their total environment than more academic zoologists would be. In his *Geographical Distribution of Animals* (1876), Wallace tried to portray peculiar animals of a given area in their typical environment. Here the fauna of East Africa are shown on the veldt. In the foreground stands a secretary bird, a serpent killer. Behind it is an aardvark. The antelopes and rhinoceros are well known. The bird on the wing is a red-billed promerops.

389

389. *The zoological landscape: New Zealand birds.* New Zealand was populated by rather strange avian forms. In the middle foreground is the owl parrot, a nocturnal burrowing parrot that climbs but cannot fly. On the right is a pair of kiwis, birds whose wings are mere rudimentary stubs. In the air are a pair of crook-billed plovers, the only bird known to have a bill bent sideways. At left is a pair of large rails with useless wings.

390

390. *The zoological landscape: the Canadian forest.* The skunk, in the foreground, is found only in America. Climbing a tree is a porcupine, while under it a jumping mouse evades the skunk. At rear stands a pair of moose.

The Science of Life

What the theory of evolution provided for the natural historian was the necessary thread by means of which he could find his way out of the increasingly confusing multiplicity of species which the new discoveries dumped on his doorstep almost daily. The pressure had begun to mount already in the eighteenth century when the sheer number of botanical species had risen so high that there appeared to be no way to keep them in order. Linnaeus had suggested a classificatory scheme based upon the structure of the flowers of plants. At first the system had worked well but it soon broke down as the number of anomalies increased beyond the ability of the system to take care of them. A later system, devised by Jussieu, insisted that all parts of the plant—leaves, flowers, vascular system, and roots—must be taken into account when attempting to put individuals into families. Again, the insight was correct but what was lacking was the ability to distinguish true families from mere coincidental resemblances. It was into this situation that Darwin's theory shone a bright and intense light. Families were tied together by the common element of genetic descent and once this was realized, a firm and sound foundation for taxonomy was available. The new science of biology could be placed solidly on the relationships that evolution revealed. Paleontology, natural history, zoology, botany, and the study of the geographical distribution of plants and animals could now be welded together into a single study, founded on observation and upon the action of natural forces observable in everyday life.

391. *The birth of biology.* The study of nature goes back to antiquity, but the analysis of living nature in scientific terms is more recent. The word describing the science of life—biology—was first used at the beginning of the nineteenth century. Around 1800 Lamarck used the term in the text of one of his treatises, and Gottfried Reinhold Treviranus in Germany was the first to use it in the title of a book.

392. *Goethe's archetypal plant.* The great German poet J. W. Goethe was fascinated by living forms, and sought to discover their basic principles. He felt that all plants had diverged or "evolved" from an original progenitor that had contained the germs of all the variations of plant forms within it. This *Urpflanze* is depicted here with its primary root, leaf, and sexual systems shown. All other plants, Goethe believed, were mere variations on this design.

391

Biologie,

oder

Philosophie

der

lebenden Natur

für

Naturforscher und Aerzte.

Von

Gottfried Reinhold Treviranus.

[1776-1837]

Erster Band.

Göttingen,

bey Johann Friedrich Röwer.

1802.

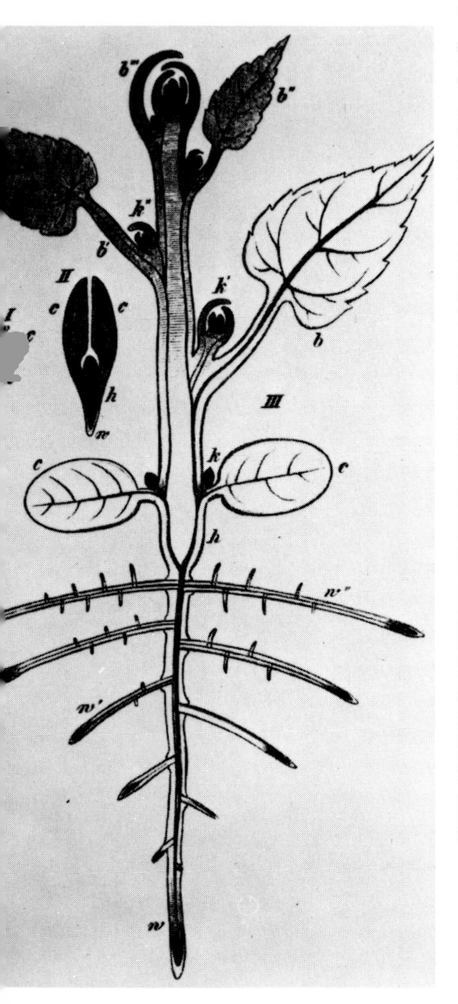

392

393a

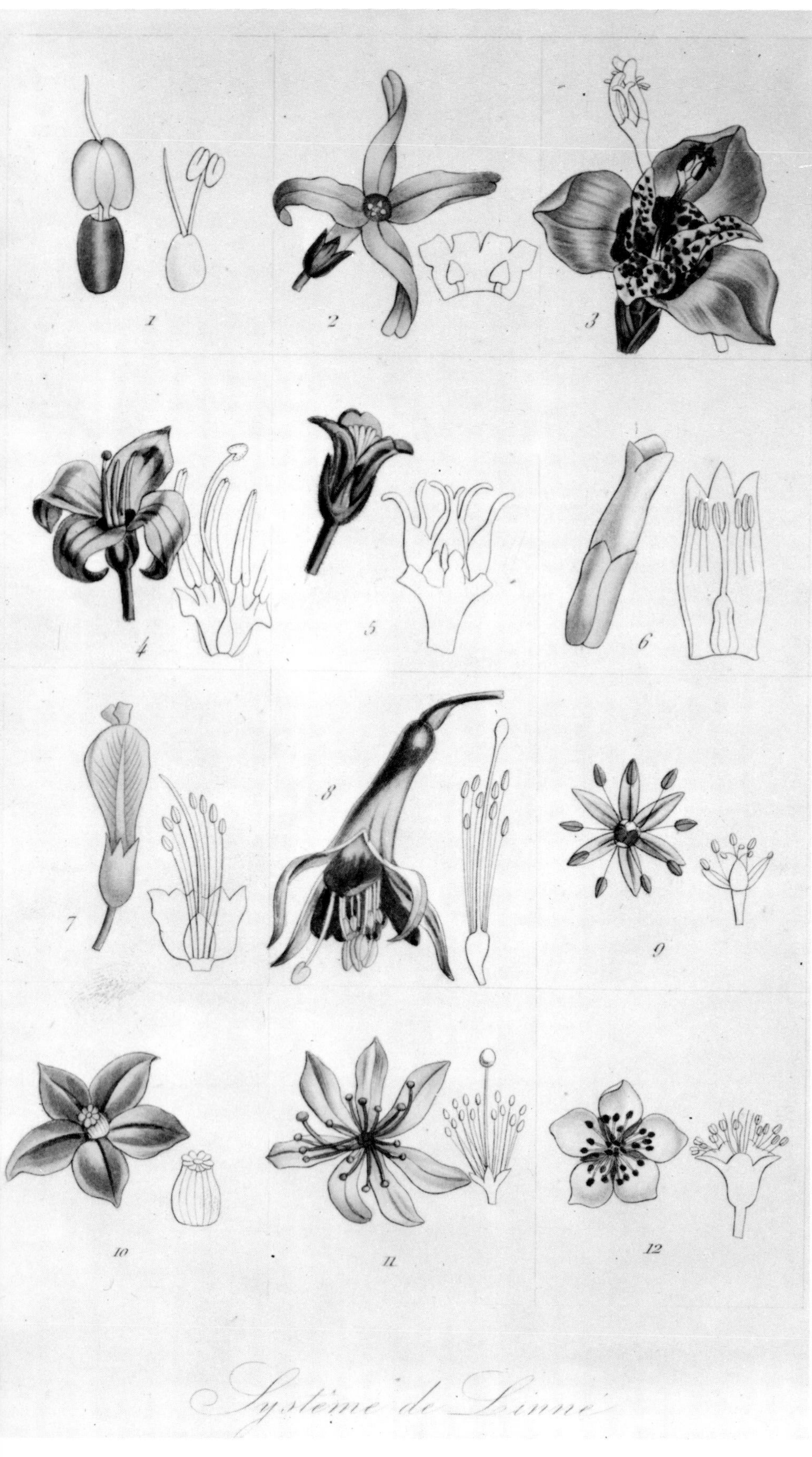

393a, b. *The Linnaean system of botanical classification.* The extraordinary variety of plant forms had put enormous strain on taxonomists before the eighteenth century. It was the Swedish botanist Carl von Linné—his name was Latinized as Linnaeus—who provided some clarification by suggesting that plant flowers, pistils, and stamens could be used as classificatory aides. These plates illustrate Linnaeus' system, which was in general use at the beginning of the nineteenth century. In figure 393a, numbers 1–12 illustrate species differentiated by the number of anthers in the flower. Figure 393b shows the more complicated flowers with which the Linnaean system attempted to deal.

13
14
15
16
17
18
19
20
21
22
23
24
Système de Linné

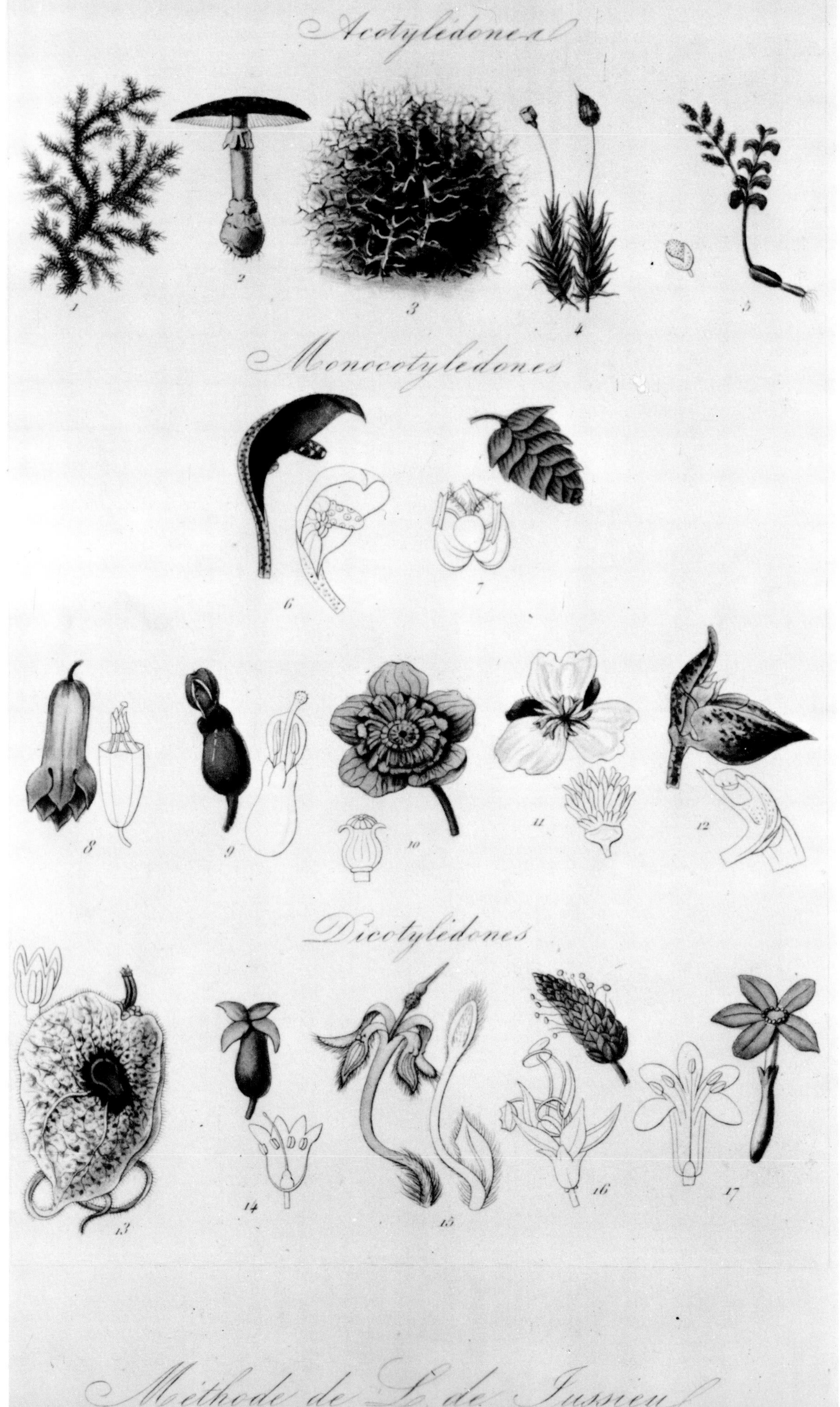
Acotylédones
1
2
3
4
5
Monocotylédones
6
7
8
9
10
11
12
Dicotylédones
13
14
15
16
17
Méthode de L. de Jussieu

394. *The system of Antoine-Laurent de Jussieu.* Linnaeus' system was not adequate to the needs of nineteenth-century botany, so other systems were invented. That of Antoine-Laurent de Jussieu attempted to classify plants according to their entire structure, not just according to their flowers. Thus leaves, stems, root systems, and other characteristics were considered, as in numbers 1–5 in the engraving reproduced here. Numbers 6–12 show some characteristics of monocotyledons; 13–17 illustrate some dicotyledons.

395. *On the Origin of Species.* The work that solved the taxonomic problem for biology, as well as other problems, was Charles Darwin's *On the Origin of Species,* which appeared on November 24, 1859. It certainly was one of the most influential books ever published.

395

ON

THE ORIGIN OF SPECIES

BY MEANS OF NATURAL SELECTION,

OR THE

PRESERVATION OF FAVOURED RACES IN THE STRUGGLE FOR LIFE.

BY CHARLES DARWIN, M.A.,

FELLOW OF THE ROYAL, GEOLOGICAL, LINNÆAN, ETC., SOCIETIES;
AUTHOR OF 'JOURNAL OF RESEARCHES DURING H. M. S. BEAGLE'S VOYAGE ROUND THE WORLD.'

LONDON:
JOHN MURRAY, ALBEMARLE STREET
1859.

The right of Translation is reserved.

396

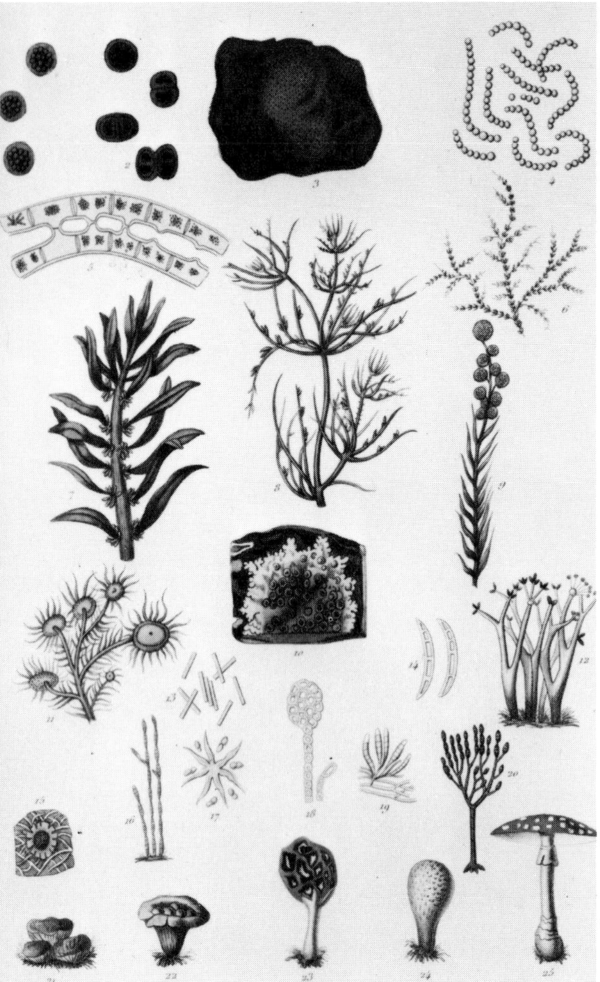

397

396. *The evolution of form.* The increasing complexity of forms within biological classes had long been known, as this picture, published in 1840 and showing various acotyledons from simple spores to complex toadstools, demonstrates. Darwin's work explained this progression genetically, as the result of evolution through natural selection.

397. *The Great Chain of Being (plants).* This illustration from a popular picture encyclopedia, published eight years before Darwin's *Origin of Species*, shows the gradations of plant structure.

398. (pp. 272 and 273) *The Great Chain of Being (animals).* A chain of increasingly complex forms could be discerned among animals as well as plants. Darwin's accomplishment was to explain how this chain had been constructed.

7
9
10
11
12
13
14
23
24
26
27
32
33
35
40
43
47
50
52
53
55
56
57
58
59

398

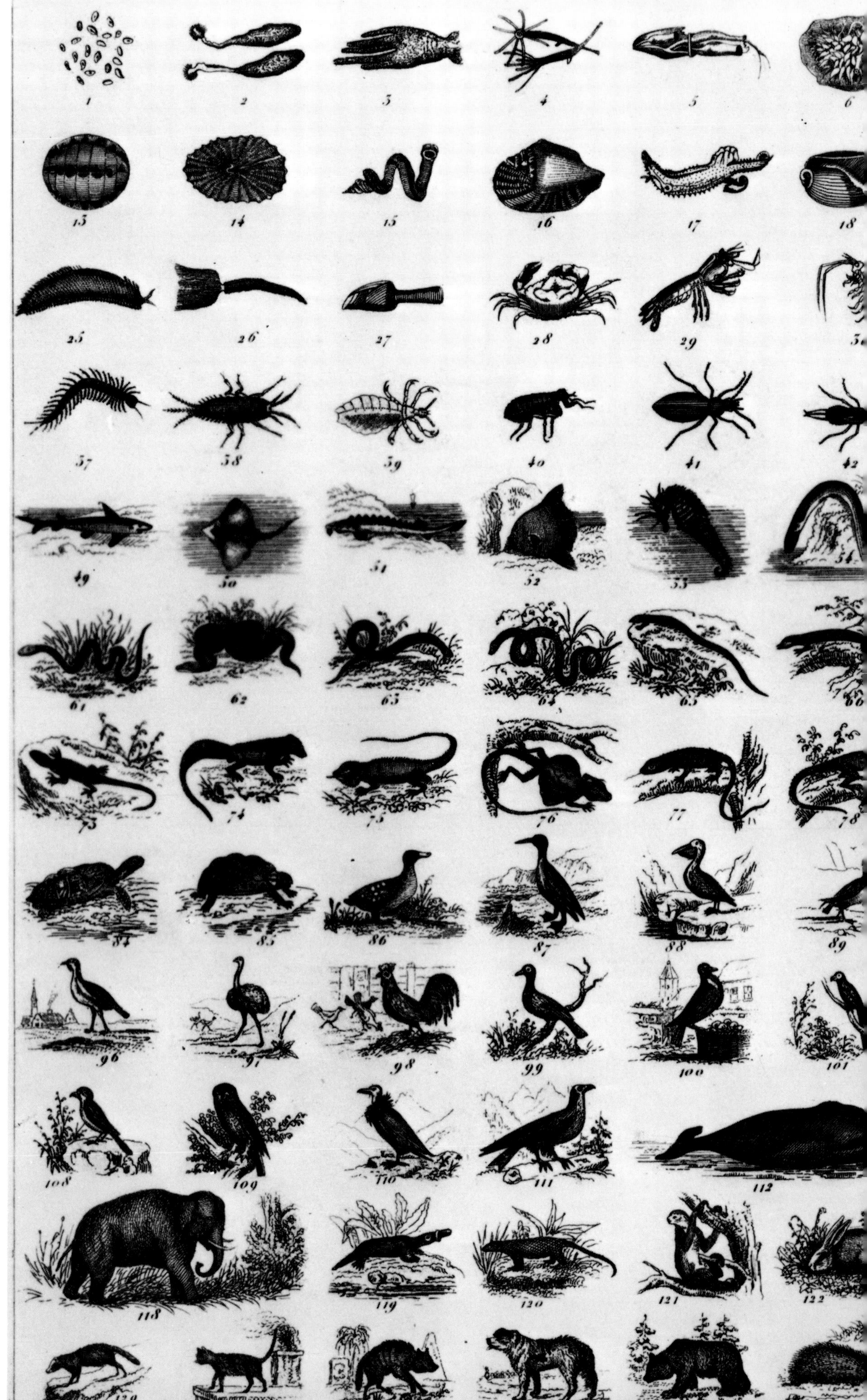
1 2 3 4 5 6
13 14 15 16 17 18
25 26 27 28 29
37 38 39 40 41 42
49 50 51 52 53 54
61 62 63 64 65 66
74 75 76 77 78
84 85 86 87 88 89
96 97 98 99 100 101
108 109 110 111 112
118 119 120 121 122
129 130 131 132

399a

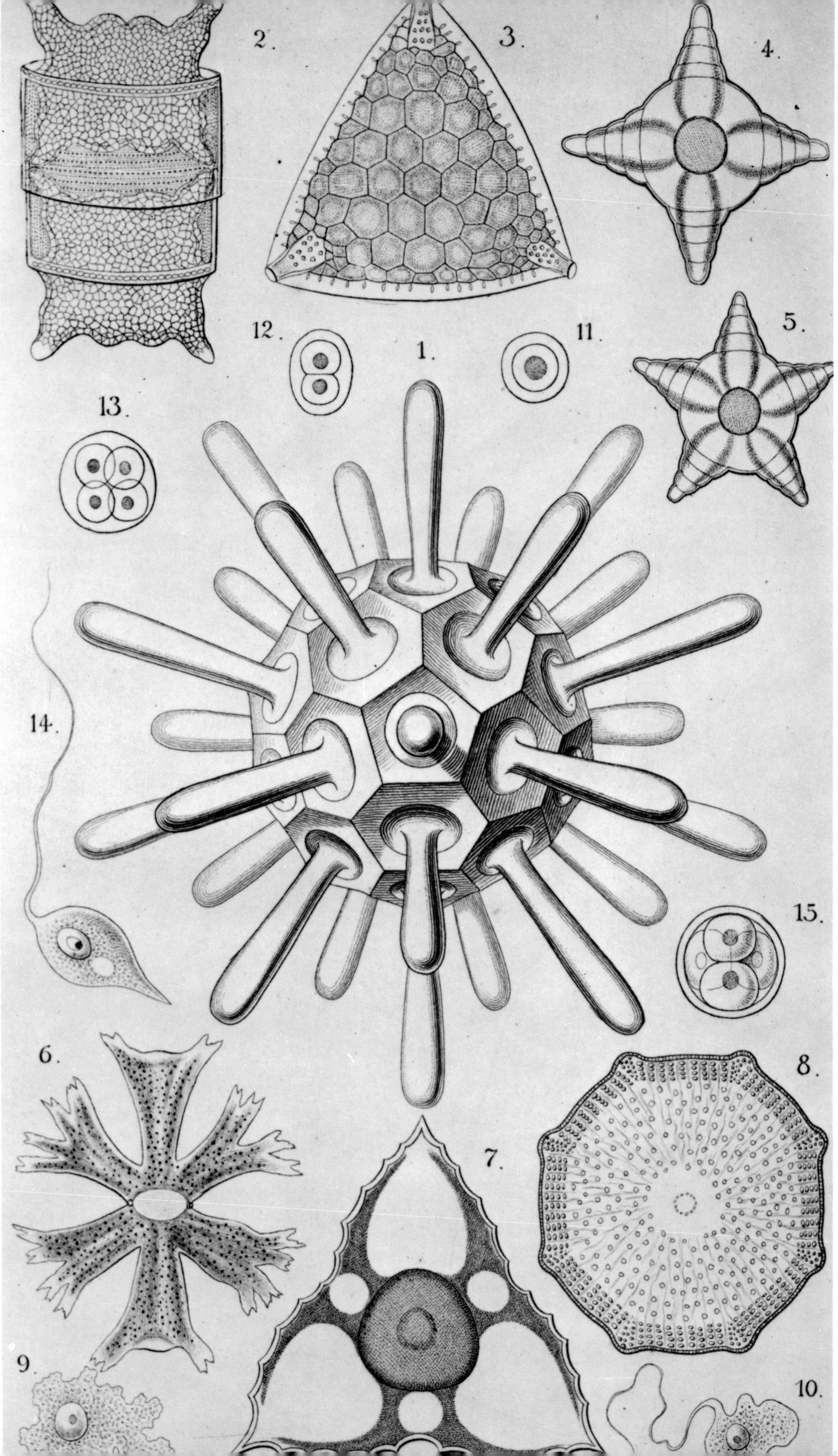

399b

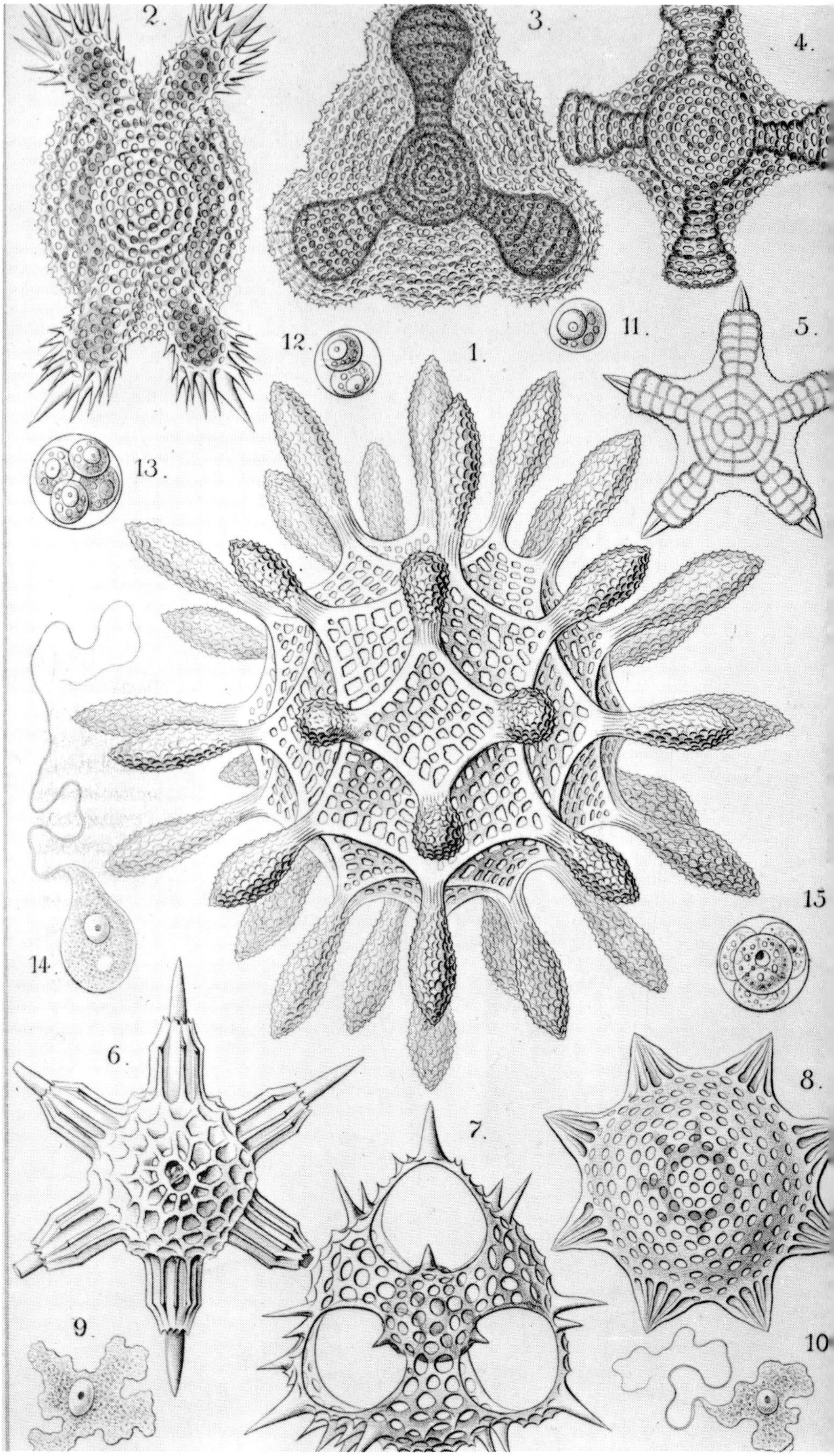

399a, b. *Basic forms of plants and animals.* Goethe's *Urpflanze* had to be replaced, after Darwin, by Ur-ancestors, which obviously must be very simple organisms. Here are Ernst Haeckel's ideas of what the basic forms of plants (399a) and animals (399b) must have been. From them, Haeckel suggested, all life evolved.

400. *How species evolve—a humorous view, 1878: "Our Village Grocer (great Floriculturist)."* "Most extr'or'nary Thing, Sir. Last year I had some bacon in my shop that went bad durin' that hot Weather, and I buried it in my garden. You'll hardly believe it, but all my Asters this season come up Streaky!"

" DARWINIAN."

400

The Microscopic World

The theory of natural selection and of evolution provided a conceptual scheme for organizing the bewildering variety of plant and animal life that pullulated on the earth's surface. It did little to help in the understanding of what life was, how it came about, and how it continued in organisms. At the beginning of the nineteenth century, the so-called imponderable fluids of electricity or heat or light were considered to be the basis of vitality, as will be seen with the galvanic experiments on corpses. Failure to turn the story of Frankenstein from fiction into fact led scientists to look elsewhere. The obvious place to look was at the microscopic level where, since the seventeenth century, it had been known that living creatures existed. Smaller almost inevitably implied simpler and so if one could just come to grips with life on the microscopic level, perhaps the basis for life would be discovered.

The problem with this approach was that there were no really good means for studying microscopic life when the nineteenth century began. There were, to be sure, microscopes aplenty but they all suffered from the same difficulties. Chromatic aberration was so severe that no sharp outlines of microscopic bodies could be observed. Everything was surrounded by a colored halo that blurred it beyond recognition. The great pathologists of the beginning of the century, men like Bichat and Magendie, carefully avoided using microscopes because they knew how unreliable they were. To this unreliability of the instrument must also be added the ignorance of the effects of substances used to prepare microscopic slides. In the eighteenth century, the great Dutch physician Herman Boerhaave had founded an entire theory of pathology on his observations that blood corpuscles could be broken down into still smaller corpuscles. He observed his blood corpuscles in aqueous solution and what he really observed was the destruction of the corpuscles by the water, not their decomposition into smaller, constituent particles. So, the investigation of the microscopic world had to await the development of microscopes that could provide accurate images and of techniques whereby microscopic objects could have their features enhanced without destroying their integrity completely.

In 1837 Giovan Battista Amici constructed a microscope with a hemispheric frontal lens that suddenly opened up the subvisible world to scientific investigation. It was both achromatic and of a much higher resolving power than any previous compound instrument. With such microscopes (and they rapidly gained popularity) it was now feasible to explore the microscopic realms with some hopes of providing scientifically reliable observations. The growth of biological microscopy from this date on was prodigious.

Among the earliest important microscopic events of the nineteenth century was the work of G. C. Ehrenberg on infusoria. Most of the unicellular organisms that had been observed previously had appeared to be those vital "monads" underlying life. What Ehrenberg did was to show that the situation was far more complicated than that. These were not Buffon's "organic molecules" but complete organisms, as complicated in structure as their macroscopic kin. From the beginning of microscopic biology, the point was made that the living cell was not a simple thing, but a highly structured and complicated entity whose anatomy and physiology were likely to be very difficult to lay bare. Ehrenberg, in the exposition of his work, also produced some of the most beautiful colored plates of infusoria ever made. They are truly works of art.

Ehrenberg's conclusions on the nature of infusoria did not prevent other microscopists from building universal theories around their microscopic observations. The most important of these was the cell theory, put forward by Matthias Schleiden, a botanist, and Theodor Schwann, a zoologist, in 1838. The cell theory stated that all living organisms were composed of cells and that cells were the basic building blocks of tissues. Although this statement is not quite correct—there are organisms such as slime molds that have no cellular structure—the cell theory had a momentous impact upon biology. By literally focusing on cells, Schleiden and Schwann provided biologists with something akin to the chemical elements in chemistry. Like the elements, different kinds of cells could be identified and their role in tissue formation was very similar to that of the chemical elements in chemical compounds. Furthermore, identification of healthy cells permitted the identification of pathological cells, thus moving the seat and causes of disease from the level of organs and tissues to that of their constituent cells. In 1859 Rudolf Virchow published his *Cellular Pathology* which put the cell theory at the very center of medicine.

It is important to note that Schleiden's and Schwann's ideas about cells were not as successful as their cell theory. Both men were convinced antivitalists and their theory of cell formation, which they considered as important as their generalization about the universality of cells in living tissue, was their declaration of independence from vitalism. Cells, they argued, were produced by a kind of *Urschleim*, or protoplasmic soup, out of which they were "crystallized." They were living monads and neither Schleiden nor Schwann took Ehrenberg seriously about their internal complexities. On this front, they were to lose. Virchow established the formula *omnis cellula e cellula* ("Every cell comes from a pre-existing cell"), and future microscopic studies were to be devoted to the mysteries of cell structure and cell function.

Ordinary observation of cells through a microscope reveals very little about them. One can make out their outlines, their nuclei (sometimes), and a few other features. To make effective use of the microscope, means must be used to bring out particular aspects of the cells under study. Cell staining was a nineteenth-century art. It tended to develop by trial and error but as it developed it enabled microscopists truly to explore the interior of cells. Some features thrust themselves upon the attention of the observer. Chromosomes (literally, "colored bodies") were so named because they took up basic dyes so readily and stood out from the rest of the cell. The perfection of the staining procedures made it possible to follow cellular processes in some detail. With the addition of the camera to the microscope, these processes could be photographed and then studied at leisure. By the end of the nineteenth century, the mystery of fertilization had been solved by actual pictures of the penetration of a single spermatazoon into an egg and the subsequent division of the egg. The process of mitosis within the cell had also been demonstrated, providing valuable but difficult clues to the mechanisms of heredity. These clues were not to bear real fruit until after 1900 when Gregor Mendel's fundamental paper creating genetics was rediscovered and applied. There is an interesting point here worth calling to the reader's attention. When Mendel did his classical work on the heredity of ordinary garden peas in the 1860's, microscopic studies of cells were still crude and such bodies as the chromosomes had not yet been distinguished. His results simply did not fall on fertile ground since neither students of inheritance, who were not expert microscopists, nor expert microscopists had any mechanism at their fingertips by which his results could be explained.

The final triumph of microscopy in the nineteenth century was, of course, the germ theory of disease. It was not essential to medicine that germs be visible to investigators before medical measures could be taken. Jenner's use of vaccination against smallpox was effective long before the smallpox virus was detected by the electron microscope. But it was essential that a germ theory of disease be based upon the solid evidence of the presence or absence of the germs supposedly causing the disease. The ability to see anthrax or tuberculosis bacilli made it possible to culture these bacteria and follow them through the course of the diseases they caused. It was only in the twentieth century that it was realized that germs were a necessary but not sufficient cause of some diseases.

By the end of the nineteenth century, the whole science of biology had undergone a fundamental revolution. Cells were not even known when the century began except as casual structures. By the end of the century, it had become obvious to those on the frontiers of biological research that the secrets of life were contained within the cell. It was the cell that controlled heredity and heredity that controlled growth. Cells were the fundamental physiological units, as well as the basic building blocks of most organisms. More importantly, it was gradually becoming clear as the century ended that cells themselves could be reduced to amazingly complex physico-chemical units. There was still no substitute for biological experimental material, but the methods that increasingly were to be used to understand this material were those provided by chemistry and physics. Molecular biology was a creation of the nineteenth century and its success lay partially in the success of the chemists and physicists of the century who had been able to penetrate with amazing subtlety into the submicroscopic world of atoms and molecules.

401

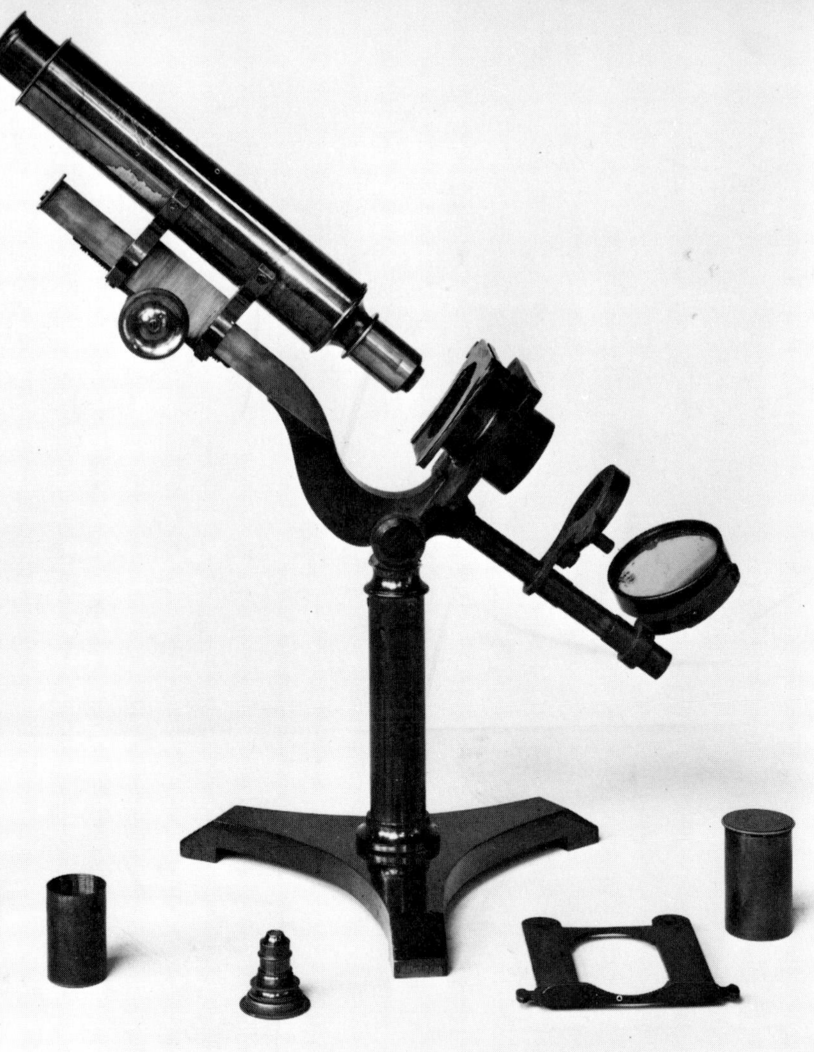

402

403

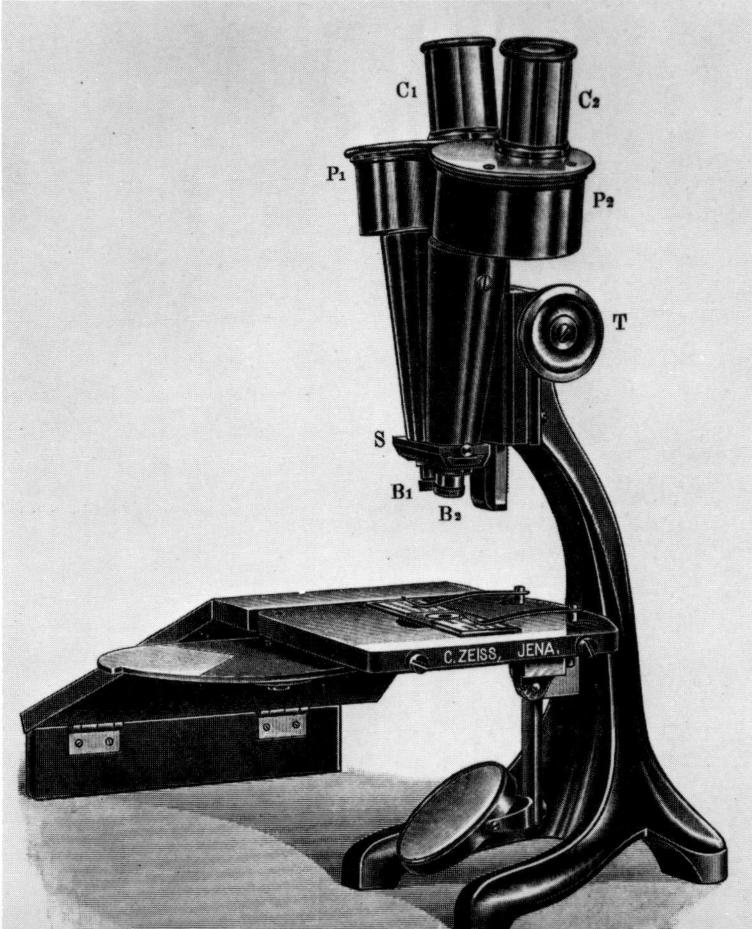

401. *Early achromatic microscope.* The most important tool for the intimate study of living things was the microscope. Until the 1830's, however, chromatic aberration was so bad that it was almost impossible to discern microscopic detail. The invention of the achromatic objective by G. B. Amici removed the aberration. This figure shows an achromatic microscope of the second quarter of the nineteenth century.

402. *A Zeiss microscope (1898).* By the end of the nineteenth century, the microscope was a precision instrument with both optical system and mechanical construction reflecting the advances of the Industrial Revolution.

403. *A binocular microscope from the end of the nineteenth century.* The binocular microscope provided depth to the microscopic field, thus making microscopic detail more easily observable.

404

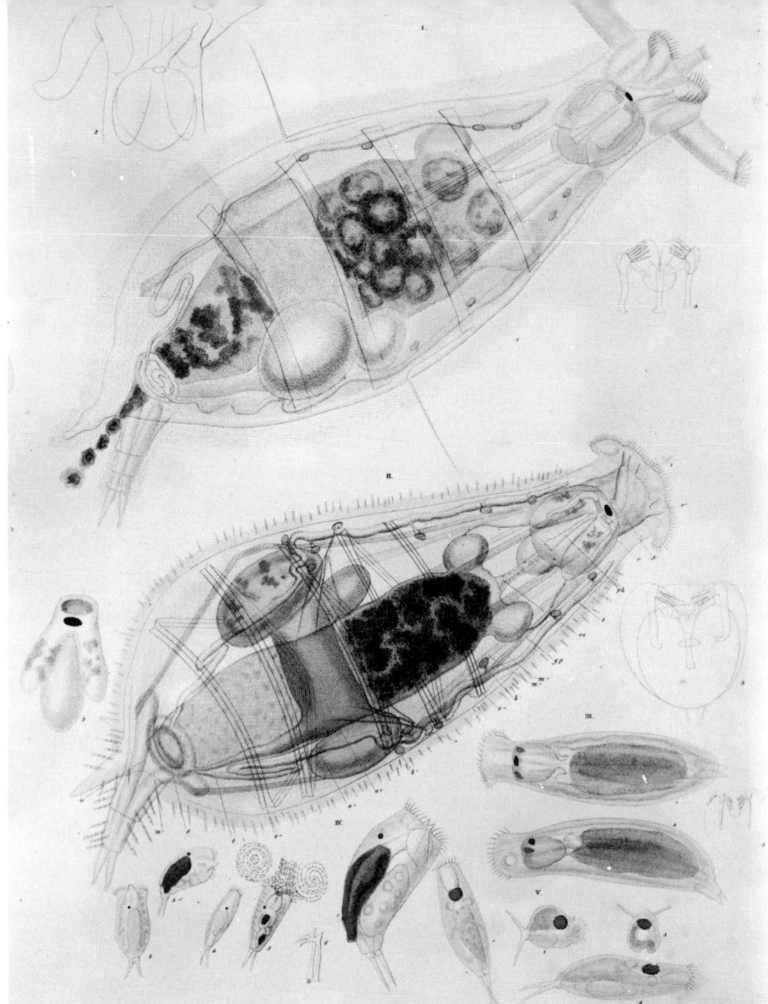

404. *Ehrenberg's infusion animalcules (1838): Notomata.* One of the great, and most beautiful, works of microscopical science in the nineteenth century was C. G. Ehrenberg's huge volume, *Die Infusionsthierchen als vollkommene organismen* ("Infusion Animals as Complete Organisms"). In gorgeous colored plates, Ehrenberg illustrated his thesis that microscopic animalcules were not simple, living monads but highly complicated organisms.

405. *Ehrenberg's infusion animalcules: Vorticella.*

406. *Ehrenberg's infusion animalcules: Stentor.*

405

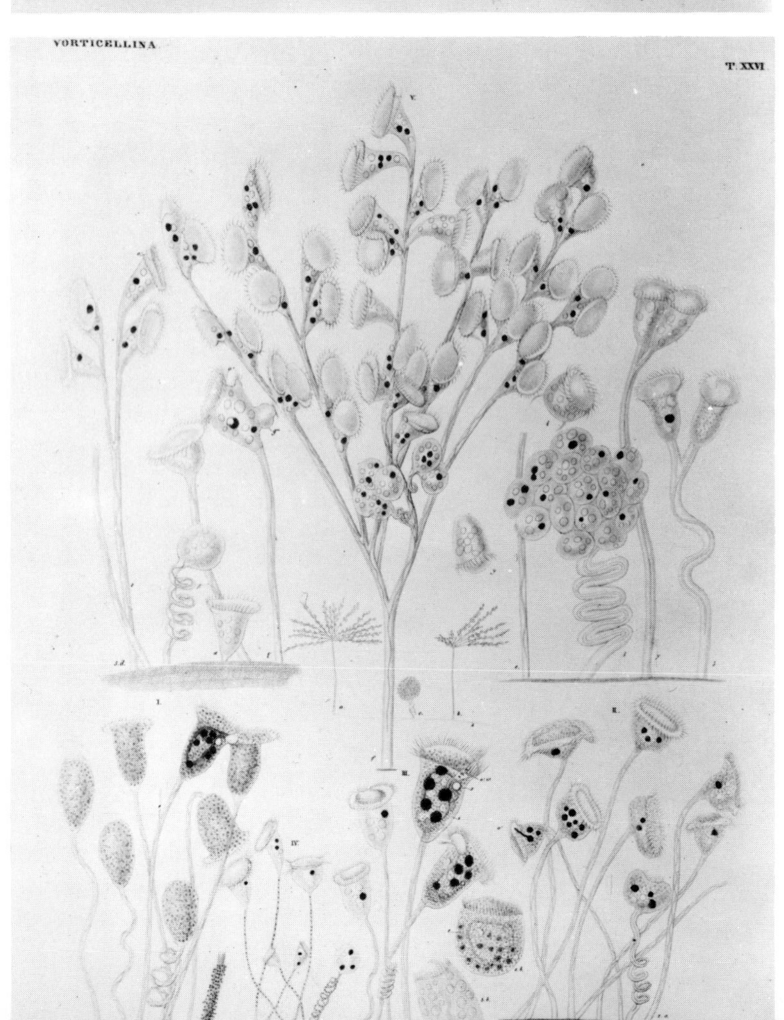

406

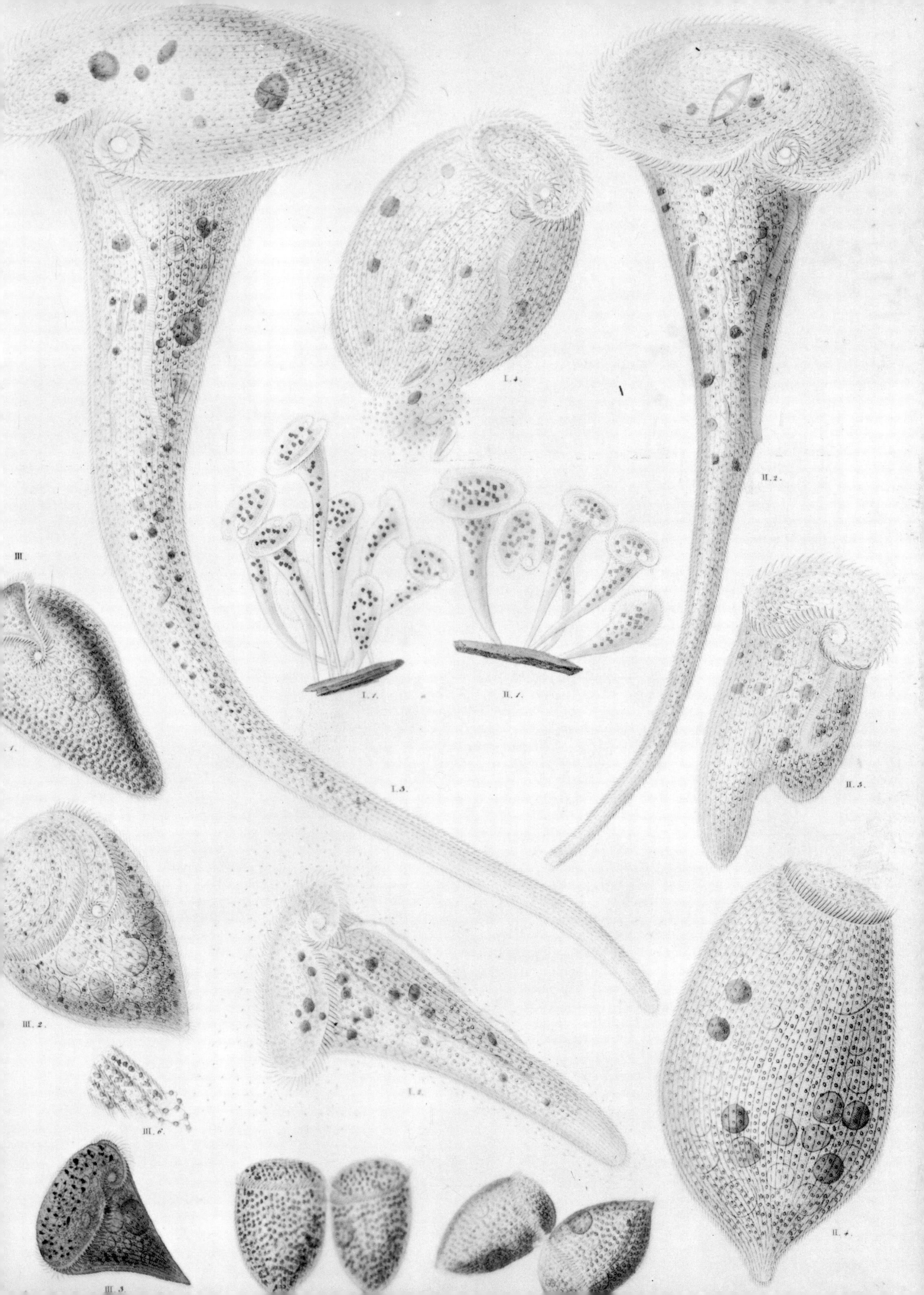
I. 4.
II. 2.
III.
III. 2.
I. 2.
II. 4.

407

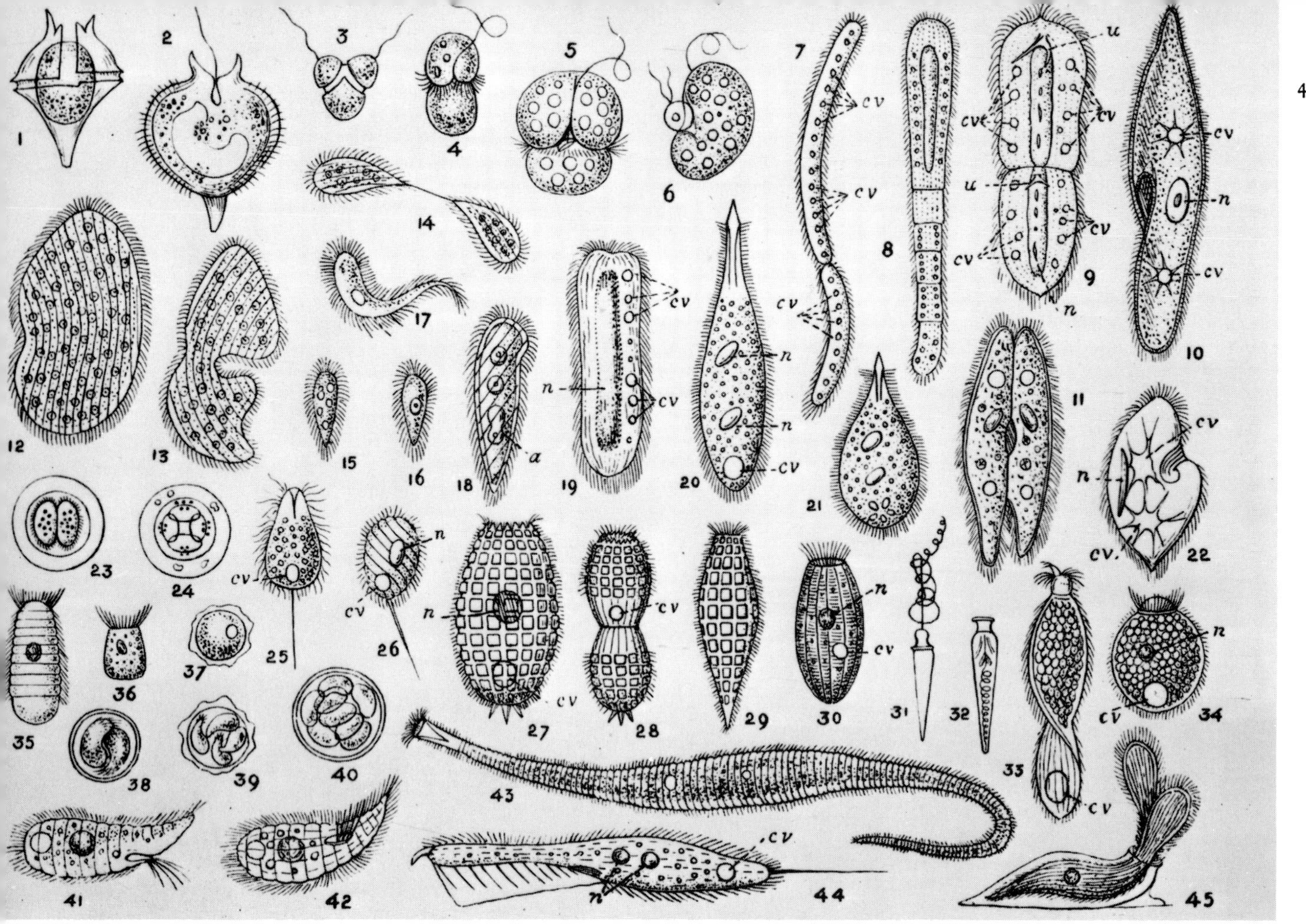

408

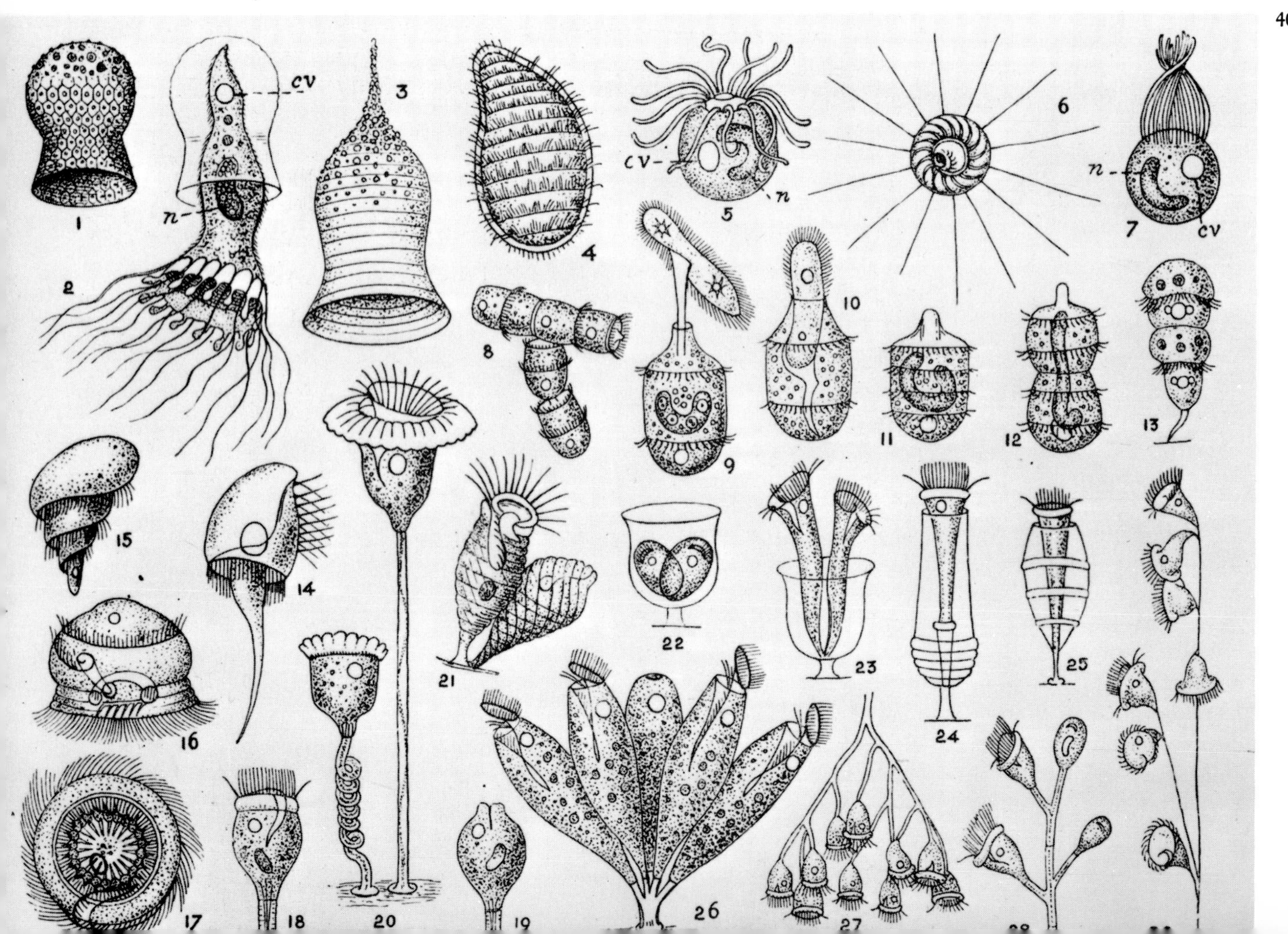

407. *Infusoria as seen at the end of the nineteenth century.* Ehrenberg's work opened up the serious study of infusoria. Other, later observers did not always see what he had seen, as these plates reveal. But, by the end of the century, the basic anatomy of most infusoria had been laid bare.

408. *Infusoria at the end of the nineteenth century.*

409. *The microscopical evidence for the cell theory.* The most important theoretical effect of the new achromatic microscopes was the enunciation of the cell theory in 1839 by Theodor Schwann and Jacob Mathias Schleiden. These drawings of animal and plant cells are from Schwann's *Microscopical Researches.* The cell theory did for biology what Lavoisier's definition of chemical elements had done for chemistry; it provided a clear concept of what could be considered the basic biological element.

410. *Leaf of ivy—a microscopical view.* The microscope revealed the intimate structure of living tissues and soon proved to be an indispensable instrument for understanding living processes. This view of an ivy leaf provides a valuable glimpse of its cellular life.

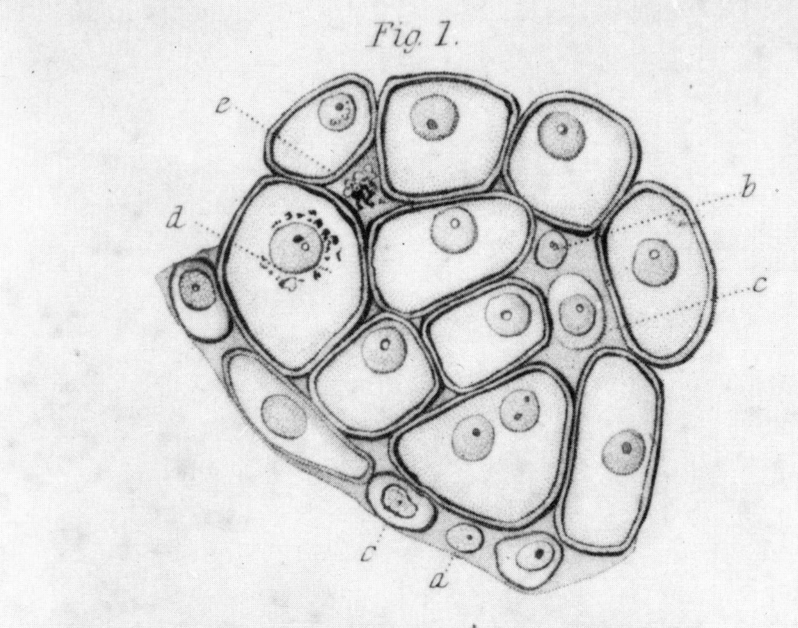

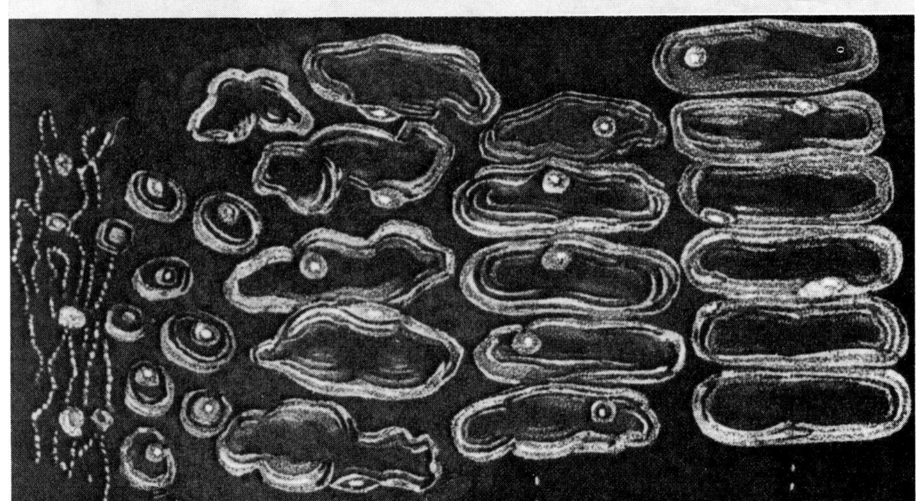

409

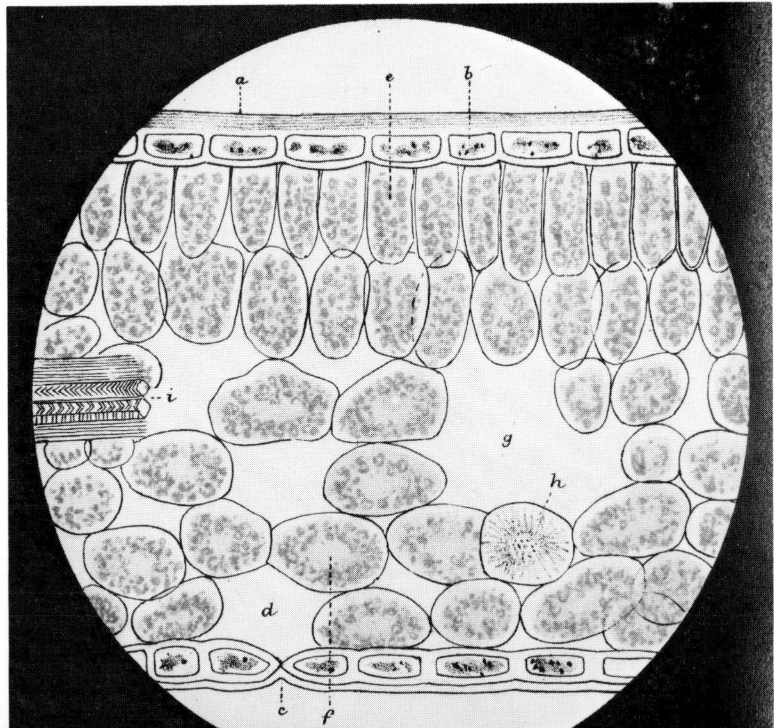

410

411

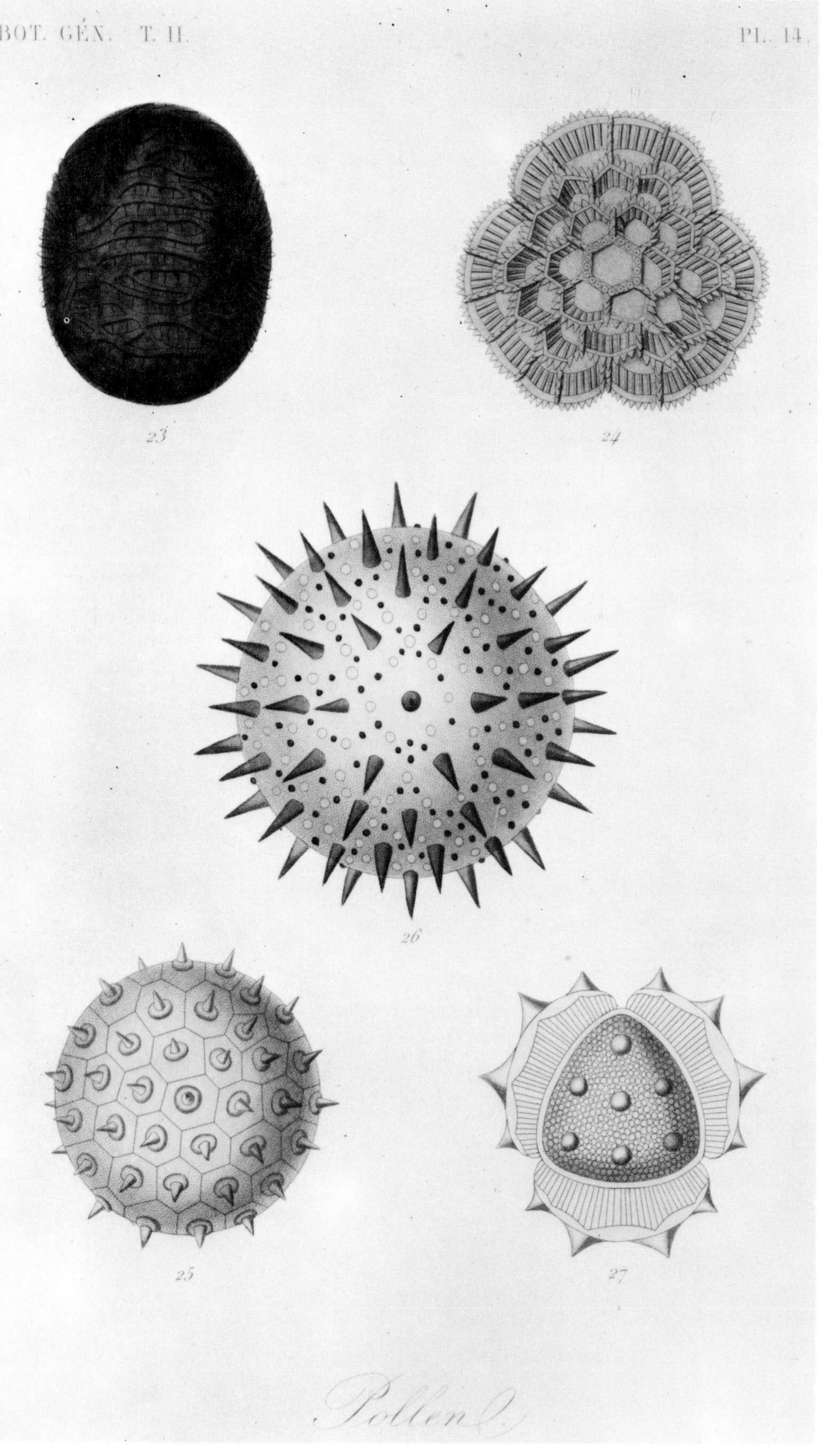

411. *Forms of pollen.* Generally speaking, the microscope revealed that microscopic nature was far more complicated than anyone had thought it to be. Pollen dust had seemed a simple botanical product; who would have thought it contained such complicated forms as these?

412. *Anthrax bacilli.* One of the great triumphs of nineteenth-century microscopy was the identification of germs and the proof of their role in disease. The anthrax bacillus was the first to be clearly proved guilty of causing a disease. In a classic paper published in 1876, Robert Koch described and illustrated its development. Figure 1 shows the bacillus in its "normal" state, together with red blood cells and spleen cells. In time the bacilli grow to form long threads (Figures 2 and 3) that are more than 100 times their original length. Then (Figure 4) the threads disappear, leaving behind spores by which the bacilli reproduce.

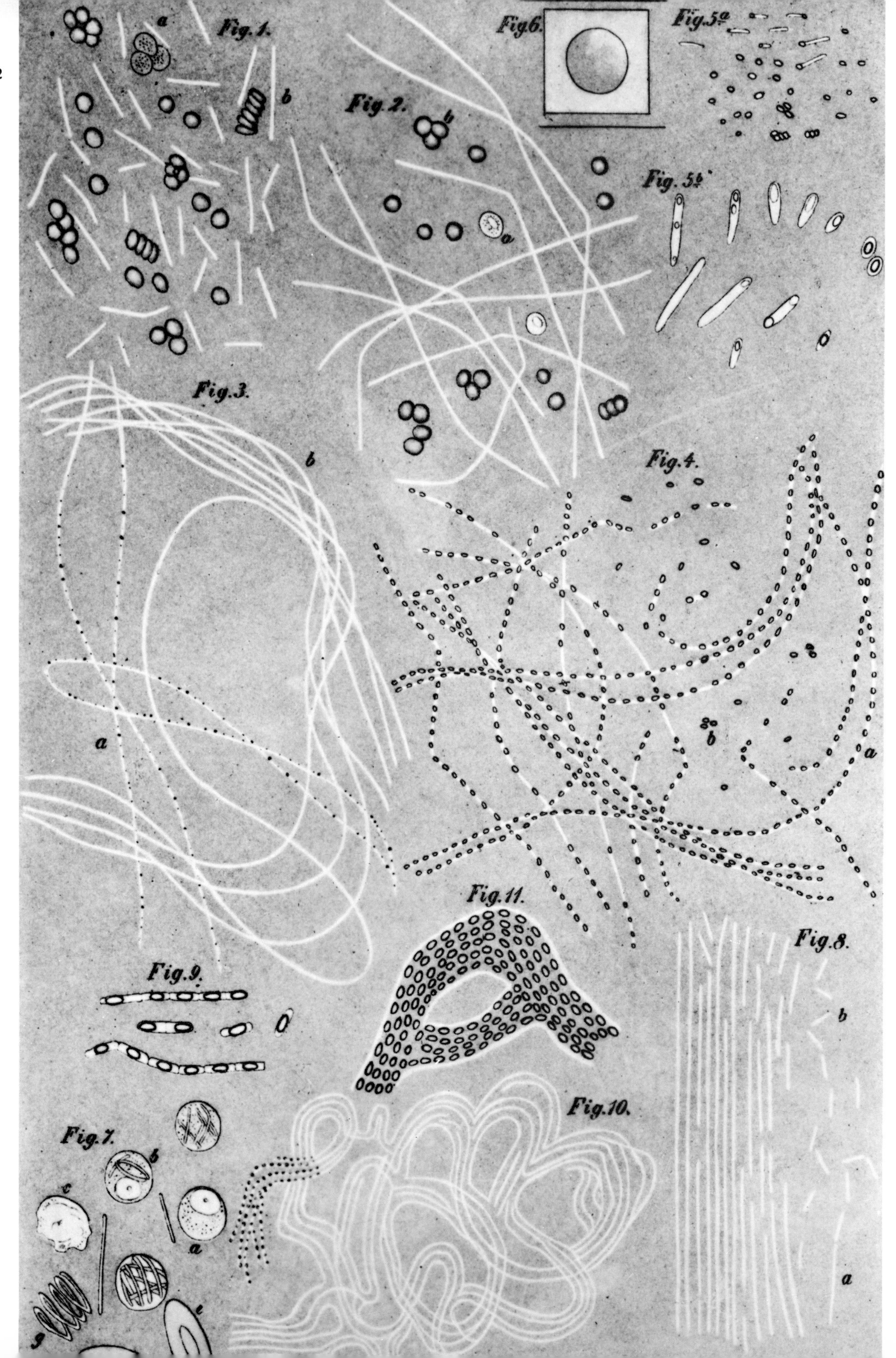
Fig.1.
a
b
Fig.2.
b
a
Fig.6.
Fig.5a
Fig.5b
Fig.3.
b
a
Fig.4.
b
a
Fig.11.
Fig.8.
b
a
Fig.9.
Fig.10.
Fig.7.
b
c
a
g

413

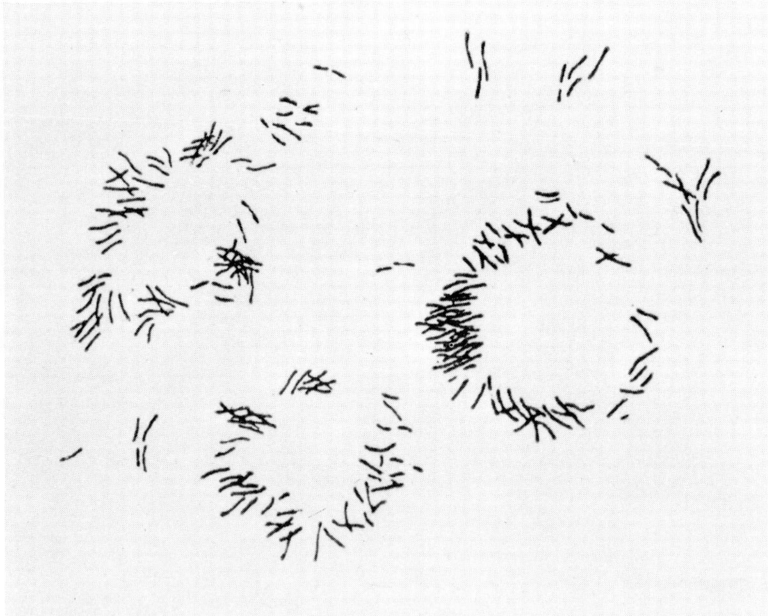

414

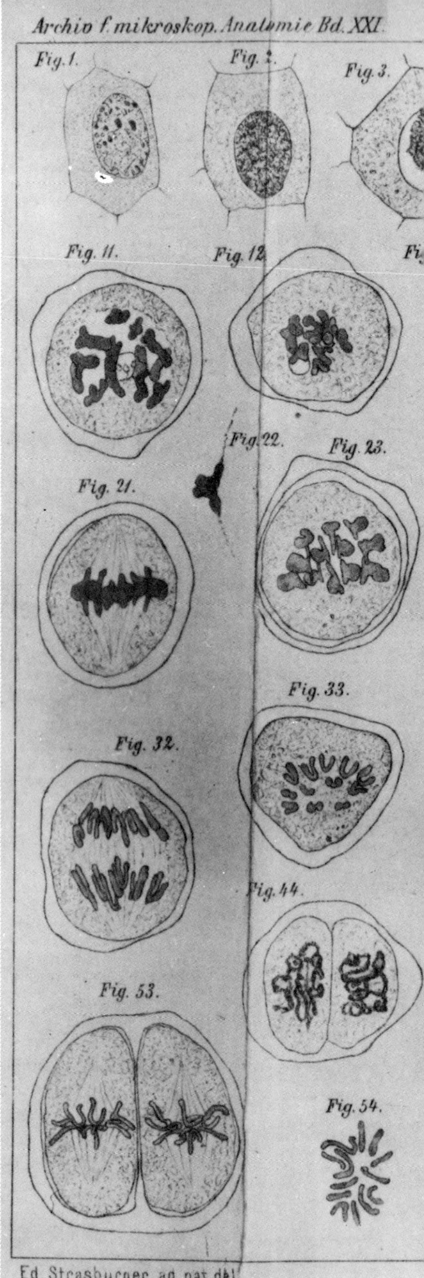

415

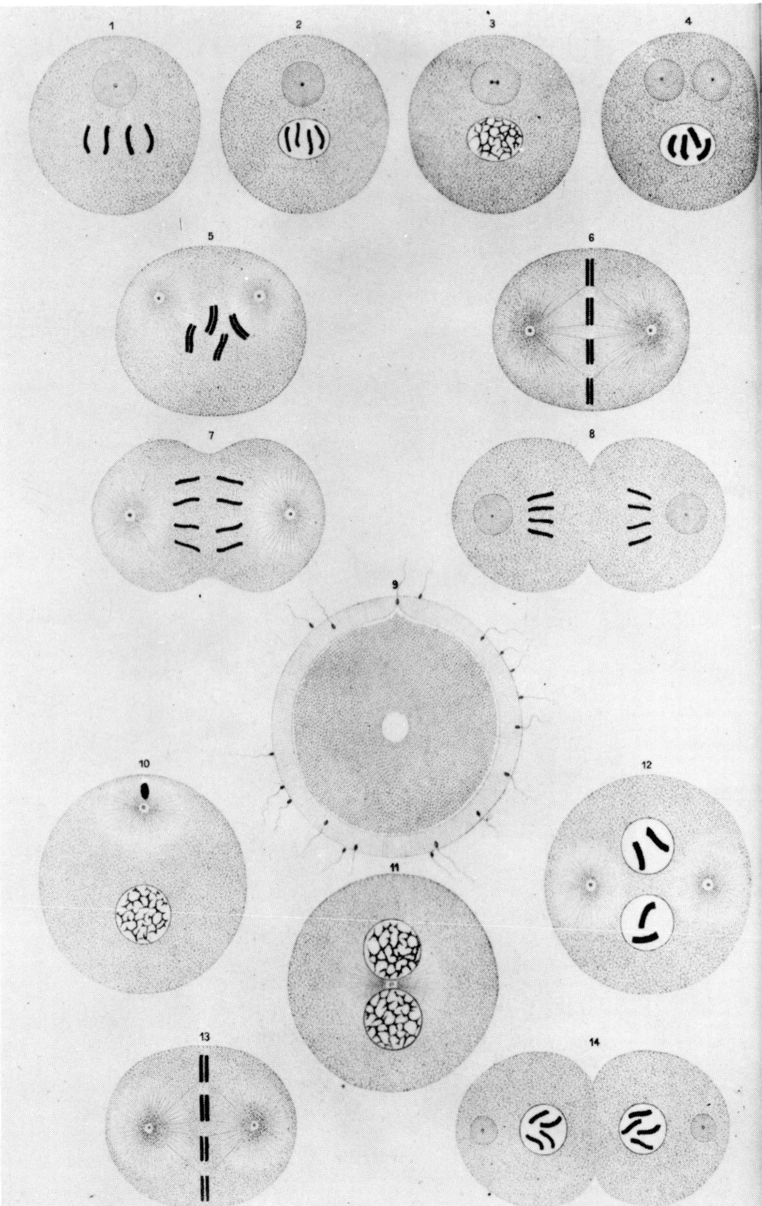

413. *Tuberculosis bacilli.* Koch was the foremost worker on tuberculosis bacilli, isolating and identifying the organism. There, linear in form, was the cause of the "white plague" that had scourged nineteenth-century Europe.

414. *The workings of the living cell: mitosis.* The development of the microscope and the evolution of staining techniques permitted the observation of the intimate processes, as well as the structures, of the living cell. In 1882, the complicated stages of cell division were depicted in this plate. The chromosomes had only recently been made visible through use of the new aniline dyes produced by the organic chemists.

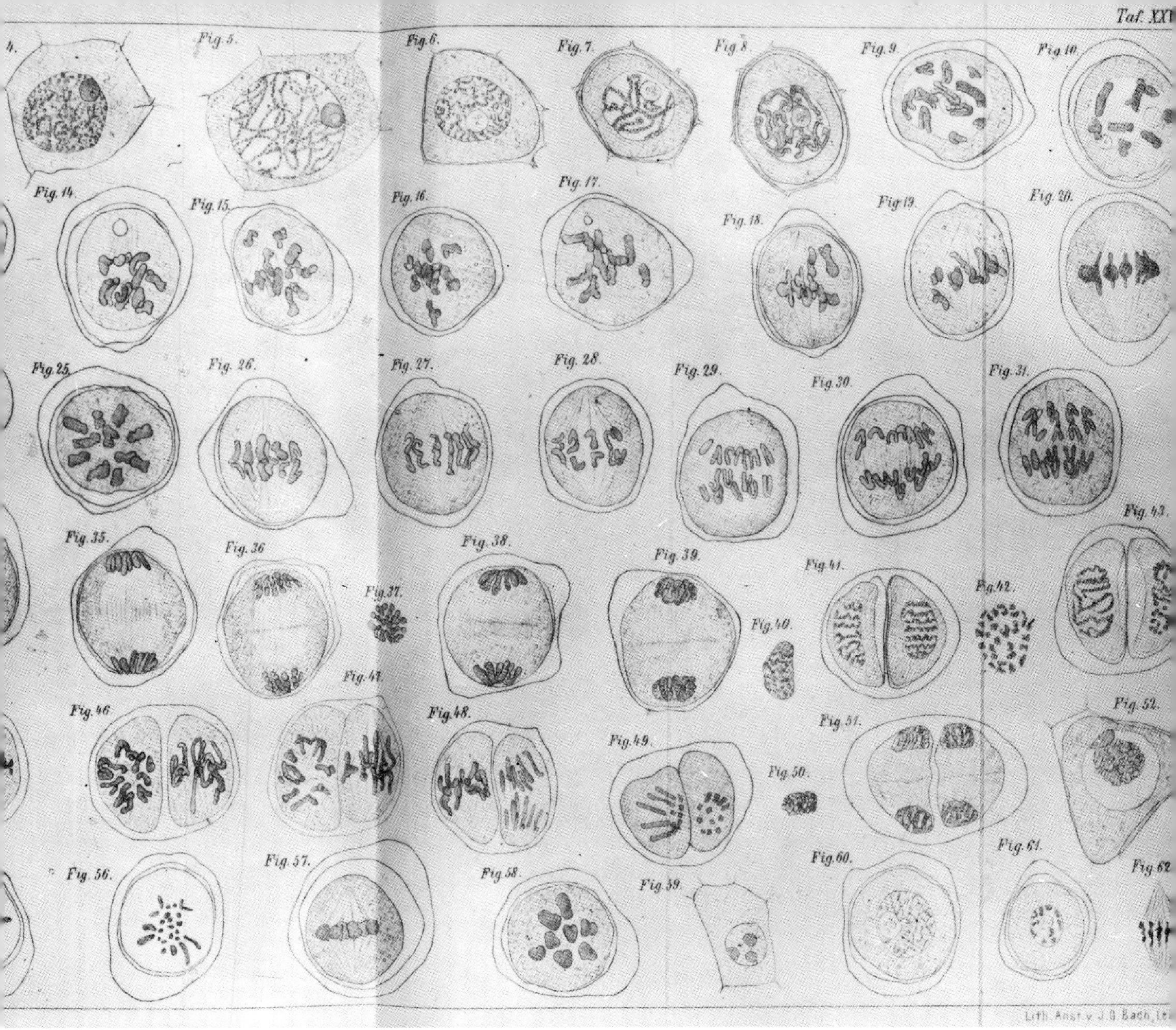

415. *Fertilization and cell division.* By the end of the nineteenth century, the mystery of sexual reproduction had been solved, as had the problem of cell division. These schematic views show the basic stages of fertilization and mitosis.

416. *Microphotographic apparatus.* The ability to take photographs of what the microscope revealed greatly furthered the study of living tissues. As in astronomy, people had tended to "see" things and draw structures that were not there. The photograph changed all that and allowed the creation of an objective, microscopical science. This rather simple apparatus made that advance possible.

416

417-424 (left to right, top to bottom)

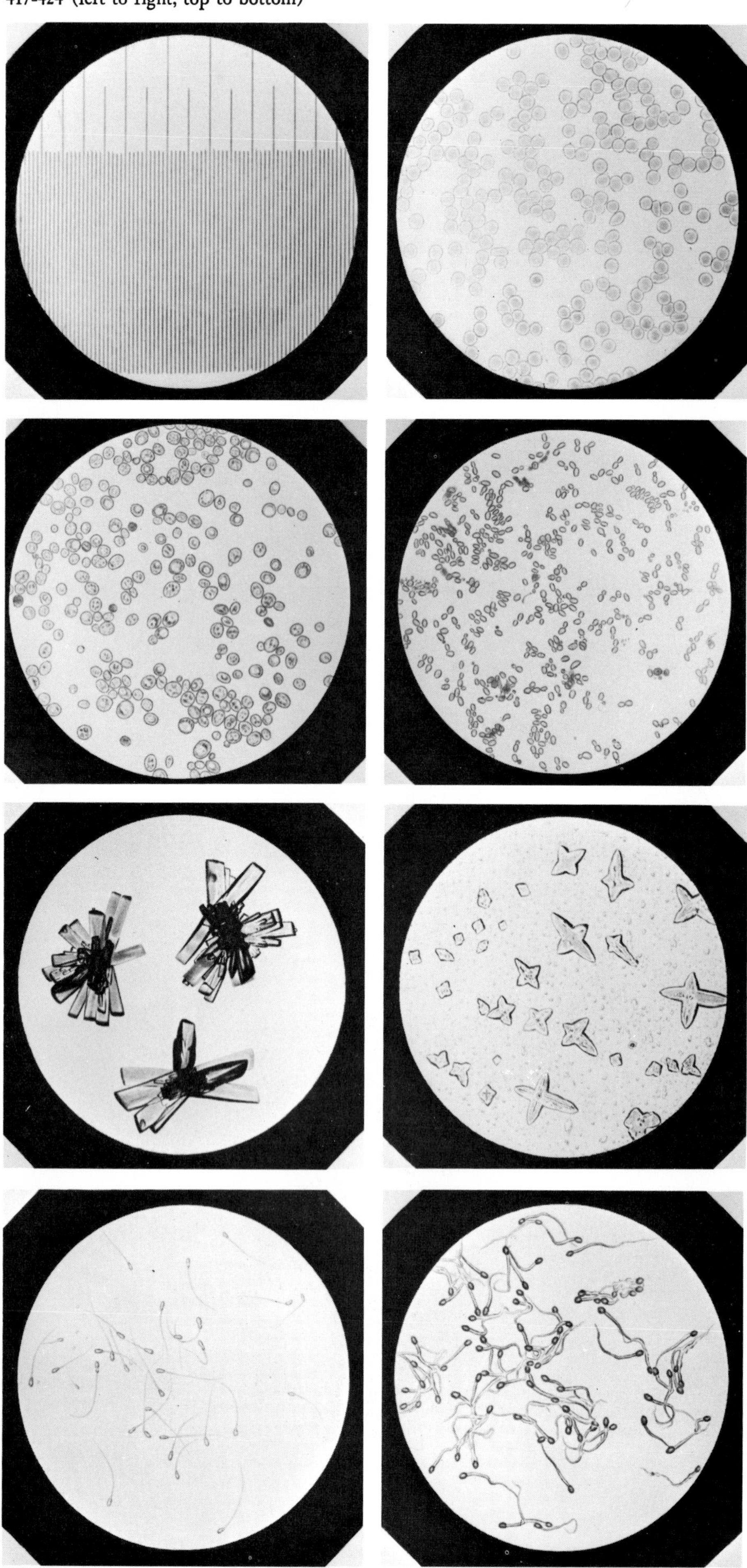

417. *Early microphotographs.* The inventive Léon Foucault was among the first to see the possibility of applying the new photographic techniques to microscopy. In 1845 his collaboration with Alfred Donné, a physiologist and physician, produced numerous pictures of microscopic objects (see Figures 418–426) that were published in an atlas to accompany Donné's course in microscopy. One advantage of microphotography is illustrated by Figure 417. A metrical standard could be created and then applied to microphotographs to provide absolute dimensions. Figure 417 shows a millimeter divided into 400 parts.

418. *Foucault's early microphotographs: human red blood corpuscles.*

419. *Foucault's early microphotographs: globules of yeast.* When these photographs were published, the nature of the agent that caused fermentation was not at all clear. The scientific world was divided between those who believed that the "ferment" was a living organism and those who considered it purely chemical. Microphotographs permitted both camps at least to agree on what the "ferment" looked like.

420. *Foucault's early microphotographs: the fermenting agent of sugared urine.* This and Figure 419 show clearly that, whatever "ferments" might be, they were different for beer and for urine. This fact supported the organic view.

421. *Foucault's early microphotographs: crystals of uric acid.* The study of organic nature was not the only discipline to profit from the application of photography to microscopy. Crystallography also benefited from the new ability to make accurate pictures of crystal forms. Since the analysis of such forms was essential for the classification and identification of crystal species, photographs greatly aided the study of microsamples and microcrystals.

422. *Foucault's early microphotographs: crystals from evaporated urine.* These are crystals of uric acid, contaminated with other components of the urine.

423. *Foucault's early microphotographs: human spermatozoa.* Compare this photograph and the two that follow with Leeuwenhoek's and Buffon's drawings in Figure 309.

424. *Foucault's early microphotographs: bat spermatozoa.*

425. *Foucault's early microphotographs: mouse spermatozoa.*

426. *Foucault's early microphotographs: rabbit's ovum.*

427. *Microphotographs of various objects.* By the 1880's, microphotographs could be inserted in books and used as illustrations. Here, from a popular German work on light, are some examples: 1. section of a child's larynx; 2. section of lower segment of a fetal leg; 3. section of a cat's tongue; 4. section of the head of a tapeworm; 5. the trichinal parasite (*Trichinella spiralis intestinalis*); 6. *Navicula lyra.*

428. OVERLEAF: *Microphotographs of various objects.* 1. section of shell of *Pleurosigma formosum;* 2. section of shell of *Pleurosigma formosum;* 3. *Arachnoidiscus Japonicus;* 4. anthrax bacilli and blood corpuscles; 5. tuberculosis bacilli in sputum; 6. bacteria in polluted well water; 7. micrococci; 8. nerve ganglion cell; 9. section of eye of a turtle, showing retina; 10. horse mange mite; 11. section of lower jaw of a rat; 12. the common head louse.

425

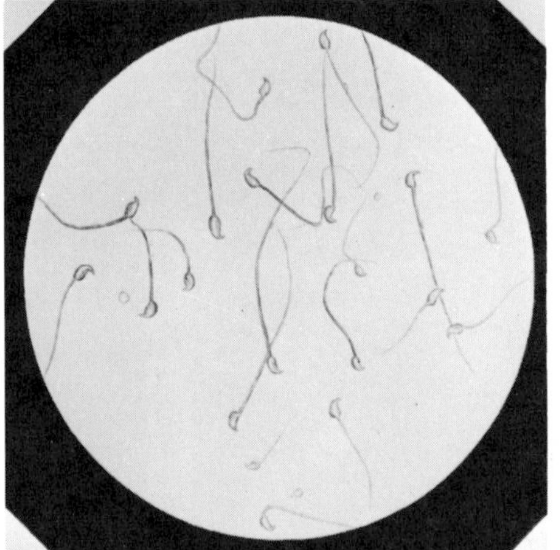

426

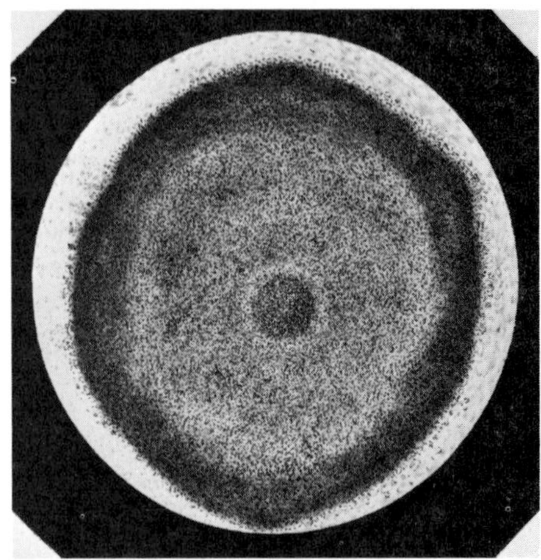

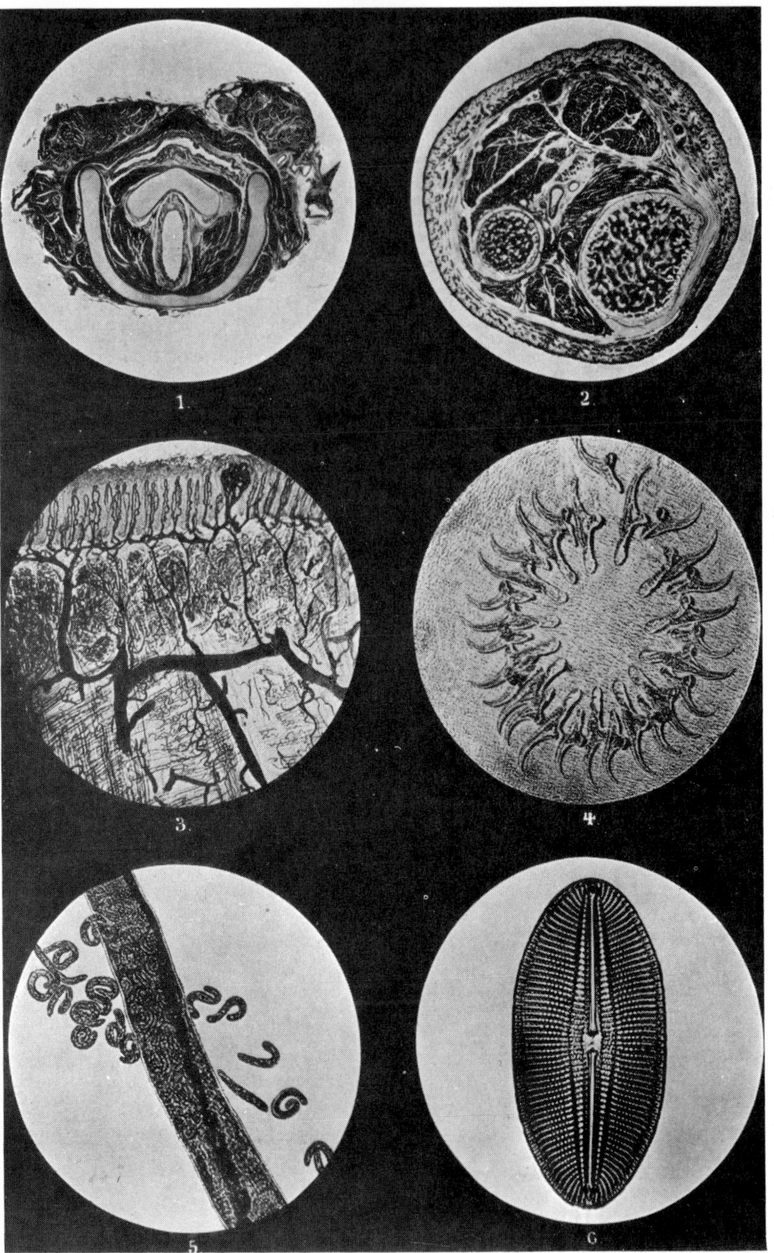

427

428

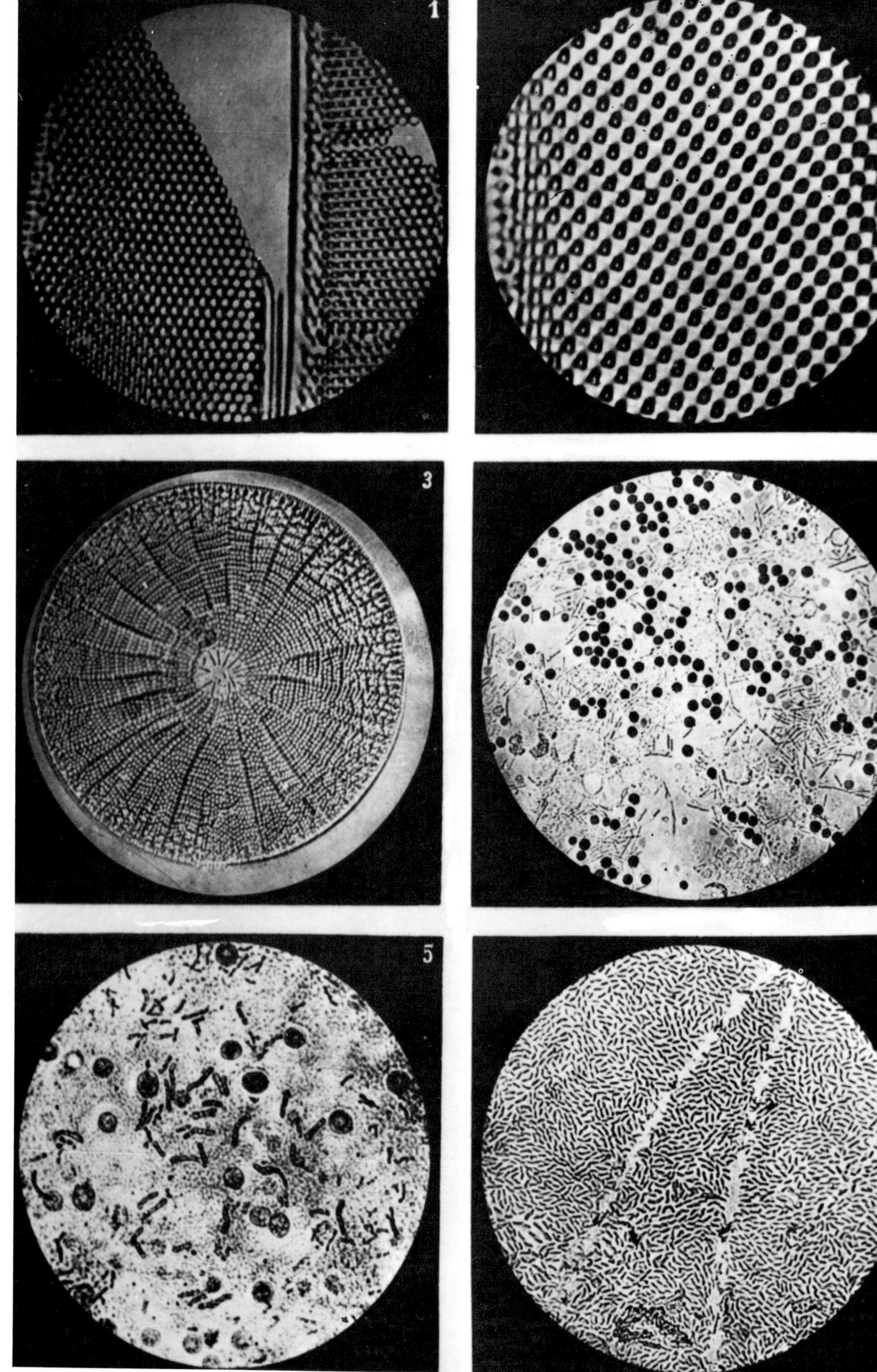

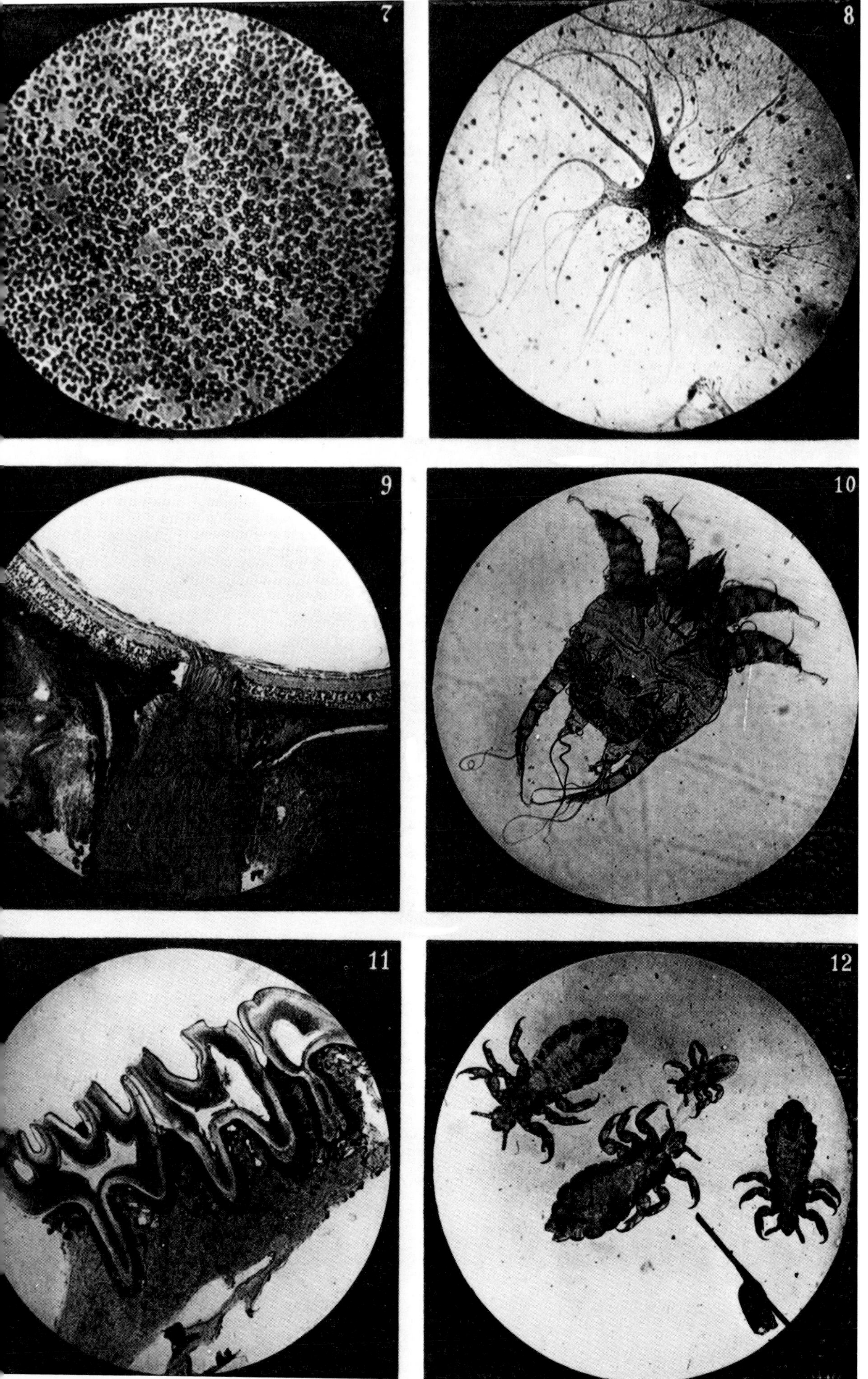
7
8
9
10
11
12

429

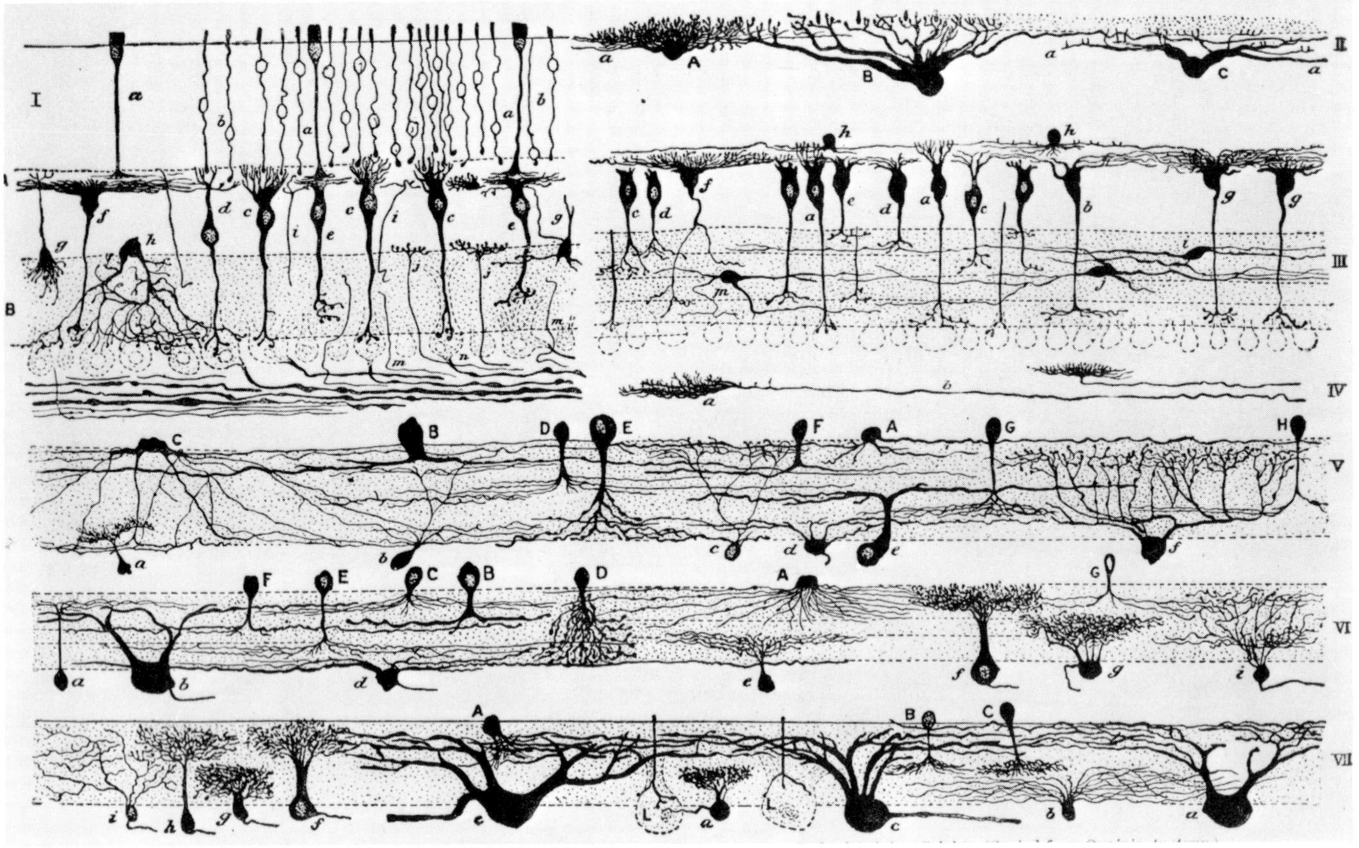

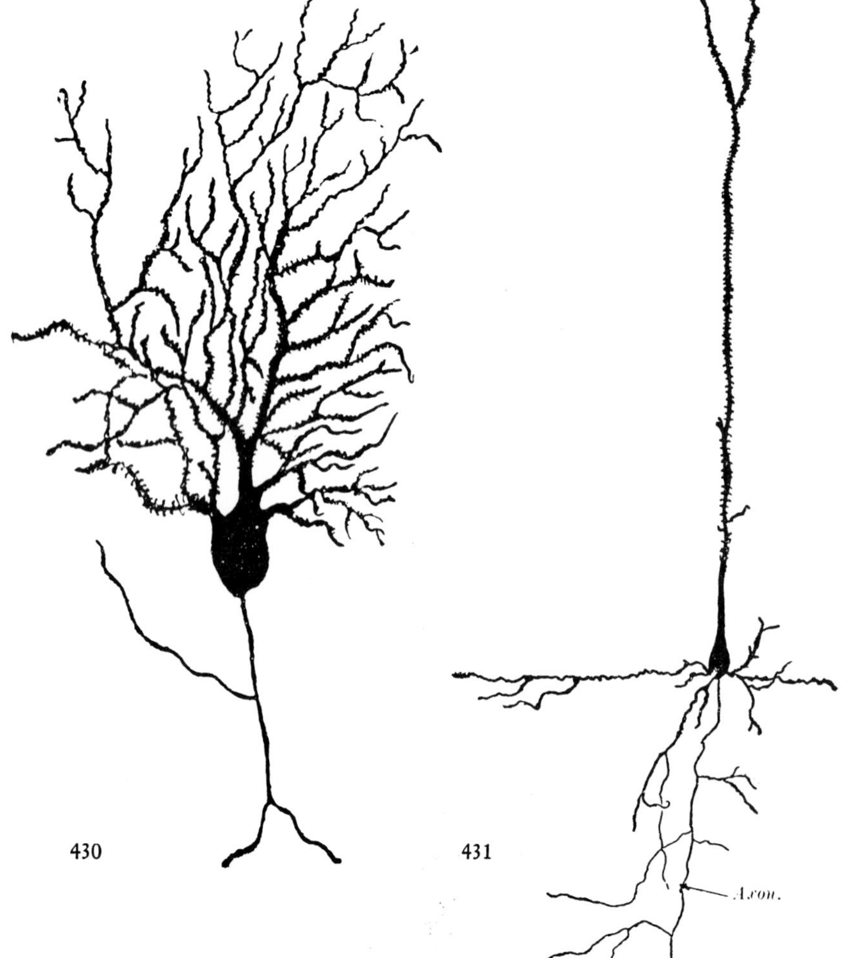

429. *Elements of the retina.* Greater detail could be revealed by combining photography with new methods of preparing the microscopic sample. Using chromate of silver, the Italian histologist Camillo Golgi was able to bring out considerable detail in the structure of the retina of mammals. Compare this with Figure 428, item 9.

430. *Cell of Purkinje.* Nerve cells, in particular, responded well to Golgi's method. Here is a Purkinje cell from the cerebellum of a cat, as photographed by the Spanish neurologist Santiago Ramón y Cajal.

431. *Pyramidal cell from the cerebral cortex of a mouse.* At the end of the nineteenth century, Ramón y Cajal was the foremost proponent of the neuron theory of the propagation of neural stimuli. This theory depended upon the ability to see the branching of nerve cells and, as this figure shows, the silver chromate method made this aspect of the neuron obvious.

432. *The details of fertilization.* Powerful microscopes and cameras permitted the recording of intimate cellular processes, such as the actual penetration of the sperm into an ovum. Comparison of this figure with the depiction of conception shown in Figure 310a will reveal how much new territory had been opened up by microscopists in the nineteenth century.

433. *Gastrulation.* Embryonic development also could be followed step by step on the microscopic level, from fertilization to the appearance of the mature embryo. Here Haeckel compares the early development of a worm (left), a frog (center), and a mammal (right).

432

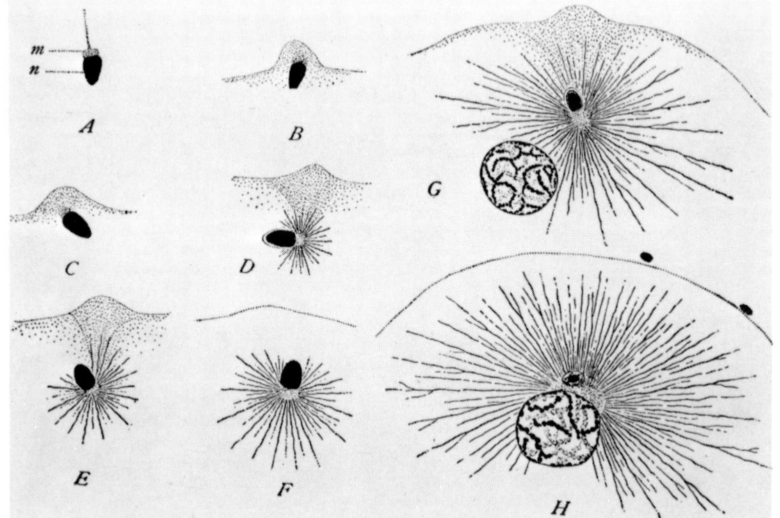

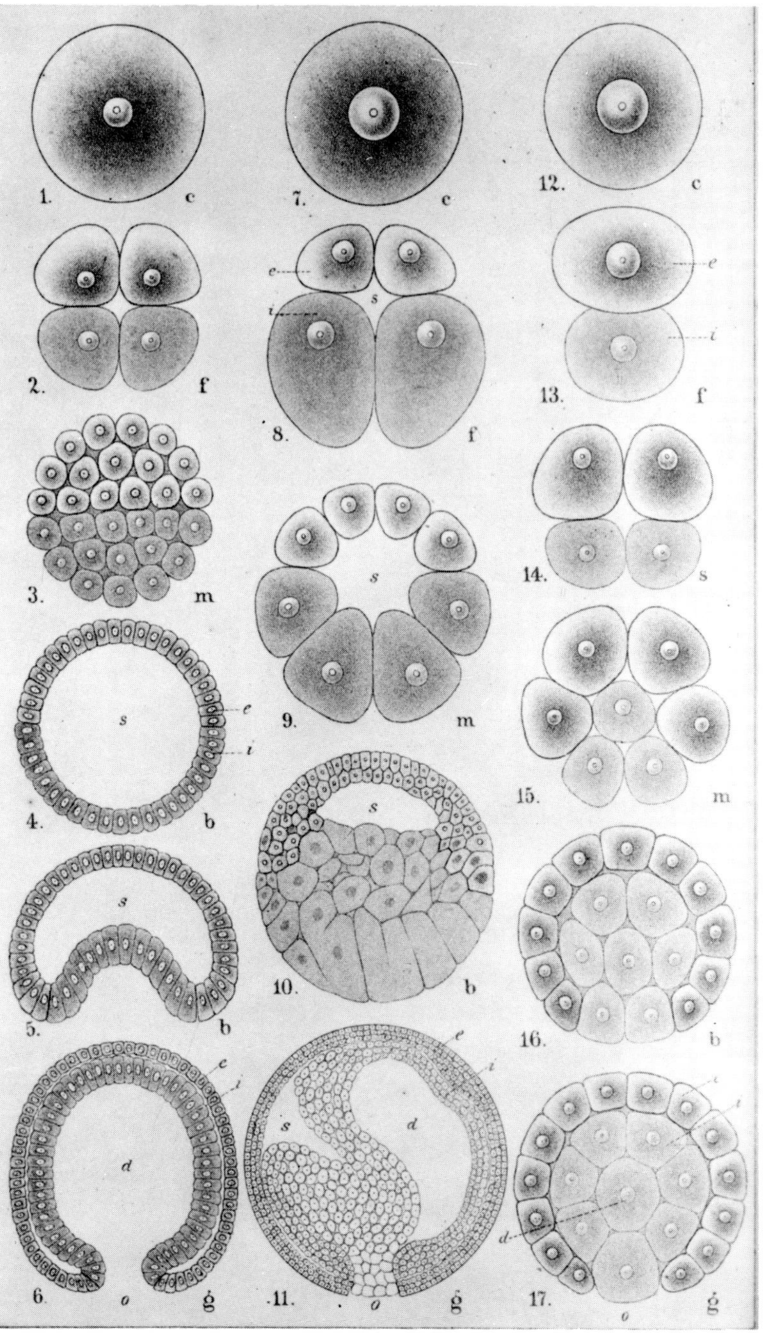

433

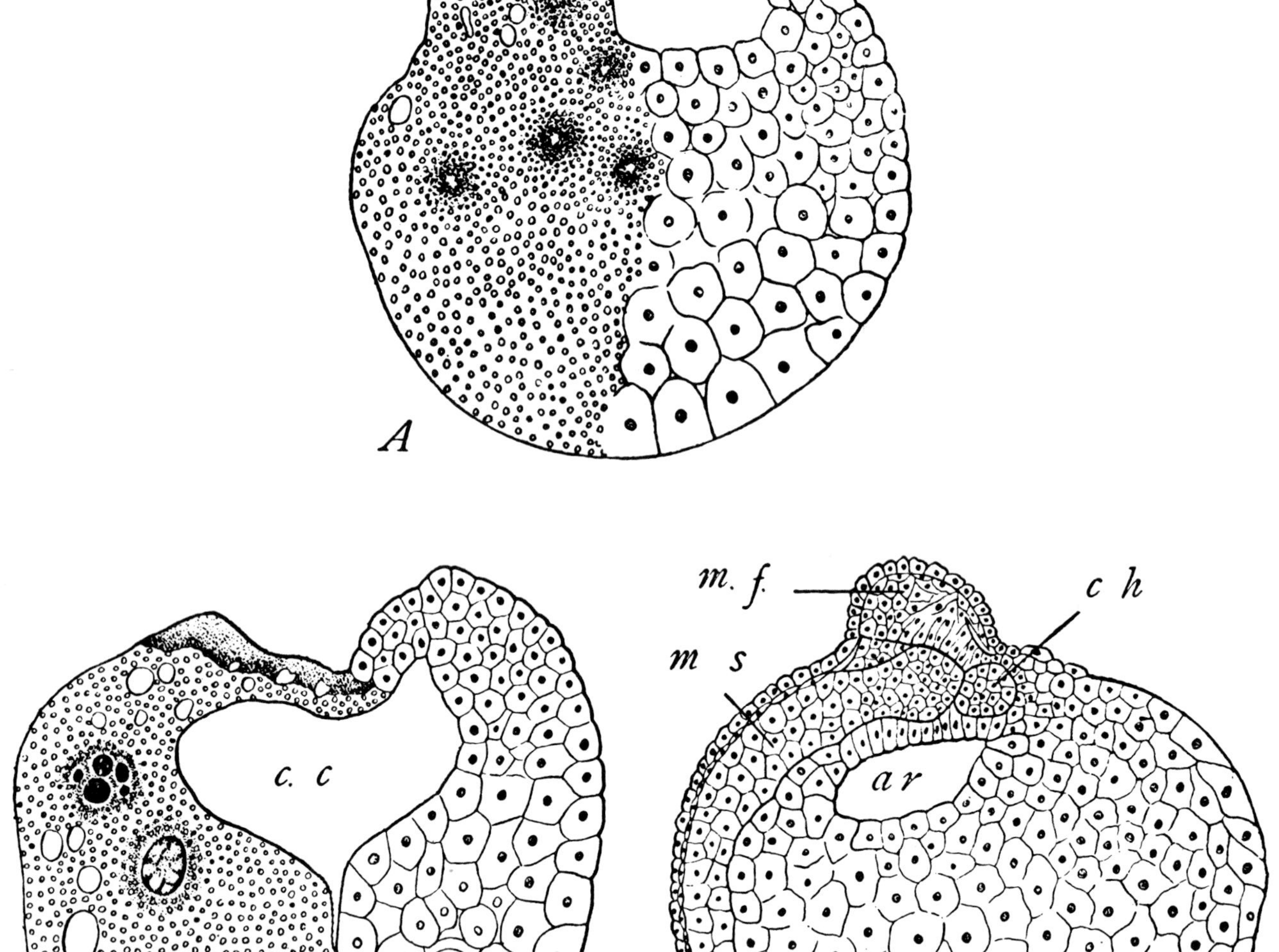

Fig. 131.—Half-embryos of the frog (in transverse section) arising from a blastomere of the 2-cell stage after killing the other blastomere. [Roux.]

A. Half-blastula (dead blastomere on the left). *B.* Later stage. *C.* Half-tadpole with one medullary fold and one mesoblast plate; regeneration of the missing (right) half in process.

ar., archenteric cavity: *c.c.*, cleavage-cavity; *ch*, notochord; *m.f.*, medullary fold; *ms.*, mesoblast-plate.

435a

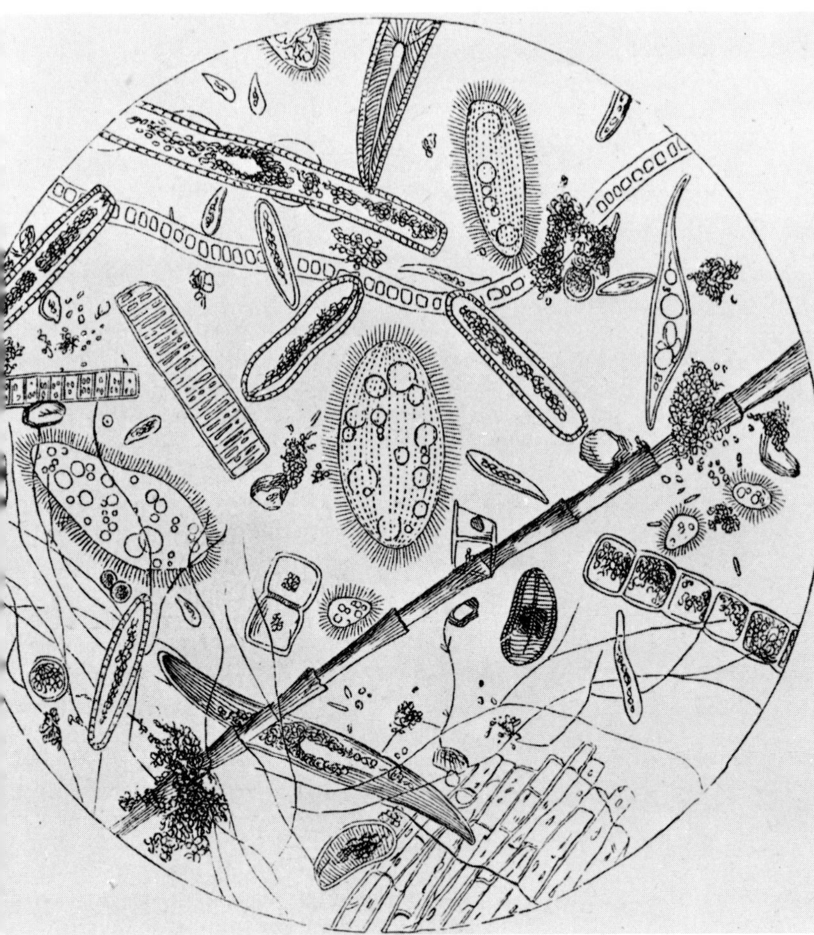

434. *A famous embryological experiment.* The ability to follow embryonic development on the microscopic level led, inevitably, to experimental intervention. In the 1880's Wilhelm Roux performed a famous experiment. Working with a fertilized frog's egg when it had divided into two cells, Roux killed one of the daughter cells with a hot needle. Only half an embryo developed, as shown here. This led Roux to suggest that the "plan" of embryological evolution was contained, like a mosaic, in the fertilized egg and that each division led to a partition of the mosaic plan. This "mosaic" theory was to be refuted in the twentieth century by Hans Driesch.

435a, b. *Microscopy and public health.* The suspicion that microorganisms might have some effect on health was relatively widespread long before the germ theory of disease was put forward as a scientific hypothesis. These examples—435a of well water and 435b of river water—show what drinking water in the nineteenth century contained.

436. *A drop of Spanish water.* This picture of the inhabitants of a *gota de agua* was presented to the Spanish public in 1840 for them to wonder at. We can wonder at the intestinal fortitude of people who could drink such water without being seriously affected. Compare these infusoria with those drawn by Ehrenberg only a few years before (Figures 404–406).

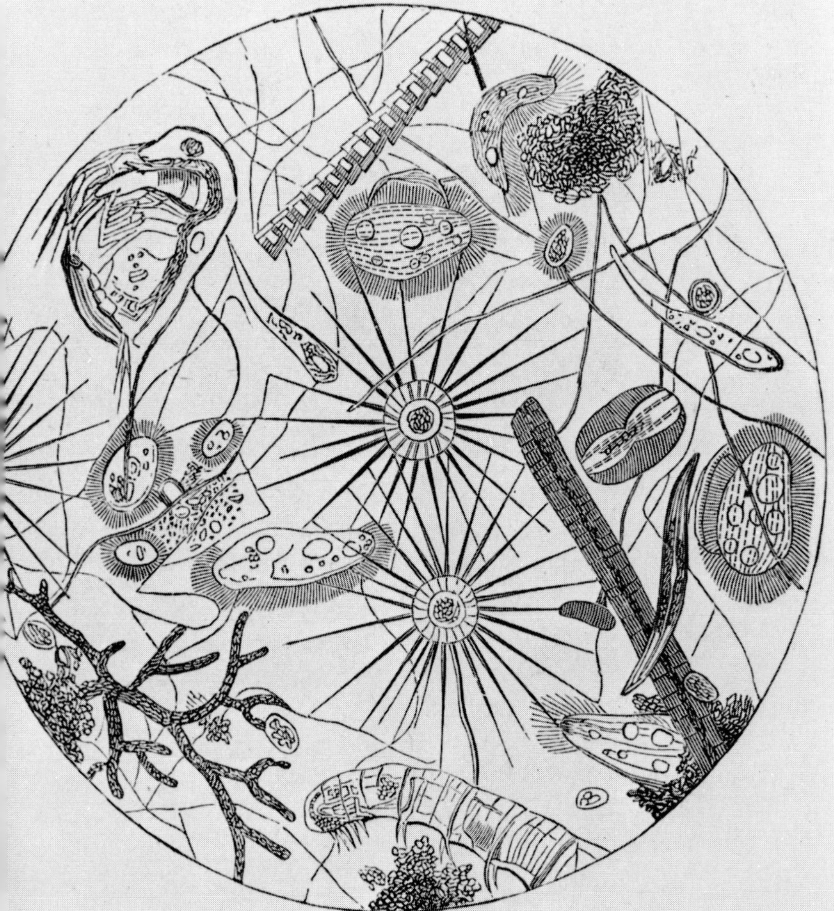

435b

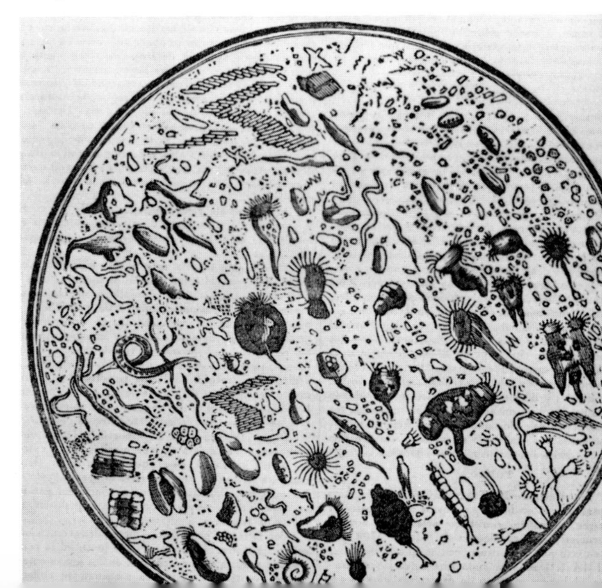

436

437. *Parliament under the satirical microscope. Punch* could not resist caricaturing the members of Parliament in terms of the new microscopical science. These strange political life forms could be as puzzling as germs.

V

Atoms and Molecules and Forces

The Laboratory and Its Instruments

The business of chemistry when the nineteenth century began was the separation of the elements from their compounds, and their purification and description. Compounds themselves were to be analyzed into their constituents so that their empirical formulae could be determined and then the compounds could be classified according to the scheme laid out by Lavoisier in his great *Treatise on Chemistry*, published in 1789. In short, chemistry was an analytical and taxonomic science that had little to do with either synthesis or explanation.

In order to analyze and classify, chemists had to have tools and a place to work. The chemical laboratory had existed, in one form or another, since late antiquity, but it was not until the nineteenth century that it became the standard place for chemists to work. The instruments tended to be rather simple when the century began, with the chemical balance probably the most accurate. It remained the essential tool of the chemist throughout the period. Most chemical processes used some kind of distillation at some point. Here the basic problem to be faced was that of adequate sealing of vessels and joints between tubes. Luting, as it was called, had become an art, involving all kinds of glues and pastes to prevent the escape of the vapors produced during distillation. Because it was almost impossible to achieve complete success through luting, the tendency was to avoid joints as much as possible and rely, instead, on the ingenuity and skill of the glassblower to make a single vessel in which the distillation could be carried out. Scientific glassblowing became a commercial art and chemists of the nineteenth century could buy apparatus ready made. One of the major advances in the construction of simple apparatus came with the ability to use rubber stoppers and tubing. With little trouble, rubber joints could be made airtight and more complicated apparatus could, then, be put together. The advance of industrialization also had its effect. The gas industry made the Bunsen burner possible and anyone who has ever spent any time in a chemical laboratory will appreciate what life must have been like before that instrument was available. Motors and pumps for creating either pressures or vacuums opened up new environments in which chemists could operate. Even the materials out of which utensils were made changed so that chemists late in the century could do things that earlier chemists could not. Henri Moissan, for example, used platinum distilling apparatus to isolate fluorine. The process of refining and working platinum so that it could be turned into usable forms was discovered in the early nineteenth century.

438

439

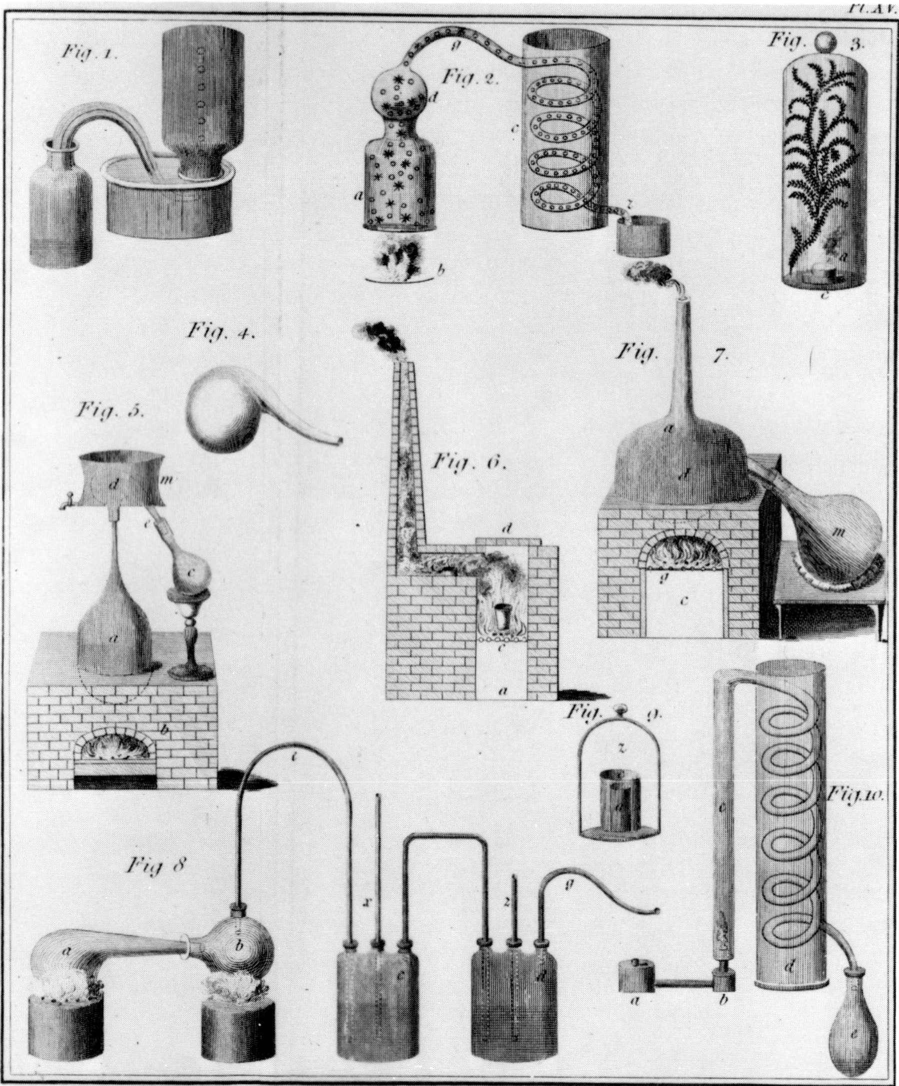

438. *The chemical balance.* The chemical balance remained one of the more accurate instruments available to the nineteenth-century scientist. Samples could be accurately weighed to at least 1/1,000th of a gram; this precision permitted chemists to analyze substances and have confidence in the results. The balance was the most important tool for probing the molecular world.

439. *Separation apparatus.* The chemist dealt practically with combinations and separations of substances. Some of the apparatus available in 1800 for separating constituents of compounds is shown here. Figure 1 is a simple pneumatic trough for the collection of gases. Figure 2 is a distillation apparatus; in it wine is heated (*a*) and the resulting alcohol, passing into a cooling coil surrounded by water, condenses into the receiver (*z*). The diagram nicely represents one contemporary theory of the nature of heat: the stars are fire particles that vaporize the wine by forcing apart the wine particles, here represented by circles. Figure 3 shows the sublimation of benzoin gum onto the surface of the enclosing glass jar. Stronger heats are used in Figures 5, 6, and 7 for separating the elements of various compounds. Figure 8 illustrates Peter Woulfe's apparatus for the separation of mixed gases. If the three-necked jar (*c*) is half filled with dilute limewater and the second jar (*d*) with alkaline solution, then carbon dioxide will be removed in the first jar and "acid" vapors in the second. Figure 10 is an apparatus for burning alcohol to water and then condensing the water; this illustrates weight gain through combustion.

440

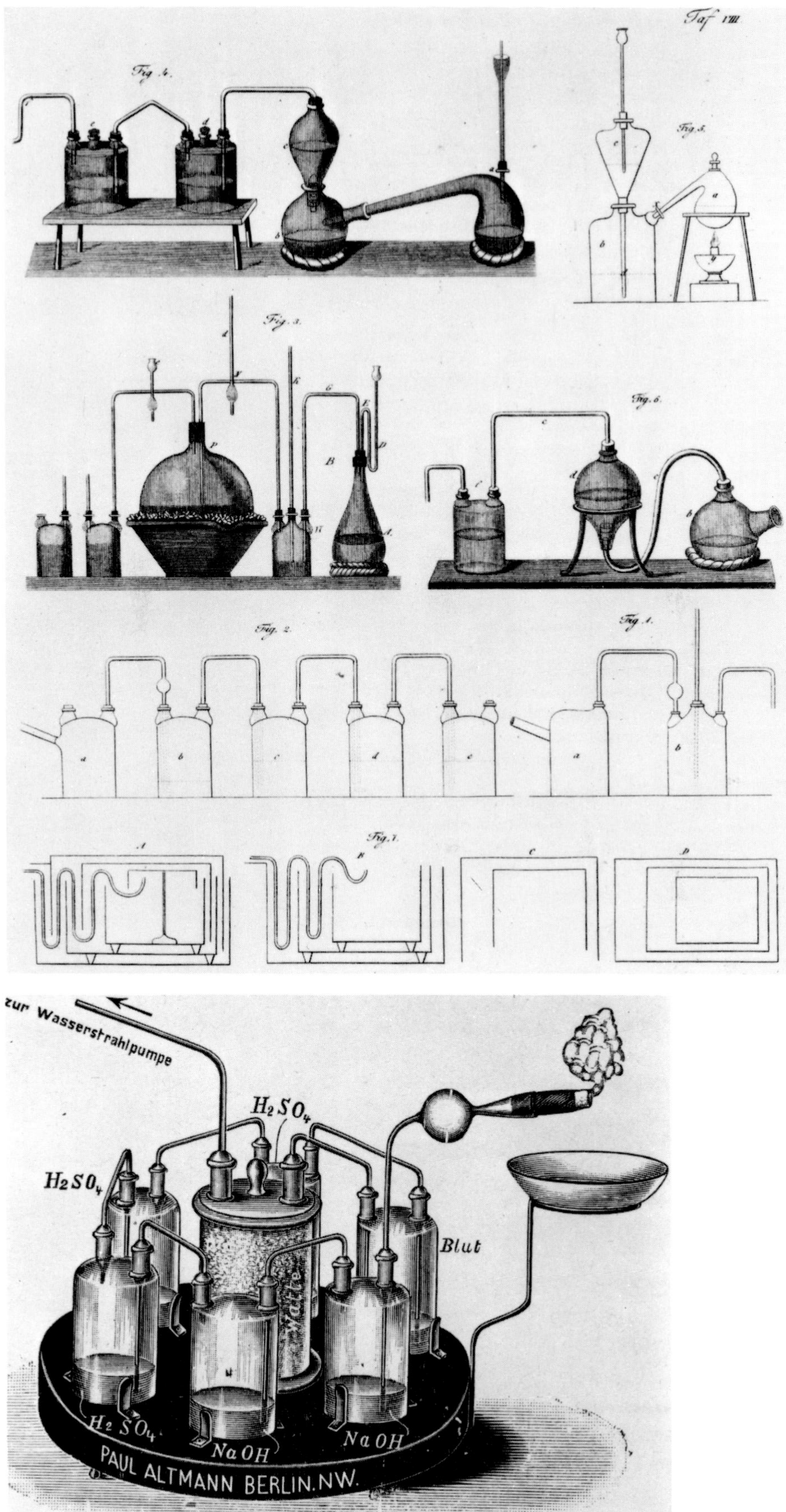

441

440. *Distillation apparatus (1825).* The basic principles of the distillation apparatus remained the same, but the "sensitivity" of the apparatus increased. By 1825, as this illustration from a journal devoted to the description of new laboratory equipment shows, the resolving power of the various stills had improved. Fractional distillation, shown in Figures 3 and 4, permitted the separation of more than two constituents. Trains of jars (Figure 2) could be set up for removing specific substances; and the various baffles shown in Figure 1 permitted chemists to work closely with gaseous mixtures.

441. *The analysis of tobacco smoke.* Ease in handling gases led to the investigation of all aspects of the gaseous state. The cigar was a nineteenth-century invention, and it seems only proper that it undergo a nineteenth-century analysis. The smoke was passed through sodium hydroxide and sulfuric acid, then through blood to determine its effects on the latter. There is a note of Victorian neatness in the ashtray held expectantly under the cigar.

442. *The Bunsen burner.* The Bunsen burner, invented by Robert Bunsen, took advantage of the new gas facilities that were being used to light the homes and streets of European cities. It consisted of a tube, a variable air intake, and a gas source. By twisting the tube, the mixture of air and gas could be varied until the flame was just right. This simple invention had an incalculable effect on the progress of chemistry; it made any warming or heating operation as simple as striking a spark.

443. *Electrolytic apparatus.* The discovery that an electric current could decompose water started chemists exploring ways to use static electricity for the same purpose. The electrolytic apparatus shown above could equally well have been used to electrolyze solutions with a steady current; all that would have to be changed was the electrodes.

444. *Apparatus for determining CO_2 content.* As the nineteenth century progressed, apparatus of increasing complexity could be built. This is the setup used in 1887 to determine the CO_2 content of complex substances such as foodstuffs. The glassblower's art has progressed since the beginning of the century, and rubber stoppers and tubing made it possible to seal and connect chemical vessels easily; previously, complicated luting methods had been required.

445. *Henri Moissan's apparatus for producing fluorine.* Peculiar problems, such as the isolation of the element fluorine, required peculiar apparatus. Fluorine is such an active element that special arrangements have to be made in order to collect it and keep it from combining with other substances. Since it combines readily with the silicon in glass, utensils of that substance could not be used. Only platinum was sufficiently inert and strong to serve the purpose. It was in this apparatus that Henri Moissan first isolated fluorine in 1886.

442

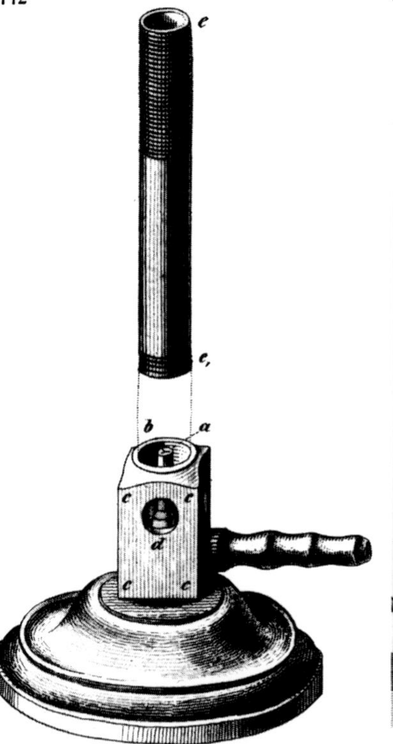

444

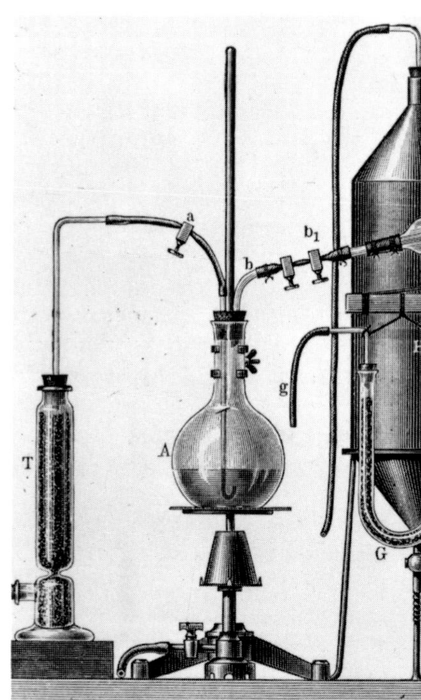

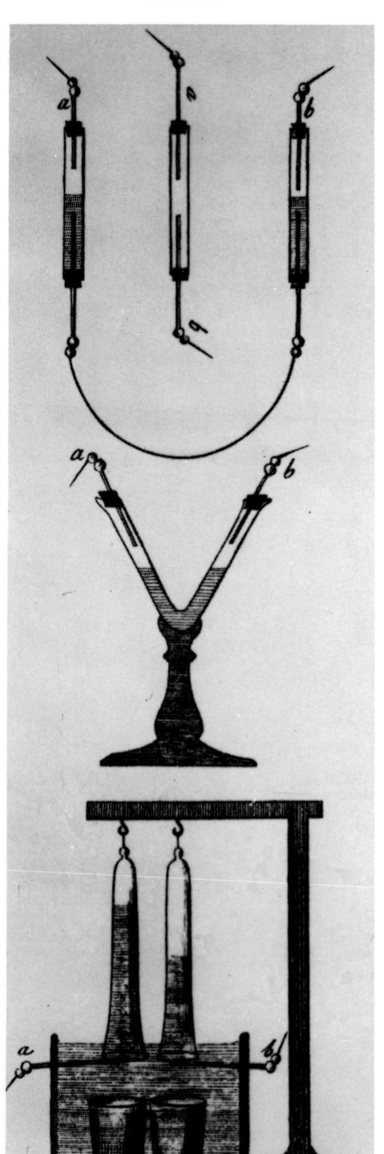

443

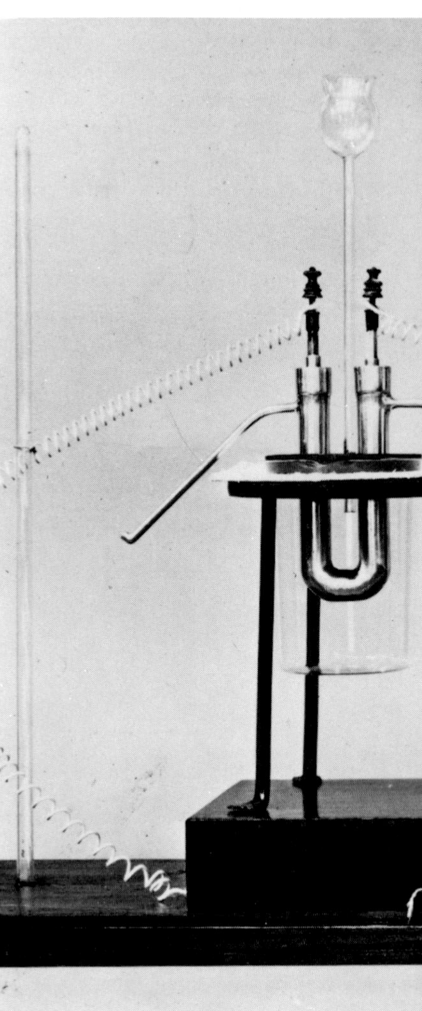

445

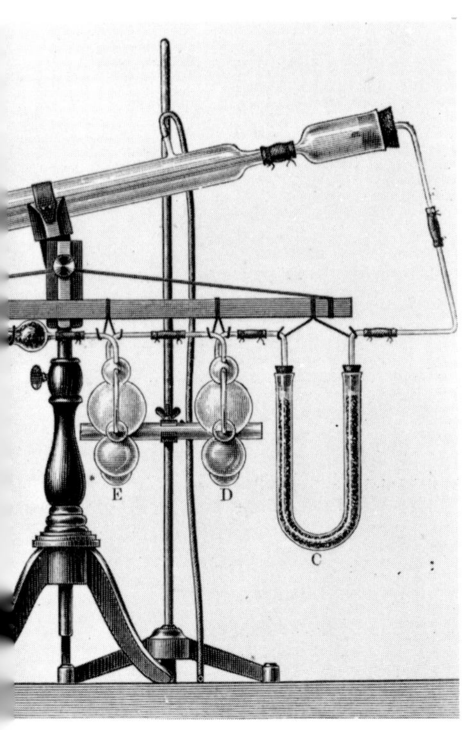

446

446. *The synthesis of acetylene.* Advances in the synthesis of substances lagged far behind developments in chemists' ability to decompose compounds. It was only in the second half of the nineteenth century that the art of synthesis matured. Probably the most dramatic moment in this maturation came with Marcellin Berthellot's 1862 synthesis of acetylene in this apparatus, using hydrogen gas and the carbon in carbon-arc electrodes. From acetylene, he could make other "organic" substances such as alcohol, thus destroying forever the notion that "organic" chemicals required some kind of vital force for their creation.

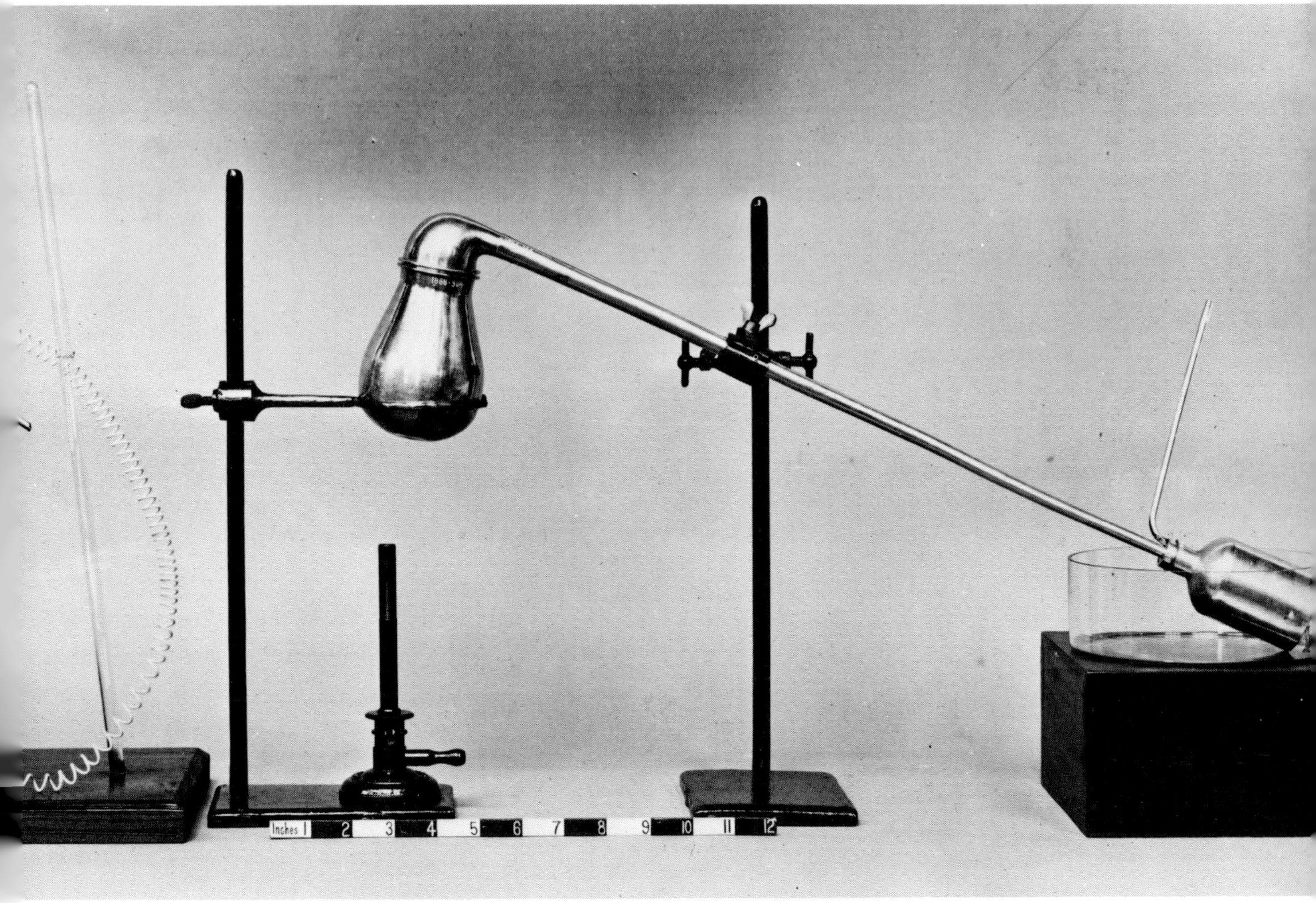

Atoms and Molecules

When the nineteenth century began, atoms and molecules had been exiled from chemistry as metaphysical entities that could not properly be studied in the laboratory. Lavoisier wrote that it is probable that we know and could know little or anything about the ultimate particles of matter. Better, he advised, to stay close to observables for only then could chemistry become a precise and accurate science. He followed his own advice when he gave his definition of a chemical element. It was, he suggested, the last point of a laboratory analysis. When you had decomposed a body as far as it would go, then what you were left with were the elements out of which it had been made up.

Given the elements, and given the methods of analysis, it ought to be possible to make precise tables of the elements and their combinations. This is what Lavoisier and his followers set out to do. Lavoisier recognized that two elements could combine in more than one way and it was he who devised the endings of *-ic* and *-ous* to indicate the degree of saturation of one element for another in a compound. There was no theoretical limit, however, to the number of different compounds that two elements could form; the job of the chemist was simply to discover as many as possible and put them in the right order. Nor did Lavoisier's chemistry try to deal with the causes for compound formation or with the reasons for some elements joining together and for others remaining aloof.

Fortunately for chemistry, the tendency to conceive of the elements in atomic terms had persisted in scientific thought as a quasi-philosophical doctrine. Isaac Newton, after all, had given his stamp of approval to the atomic doctrine and had even attempted to justify it philosophically. In his Third Rule of Reasoning, in the *Principia,* Newton had postulated that the properties of sub-sensible entities like atoms could be apprehended through experiment. He argued that those qualities of gross bodies that do not change with circumstances, such as inertia, mass, impenetrability, and so on could be considered the properties of all bodies, even atoms. That being the case, Newton insisted, it was both legitimate and necessary for natural philosophers to consider atoms with these properties in their natural philosophies. It was a persuasive argument and one of Newton's more ardent disciples of the early nineteenth century, John Dalton (1766–1844), took it very much to heart. Dalton's work in meteorology and on the solubility of gases led him to hypothesize that the gaseous elements differed from one another by one essential quality, their weight. When this hypothesis was made, then a number of otherwise inexplicable facts could be easily understood and Dalton enthusiastically extended his "discovery" from the gases to all elements.

Dalton's atoms were hard, billiard-ball-like objects surrounded by an atmosphere of heat, or caloric, that clumped together to form compounds. This did account for the observed facts, but it also left untouched some of the questions about the nature of chemical composition. That problem, together with a number of other chemical questions of some subtlety, could be attacked if one adopted another atomic theory that enjoyed some popularity in the nineteenth century. This theory argued that atoms were mere mathematical points surrounded by "shells" of attractive and repulsive forces. When two or more of these atoms joined together, their resultant force field was a complicated one and it was this complication that could be used to explain elective affinities, or the willingness of an element to combine with one element but not with another. This atomic doctrine had been devised in the eighteenth century by R. J. Bošković and provided some of its adherents in the nineteenth century with considerable insights into chemical mechanics.

Not everyone was immediately persuaded

of the truth of Dalton's (or Bošković's) atomic doctrine and it took some time for it to become a central part of chemistry. Part of the reason for hesitation, as far as Dalton's atomic theory was concerned, was that it did not address itself to the question of elective affinities. Dalton relied upon gravity to hold his compounds together but since gravity is both universal and undiscriminating, it could not explain why some reactions "go" and others do not. The problem did not appear insoluble to a follower of Dalton. The Swedish chemist Jöns Jacob Berzelius (1779–1848) simply added the two electrical "fluids" of + and − electricity to Dalton's atoms and then insisted that chemical combination was the result of electrostatic forces. This accorded well with the new electrochemistry and also provided a working and workable model to account for the mechanics of chemical combination. Berzelius' influence was to be deep and abiding throughout the century and chemical bonding was to be considered always as an electrical phenomenon bearing some resemblance to Berzelius' early theory of it.

As the century progressed and chemists became more and more subtle in their ability to manipulate matter, one point began to come to the foreground. It was not enough to know the exact composition of a body to know its chemical nature. Compounds were discovered that contained exactly the same number of the same atoms, but which had different chemical characteristics. Such isomers were to be found particularly in organic chemistry and they embarrassed organic chemists a good deal. Indeed, organic chemistry was a rather embarrassing study. It was not that one could not know anything about organic chemicals because of some "vital force" excluding them from the conceptual scheme of chemistry as a whole. Before many years of the century had passed there were few organic chemists who let that belief play a role in their research or hold it back. It was that there was such a bewildering array of chemical compounds that could be made out of so few elements. Just carbon and hydrogen could produce a long list of compounds and it was difficult, if not impossible, to understand how this could be so. If, as Berzelius insisted, chemical combination was a matter of electrostatic attractions, then the carbon and hydrogen atoms ought to run out of excess charges before their combination grew into large molecules. Unless atoms of similar electrical charge could somehow be attached together, there seemed to be no way out of the theoretical impasse that the progress of organic practice had created. This was precisely the leap that August Kekulé took in the early 1860's when he suggested that two carbon atoms could be attached to one another by a bond, and that the carbon atom had four bonds by which it could tie on to other elements.

Kekulé's hypothesis came just when other chemists realized that Berzelius' theory had to be wrong, particularly in the realm of gases. Avogadro had suggested in 1811 that the "particles" of a gaseous element were composed of a group of atoms of the same element combined into a single molecule. Thus, oxygen gas was really made up of molecules containing two atoms of oxygen (O_2) instead of one atom (O). Berzelius on the other hand had rejected this hypothesis as electrostatically impossible. In 1860 at a chemical conference in Karlsruhe, Germany, Stanislao Cannizzaro had persuaded his colleagues that Avogadro had been right. The consequence of this was an important new means for determining accurately the atomic weights of gases since Avogadro had shown that equal volumes of gases at standard temperature and pressure must contain equal numbers of molecules. From this fact, it was easy to deduce that 22.4 liters of a gas must weigh, in grams, the molecular weight of the gas. Armed with this new insight, chemists could and did accurately measure the atomic weight of carbon, a particularly hard element to deal with in this regard, and were thereby able to clarify large numbers of empirical formulae in organic chemistry.

Given the newly determined atomic weight of carbon as 12 and the carbon-carbon bond, men such as Kekulé were able to put together organic molecules by using models to illustrate them. Kekulé and his contemporaries insisted that these models were not to be taken seriously, but they had to be since they were immediately found of enormous help in understanding organic reactions and the properties of organic compounds. The reality of the model became inescapable in 1865 when Kekulé solved the problem of benzene. Benzene had the empirical formula C_6H_6 and if constructed as a chain hydrocarbon, would contain a number of double and triple bonds that do not accord at all with its chemical activity. Kekulé, in a dream, saw the benzene molecule as a ring and when he applied this idea in the laboratory, discovered that it worked amazingly well to explain the properties of benzene. Whatever else was false in this model, the ring structure, Kekulé was certain, had to be true.

The idea of structure being responsible for chemical properties was picked up immediately and applied in rather wild and speculative ways. It became clear that more had to be done than simply to make up models by using the imagination and then showing that these molecules would be able to explain certain chemical facts. There had to be evidence for every step in the model building or else the models would become metaphysical fancies of little or no scientific use. It was this method of showing how and why certain structures had to be true that J. H. van't Hoff and Joseph Le Bel introduced into chemistry in the 1870's. Stereochemistry was to become as sound and as hardheaded as any other branch of the science in the hands of such masters as van't Hoff and Johannes Wislicenus. By the end of the century, only a few chemists could still doubt the existence of atoms and molecules. Most were firmly convinced of their reality and most, in fact, had become exceedingly skillful in working with them, building compounds out of wooden balls and wire before attempting to synthesize them in the laboratory. For most chemists, atoms and molecules were here to stay.

The idea that structure might give rise to properties was not restricted to chemical compounds. What about the elements, themselves? Was there any sufficient reason to insist that every element was a hard, impenetrable, elastic sphere that differed from other elements only in its weight? As the number of known elements increased by leaps and bounds during the century, the temptation became increasingly strong to answer with a loud No, and look around to find some way to justify this stand. The first hypothesis that implied a structure to the elements was that put forward anonymously by William Prout in 1815. Prout suggested that all elements were merely compounds of hydrogen and that, therefore, all atomic weights would be found to be integral multiples of that of hydrogen. When Berzelius showed that the atomic weight of chlorine (compared to hydrogen as 1) was 35.5 that seemed to sound the death knell of that speculation. Yet, it kept bobbing up again and again throughout the century. It did so for a very good reason. The elements were not separate and completely different from one another; their properties tended to repeat in other elements. Chlorine, iodine, and bromine, for example, underwent similar chemical reactions and acted very much alike under similar chemical conditions. Other elements also fell into natural families and there seemed to be good reason to try and classify elements according to these repeating properties. The early attempts at putting the elements into some natural order all failed but even in their failure they did call attention to the periodic nature of chemical elements. When Dmitry Mendeleev finally composed his periodic table in 1869, he knew enough to realize that gaps had to be left for elements as yet undiscovered. It was this fact that allowed him to classify the elements more accurately than his predecessors and that made his table a research guide as well as a handy summary. He actually predicted the properties of the undiscovered elements by closely considering their neighbors. When the gaps were filled in by other chemists and physicists, the reality of his classification was recognized, for the properties were as predicted. And, of course, the whole classification implied that elements were structured. The periodicity of properties almost necessarily entailed periodicity of structure. But the question of *how* atoms were structured baffled the minds of the nineteenth century. It was, in fact, a problem that was only to be solved in the twentieth but it did raise a new issue in the nineteenth. If atoms had a structure, they must be made of something smaller than atoms. What could such subatomic particles be? And, could they be detected in the laboratory or were they destined to remain "metaphysical" entities used by man's mind to solve problems that the senses could not deal with?

447. *Atoms and forces: Bošković's curve.* In mideighteenth century, a Jesuit, Roger Joseph Bošković, published a large folio volume based upon the curve shown here. The curve represented the forces of attraction (below the line *c-c′*) and repulsion (above *c-c′*) encountered if a test particle were brought toward an "atom," which is indicated by the mathematical point center of these forces at *A*. On the right side of the curve (*S-V*) is the hyperbola of universal attraction, decreasing as the square of the distance from *S*. Between *S* and *F*, there are alternating repulsive and attractive forces, which Bošković suggested could account for chemical combinations, explosions, and a host of other chemical phenomena. From *F* to *D*, the repulsive force increases, becoming infinite at *A* and thereby preserving the essential "atomic" quality of impenetrability. Bošković's "force atom" was well known in the nineteenth century and influenced the thinking of such chemists as Humphry Davy, Michael Faraday, and William Thomson (later Lord Kelvin).

447

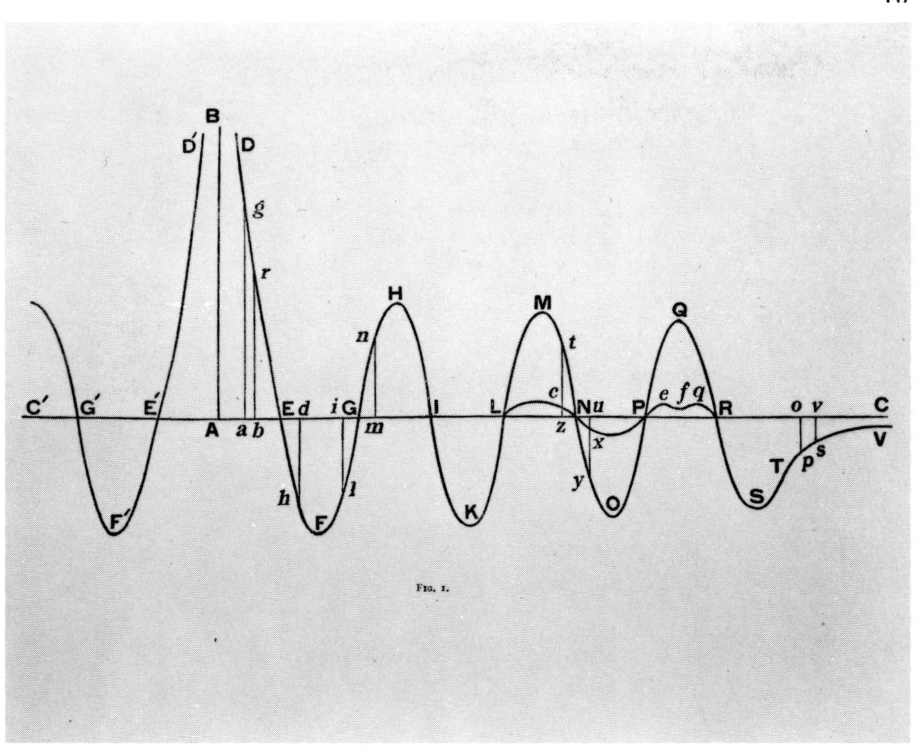

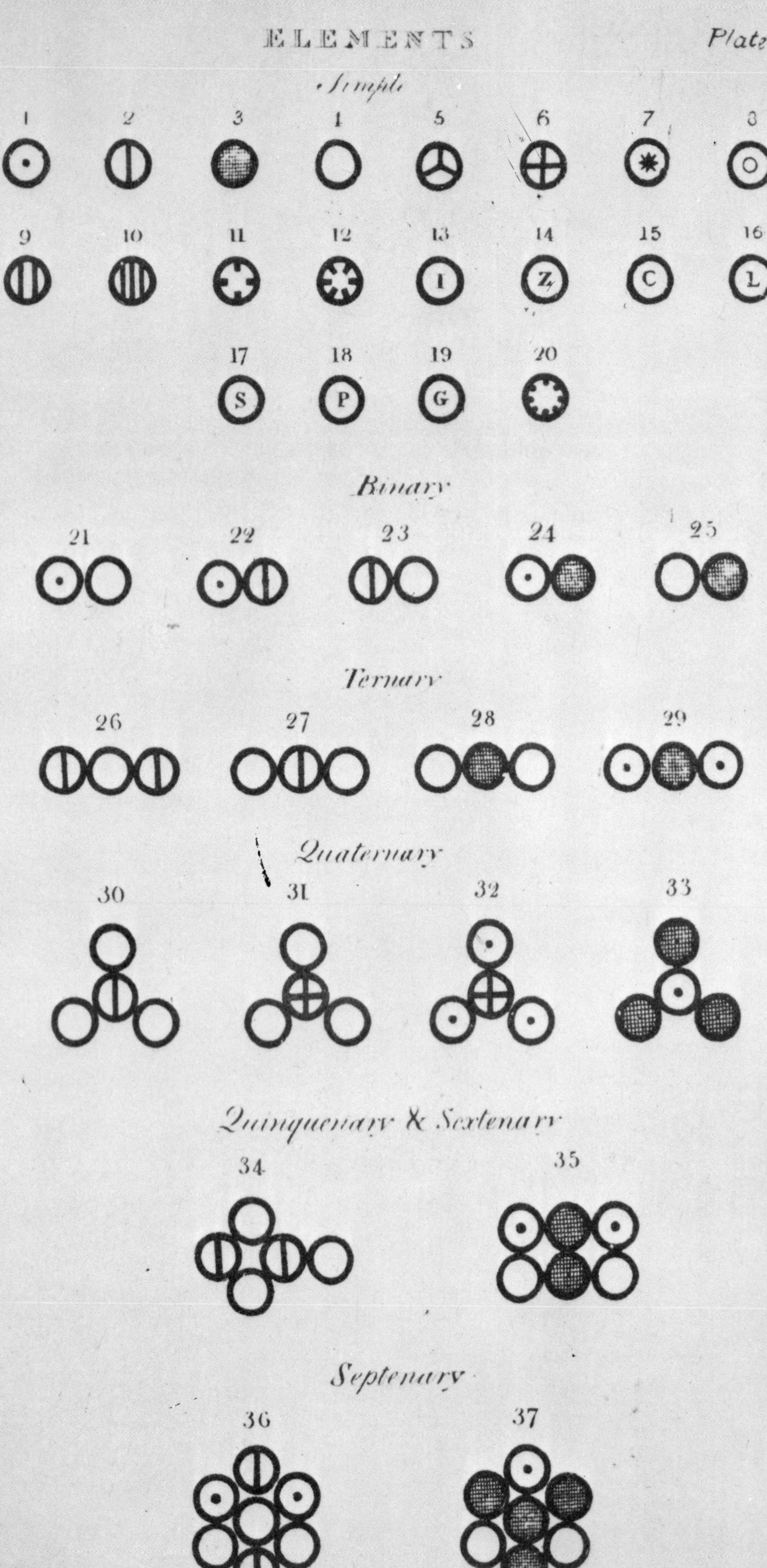

448a, b. *John Dalton's Atoms and Molecules.* When atoms were spoken of in the early nineteenth century, they were generally regarded as similar to very small billiard balls. Because of their small size, most chemists felt that little or nothing could be known of them. The English Quaker, John Dalton, disagreed. In *A New System of Chemical Philosophy*, he argued in 1808 that chemical elements differed in weight, and that their relative weights could be determined in the laboratory. It then seemed only natural to Dalton to represent substances by simple pictures of their atoms (at top, 448a). In numerical order from 1 to 20, these are hydrogen, nitrogen, carbon, oxygen, phosphorus, sulphur, magnesia (the oxide, for magnesium had not yet been isolated), lime (the same is true for calcium), soda, potash (Humphry Davy had just isolated sodium and potassium, but not before Dalton had written), stronties, barytes, iron, zinc, copper, lead, silver, platina, gold and mercury. Compounds could then be represented, as at bottom, simply by the juxtaposition of the symbols for the elements. Dalton assumed that the formula for the compound of two elements that did not form any other compound must be simply one atom of each. Thus water (21) was HO, ammonia (22) was HN, and so on. Beginning with number 23, the compounds shown here are ethylene, carbon monoxide, nitrous oxide, nitrous acid, carbon dioxide, methane, nitric acid, sulfuric acid, hydrogen sulphide, alcohol, N_2O_3, acetic acid, nitrate of ammonia, sugar. These representations were not intended by Dalton to depict anything but the composition of the compounds; but the use of symbols such as these over the course of years led almost inevitably to the idea that chemical formulas also might be used to represent the actual structure of molecules, as seen in figure 448b. It was to take almost sixty years for that chemical dream to come true.

448a

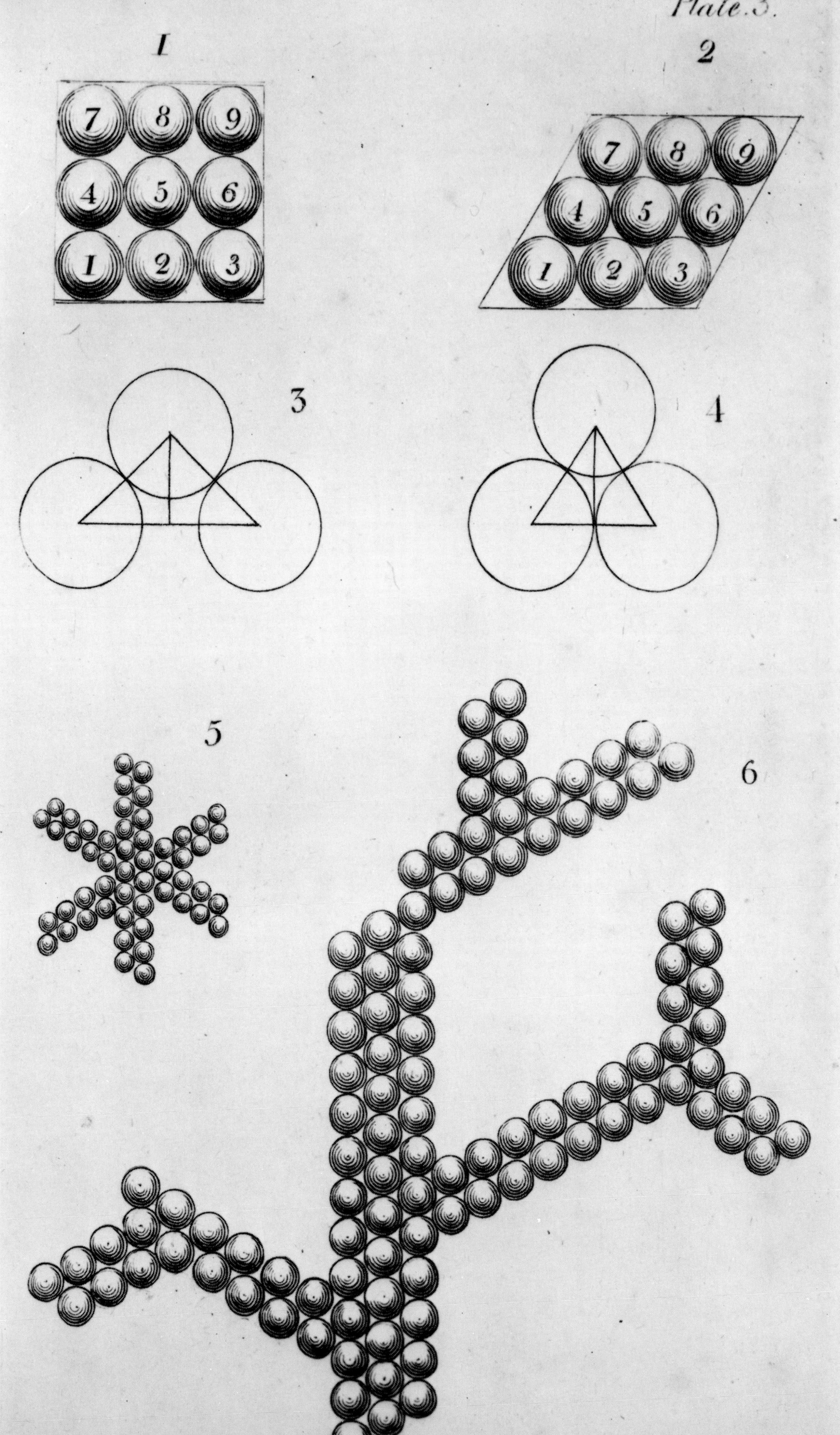

448b

449

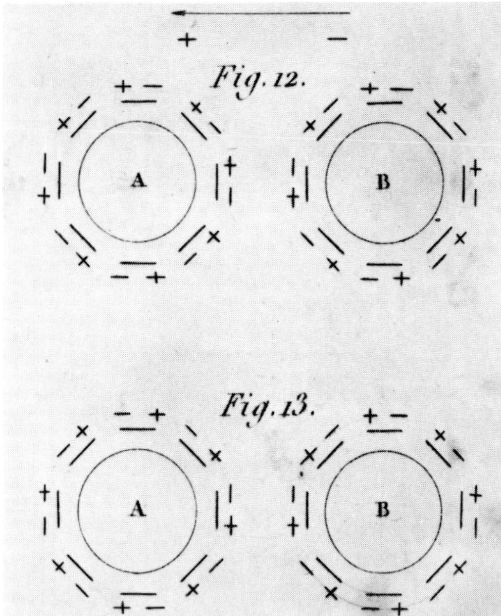

449. *Electrochemical atoms (after Berzelius).* When John Dalton developed his new atomic system, he paid no attention to the recent discoveries in electrochemistry. But electrochemistry could not be ignored. If an atomic doctrine were to make headway, it would have to account for the role of electrical charges in atomic dynamics. This was done by Jöns Jacob Berzelius, who illustrated the role of electricity in chemistry in a treatise published in 1818. When, as in Figure 12, the distribution of electrical charges was such that opposite charges faced one another, atoms combined; but when, as in Figure 13, similar charges faced one another, they repelled each other and no reaction occurred. Berzelius did not try to explain the distributions of charges.

450. *Berzelius' chemical symbols.* In a table, published in 1818 separately from his treatise on chemistry, Berzelius described the composition of a large number of compounds, using his new chemical symbolism that made chemistry much simpler and easier to understand. (Compare Berzelius' compounds, for example, with Dalton's in Figure 448b.) Berzelius' can be written in a moment; Dalton's take some time. Furthermore, Berzelius' symbols make no statement about molecular structure.

450

Exempel af några Dubbelsalters sammansättning.

Namn.	Formel.	Partikelns vigt.	Starkaste basis	Svagare basis ell. syra	Starkaste syran.	Vatten.
Carbonas Magnesico Calcicus, Bitterspat, Dolomit . .	$\ddot{Ca}\ddot{C}^2+\ddot{Mg}\ddot{C}^2$	2330,10	30,56	22,18	47,26	
Fluosilicias Ammonicus . .	$(3\dot{N}H^6+2\dddot{Si})+(3\dot{N}H^6+3\ddot{F})$	3310,57	38,88	36,13	24,99	
Hydricus . .	$3\ddot{F}Aq^2+2\dddot{Si}^2\ddot{F}^3$	5540,55		43,06	44,67	12,27
Kalicus . . .	$3\ddot{K}\ddot{F}+2\dddot{Si}^2\ddot{F}^3$	8400,44	42,13	28,41	29,46	
Oxalas Ammonico-Cupricus .	$2\dot{N}H^6\overline{O}Aq^2+\ddot{Cu}\overline{O}^2Aq^2$	3908,95	11,02	25,36	46,23	17,39
biammonico-Cupricus	$(3\dot{N}H^6+2\overline{O})+2\ddot{Cu}^3\overline{O}^2+Aq$	9418,55	6,86	63,16	28,78	1,20
triammonico-Cupricus	$2(3\dot{N}H^6+\overline{O}Aq^3)+\ddot{Cu}^3\overline{O}^2Aq^6$	7433,09	17,39	40,01	24,31	18,29
Oxalas Kalico-Cupricus, Var. 1:ma	$\ddot{K}\overline{O}^2Aq+\ddot{Cu}\overline{O}^2Aq$	4204,83	28,05	23,58	42,98	5,39
Var. 2:da	$\ddot{K}\overline{O}^2Aq^2+\ddot{Cu}\overline{O}^2Aq^2$	4431,37	26,63	22,37	40,78	10,22
Natrico-Cupricus .	$\ddot{Na}\overline{O}^2Aq+\ddot{Cu}\overline{O}^2Aq$	3806,84	20,54	26,04	47,47	5,95
Murias Ammonico-ferrosus .	$2\dot{N}H^6\ddot{M}+\ddot{Fe}\ddot{M}^2$	3376,19	12,76	26,02	61,22	
Hydrargyricus	$2\dot{N}H^6\ddot{M}Aq+\ddot{Hg}\ddot{M}^2$	5455,89	7,90	50,07	37,88	4,15
Platinicus .	$2\dot{N}H^6\ddot{M}Aq+\ddot{Pt}\ddot{M}^2$	4139,52	10,41	34,19	49,93	5,47
Kalico-platinicus .	$\ddot{K}\ddot{M}^2+\ddot{Pt}\ddot{M}^2$	4661,94	25,30	30,36	44,34	
Natrico-Platinicus .	$\ddot{N}\ddot{M}^2+\ddot{Pt}\ddot{M}^2$	4263,95	18,34	33,19	48,47	
Hydro-Carbonas Cupricus .	$\ddot{Cu}Aq^2+2\ddot{Cu}\ddot{C}^2$	4302,02		69,13	25,60	5,27
Magnesicus . (v. Magnesia Alba).	$\ddot{Mg}Aq^8+3\ddot{Mg}\ddot{C}^2$	4625,00	44,69		35,72	19,59
Zincicus .	$\ddot{Zn}Aq^6+3\ddot{Zn}\ddot{C}$	5531,39	72,78		14,93	12,29
Muriocarbonas Plumbicus .	$\ddot{Pb}\ddot{M}^2+\ddot{Pb}\ddot{C}^2$	6813,96	81,86	8,08	10,06	
Sulphas Aluminico-Ammonicus	$\dot{N}H^6\dddot{S}+\ddot{Al}\dddot{S}^3$	2862,40	7,53	22,44	70,03	
Kalicus . (Alumen).	$\ddot{K}\dddot{S}^2+2\ddot{Al}\dddot{S}^3$	6473,75	18,23	19,84	61,93	
c. Aqua	$\ddot{K}\dddot{S}^2+2\ddot{Al}\dddot{S}^3+48Aq.$	11910,60	10,15	10,54	33,66	45,65
Natricus .	$\ddot{Na}\dddot{S}^2+2\ddot{Al}\dddot{S}^3$	10085,04	7,75	12,74	79,51	
Ammonico-Cupricus .	$2\dot{N}H^6\dddot{S}Aq^2+\ddot{Cu}\dddot{S}^2Aq^{10}$	7017,30	6,15	14,13	57,13	22,59
triammonico-cupricus (Cuprum Ammoniacum).	$4(3\dot{N}H^6+\dddot{S})+\ddot{Cu}^3\dddot{S}^2Aq^6$	9246,01	27,96	32,17	32,52	7,35
Ammonico-Kalicus .	$\ddot{K}\dddot{S}^2+2\dot{N}H^6\dddot{S}Aq^2$	6073,06	19,43	7,09	66,02	7,46
Magnesicus .	$2\dot{N}H^6\dddot{S}Aq^2+\ddot{Mg}\dddot{S}^2Aq^{10}$	6542,63	6,59	7,90	61,28	24,23
Calcico-Natricus . (Glauberit).	$\ddot{Na}\dddot{S}^2+\ddot{Ca}\dddot{S}^2$	5503,18	14,21	12,94	72,85	

X

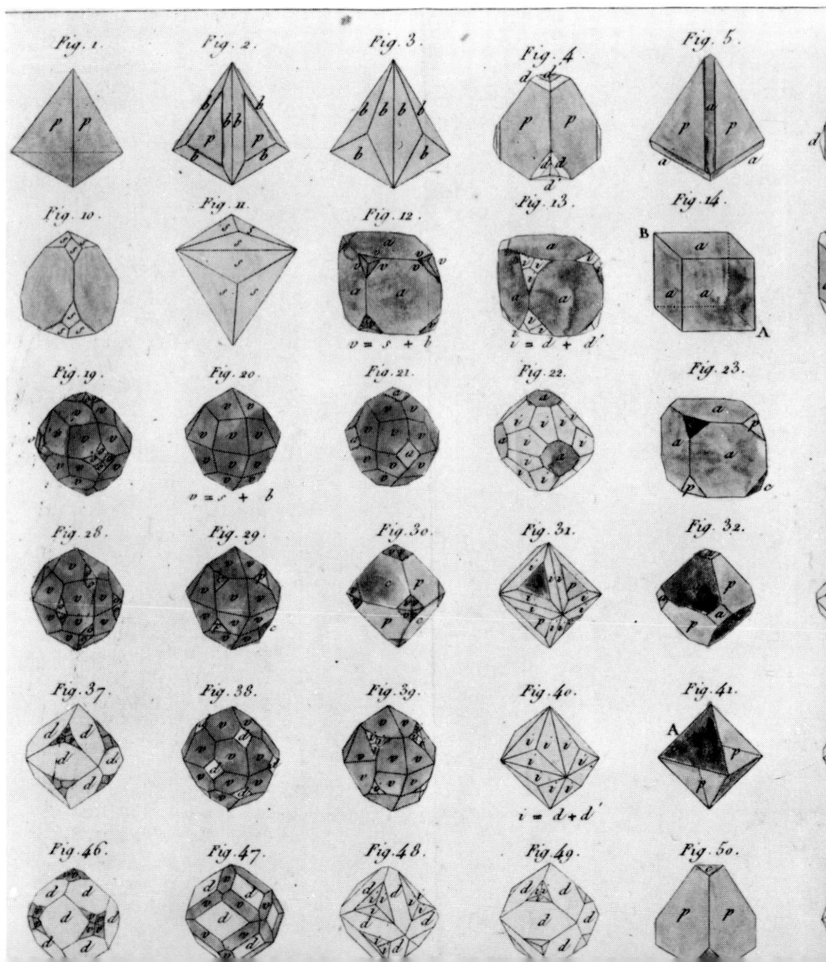

452

451. *Crystallography and form.* Chemists might be able to ignore structure; crystallographers and mineralogists could not. They shared a deep-seated intuition that observable forms reflected the order that existed on the molecular or atomic level. Here some of the manifold crystal forms are illustrated in a plate from one of the early standard treatises on mineralogy.

452. *Hessel's crystallometry.* The variety of crystal forms seemed endless. In 1830, however, J. F. C. Hessel proved mathematically that there could be only 32 crystal classes and that only two-, three-, four-, and six-fold axes of symmetry could occur. In this illustration from Hessel's work, crystals are depicted as ordered aggregates of atoms whose arrangements are confined to those Hessel had derived mathematically.

451

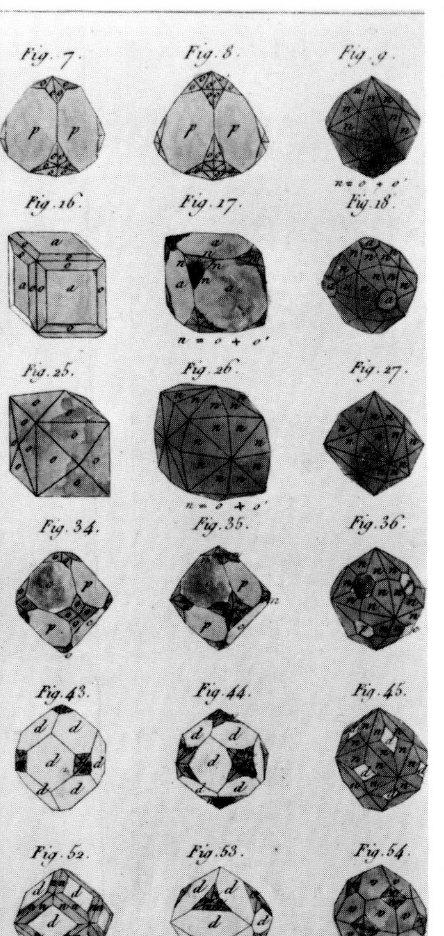

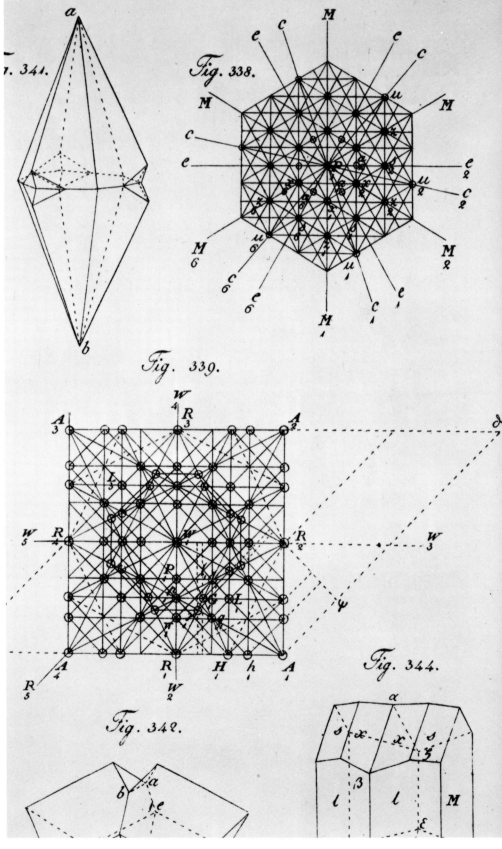

453. *August Kekulé's molecular models.* It was possible to do simple inorganic chemistry without worrying very much about the arrangement of atoms in compounds. It was almost impossible *not* to think of these arrangements in organic chemistry, however. In his great *Lehrbuch der organischen chemie* ("Textbook of Organic Chemistry"), August Kekulé made considerable use of molecular models to illustrate his points. In this figure from the book, models represent, from left to right: CH_4 (methane), CH_3Cl (methyl chloride), $COCl_2$ (phosgene), CO_2 (carbon dioxide), and HCN (hydrocyanic acid). The peculiar sausage-like appearance of the carbon, oxygen, and nitrogen atoms was designed to show their binding power or valence, which had recently been discovered.

454. *Other models by Kekulé.* From left to right, these compounds are, in Kekulé's nomenclature: HCl (hydrochloric acid), H_2O (water), NH_3 (ammonia), O_2 (oxygen), SO_2 (sulfuryl radical), SO_2Cl_2 (chlorosulfuric acid), and $SO_2(OH)_2$ (hydrated sulfuric acid, which is, of course, H_2SO_4, our sulfuric acid). In 1861, when this text was written, sulfuric acid was considered to be SO_2, so the extra 2H and 2O had to be added as a hydrate. This is a good example of the way in which molecular models could obscure chemical truths. At the far right is HNO_3 (nitric acid).

455. *Chemical dynamics, after Kekulé.* The ease with which simple molecular models could illustrate simple chemical processes is here made manifest. When compound *aa′* meets *bb′* under proper conditions, the two compounds disintegrate and recombine to give *ab* and *a′b′*. How easy it all seems when put in pictures!

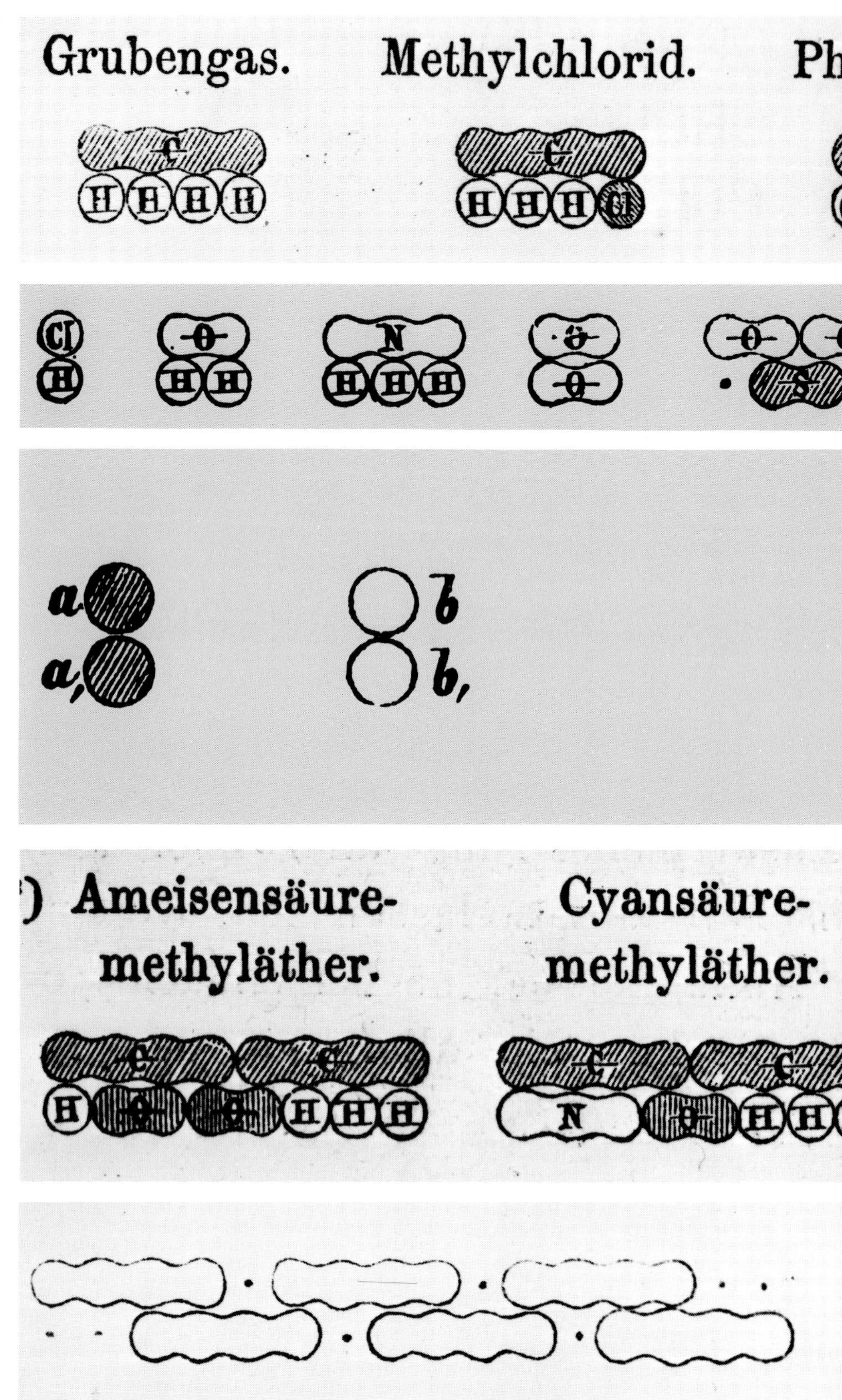

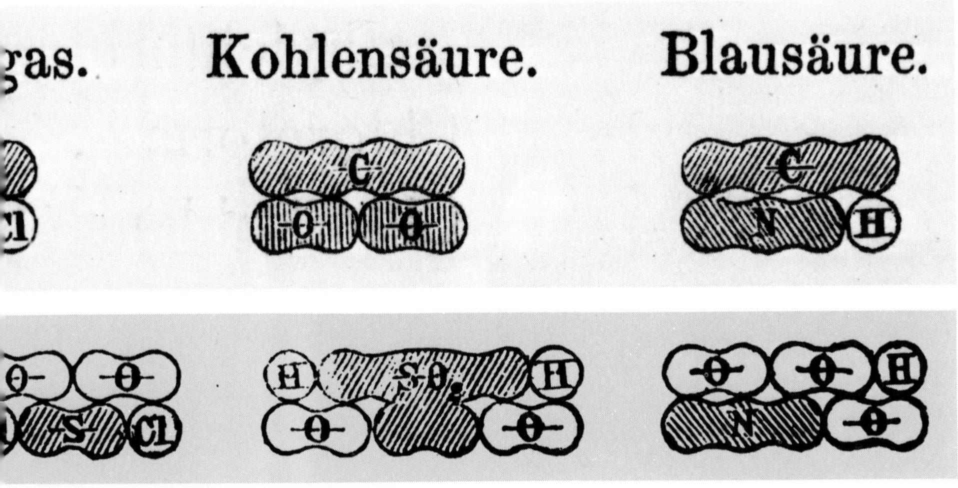

gas.
Kohlensäure.
Blausäure.

Cyanmethyl.
Essigsäure.

geschlossene Kette.

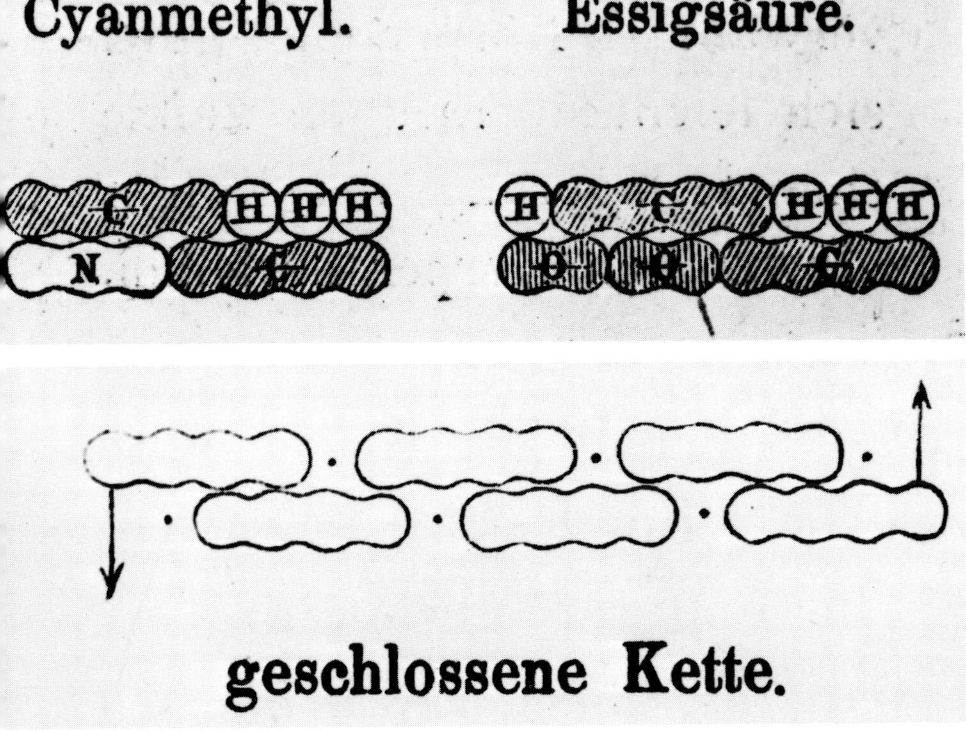

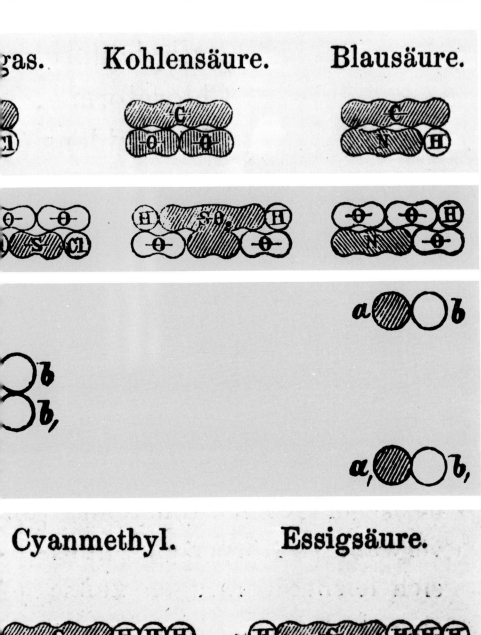

456. *More complicated molecular models, after Kekulé.* Kekulé's models were capable of representing some of the more complicated organic molecules of the time. But they had a weakness that appears here: although structure is implied in the very idea of a model, Kekulé's models make no attempt to incorporate substructures into the overall molecule. All of them look alike, yet they are all different chemically. Because the models do not reflect differences, they are of limited use.

457. *Kekulé's closed ring.* In 1865, Kekulé announced the solution to the great riddle of the benzene molecule. All would be explained if the six carbon atoms were considered to be a closed ring instead of an open chain. On the left above is a C_6 chain; on the right, Kekulé's representation of a closed ring, a model that makes no serious attempt to represent structure.

458

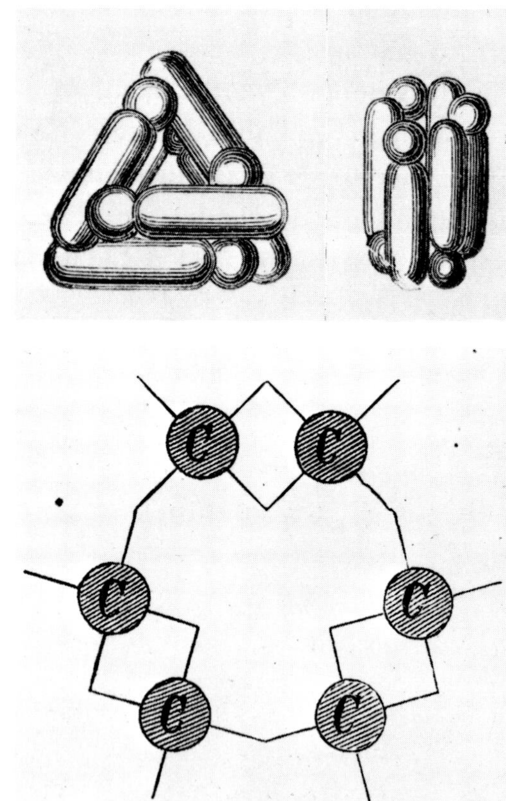

459

458. *Havrez's model of benzene (1865).* It was not immediately obvious *how* the carbon atoms of the benzene ring could be arranged. One of the earliest attempts at solving this problem was that by the French chemist Paul Havrez, who was the first to wrestle with the physical implications of Kekulé's theory. The long sausage-shaped atoms represent carbon; the balls represent hydrogen. The problem was to satisfy all the carbon valences with only 6 hydrogen atoms and neighboring carbons.

459. *Kekulé's benzene ring.* This solution of the benzene problem appeared in the second volume of Kekulé's textbook on organic chemistry, published in 1866. The tetravalence of carbon is preserved by creating double bonds between carbon atoms. These bonds implied the existence of extra energies; but the latter could not be found, and it was soon realized that the model was impossible.

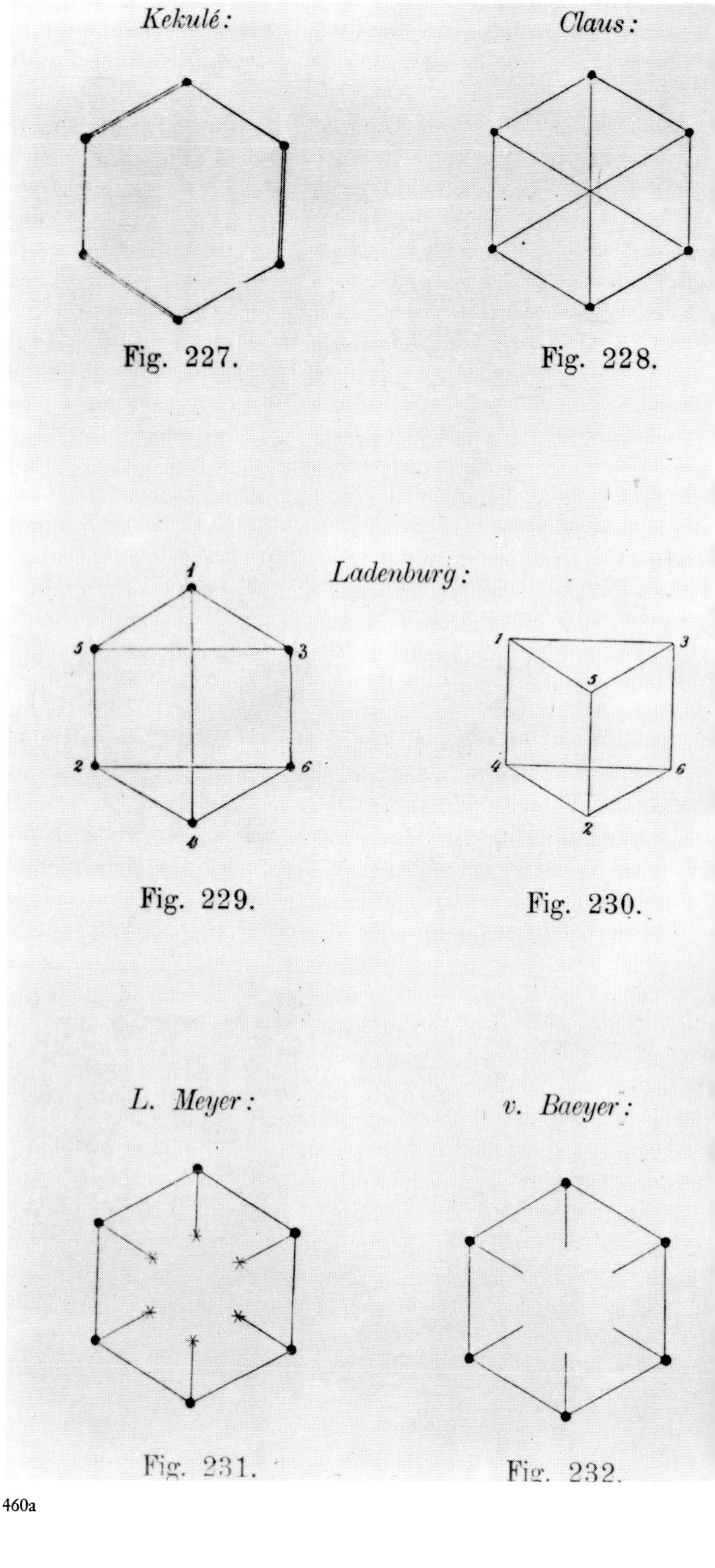

Kekulé: Fig. 227.

Claus: Fig. 228.

Ladenburg: Fig. 229. Fig. 230.

L. Meyer: Fig. 231.

v. Baeyer: Fig. 232.

460a

460a, b. *Some later attempts to solve the ring mystery*. The fact that Kekulé's ring model of benzene was quickly proved inadequate accounts for the rash of alternatives that appeared in the latter decades of the nineteenth century. As these diagrams show, the problem was how to account for all the available bonds while preserving the relative chemical inactivity of benzene. The favorite solution was to direct the extra bonds into a ring, where they were no longer available for combination. But this solution, too, proved incapable of explaining all the facts. The answer had to wait until the twentieth century and the advent of quantum mechanics.

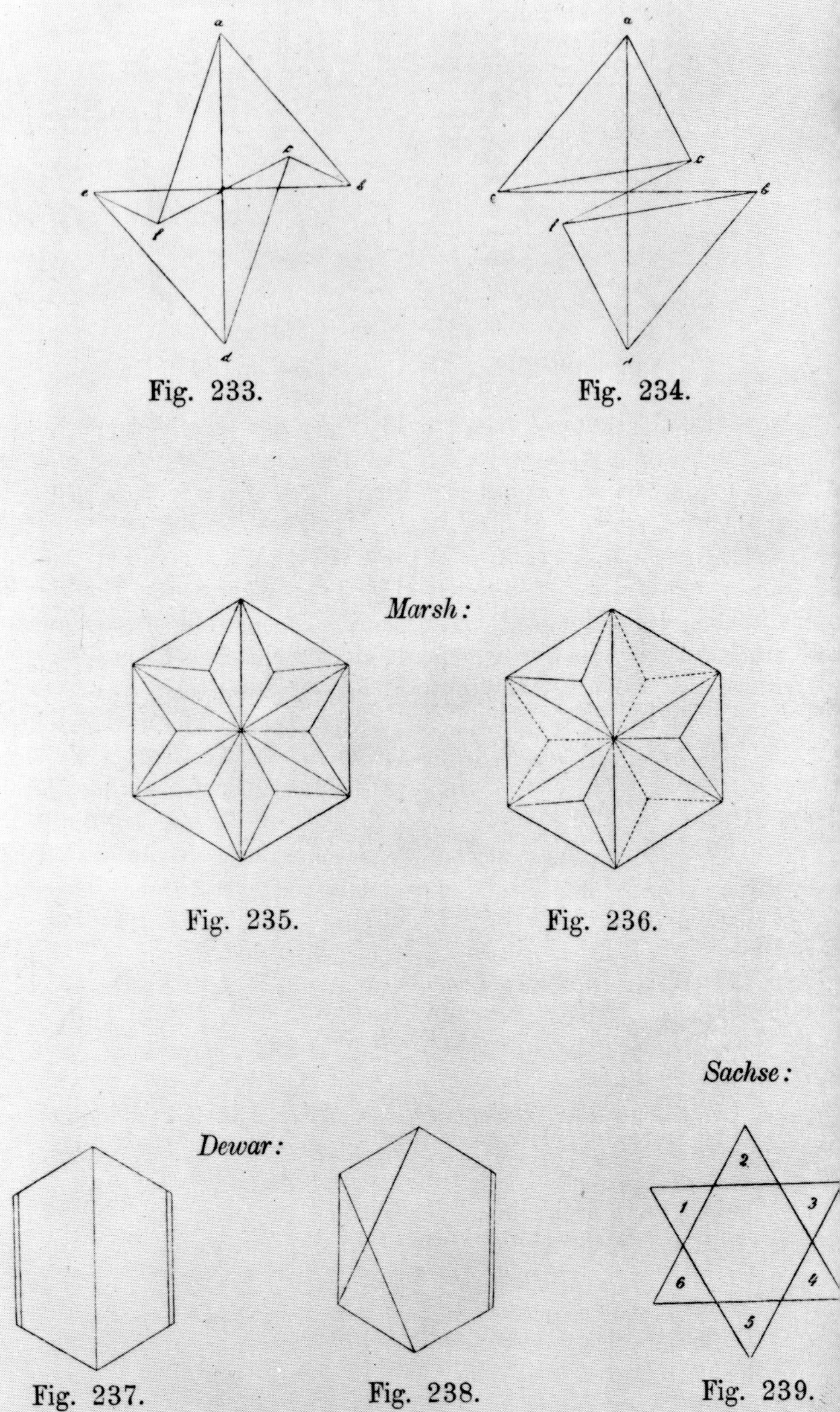

460b

461

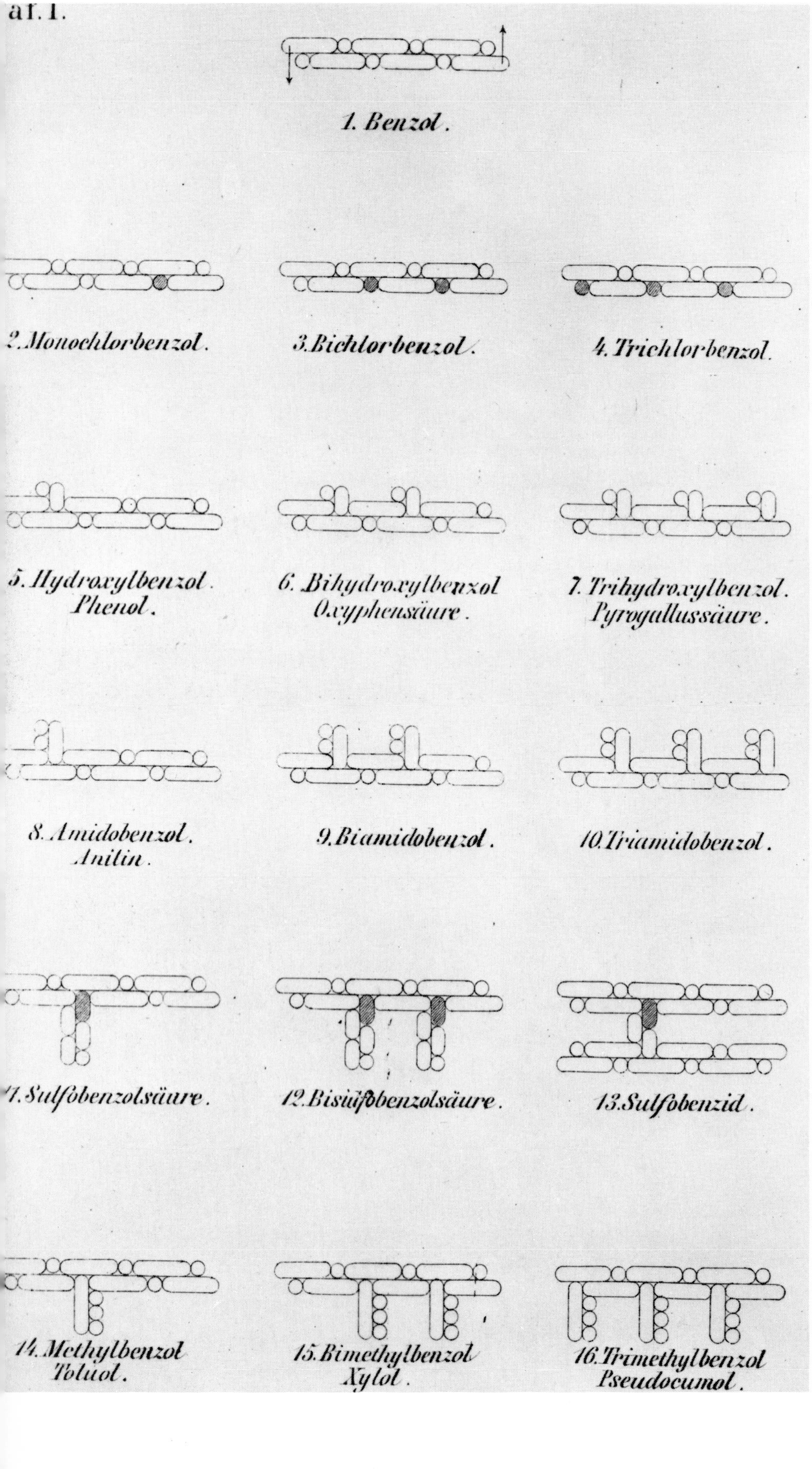

af. 1.

1. Benzol.

2. Monochlorbenzol. 3. Bichlorbenzol. 4. Trichlorbenzol.

5. Hydroxylbenzol. Phenol. 6. Bihydroxylbenzol Oxyphensäure. 7. Trihydroxylbenzol. Pyrogallussäure.

8. Amidobenzol. Anilin. 9. Biamidobenzol. 10. Triamidobenzol.

11. Sulfobenzolsäure. 12. Bisulfobenzolsäure. 13. Sulfobenzid.

14. Methylbenzol Toluol. 15. Bimethylbenzol Xylol. 16. Trimethylbenzol Pseudocumol.

462

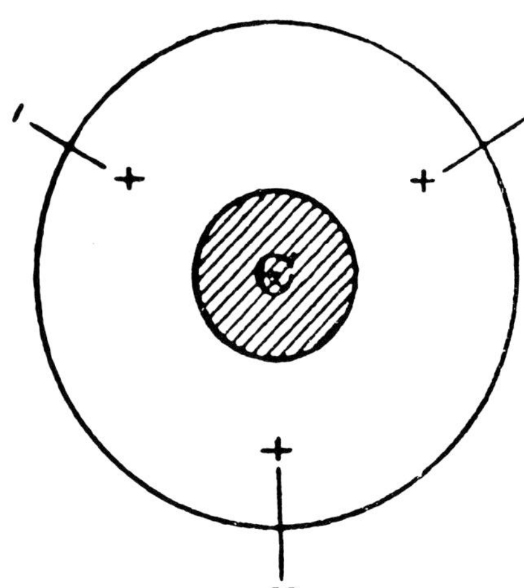

461. *Some heuristic aromatic models.* Without agonizing over the fine structure of his new concept, Kekulé proceeded to use it to represent aromatic compounds. His "sausage" chains were simply understood to be closed, as in the top example, which shows benzene. It was then rather easy to represent a host of compounds simply by substituting elements or radicals between the sausages. These models are totally worthless as guides to further research, for they bear no relationship to the real structure of the molecules they supposedly represent.

462. *The nature of the chemical bond.* The problem of chemical bonding became crucial with the acceptance of the ring structure of benzene and the further development of organic chemistry. In stereochemistry particularly, the direction of the bond was important. This figure of a carbon atom appeared in a handbook of stereochemistry published in 1894. The atom is shown surrounded by an electrical atmosphere (much like Berzelius' model of some 75 years before) whose character will determine its bonding. Almost all the elements of the model were arbitrary and could not be explained by contemporary science. Once again, the explanation had to wait until the advent of quantum mechanics.

463a, b. *The structure of diatomic molecules.* One of the earliest proponents of the theory that chemical properties depended largely upon the organization of atoms within molecules was the French chemist M. A. Gaudin. In 1873 Gaudin published a book in which he laid out his ideas on chemical structure. He began by illustrating Avogadro's hypothesis that the molecules of ordinary gases such as oxygen (marked *a* in Figure 463a), hydrogen (*b*), chlorine (*c*), and nitrogen (*d*) were composed of two atoms of the same element. These two atoms circled one another (as in Figure 463b) "exactly like double stars" and were moved by the same force that moved double stars. This was one of the earliest dynamic models of a molecule, but it was rapidly abandoned, even by Gaudin.

464. *A stereochemical representation of a chemical reaction.* Gaudin used simple figures to illustrate his thesis. The molecule on the far left is one of the sesquioxides, with the general formula A_2B_3, in which A represents a metal and B represents oxygen. The reaction shown involves the decomposition of a more complex molecule, seen on the right, into its components—two simple elements plus the sesquioxide whose properties, in part, are explicable by its geometrical form.

465. *A complex chemical molecule.* This rather beautiful molecule was meant by Gaudin to represent monopotassium stearate. In the accompanying text, he elaborates on the reasons for assigning the compound this composition and structure, again making the point that form determines properties.

463a

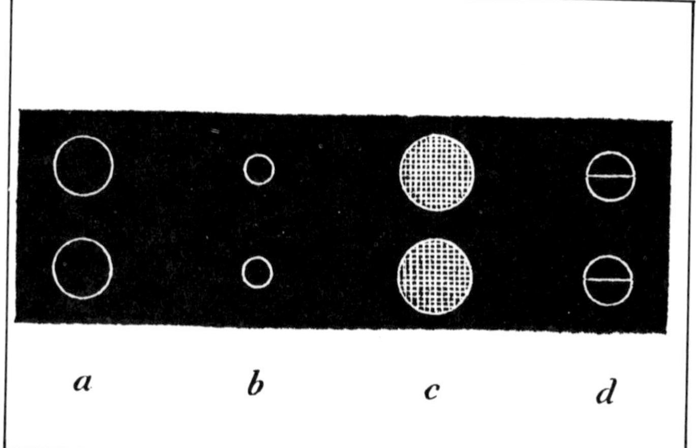

463b

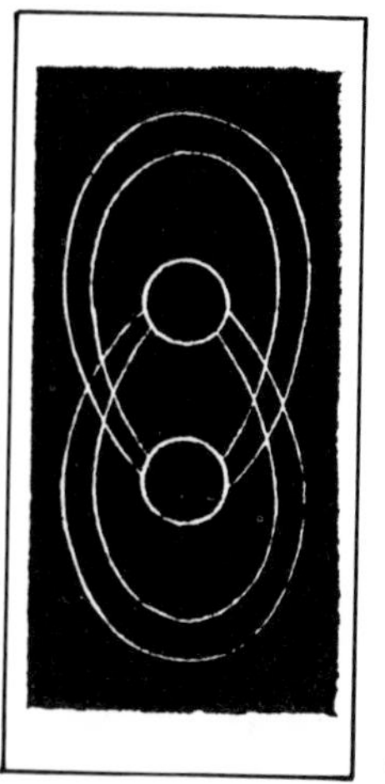

464

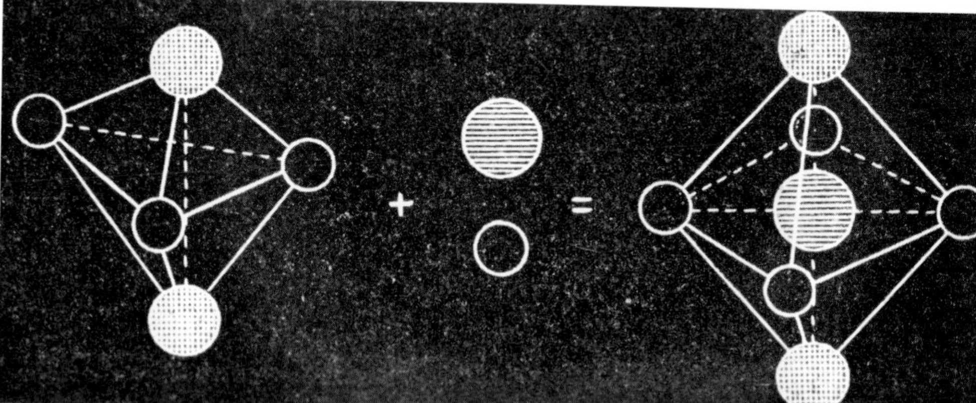

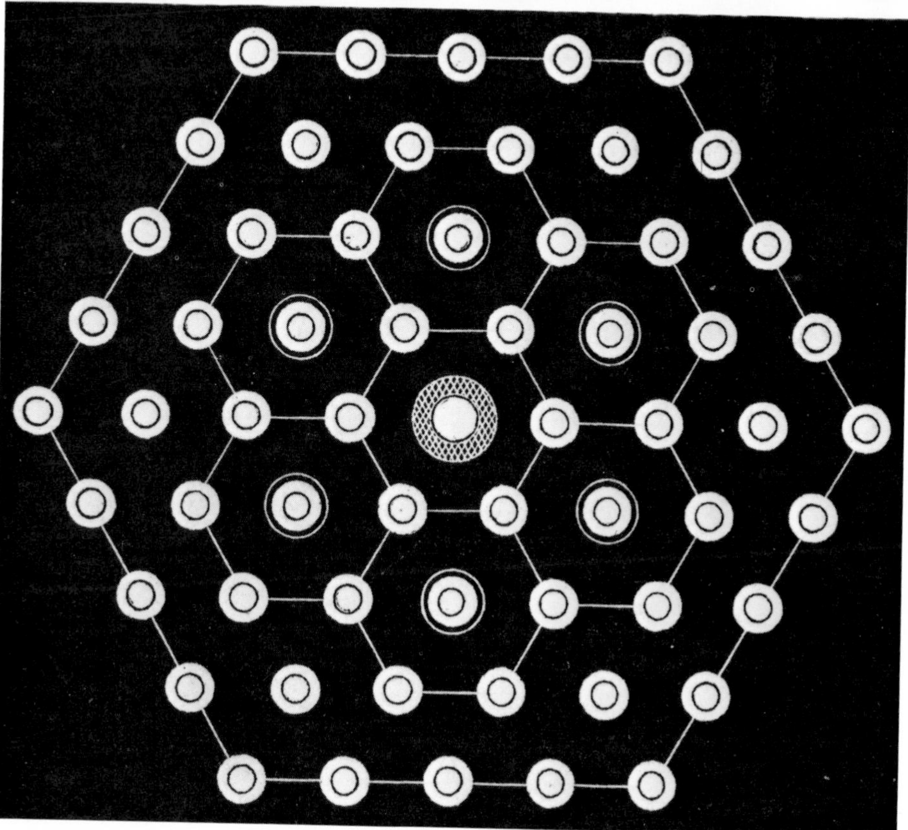

465

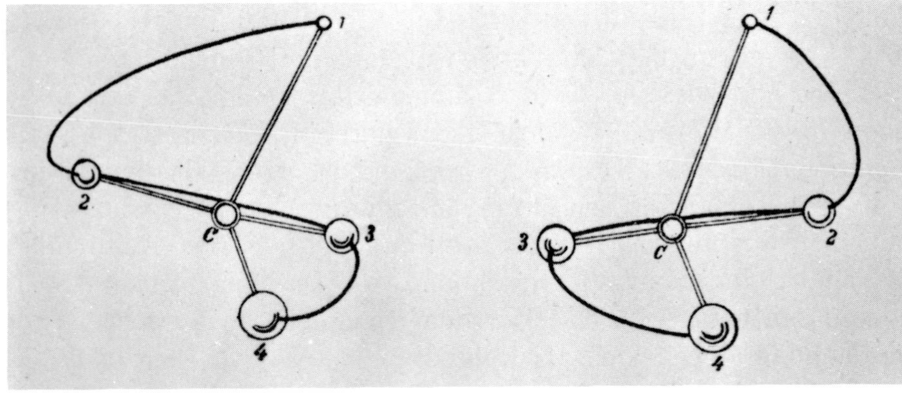

467

Fig. 74. Fig. 75. Fig. 76. Fig. 77.

HOCO H COOH H HOCO H Br Br H COOH H COOH

$+ Br_2 =$ Drehung: $= HBr +$

H COOH H COOH H COOH Br COOH

Br Br

Fumarsäure Dibrombernsteinsäure Brommaleïnsäure

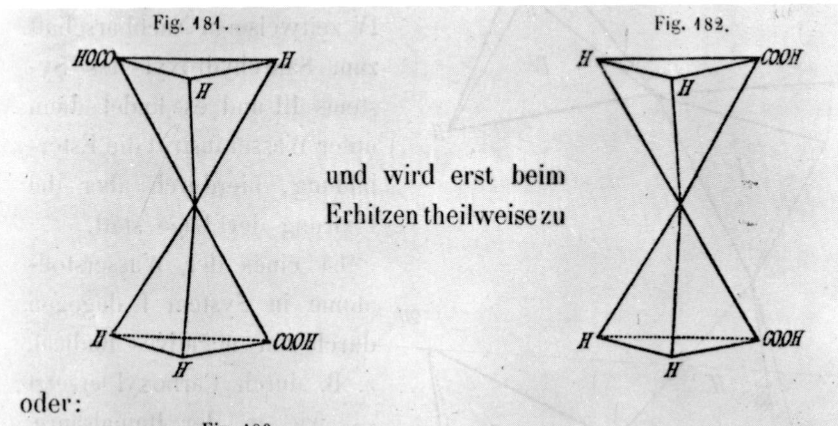

oder:

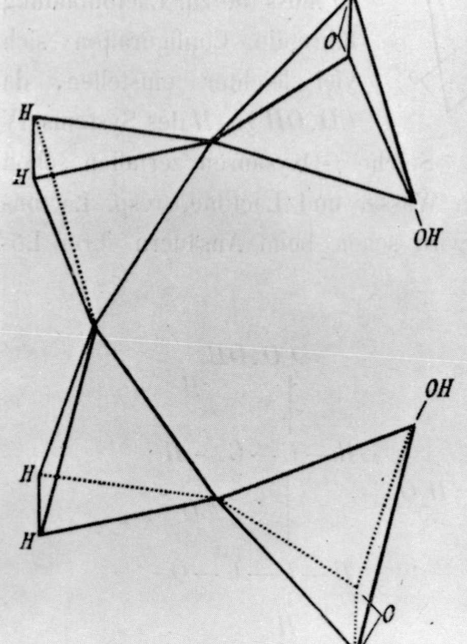

Ist die Anhydridbildung in dieser Lage einmal vollzogen, so kehren die Systeme in die an sich bevorzugtere Configuration nicht zurück, da die neue jetzt fixirt ist und erst dann wieder durch Drehung verändert werden kann, wenn die Schliessung des Ringes durch Hydratbildung gelöst wurde.

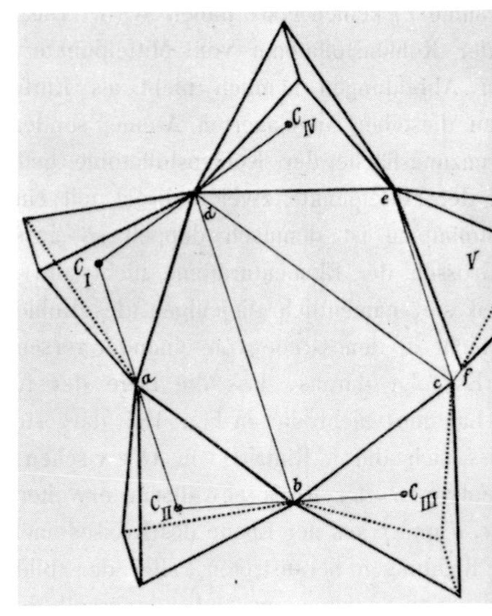

466. *A later stereochemical model.* By the end of the nineteenth century, Gaudin's fancies could be replaced by more precise determinations of molecular form. Here, the positions of the constituent atoms are fairly well represented. The figure shows two optical isomers, molecules with the same components but with structures that are mirror images of each other.

467. *Stereochemistry and chemical qualities, after Wislicenus.* Having obtained enough theoretical background to build hypothetical models of molecules, and having developed refined laboratory techniques to check their hypotheses, chemists could now explain a host of chemical reactions by correlating them with differences in atomic arrangement. One of the early masters of this art was Johannes Wislicenus, whose models, shown here, reveal how powerful the new perspective was in clarifying certain aspects of chemistry. The reaction shown, between fumaric acid and bromine, yields dibromosuccinic acid. That, in turn, under pressure gives off hydrobromic acid and forms bromomaleic acid. Note the alterations in geometrical form that accompany the chemical changes.

468a, b

468a, b. *Stereochemistry and chemical reactions.* Stereochemistry could be used to elucidate the mechanism of chemical reactions. Figure 468a illustrates how stereochemical molecules could bond together to form more complex molecules. Three molecules like the onc shown at the upper left in Figure 468a could be put together to form the star in Figure 468b.

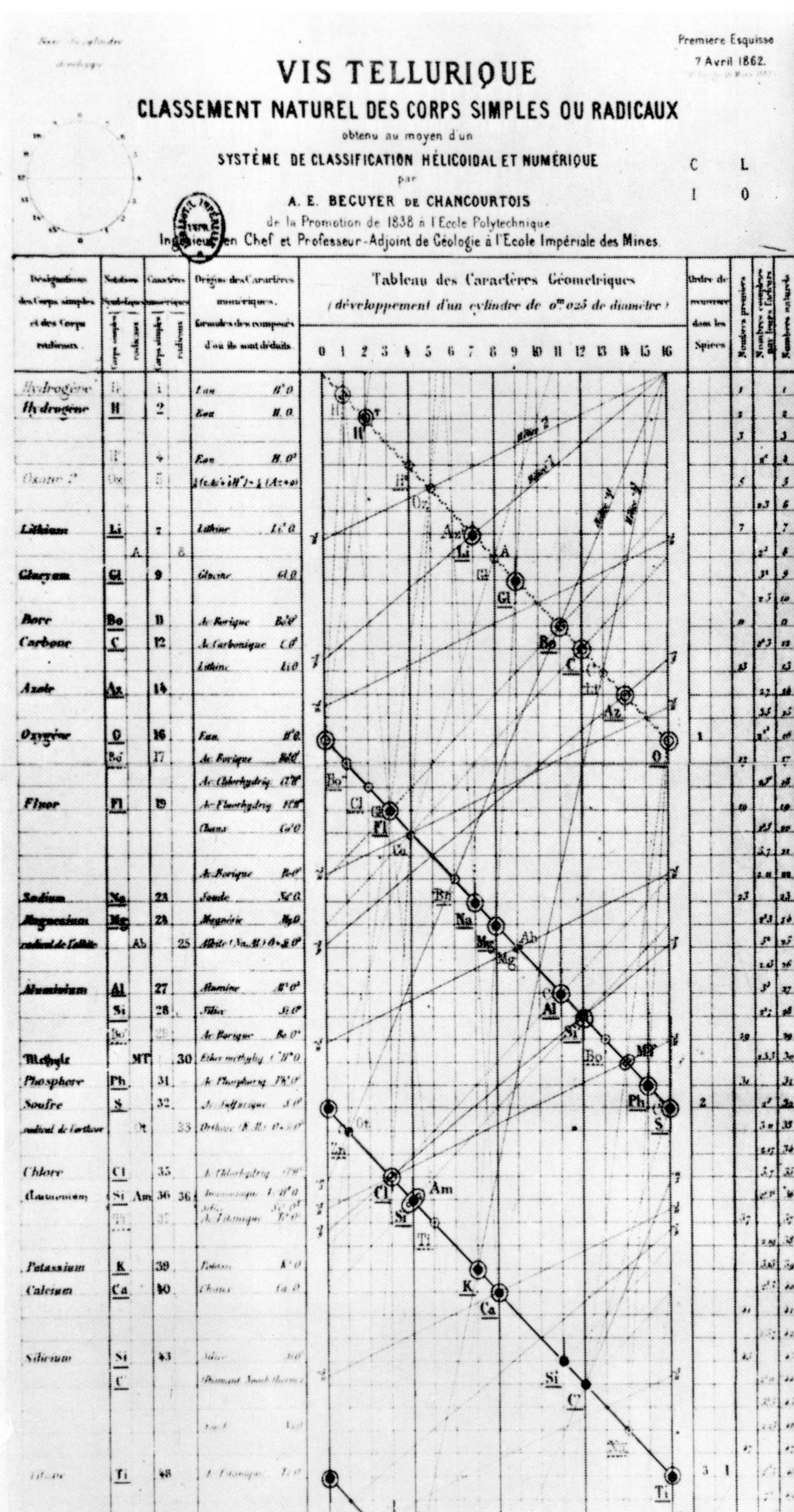

469. *The classification of the elements by analogy.* The number of known chemical elements almost tripled in the nineteenth century. It seemed impossible that each element was totally unrelated to all the others and, indeed, elementary chemistry revealed similarities that could not be ignored. Sodium, potassium, and lithium, for example, acted very much alike, as did chlorine, bromine, and iodine. Such similarities appeared to provide a basis for classifying the elements into groups with similar properties. One of the earliest attempts to do this was that of A. E. Béguyer de Chancourtois, who in 1862 arranged the elements in a spiral around a cylinder in such a way that analogous elements fell on the same vertical line. A close inspection of what Béguyer de Chancourtois called his *vis tellurique* (telluric screw) will reveal a number of relationships among the elements. However, the *vis* served better as a shorthand aid than as a window looking deeper into chemical reality.

470a

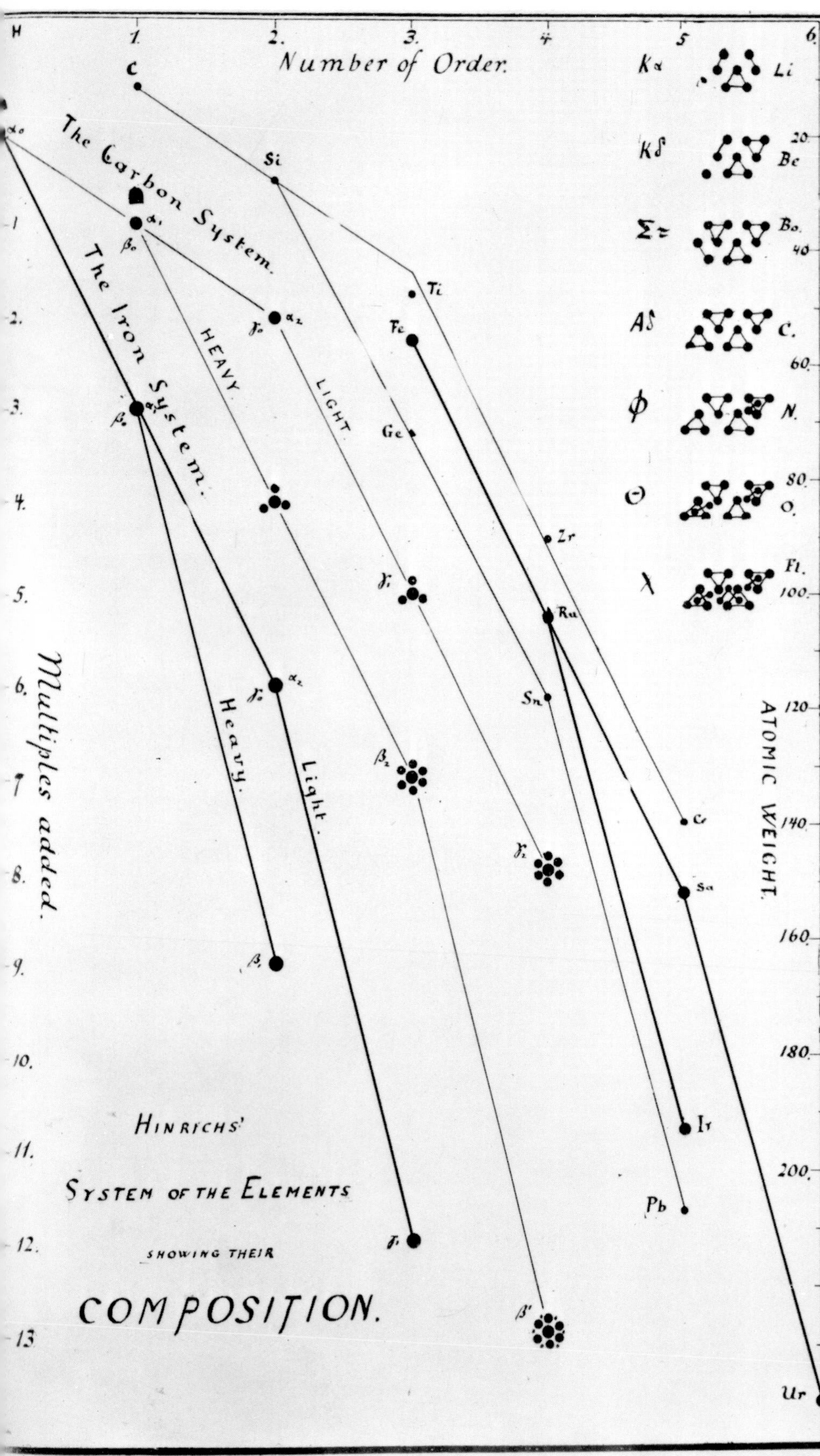

470a, b. *Classification of the elements by structure.* In 1815, Dr. William Prout in England had suggested that all elements were but compounds of hydrogen. Prout's hypothesis haunted the nineteenth century and kept popping up even after it had been conclusively proved that elementary atomic weights were not simple multiples of the atomic weight of hydrogen. G. D. Hinrichs of Iowa State University revived Prout's ideas in the 1860's and attempted to classify elements according to their structure, on the assumption that all the elements were compounds of a simpler matter. His system is shown in Figure 470a. Figure 470b, a page from the manuscript of his lecture notes, shows how Hinrichs used atomic structure to account for the properties of elements.

470b

87. We will at least mention here the
Imid- and Nitrid bases, and the Ammon 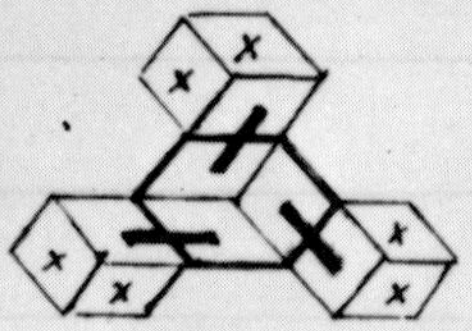
oxide base. The adjoining
plan of a Tetramin must
The 4 N are represented by the
agonal atomares; the diatomic
als by ▬; the monatomic by x. Take into
siderat. the base of O. 2.2².

88. Six atoms may, it is true, find
with one N; more readily, howev
those designated in the diagram by O. 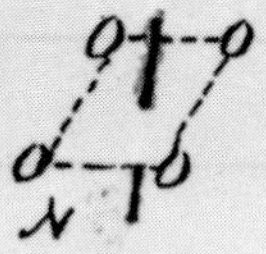
This latter gives Am. = NH_4, and
derivatives. The combinations
espec. Haloid combinats.- result
(See the lower annexed perspectiv
Nx forms the axis; 4 H an equator.
It is the prototype of the Spinel, if
were tessular.

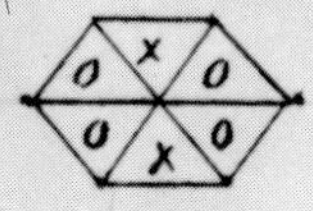

N

, x ,

' N '

89.

Am.

Ammonium represented a
compared with Kalium.

Ka.

90.

Cy.=CN. Cl.

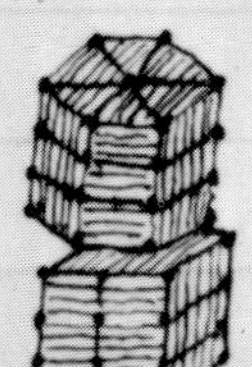

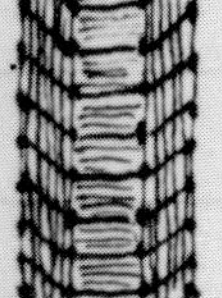

Cyanogen represented
compared with the Chlor

471. *Baumhauer's spiral classification (1870).* In 1869, Dmitry Mendeleev published his classification of the elements, which forms the foundation for the modern periodic table. This classification did not satisfy those who were convinced that periodicity was the result of inner structures that reflected simple combinations of some elementary atomic particle. Heinrich Baumhauer tried, like Hinrichs, to illustrate this concept by arranging the elements in a spiral so that elements of the same group would appear in analogous positions and yet remain in order according to their atomic weights. The result was more elegant than Mendeleev's table, but not half so useful in predicting new elements or in correlating the properties of different elements.

472. *William Crookes's model of the elements.* William Crookes's fertile imagination was stimulated by the similarities among the elements, which were being seized upon as the basis for their classification; and he built a model that grouped elements with similar properties. It is purely coincidence that the model, as a whole, makes a figure 8; it was not then known that 8 is the basic repeating number in the periodicity of the inert gases. Note the blank disks where elements ought to exist but had not yet been found.

473. *The genesis of the elements.* Only one thing remained to be done to the newly developed periodic system to make it completely reflective of nineteenth-century thought, and that was to fit it into an evolutionary framework. This William Preyer did at the end of the century, ending his book *Das genetische System der chemischen Elemente* ("The Genetic System of the Chemical Elements") with the chart reproduced here, which is supposed to show how the elements had grown out of one another through evolution.

471

472

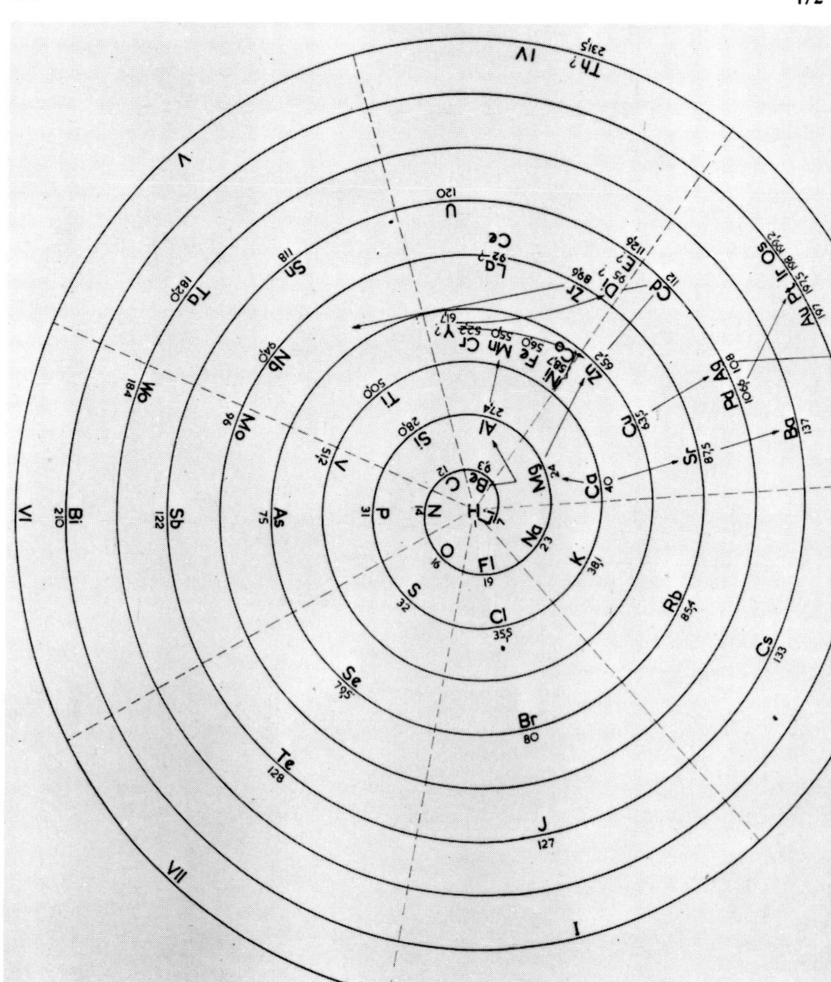

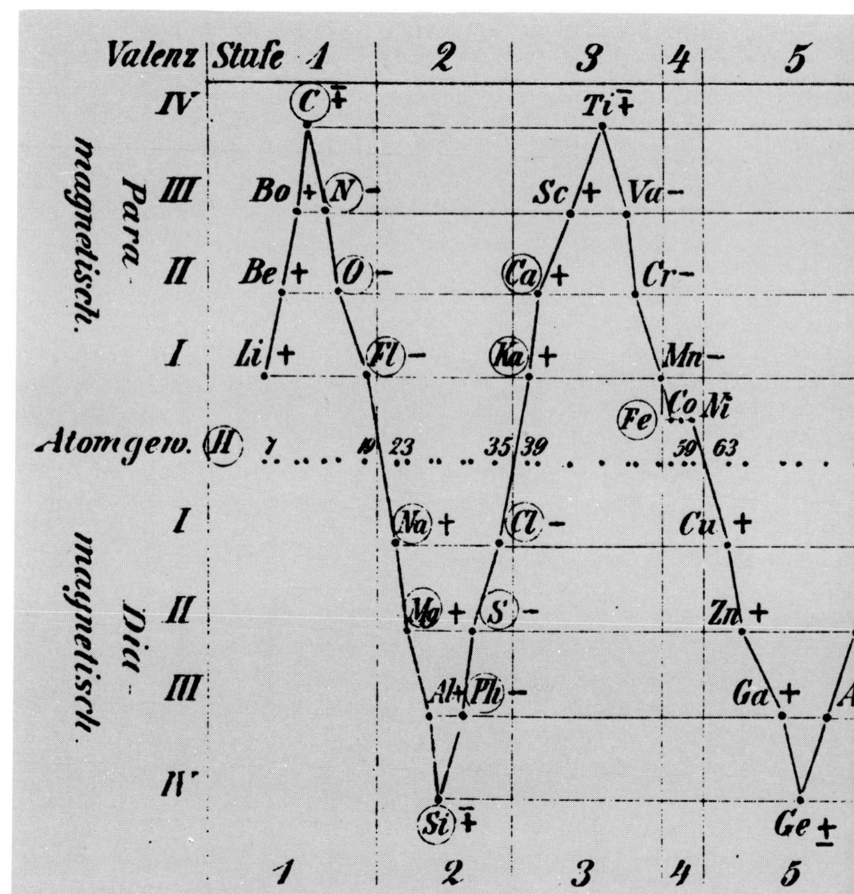

473

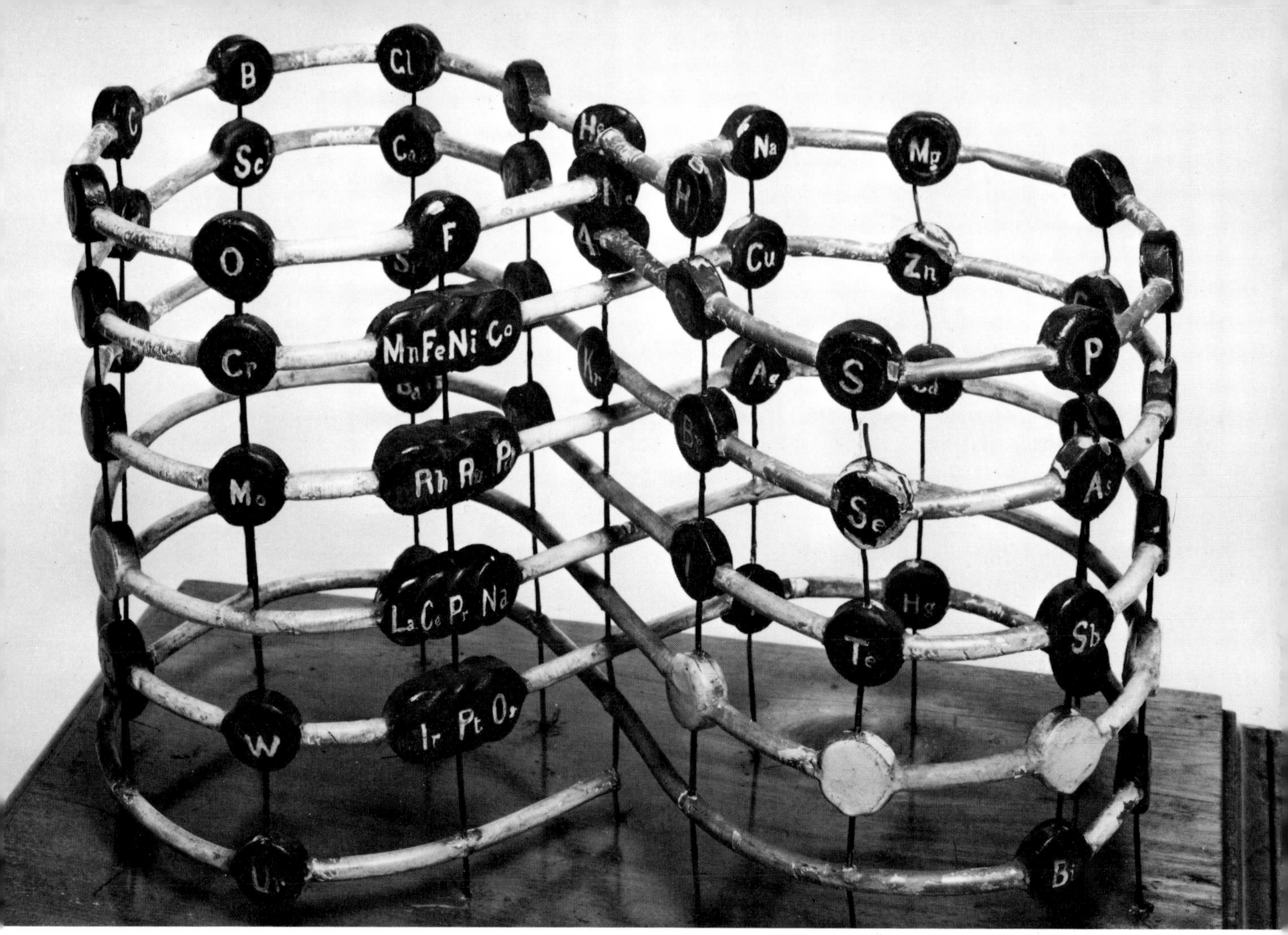

B
Cl
C
Sc
Ca
Hg
Na
Mg
H
O
F
Cu
Zn
MnFeNiCo
Cr
Kr
Ag
S
P
Mo
RhRuPd
Se
As
LaCePrNa
Hg
Te
Sb
W
IrPtOs
Bi

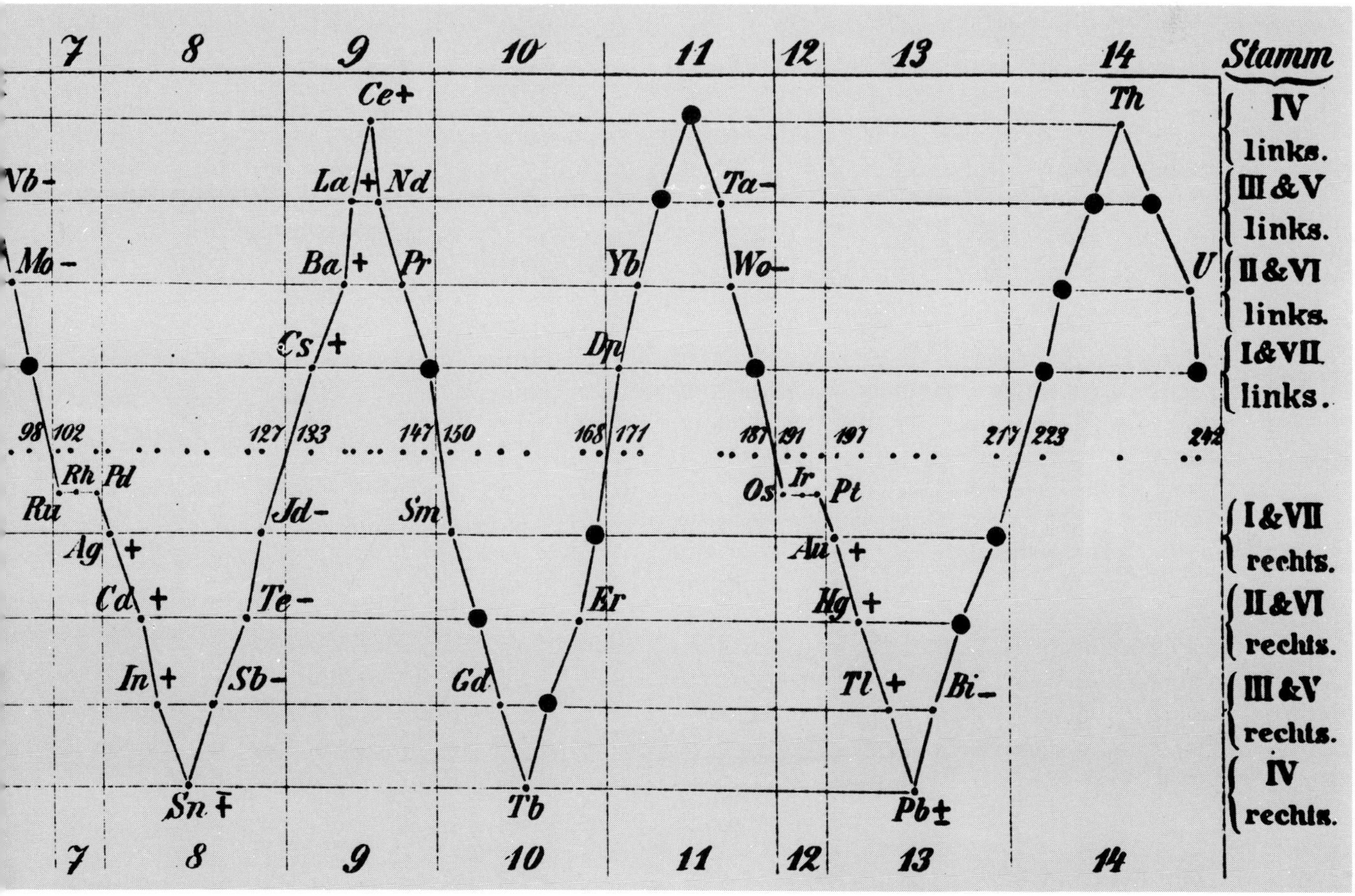

7 8 9 10 11 12 13 14
Stamm
Ce+
Th
IV links.
Vb-
La + Nd
Ta-
III & V links.
Mo-
Ba + Pr
Yb
Wo-
U
II & VI links.
Cs +
Dp
I & VII links.
98 102 127 133 147 150 168 171 187 191 197 217 223 242
Rh Pd
Ru
Os Ir Pt
Jd-
Sm
Ag +
Au +
I & VII rechts.
Cd +
Te-
Er
Hg +
II & VI rechts.
In +
Sb-
Gd
Tl +
Bi-
III & V rechts.
Sn ∓
Tb
Pb ±
IV rechts.
7 8 9 10 11 12 13 14

Electricity and Magnetism

When the nineteenth century began, there were subatomic particles already assumed in the orthodox science of the day. The phenomena associated with static electrical discharge, electrostatic attractions and repulsions, magnetic attractions and repulsions, heat and light were all considered to be caused by "subtile" fluids whose particles were the centers of various kinds of forces. Thus there were two fluids of electricity whose particles of opposite sign attracted one another, while particles of the same sign repelled one another. The same analysis seemed valid for magnetism, where two fluids acted in precisely the same way as did the electrical fluids. The two kinds of fluids, electrical and magnetic, were absolutely distinct for, as Coulomb had shown in the 1780's, the laws of attraction and repulsion governing electrical and magnetic action were identical in form, but substantively different. These laws closely resembled the Newtonian law of universal attraction for ponderable matter, being dependent upon the product of the electrical or magnetic "masses" and inversely proportional to the square of the distance separating the charged or magnetic bodies. Light and heat differed from electricity and magnetism in that the laws of their action were not so precisely known.

Qualitatively, electricity and magnetism also differed from one another. Electricity appeared to be associated with all kinds of ponderable matter. It either could be accumulated on material objects like glass or pitch, or could be conducted through substances such as copper wire. Magnetism, on the other hand, occurred only in iron, nickel, and cobalt. Electricity seemed to be everywhere, literally falling from the skies in lightning and capable of extraction from so homely an object as a glass rod. Magnetism, on the other hand, could only be made manifest if a magnetic body were present. Thus magnets were made by stroking iron (or nickel or cobalt) with a magnet, or by striking iron in the earth's magnetic field, or by heating iron and allowing it to cool slowly in a magnetic field. Already by the nineteenth century, it was believed that the magnetic fluids were contained within the molecules of iron and that magnetization consisted of the alignment of these magnetic dipoles. Electricity, on the other hand, appeared to be able to move with varying degrees of freedom between the molecules of bodies.

The nineteenth century was well supplied with the relatively simple instruments used to generate static electricity. Both the Leyden jar and the static electricity generator had been invented in the eighteenth century. They remained important research instruments in the nineteenth. To give but one example, Faraday used a battery of Leyden jars much like the one in Figure 475 in his researches that led to the enunciation of his two laws of electrolysis. The Wimsatt electrostatic generator was also a popular instrument for public lectures since its effects were spectacular and guaranteed to bring ooohs and aaahs from the public.

One of the greatest discoveries in electrical science came as the new century started. In 1800 Alessandro Volta announced his invention of the voltaic pile in a letter to the Royal Society of London. The pile produced steady electrical currents (a term used by Volta to describe the effect) instead of transient sparks. Almost immediately after this announcement, it was found by Nicholson and Carlisle in England that such currents could cause water to decompose. Electricity obviously had a much more intimate relationship to chemistry than had hitherto been suspected. Indeed, the chemical effects of electricity showed that electricity was intimately bound up with matter and was not merely a substance that existed between the pores of material bodies. In the course of the nineteenth century, a good part of this relationship was to be explained for, as we have seen, the first suggestion that electricity was the binding glue in chemical combination was made by Berzelius early in the century. This idea was to be refined and developed until a reasonably coherent theory of valency and of the chemical bond existed by the turn of the twentieth century. It should be noted, once again, that a really satisfactory theory of the nature of the chemical bond had to await the advent of quantum mechanics.

The discovery of electrical currents and their physiological effects immediately raised hopes that current electricity was the secret of vitality itself. Volta had been led to his discovery by Galvani's observation that electrical stimulation of a frog's leg by discharge of static electricity through it caused the leg to jump as if it were alive. Electrical currents were now passed through corpses with amazing results. Breathing, general muscular contraction, even expression of the emotions could be caused by simply moving electrical contacts to various portions of the body and permitting a current to flow. Although these hopes of finally capturing the essence of life were to be disappointed, the electrical effects on tissues did offer the opportunity to explore the role of electricity in the animal organism. Electrophysiology became a respectable science in its own right. With the measurement of the speed of a nerve impulse by Helmholtz at midcentury, the means were provided for using electricity as a physiological tool. Furthermore, the researches of Emil du Bois-Reymond and of Matteucci were to reveal that many physiological processes were intimately tied up with electrical currents in the body. This knowledge laid the foundation for the later development of the electrocardiograph and the electroencephalograph.

Throughout the nineteenth century there was disagreement over what the electrical "current" might be. Volta had simply assumed that the steady effect achieved by his pile was a flow of electrical fluids from one point to another. It was an unjustified assumption and it served to cloud ideas for some time. For example, if that was what a current was, then there was no reason to suspect that mere motion of the electrical particles should have anything whatsoever to do with other forces in nature. Fortunately, there was another school of thought that rejected the materiality of electricity for philosophical reasons. Oersted believed, with Kant, that all the forces of nature were related to one another and were mutually convertible, one into the other. These forces existed independently of any substance. It was this conviction that led Oersted to look for the conversion of electricity into magnetism. In 1820 he was successful in detecting the magnetic effect of a current-carrying wire. One would think that this discovery would have destroyed the opposing theory of the materiality of the electrical fluids for it is literally inconceivable *how* mere motion of an electrical particle could generate a magnetic force. Yet in spite of this apparent inconsistency André-Marie Ampère was able to construct the whole edifice of electrodynamics on the supposition of two electrical fluids whose motion created a magnetic field. He did this simply by making it an axiom that electricity in motion generated magnetic force. Since this was amply confirmed by experiment, he did not have to go into lengthy explanations of why this occurred.

One of the consequences of Oersted's discovery was the invention of the galvanometer. Since the strength of the magnetic field created by a current was dependent upon the strength of the current, it was possible to devise instruments to measure the magnetic effect and, thereby, give a measure of the electrical current passing. Faraday's work on electrolysis made it possible to convert galvanometer readings into absolute quantities of electricity that could be measured by the weights of substances deposited on electrodes during electrolysis. This is how the coulomb is defined today.

Faraday's work went far beyond such practical things as the laws of electrolysis or the laws governing the generation of currents from magnets, one of his more important discoveries and the basis for the modern dynamo and electric motor. Faraday was deeply interested in the nature of electricity and could not believe that it was composed of little charged particles. His researches led him to see electricity as a force, not a substance, that was transmitted by the molecules of ordinary matter which electrical forces put into a state of tension. Faraday used Bošković's atomic model to advantage and was able to show how molecules made up of Bošković's atoms could be distorted and strained by the electrical forces. He was then able to unify the fields of static, dynamic, and chemical electricity. Substances whose molecules could take a lot of strain before "snapping" were insulators that could be "charged" by placing their molecules under strain. Electrochemical decomposition took place when the intermolecular strain in solution was sufficient to move the two constituents in a compound in opposite directions. The laws of electrolysis followed from the fact that this force was equivalent to the chemical force of affinity binding the substances together. An electric "current" was not the passage of a substance through the wire, but the passage of a vibration caused by the buildup and breakdown of intermolecular strain in the wire itself. Conductors were bodies that could not take much strain, so the buildup and breakdown of the strain were extremely rapid. Thus, for Faraday, electricity was a "force" spreading throughout space, wherever material particles could carry it. If that were true, then magnetism must be similar. Here Faraday was forced to admit that magnetic force could apparently exist where there were no material particles, i.e., in a vacuum, but Faraday was able to accept this by the simple expedient of permitting space itself to carry the magnetic strain. Thus, paradoxically enough, the strength of a magnet is not in the iron bar but in the space surrounding it. The iron bar, as Faraday put it, was merely the habitation of magnetic lines of force.

This development of Faraday's ideas took place over the years from 1831 when he first discovered electromagnetic induction until the late 1850's when his creative powers waned. He was led away from the standard concepts of forces associated with particles to the basic idea of field theory which places the energy of a system in the system as a whole, rather than in the particles of which it may be composed. This idea excited James Clerk Maxwell and stimulated him to investigate Faraday's ideas mathematically. Calling upon the ether to carry the strains that Faraday had envisioned in space itself, Maxwell developed his famous equations that described most electric and magnetic phenomena. Electricity and magnetism were strains of various kinds in the ether, and these strains could be put into mathematical form. One of the great advantages of doing this is that the mathematics can often lead to further deductions and that was the case with Maxwell's equations. Maxwell had noticed that the conversion factor between electrostatic units (electricity at rest) and electromagnetic units (electricity in motion) was very close to the known speed of propagation of light. Perhaps, Maxwell mused, light was merely an electromagnetic disturbance, propagated through the ether. If that were the case, then one ought to be able to produce such disturbances at will and measure the wavelength that resulted. This was what Heinrich Hertz did in 1886, apparently providing the last element of experimental proof for the Faraday-Maxwell theory of the field. Physical reality was nothing but the ether; electricity, magnetism, and even matter itself could be viewed as mere local disturbances in the ether. The old imponderable fluids could now join phlogiston, animal magnetism, and other false ideas on the scrap heaps of the history of science.

At the same time that Maxwell and Hertz were establishing field theory, other researchers were finding rather solid evidence for the materiality of electricity. The invention of the mercury pump by Geissler in the 1850's made it possible to achieve fairly hard vacuums in sealed glass tubes. The passage of electricity through these tubes was both spectacularly beautiful and fascinating. For a whole generation, no one knew what was happening and about all that could be done was to observe and describe what one saw. Gradually, however, it became possible to draw some conclusions about the nature of the strange rays that were finally recognized as coming from the cathode. William Crookes, in the 1870's and 1880's was particularly clever in devising simple but illuminating experiments to confirm his belief that cathode rays were matter in an extremely tenuous form. The rays did seem to have all the properties associated with matter, but there were many skeptics who preferred to see the rays as disturbances in the ether. J. J. Thomson, at the end of the century, undertook to bring all the evidence for the materiality of cathode rays together in what he hoped would be an absolutely conclusive argument. This he did in 1897 when he presented his measurements of the ratio of the charge to the mass (e/m) of the supposed particle of which the cathode rays were composed. If one believed in particles, then this was a particularly nice measurement to have but it did not conclusively prove Thomson's point. Nevertheless, Thomson went ahead as though it had and adopted the name of "electron" for this particle of electricity that he went on to show was about 1/1800 the weight of a hydrogen atom.

There were skeptics and their skepticism was undoubtedly increased by the appearance of a new phenomenon from cathode ray tubes. While experimenting with these tubes, Wilhelm Röntgen noticed in 1895 that a nearby phosphorescent screen glowed when the cathode ray tube was on. When he investigated the peculiar and unknown rays (hence their name, X rays) that emanated from the tube, he discovered to his amazement that they passed through flesh and could be used to take photographs of living bones. X rays were the final surprise* in a century so filled with scientific surprises that one would have thought that people would have become accustomed to them. But, after all, that was what the nineteenth century expected from science. It had served up one new discovery after another and there was something both exhilarating and exciting about this latest one. After a brief flurry of concern over maidenly modesty, it was soon recognized what a boon the X rays could be to medicine. This was just another example of the beneficent effects of science. On the whole, although X rays caused some alarm among theorists who found them difficult to fit into their theories, their discovery opened up an enormously important field of scientific research for the next century.

* We shall here omit the discovery of radioactivity whose major effects were to be felt in the twentieth century.

474

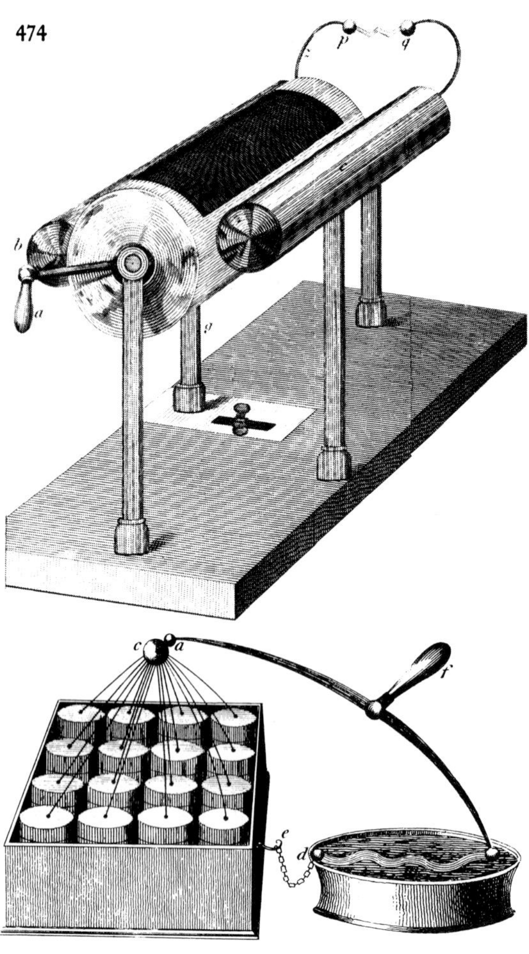

475

476

474. *The common electrical machine at the beginning of the nineteenth century.* The machine commonly used for producing electrostatic effects at the beginning of the nineteenth century was the one pictured here. It is, basically, a mechanism for rubbing glass rapidly with a cloth; this was done by turning the crank marked *a*. In the machine, the large size of the glass cylinder—18 inches long and 14 inches in diameter—made it possible to develop fairly impressive discharges.

475. *An electric battery before the invention of the voltaic cell.* Before the discovery of current electricity, an electric battery consisted of a series of charged Leyden jars, all of whose inner coatings were connected to a single terminal and all of whose external coatings made contact with an enclosing metal box. When a conducting path was made available over the surface of water, as shown here, a fairly spectacular discharge occurred.

477

476. *The Wimshurst electrostatic generator.* This machine was the most efficient producer of static electricity in the nineteenth century. Still used today in elementary science courses, it generates electricity through the friction of two glass plates revolving in opposite directions.

477. *Large-scale electrostatic generation.* Considerable energies could be attained simply by making the electrostatic generator larger. Faraday used a machine like this in his experiments on the electrolysis of compounds by static electricity. This sort of machine also was used for public lectures when spectacular effects were desired.

478. *The electric spark.* Ever since Benjamin Franklin had demonstrated that electricity and lightning were the same, the large electric spark had fascinated both professional scientists and laymen. It was, after all, through the spark that the strange substance or force called electricity became visible and "real."

479. *The electric brush.* Part of the fascination with electricity came from the beauty of its effects. The electric brush discharge was of great aesthetic as well as scientific interest.

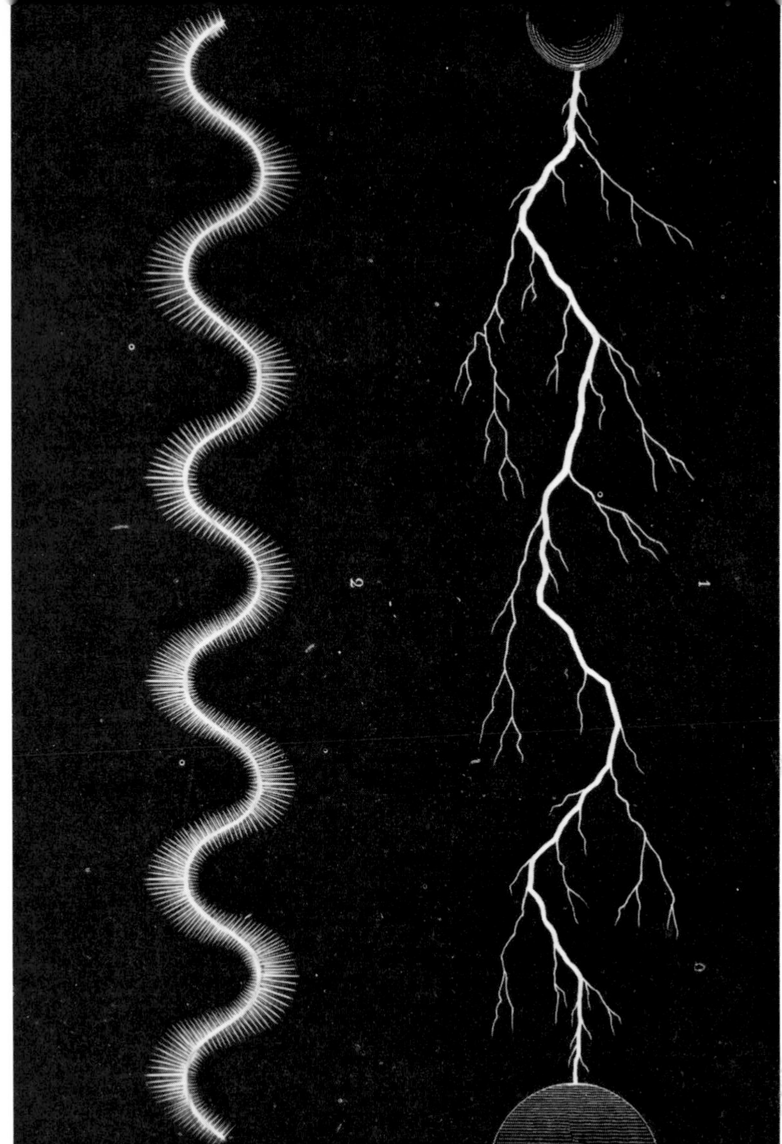

478

479

480

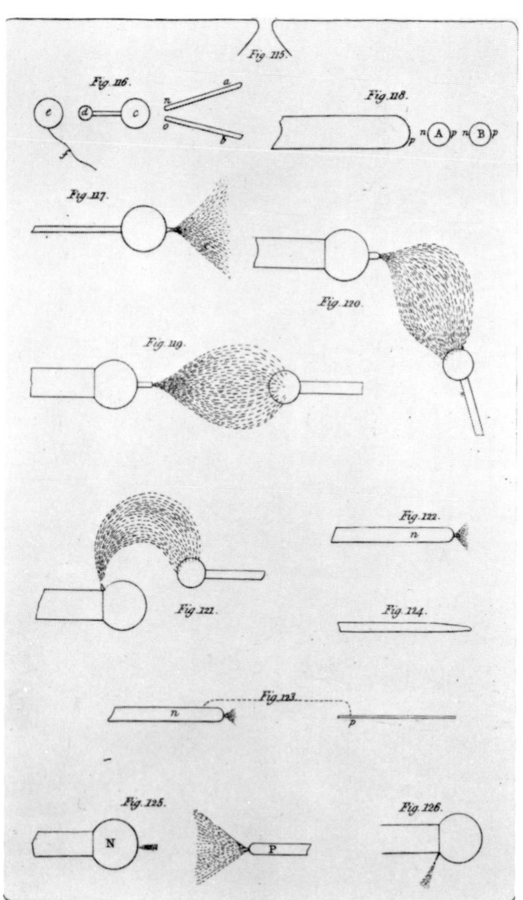

481

480. *The forms of the electric discharge.* The various forms of static electrical discharge had long been known when Michael Faraday took up their study. To Faraday, the fact that this discharge almost always occurred along curved lines indicated that it resulted from the release of intermolecular strain, rather than from the flow of a material substance. These humble illustrations provided one of the cornerstones of Faraday's field theory.

481. *The voltaic pile.* These piles were to nineteenth-century physics what the cyclotron has been to the physics of the twentieth century. Disks of copper, zinc, and blotting paper soaked in brine or dilute acid and piled on top of one another produced a steady electrical current rather than a transient spark. This current did more than anything else in the nineteenth century to open up the study of the intimate nature of matter. The invention of the pile was announced by Alessandro Volta in 1800.

482

482. *The galvanic trough.* Within a few years of the invention of the voltaic pile, its power was vastly increased by the discovery that the immersion of copper and zinc plates in dilute acid produced a far stronger current than a pile of metal disks and damp paper did. The galvanic trough, as pictured here, became a fundamental tool of electrochemical research.

483. *The great galvanic battery at the Royal Institution.* Humphry Davy at the Royal Institution of Great Britain was one of the first chemists to realize the potential offered by large galvanic batteries. An instrument such as the one pictured here could provide the currents required for Davy's epoch-making electrochemical researches that culminated in his isolation of sodium and potassium.

484. *A Wollaston battery.* Intense chemical action in the galvanic battery necessitated frequent changes of the acid solutions that filled it. One of Davy's contemporaries, William Hyde Wollaston, devised a new system for increasing the convenience and decreasing the volume of batteries. All plates could be lifted out together while the acid was changed. The arrangement of the plates, with a zinc core surrounded by copper sheeting, gave the maximum active area for the generation of electricity.

485. *Postmortem effects of galvanic stimulation.* To some savants electrical current appeared to be the principle of life itself, so it was only natural for current to be applied to corpses in order to discover its effects, if any. A golden opportunity appeared on November 21, 1803, in the city of Mainz, when a famous brigand and nineteen of his accomplices were executed. A group of medical doctors secured permission to seize both the bodies and the heads of the criminals after they had been guillotined and to experiment galvanically on them. The scene portrayed here is the collecting phase of the experiment.

483

484

485

486. *Dr. Andrew Ure and electrostimulation of a corpse.* The "experiments" of the doctors at Mainz had not revealed any of nature's fundamental secrets, though the cadavers had twitched when enough current was applied. A more sophisticated investigation was conducted on the corpse of a murderer by Dr. Andrew Ure of Glasgow in 1818. The spinal cord of the subject was exposed near the head and an incision was made to lay bare the sciatic nerve in the left thigh. When wires were touched to these points, the corpse convulsed. Other connections made the corpse "breathe." Finally, the muscles of the face were electrically stimulated and facial expressions denoting rage, anguish, happiness, and despair were produced by moving the contacts. It was from earlier experiments resembling these, reported in the popular press, that Mary Shelley drew her inspiration for *Frankenstein,* which was published in 1818.

487. *The founding of electrophysiology.* After the first wave of sensationalism had passed, physiologists turned to the serious study of animal electricity. The development of increasingly sensitive instruments for the detection of electrical currents, such as the astatic galvanometer shown here, made it possible to begin to trace electrical effects in living tissue.

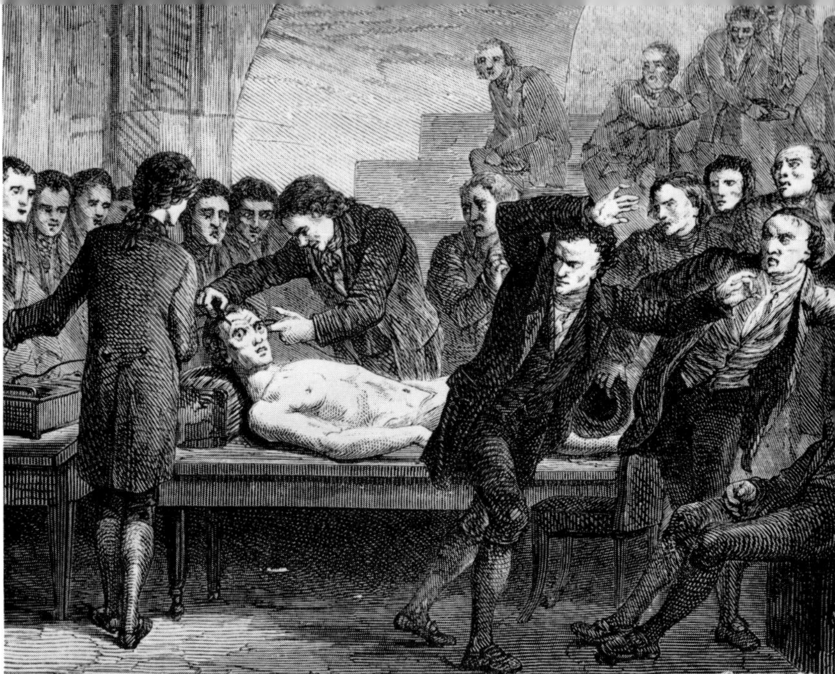

486

487

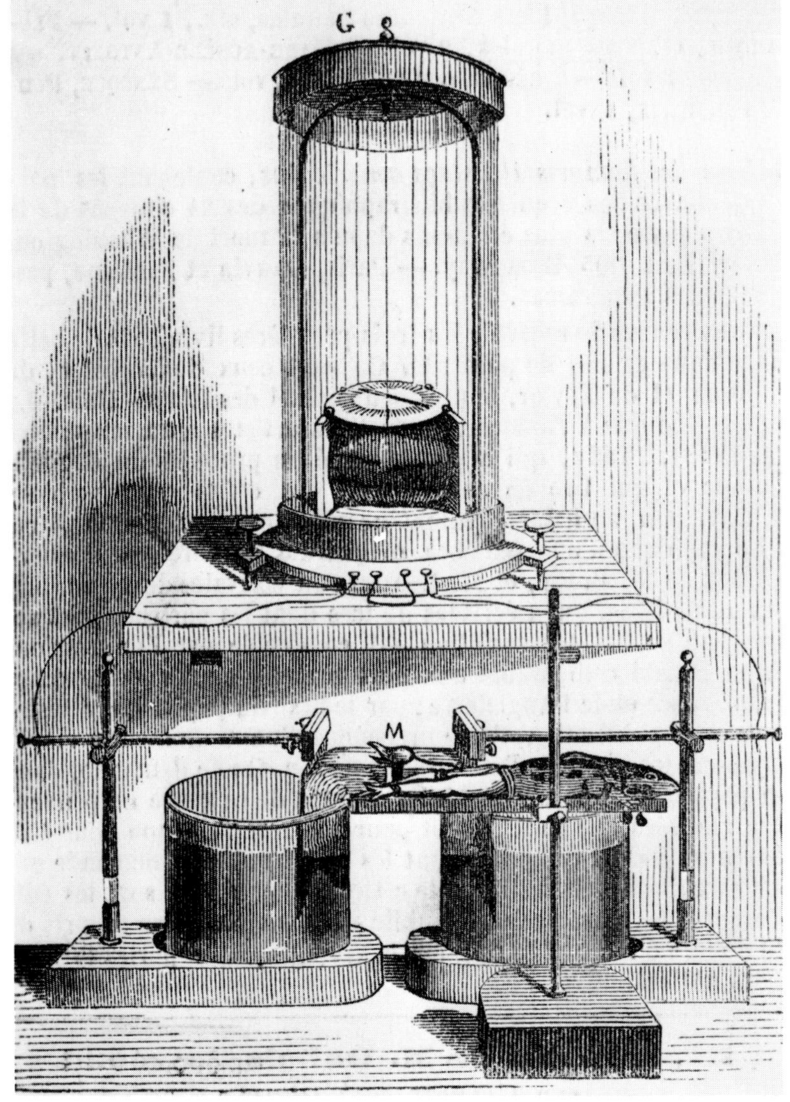

488

488. *Electrical measurement and physiology*. In the 1840's, Carlo Matteucci and Emil du Bois-Reymond began systematic investigation of the physiological effects of electricity and the electrical aspects of physiology, observing the effects produced when carefully measured intensities and quantities of electricity were passed through living tissue. From the carefully controlled quantitative biological experiment on a frog that is shown here, conclusions could be drawn that were as precise and as sound as those drawn from chemical or physical experiments.

489. *Man as an experimental animal in electrophysiology*. The study of the physiological effects of electricity relied heavily on the scientist's use of himself as an experimental object. Faraday estimated the voltage generated by electric eels by grabbing them and taking the shock through his own body. Here, the physiological effects of gradually increased currents are "recorded" by a scientist who is his own subject.

490a, b. *The discovery of electromagnetism.* One of the more dramatic discoveries of the early nineteenth century was that an electric current is accompanied by a magnetic field. The discovery, which came at the end of a lecture given by Hans Christian Oersted in 1820, was often "recreated" by popular artists. Of the two imaginary reconstructions of the event reproduced here, the first (490a) is highly unlikely because the current generated by the pile shown could not be strong enough to affect a compass placed so far away. The second version (490b), on page 334, is more realistic.

489

490a

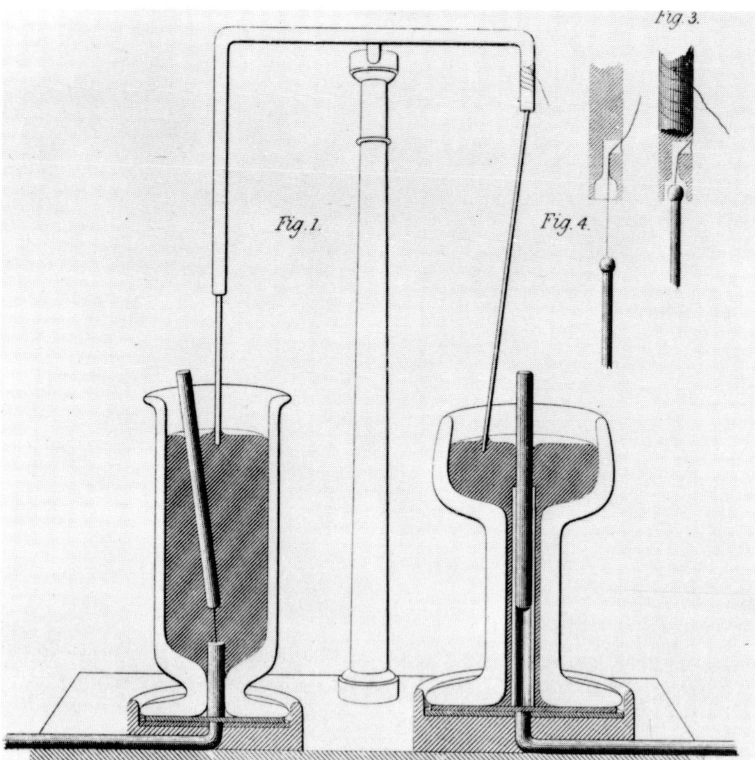

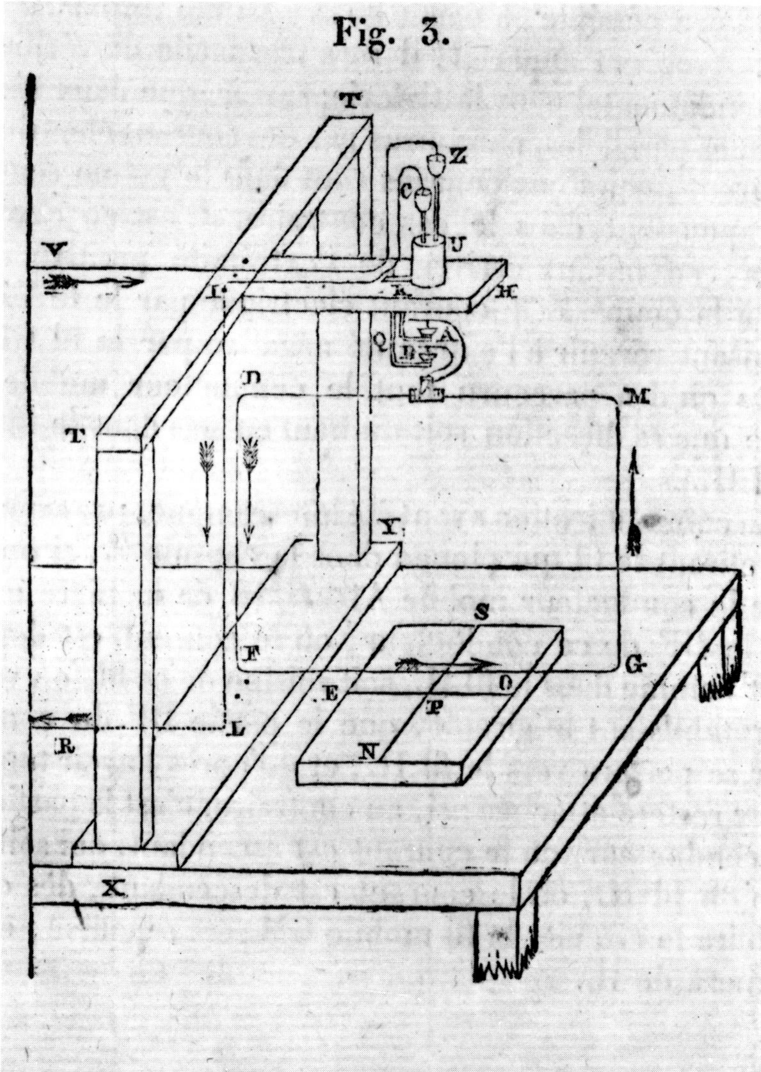

491. *The first electric motor.* The year after the discovery of electromagnetism, Michael Faraday in England devised the apparatus shown here, which converted electrical energy into mechanical rotation. In Figure 1, a bar magnet, pivoted at its lower end and immersed in mercury, rotates freely around a current-carrying wire that also dips into the mercury. In Figure 4, it is the wire that is pivoted and that rotates.

492. *Ampère's apparatus illustrating electrodynamic actions.* In France, news of Oersted's discovery stimulated André-Marie Ampère to create the science of electrodynamics. One of his first discoveries was that two current-carrying wires attract or repel one another, depending on whether their currents are going in the same direction or in opposite directions. The apparatus shown here was intended to illustrate this fundamental fact. Because the two currents, indicated by arrows, flow in the same direction, the loop, marked *MnFG*, is pulled toward the fixed wire, *IL*. When the current in the loop is reversed, the loop is repelled by the wire.

492

493. *Ampère's apparatus for illustrating the fact that electrodynamic action is dependent only on the distance through which the action takes place.* It was mathematically obvious to Ampère that the total force of attraction or repulsion between two current-carrying wires was a function of the distance through which the two wires interacted and not of the length of the paths over which the current traveled. The apparatus pictured here was intended to illustrate this fact. The wire *AD* is much longer than the wire *GH*, but the forces between the two are the same as if *AD* were equal to *GH*.

494. *Faraday's induction apparatus.* The study of current electricity served to stimulate new interest in static electricity, for it was difficult to understand the relations between the two phenomena. Michael Faraday was the first to put forward a unified theory of electricity. Static electrical induction, he argued, strained and deformed the molecules of the medium between the inducing body and the body upon which a charge was induced. This strain, the initial state in *all* electrical phenomena, preceded the passage of a current which brought about the breakdown of the strain. It became important for Faraday to map the lines of electrostatic action. This he did with the apparatus which is shown here in three views. His discovery that lines of electrostatic action were curved convinced him of the truth of his theory of intermolecular strain.

493

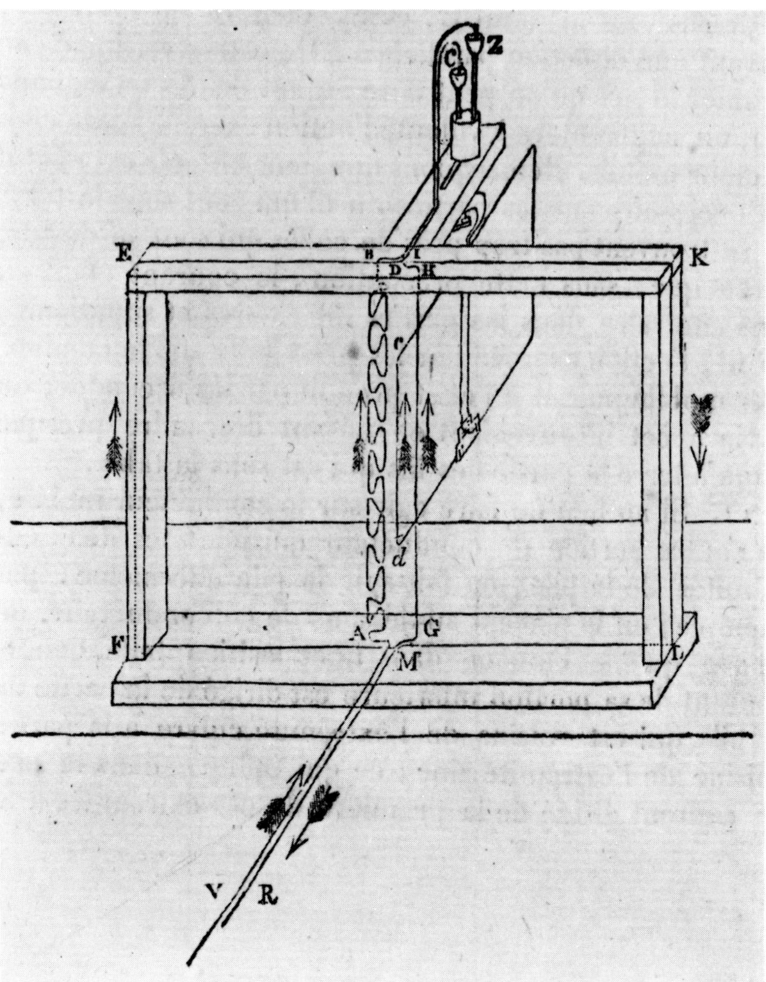

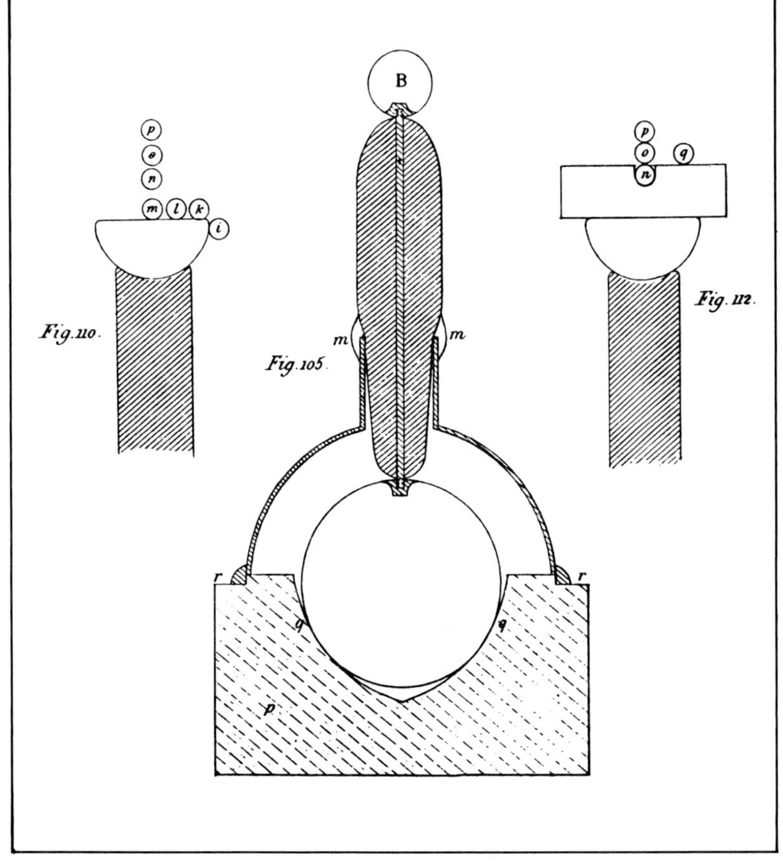

494

495. *The detection of electrical currents.* The crudest detector of an electrical current was the ordinary compass needle. As this diagram shows, it would swing to stand at an angle to the plane of a loop inside which it had been placed whenever a current passed through the loop. Because the needle also was attracted to the North Pole of the earth, the current in the loop had to be strong enough to displace the needle from the magnetic meridian.

496. *Schweigger's multiplier.* The sensitivity of the compass needle could be increased by coiling the current-carrying wire, thereby strengthening the magnetic field produced by the current. First done by Johann Schweigger in 1820, this procedure permitted fairly accurate measurements of current strength.

497. *Leopoldo Nobili's astatic galvanometer.* It was still difficult to measure feeble currents. The earth's magnetic field could overpower the electrodynamic effect, and often it was impossible to overcome this factor by increasing the number of coils because that procedure increased the resistance. To avoid this problem, the astatic galvanometer was invented. Two magnetic needles, one only slightly stronger than the other, were hung horizontally on the same silk thread. They were placed in such a way as to have oppositely charged poles above one another. Both needles were now held along the magnetic meridian by a very small force. This force was easily overcome by the magnetic force from the current, which made the more weakly magnetized needle deviate. The deviations could be read directly off a scale, thus measuring the strength of the current.

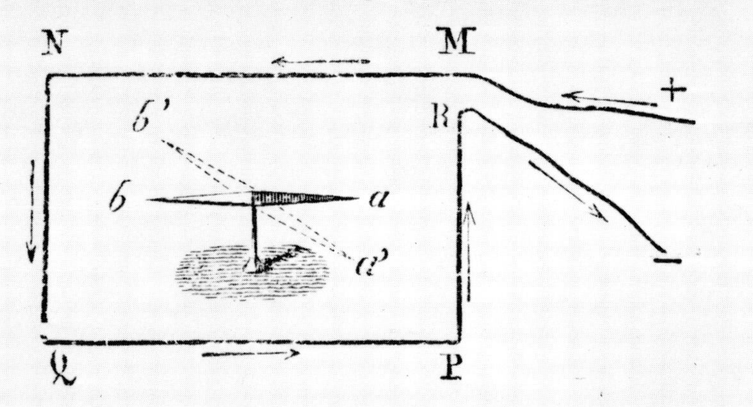

495

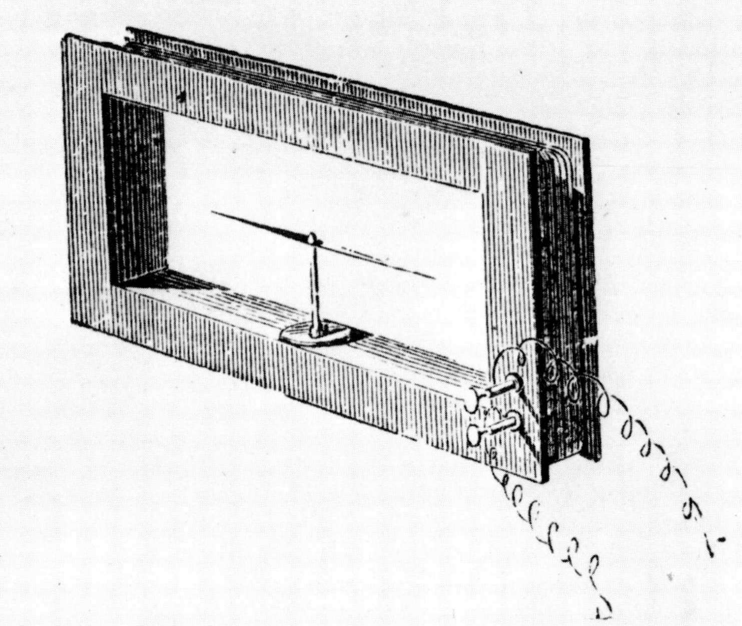
496

497

498

498. *Charles Wheatstone's astatic galvanometer*. Charles Wheatstone was one of the pioneers of electrical telegraphy. In the course of his telegraphic researches, he found it necessary to make precise measurements of currents in order to determine the resistances of circuits. Out of this work came the famous Wheatstone bridge, which accurately measures electrical resistance. Pictured here is the galvanometer he used in his work, with a microscope on top for reading the dial.

499. *Heinrich Hertz's radio apparatus*. Electromagnetic waves had been predicted by James Clerk Maxwell in his magisterial *Treatise on Electricity and Magnetism* (1873), and Heinrich Hertz in Germany, one of Maxwell's ardent disciples, set out to detect and study them. He succeeded in detecting radio waves in 1886 and began to study them in some detail. The apparatus shown at the top here is the spark gap used for detecting waves. Below it are two metal parabolic mirrors with spark gaps at their center, which were used for sending out waves.

499

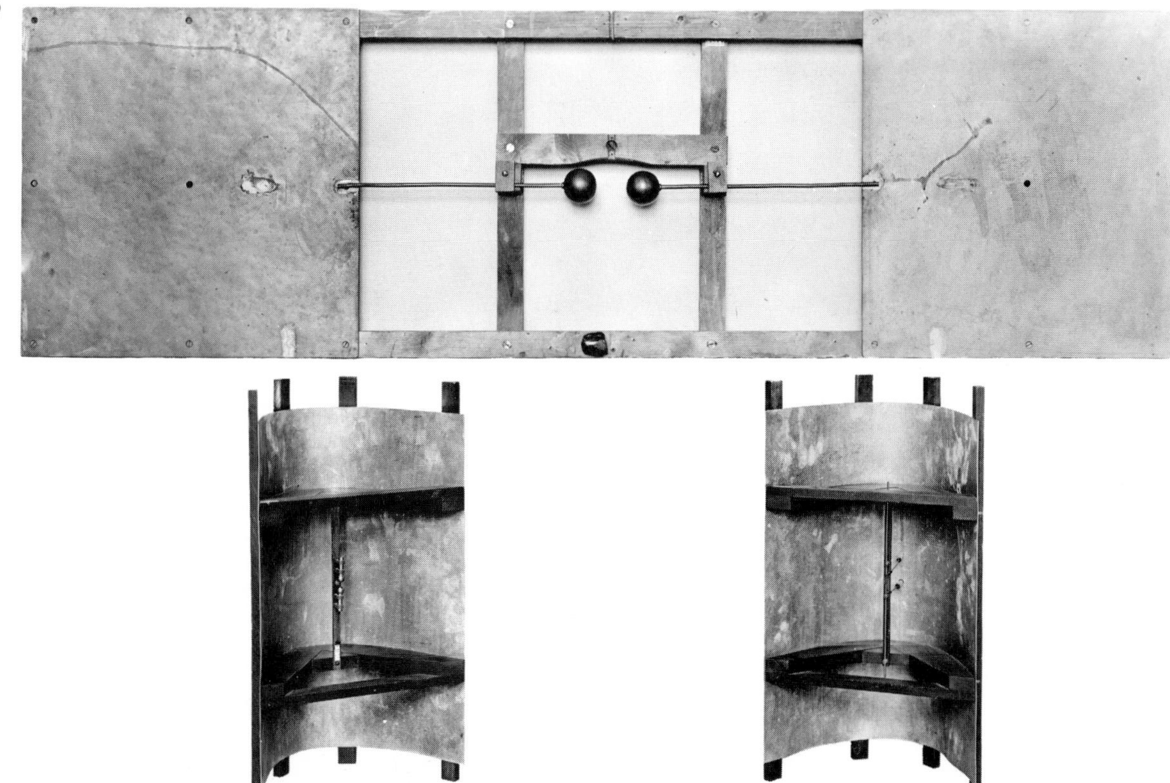

500

501

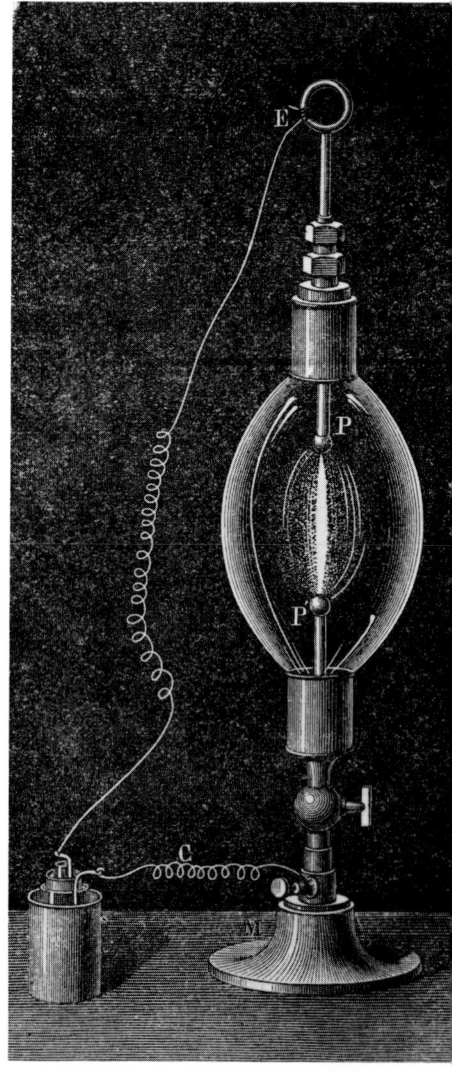

500. *Hertz's apparatus for the study of radio waves.* Radio waves, Hertz discovered, acted very much like light waves if one took into account their length of about 10 meters. Here, at the left, is a prism made of hard pitch that acted the way a glass prism does on light rays. The "grid" of wires served as a reflector.

501. *Humphry Davy's electrical light.* Electricity could generate considerable light, as the carbon-arc lamp showed. Ordinary carbon-arc lamps, however, required large batteries to provide high voltage and needed frequent adjustment as the carbon electrodes burned away. Davy resolved both difficulties by placing carbon electrodes in an evacuated glass bulb, thus lowering the voltage required and partially protecting the carbon from oxidation.

502

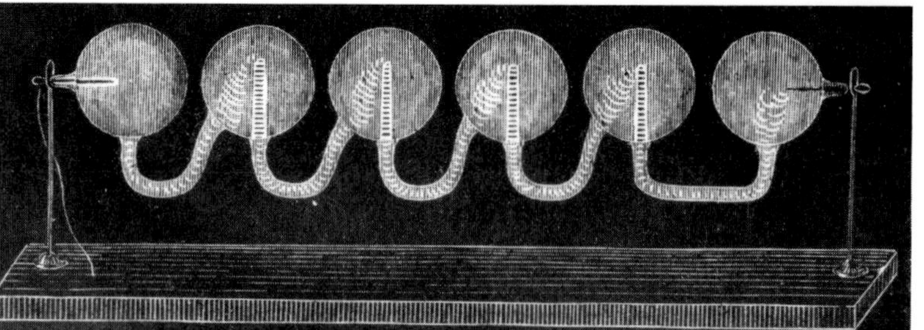

503

502. *Geissler tubes.* In 1855, Heinrich Geissler of Bonn invented the mercury air pump, which created much "harder" vacuums in tubes than had previously been obtainable. Such tubes were very small; those illustrated here had a diameter of about 1 centimeter.

503. *Rühmkorff's induction coil.* In 1851, Heinrich Rühmkorff invented an induction coil that produced fairly high voltages. The combination of the Rühmkorff coil with Geissler tubes made possible the serious study of electrical discharge in a vacuum.

504. *Electrical discharges in vacuo.* The incredible complexity and beauty of the discharges that could be obtained by varying the voltage or gaseous content of evacuated tubes is illustrated in this series of pictures. Until very late in the nineteenth century, it was difficult, if not impossible, to discover any unifying principle to explain these variations.

505. *Cathode rays.* Twenty years after Geissler and Rühmkorff had made it possible to study electrical discharges in high vacuums, it was realized that these discharges emanated from the cathode (negative electrode) rather than from the anode (positive electrode). Once scientists understood that, they began to vary the form of the discharge by changing the shape of the cathode or by placing the anodes in such a way that the "rays" passed to them along peculiar paths. In this illustration, the anodes in the tube at left cause the "rays" to move in curves. In the tube at right, the "rays" are brought to a focus as a result of the curvature of the cathode. But what these "rays" were was still a mystery.

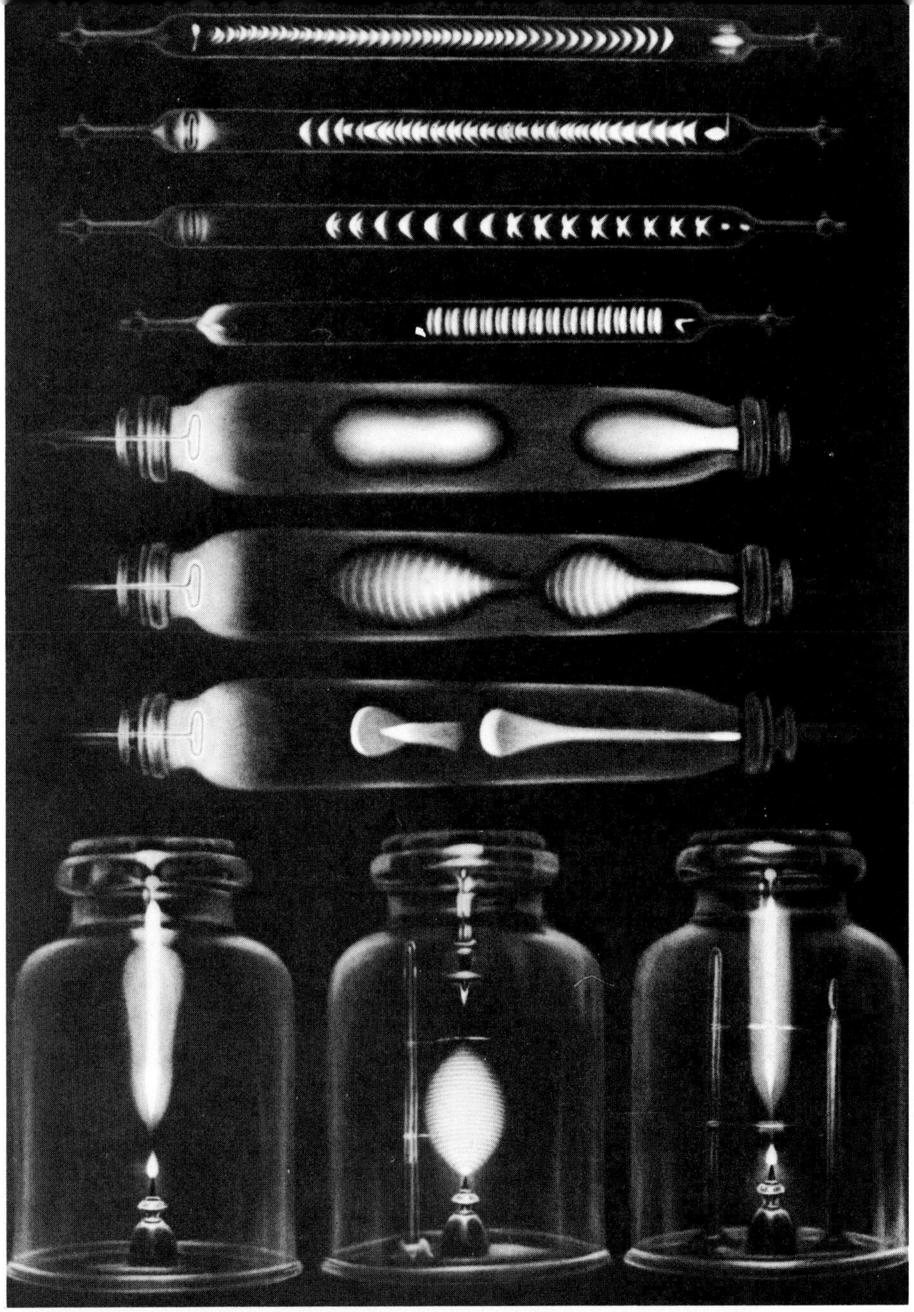

504

505

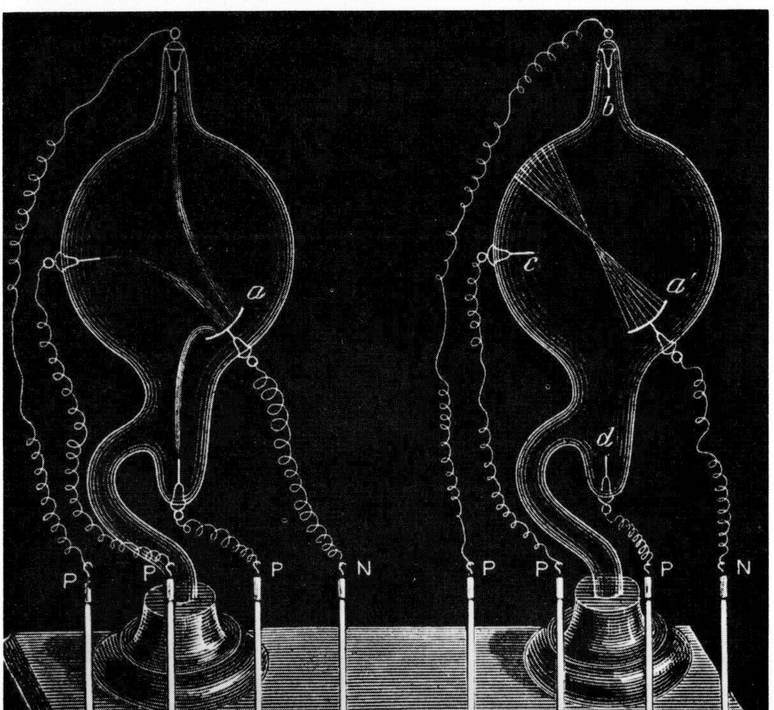

506

507

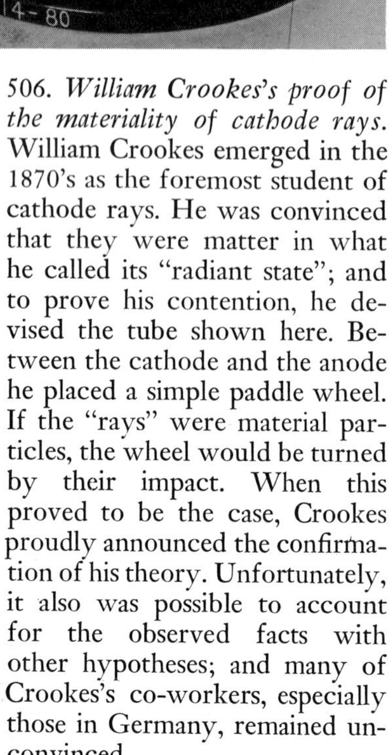

506. *William Crookes's proof of the materiality of cathode rays.* William Crookes emerged in the 1870's as the foremost student of cathode rays. He was convinced that they were matter in what he called its "radiant state"; and to prove his contention, he devised the tube shown here. Between the cathode and the anode he placed a simple paddle wheel. If the "rays" were material particles, the wheel would be turned by their impact. When this proved to be the case, Crookes proudly announced the confirmation of his theory. Unfortunately, it also was possible to account for the observed facts with other hypotheses; and many of Crookes's co-workers, especially those in Germany, remained unconvinced.

507. *Various Crookes tubes.* Crookes was an ingenious experimenter who devised a series of different kinds of evacuated tubes for the study of cathode rays. The tube on the left, for example, has two cathodes by which the attraction of two "rays" could be illustrated. From it one could infer that the "rays," whatever they were, behaved very much like ordinary electrical currents, at least in this respect.

508. *J. J. Thomson's "Crookes tube" for the measurement of e/m of the electron.* Crookes's conviction that cathode "rays" were material particles, and his illustration that the "rays" were identical with an electrical current led naturally to the supposition that the "rays" were negatively charged particles. From that hypothesis, it was a short step to attempt to measure the ratio of charge to mass in such particles. This J. J. Thomson succeeded in doing with the apparatus shown here. The cathode is at the far left. At farthest right is a scale, and in the middle are two plates that could be read on the end scale; thereby e/m could be calculated. The determination of this figure, together with other electrical phenomena, led Thomson to suggest in 1897 that there existed a fundamental particle of negative electricity. Thomson is usually given credit for the discovery of the electron.

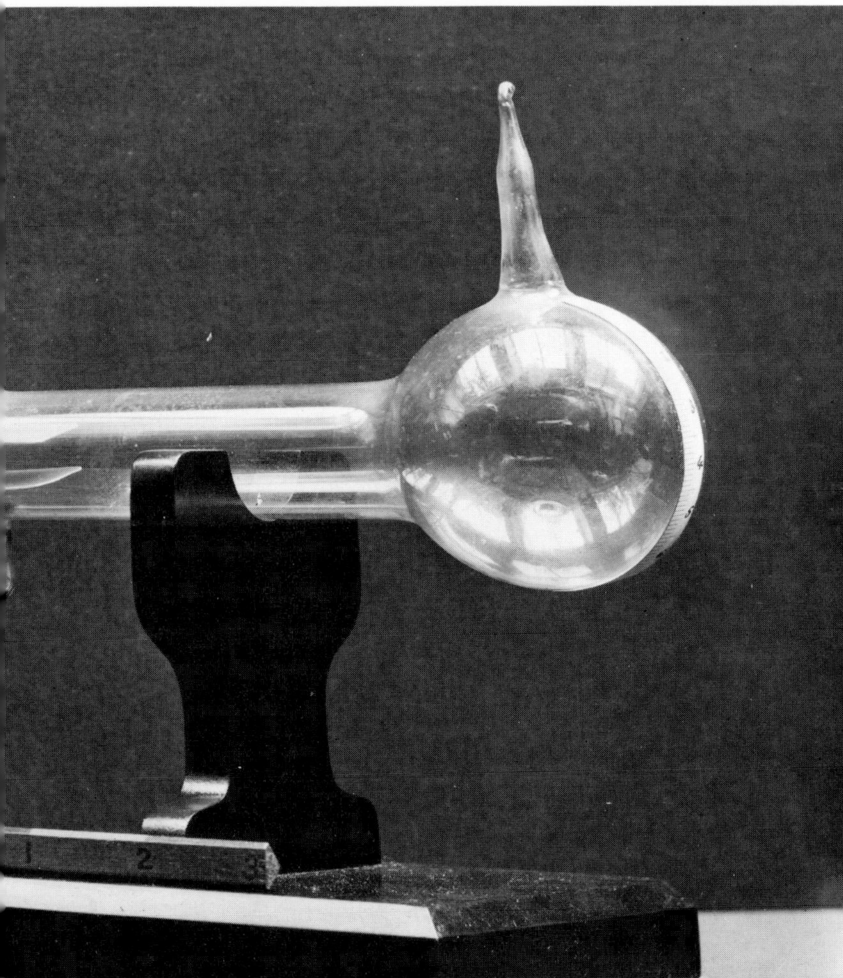

508

509

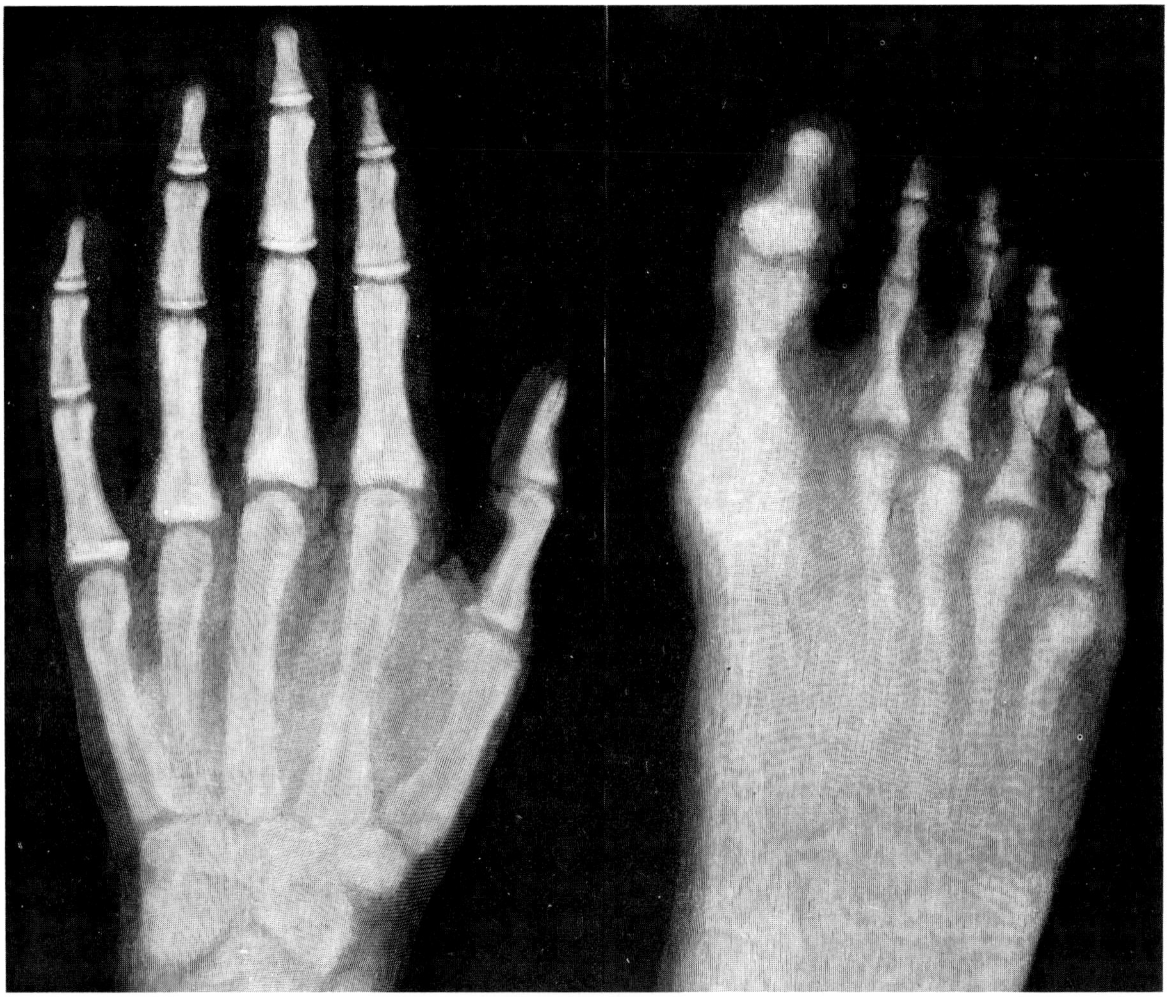

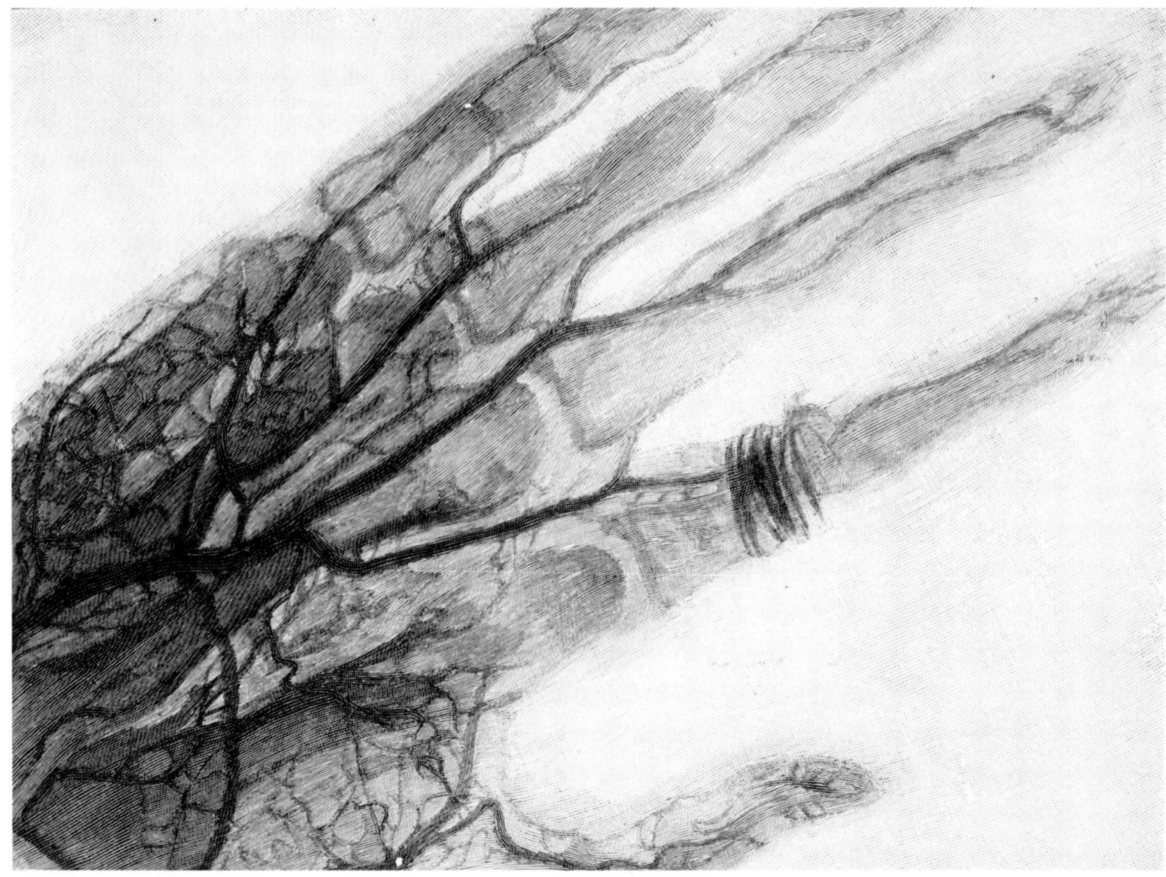

510

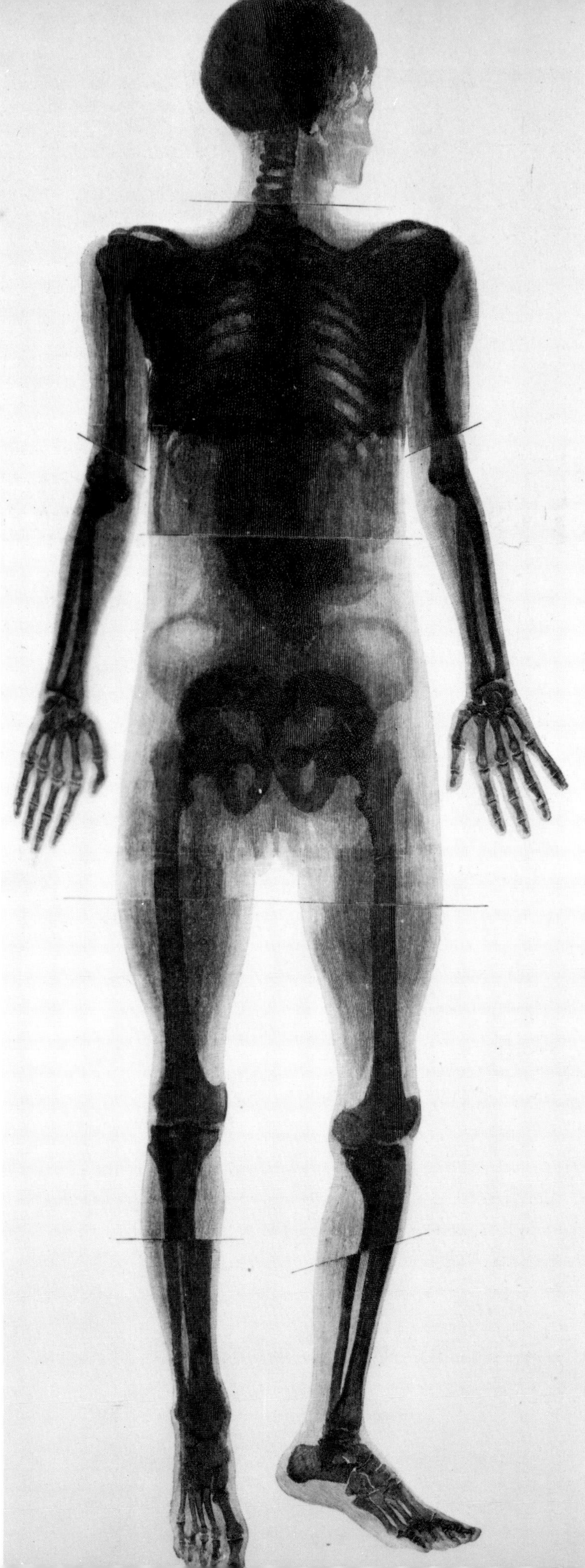

509. *Röntgen rays and Röntgen photographs.* In 1895, Wilhelm Röntgen noticed that a nearby phosphorescent screen glowed when one of his cathode ray tubes was in operation. While investigating the strange rays given off by the tube, he discovered that these rays passed through flesh and made it possible to photograph bones. Photographs like those shown here, which appeared in a German illustrated weekly in 1896, startled the general public and caused a sensation in the scientific world.

510. *X rays and anatomical structure.* More than bones could be photographed by means of the new unknown rays, or X rays, as Röntgen called them. The vascular system also could be laid bare if it were injected with suitable materials, as this contemporary photograph shows. X rays obviously offered new opportunities for diagnosis and medical treatment.

511. *The skeleton of a living human being revealed.* For the nineteenth century, this picture was the ultimate in anatomical illustration. Where, it might well be asked, would science stop when it could look inside the living body?

511

512

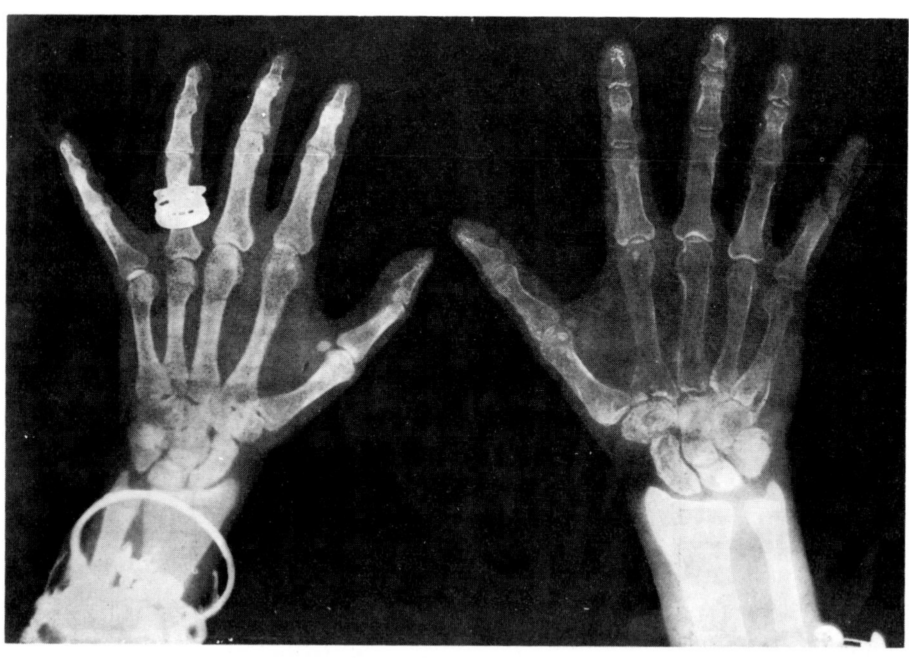

512. *Royal X rays.* The *Illustrated London News* undoubtedly sold many copies in July 1896 by printing this X-ray photograph showing the hands of the Duchess and Duke of York. The picture certainly presented an intimate and unusual view of royalty.

513. *X rays and the general public—a humorous view.* Popular newspapers could hardly resist developing the implications of the X rays. No doubt many of the scientifically naïve believed that the new rays could enable others to look through them. After all, only a few generations later, youngsters would be reading about Superman's X-ray vision.

513

THE NEW PHOTOGRAPH

Paterfamilias (peering through the floor): "Is he a burglar? But he has nothing in his hand—unless, possibly, an aluminium revolver"

Eminent medical man disregards beauty that is only skin deep, but is struck by a lady with a splendid skeleton

The baffled Kodaker: a tale of the future snap-shootists

514

514. *X rays and medicine.* The chest X ray for the diagnosis of tuberculosis was an obvious and almost immediate application of the new discovery. In this illustration, the doctor would receive a full blast of X rays while examining his patient. It is ironic that the new rays, which brought hope and life to many, were instrumental in killing most of those who studied or used them intensively.

"I can see thro' 'is back, Bill. 'Honly a halimminnum ticker"

Sound and Light and Heat

The fact that sound was a vibration in the air had been suspected since classical antiquity. Newton had used his new mechanics to compute the velocity of sound based upon the supposition that the particles of air vibrated according to the conclusions reached in the *Principia*. It was, therefore, necessary to make the measurement of the speed of sound as accurate as possible. Furthermore sound was one of the few processes that could be applied experimentally to the theory of elasticity. The speed of sound in media besides air gave important information about the molecular elasticity of these media. The measurement is not that difficult to make. Basically, it involves a visual signal tied to a simultaneous emission of sound by the sender so that the receiver can start his timing mechanism. This mechanism is then stopped when the signal is received. The signal has passed over a measured course and its speed is simple to compute. By the early nineteenth century, the speed of sound had been determined with sufficient accuracy to suit those who needed to know it.

A more surprising aspect of sound was revealed at the beginning of the century. Chladni in Germany and Savart in France showed that sound vibrations could create peculiar and beautiful permanent forms in matter. If a metal plate were strewn with a light powder, and the plate were then bowed with a violin bow, the particles of the powder arranged themselves in rather complicated but symmetrical patterns. At first, these patterns were mere scientific curiosities, but they soon assumed a larger importance. The adherents of the philosophy of forces had to rely upon vibrations for the transmission of force from one place to another. That vibrations could also create formal and permanent patterns added a whole new dimension to their ideas. This property of sound became a possibly important clue to the nature of ultimate reality and men like Faraday spent considerable time and energy in investigating Chladni figures. Their ultimate hopes of discovering the basic secrets of matter were disappointed but their time was not wasted. It has been suggested, for example, that Faraday was led to his belief that one electric current must induce another electric current in a neighboring wire by the fact that a vibrating plate can induce similar vibrations in a nearby plate. More importantly, the popularity of studying sound vibrations stimulated experimental and theoretical work on waves and on elastic media that were to be of fundamental importance, not only for the theory of sound, but for the theory of light.

Newton had concluded in his *Opticks* that light must be particulate and he had shown how light particles react to gross matter. He could not discover the laws of repulsion and attraction that accounted for reflection and refraction respectively, but he was convinced that these laws must be like those governing gross matter. Much effort was expended in the eighteenth century in the search for these laws, but to no avail. About the only theoretical advance made over Newton was the ability to discover the relative sizes of the light particles. Since all colors of light traveled at the same speed (a conclusion drawn from observations of eclipses of the moons of Jupiter in which, if the different colors traveled at different speeds, one would expect to see the moons change color at eclipse), and since the different colors of light were refracted through different angles, it was not difficult to find the relative sizes of red, orange, yellow, green, blue, indigo, and violet particles. About all this computation did, however, was to reinforce belief in the corpuscularity of light.

The corpuscular theory of light was a good theory, but it had some serious weak spots. One of the more obvious was its explanation of the phenomenon known as Newton's rings. If a plano-convex lens is placed on a reflecting surface with its plane side up and a beam of monochromatic light is directed vertically at it, a strange thing takes place. The lens looks like a shooter's target, with bright and dark bands concentric with the center of the lens. Newton had explained this phenomenon by combining his corpuscular theory with a wave theory in which a compressional wave in the ether preceded a light particle. If the particle arrived at a surface when the compressional wave was at its densest, the particle was reflected; if the wave were at its rarest, the particle entered the surface and was refracted. These "fits of easy reflexion and refraction" as Newton called them, then accounted for the bright and dark rings as light was either reflected or refracted from the bottom surface of the lens. This all seemed very clumsy to an English physician, Thomas Young, who had carried out some interesting experiments on sound and was thoroughly familiar with the behavior of sound waves. Would it not be simpler, he argued, just to assume that light was a wave, not a particle? Waves could account for all the phenomena that particles could, plus the occurrence of interference, as in Newton's rings, where the bright and dark rings could be viewed simply as the reinforcement or interference of waves when crest met crest or their annihilation when crest met trough. Young's ideas were immediately laughed out of court and he abandoned his research. The idea, however, was not to die. Augustin Fresnel, an engineer in Napoleonic France, picked up optics as a subject to keep from being bored to death in the provinces and was soon led to the same conclusion as Young. Fresnel was an excellent mathematician and was able to put his theory into mathematical form. Furthermore, with the idea of transverse waves instead of longitudinal waves, Fresnel was able to explain phenomena that no other theory could handle. By the middle of the 1820's the wave theory of light had displaced the Newtonian corpuscular theory.

If light were made up of waves, it became fundamentally important to determine the wavelengths of the visible spectrum as accurately as possible. It was this task of mapping the spectrum that inspired Fraunhofer to undertake his spectroscopic investigations. His methods, refined by later investigators such as Ångstrom, led to unparalleled precision in the determination of lengths, to the extent that the standard scientific length today is based upon such measurements of light waves.

A simple plan to measure wavelengths uncovered an unusual thing. The solar spectrum was not, as Fraunhofer and everyone else had expected, a continuous band of light from the red to the violet. It was, rather, broken up by hundreds of dark lines. Fraunhofer had no idea what these lines were, but he carefully mapped them and, indeed, found them rather useful. By labelling the more prominent of them with letters of the alphabet, it was possible to use them as optical reference points. One of the more prominent of these dark lines was one that occurred in the yellow band of the spectrum and was labelled *D*.

It was forty years after Fraunhofer's mapping that Kirchhoff and Bunsen provided the basis for an explanation of the dark solar lines. In 1859, they announced their "law" that elements excited in a flame emitted characteristic spectra permitting their easy identification. When light was passed through the vapors of elements, the light that was absorbed had exactly the same wavelengths as these elements emitted when excited. The dark lines in the solar spectrum, therefore, were absorption spectra, caused by the presence of the vapors of elements in the solar atmosphere. The *D* line, for example, coincided exactly with the yellow line emitted by excited sodium vapor. Hence, it could be concluded that sodium existed on the sun. Spectrum analysis permitted astronomers to determine the chemical composition of stars. A new era in astrophysics, cosmology, and cosmogony was begun.

The advances of spectroscopy also raised new and insoluble problems for the theoretical physicist. For a wave to be emitted, something has to wave. This means an oscillator to initiate the wave, and some substance to carry the vibration. The ether would do for the latter, but what could the oscillator be? Light waves are short in length; therefore, the oscillator had to be of the same order of magnitude and no such material particle was known to exist. Furthermore, the emission of distinct lines, rather than broad bands of light, implied minute structure to the atoms emitting the light. From spectroscopy there came strong impetus to build an atomic theory that could account for spectral lines. No theory in the nineteenth century successfully accomplished this. It was not until Max Planck had put forward his quantum theory in 1900 that the vital clue to atomic dynamics was made available, to be picked up by Niels Bohr. Planck was led to his quantum of action by a study of heat radiation.

Heat, like light, was considered at the beginning of the nineteenth century to be an imponderable fluid composed of mutually repulsive particles that were capable of combining chemically with material atoms. This appeared to be the only way one could explain such facts as the latent heat of fusion or of boiling. In this theory, the particles of heat entered a system, say a pot of water, and moved between the particles of the water. The mutual repulsion of the particles of heat caused expansion and a certain amount of "thermal pressure." A thermometer detected this pressure. When the pressure mounted sufficiently (i.e., when the temperature was high enough), the particles of heat were forced into combination with the water to form a new "compound," steam. At this temperature, every particle of heat that was added to the system forced another particle to combine with the water so that the total "thermal pressure" remained the same. Thus, at the boiling point, temperature remained constant as the water was turned into steam. When the steam condensed, these particles of heat were set free to yield the "latent heat" (latent because it was undetectable by the thermometer). This rather elegant theory accounted for the facts in a simple and straightforward way. It carried the day in the eighteenth century and was not overthrown until the nineteenth. Central to it was a measurement of the quantities of heat involved in various processes. Since heat was viewed almost exactly as another chemical substance, it was no coincidence that the best instrument for the measurement of heat, the ice calorimeter, should be invented by a chemist, Lavoisier. Until the advent of electrical methods based upon the change of resistance with heat, this remained the most accurate heat-measuring instrument of the century.

Thermometers to measure the "pressure" of the particles of heat, or caloric as Lavoisier christened it, had been developed to a fairly high degree of accuracy in the eighteenth century. The main improvements were in the art of forming the glass containers for the mercury that was used. Pyrometers for measuring high temperatures depended upon the differential expansion of two dissimilar substances, usually a metal and graphite. The thermocouple, a junction of two dissimilar metals that generated a current of electricity when there was a temperature difference between the two metals, could be used for detecting small temperature gradients.

The most dramatic development in the theory of heat in the nineteenth century does not lend itself easily to illustration. This was the creation of the science of thermodynamics, one of whose strengths is that it makes no assumption whatsoever about the nature of heat. It deals only with heat differences and is, therefore, as abstract as mathematics. It was to become one of the basic sciences, for from its principles flowed conclusions of universal validity in all the sciences. The first principle was that of the conservation of energy; the second was that in natural processes, the degree of disorder in a system tends always to increase.

The enunciation of the principle of the conservation of energy owed much to considerations of the nature of heat in the first half of the century. Of particular importance was the theoretical work done on heat engines by men such as Sadi Carnot and Séguin in France, Colding in Denmark, and Joule in England. Heat engines appeared to offer a direct analogy with hydrodynamic machines. Just as the fall of water turned a water wheel, the fall of heat through a temperature gradient could be used to account for heat engines. Implicit in both of these models was the idea that the total amount of water and heat was the same at the end of the process as at the beginning. As the study of heat engines became ever more refined, doubts began to enter as to the conservation of heat in mechanical processes. Joule developed these doubts into a systematic attack on the caloric theory.

Using simple but clever apparatus, he was able to demonstrate that heat could be both created and destroyed when mechanical work was performed. Moreover, he was able to measure with some precision the mechanical equivalent of heat, that is, just how much work had to be done to create a specific amount of heat and vice versa. With this, the caloric theory of heat was destroyed. In its place grew up the kinetic theory which attributed heat to the motion of the molecules of bodies. Once again, this was a highly mathematical theory of striking success but austerely bereft of striking illustrations. From kinetic theory and thermodynamics were to come the great revolutions of the twentieth century in atomic physics. In the nineteenth century, the foundations were laid.

515 

 516

 517

515. *Measurement of the speed of sound.* Sound had been studied since antiquity, but it was not until the nineteenth century that acoustics became a science. Fundamental to the new science was an accurate determination of the speed of sound. Here, at Montlhéry in France, the speed was determined in 1822 by comparing the time of the flash of a cannon and the time when the sound could be heard.

516. *The speed of sound in water—sending the signal.* The speed of sound in water was known, qualitatively, to exceed its speed in air; but it was difficult to discover by exactly how much. Much the same method was used for this measurement as for measuring the speed of sound in air. Activation of a lever simultaneously struck an underwater bell and produced a flash of light in the pot of coals.

517. *The speed of sound in water—receiving the signal.* By means of an underwater ear trumpet, the signal was received and the time of its arrival noted. Careful measurement of the distance then permitted the determination of the speed.

518a, b. *Wave patterns.* The study of sound inspired nineteenth-century scientists to devise mechanical means for observing and representing wave motions. Some of the patterns generated by the interaction of wave trains are shown in these illustrations. Figure 518a represents the superimposition of one wave train on another, both traveling in the same direction; Figure 518b shows some of the effects of intersecting waves.

519. *Patterns made by vibrating plates.* The vibrations of solid plates could create patterns of considerable complexity and beauty. The researches of Ernst Chladni in Germany and Félix Savart in France, from which these illustrations are taken, gave support to the view, prevalent in the last quarter of the nineteenth century, that static patterns could result from dynamic processes such as waves. The standing waves in the plates moved light particles of powder to the nodes, thus creating the patterns.

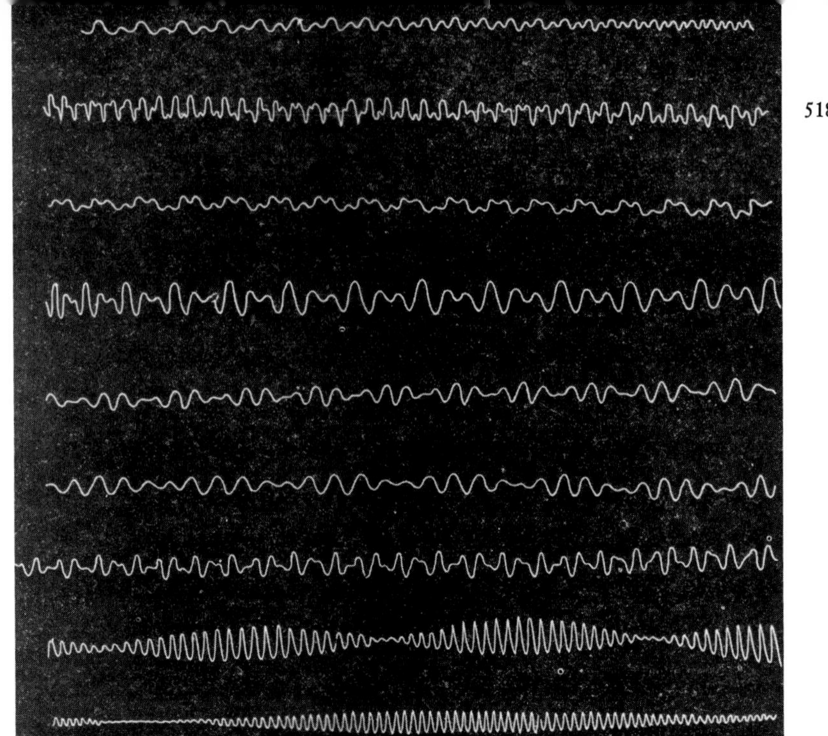

518a

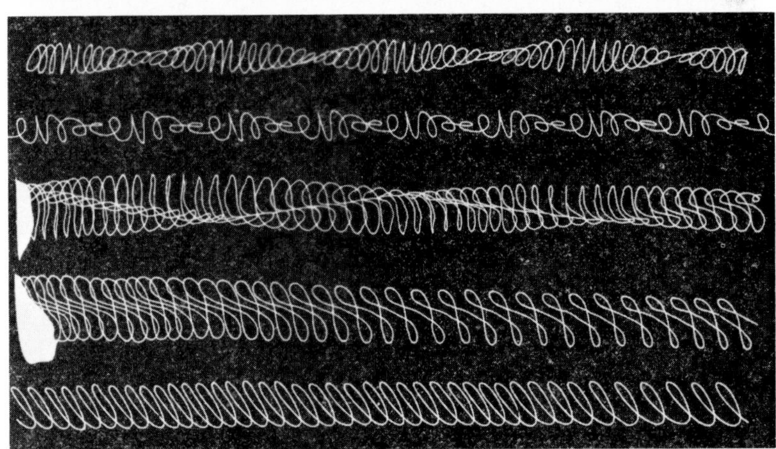

518b

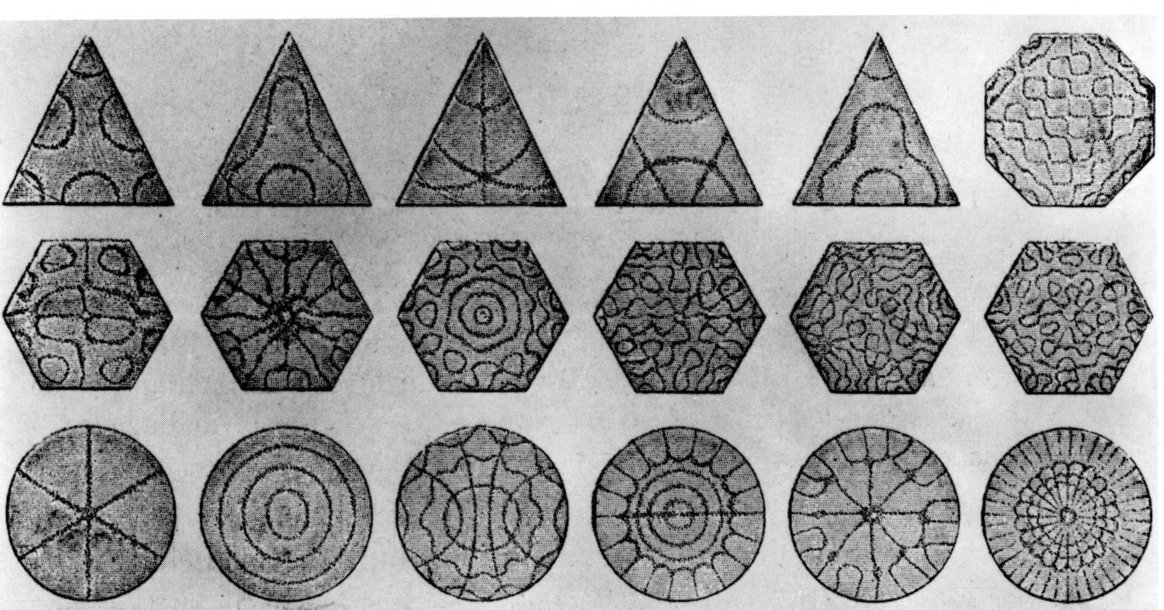

519

520

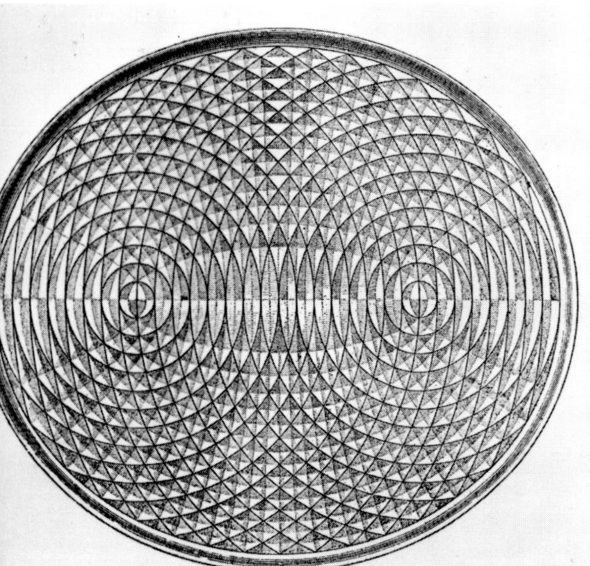

521

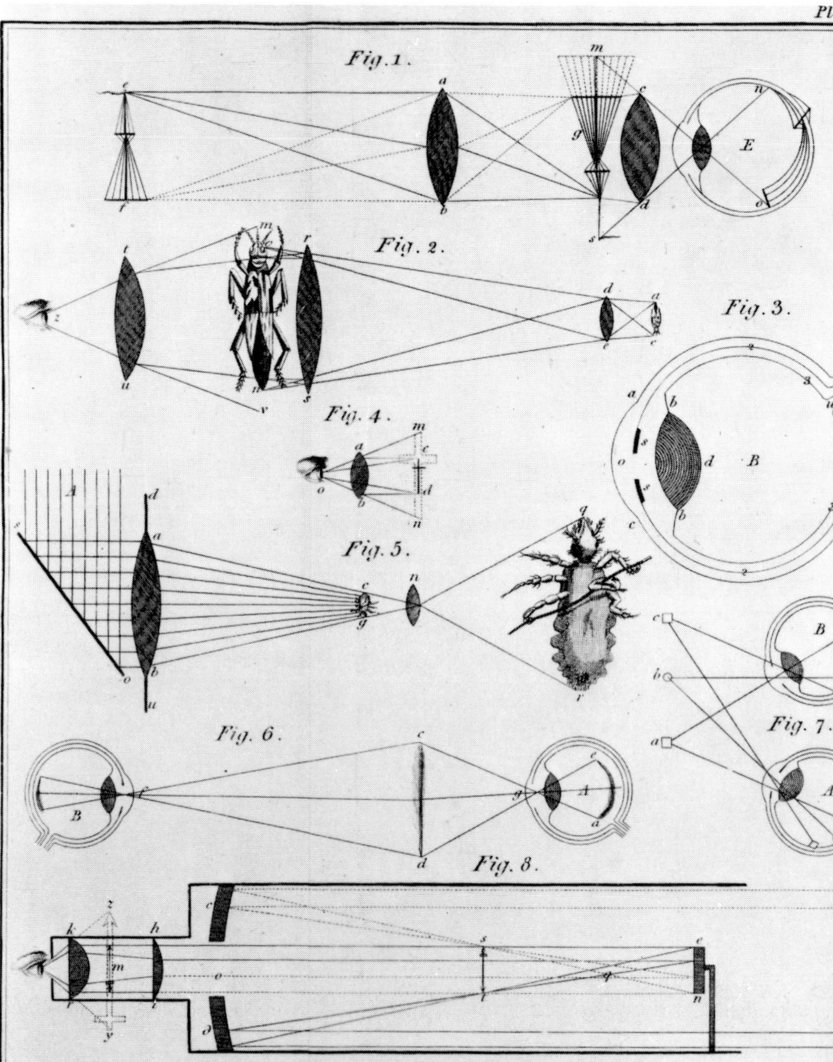

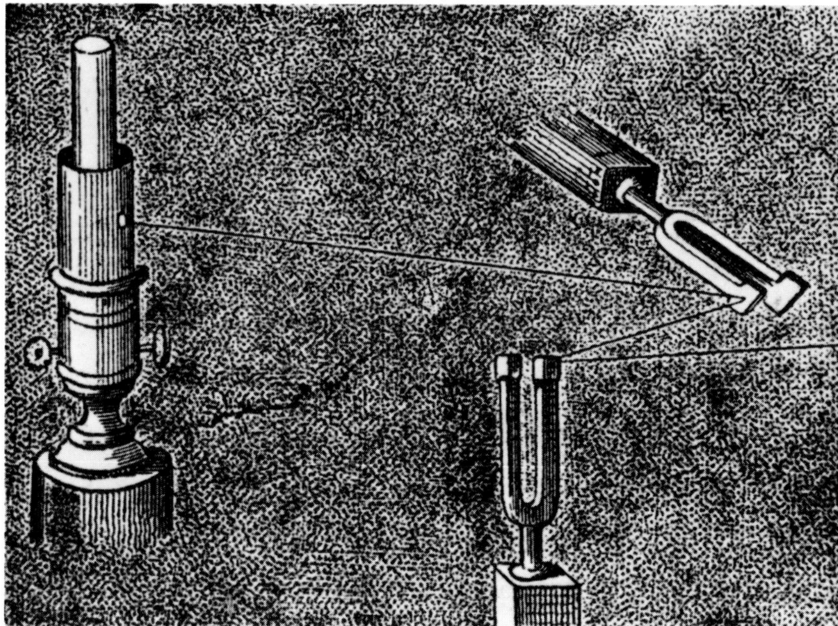

522

520. *The coexistence of waves.* In a classic demonstration, the German physicist Wilhelm Weber showed how standing waves in a dish of mercury could create a structure of great complexity. Given the omnipresence of the concept of luminiferous ether in late nineteenth-century physics, this demonstration offered a possible way to account for the complexities of the phenomenal world by explaining them as results of the various "states" of the ether.

521. *Geometrical optics.* The study of light, like the study of sound, went back to antiquity. Geometrical optics had been investigated by Euclid, and had become a part of geometry in the pre-Christian era. Only an understanding of the law of refraction made it possible for early nineteenth-century opticians to go beyond Euclid. Figure 521 shows the geometrical properties of light that were known around 1800.

522. *The demonstration of wave mixtures.* By means of this simple apparatus, it was possible to project the resultant of the vibrations of two tuning forks onto a screen. Through such instruments one could study complex wave effects.

523

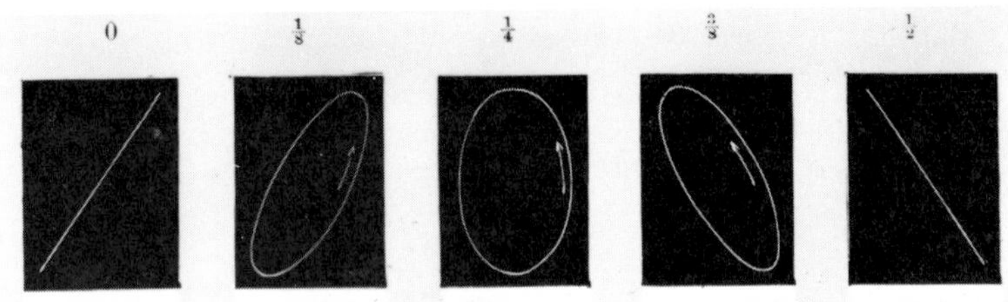

0
1/8
1/4
3/8
1/2
FIG. 136.—Optical curves representing the rectangular vibrations of two tuning-forks in unison.
0
1/8
1/4
3/8
1/2
Octave.
Fifth.
Fourth.
FIG. 137.—Optical curves. The octave, fourth, and fifth.

523. *Some optical figures of sound vibrations.* These figures represent the paths traced by light beams from the apparatus shown in Figure 523. They illustrate how vibrations from tuning forks, tuned at various intervals, could be combined to produce complex periodic phenomena.

524. *Fraunhofer's spectrometer—a demonstration.* In the early 1800's, Joseph Fraunhofer of Munich began his epoch-making investigations of the solar spectrum. The demonstration of his spectroscope is re-created here by a later artist.

524

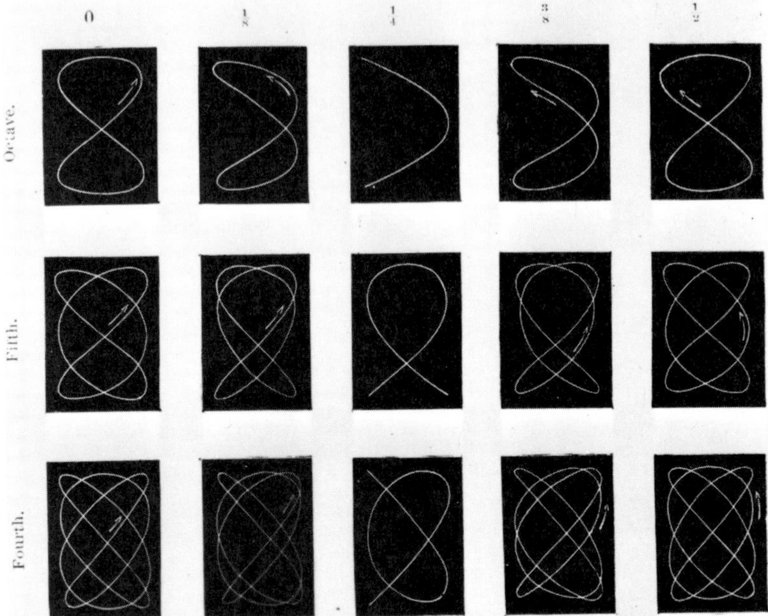

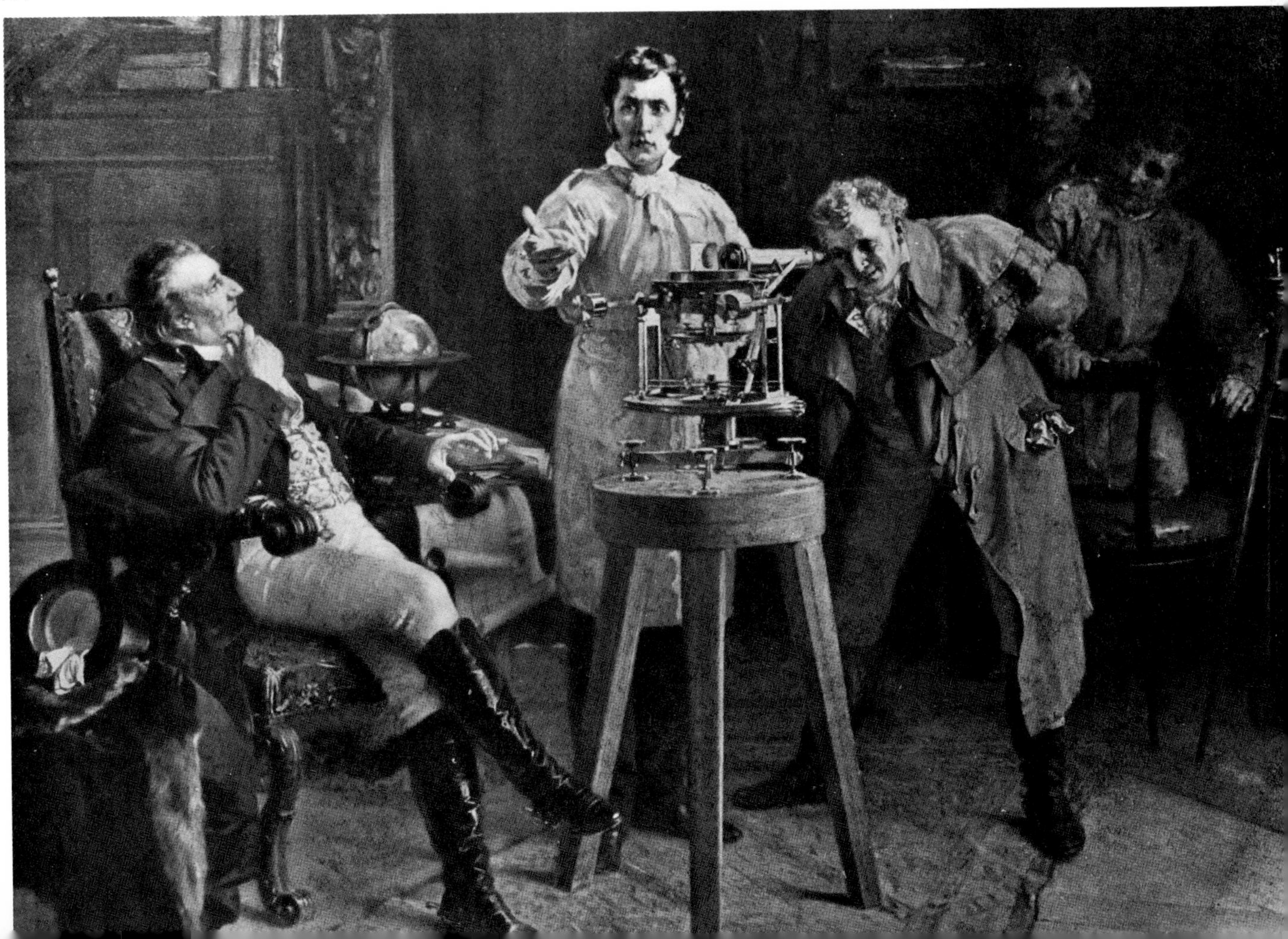

525

526

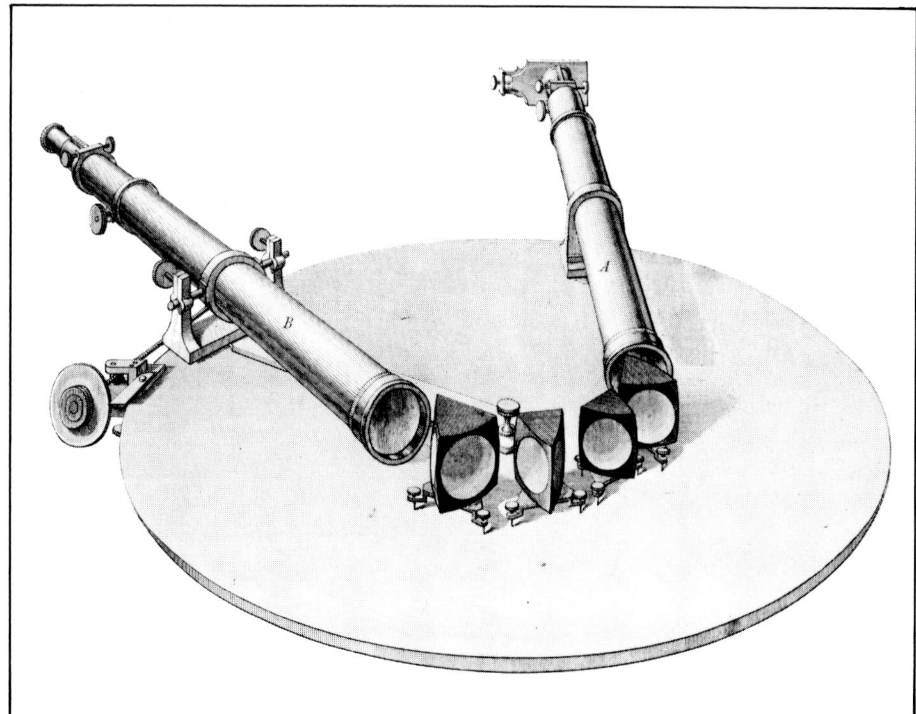

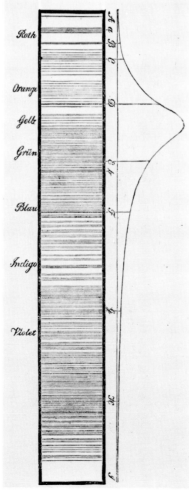

527

525. *Fraunhofer's spectrum apparatus.* Fraunhofer's spectroscope consisted essentially of a collimator for introducing the light into the prism in parallel rays, a prism to break the light into its constituent rays, and an eyepiece for viewing the result. The white light from the sun was let into the instrument through a rectangular slit. Hence the dark lines in the spectrum, visible in Figure 528, are images of that slit.

526. *Gustav Kirchhoff's spectroscope.* In 1859, Gustav Kirchhoff and Robert Bunsen announced the discovery that elements could be recognized by the characteristic line spectrum each emitted when excited. By means of its extra prisms, Kirchhoff's spectroscope permitted somewhat greater separation of the spectral lines than could be obtained with Fraunhofer's spectroscope.

527. *The Fraunhofer spectrum.* This is what Fraunhofer observed and drew. Note the double lines at *D*, the famous sodium *D* lines. Beside the spectrum is a graph showing the energy distribution of solar radiation, which was added later in the century.

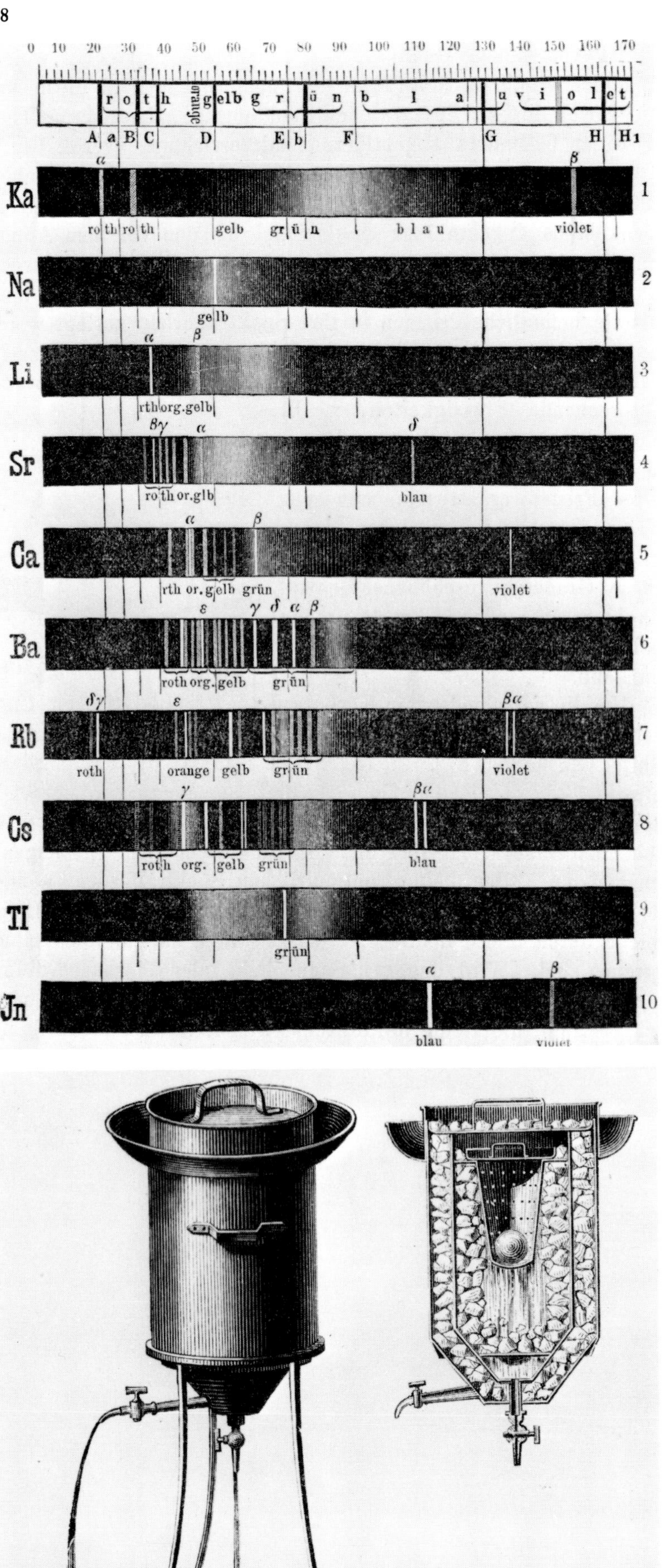

528. *Kirchhoff's and Bunsen's characteristic spectra of the elements.* Some of the spectra emitted by excited elements are shown here. The realization that bright lines in the emission spectra of the elements exactly coincided in wavelength with the dark lines in the solar spectrum indicated that the same elements that were emitting light on earth were absorbing light in the sun. This meant that the composition of the sun and the stars could be examined here on earth. Thus the science of astrophysics was born.

529. *The measurement of heat—the ice calorimeter.* The nature of heat was one of the more perplexing problems of physics and chemistry at the beginning of the nineteenth century. Was it material, or was it merely the result of molecular motions? There were facts and authorities on both sides of the question, but until midcentury there was no "crucial" experiment to decide between the two. In such a situation, the best approach was to refine observations. For this, one of the most important instruments was the ice calorimeter, first used by Lavoisier and Laplace in their classical experiments on combustion in the late eighteenth century. The calorimeter worked on the simple principle of weighing the amount of water produced by ice melting during the thermal process under study. Since the amount of heat required to melt ice was easily determined and the chemical balance was very accurate, the total amount of heat applied could be measured with some confidence.

530

530. *Thermometers for measuring temperature*. By the end of the eighteenth century, various thermometers had been devised that permitted measurement of a fairly wide range of temperature. In this illustration, depicting their evolution, Figure 1 is a not very accurate thermometer based on an instrument, made by Galileo, that reacted to both barometric pressure and temperature. Figure 2 is a thermometer invented by John Leslie that, although inconvenient and difficult to use, eliminated external air pressure by having sealed bulbs at either end. Its temperature scale measured the differential pressure between the two bulbs. Figure 3 is a simple mercury thermometer, accurate between —39°F. and 600°F. and containing mercury that expanded and contracted as the temperature changed. Figure 4 is a mercury thermometer with both Fahrenheit and centigrade scales. Daniell's pyrometer, Figure 5, was good for measuring very high temperatures by indicating the difference in the expansion of iron and platinum rods inserted in a carbon tube.

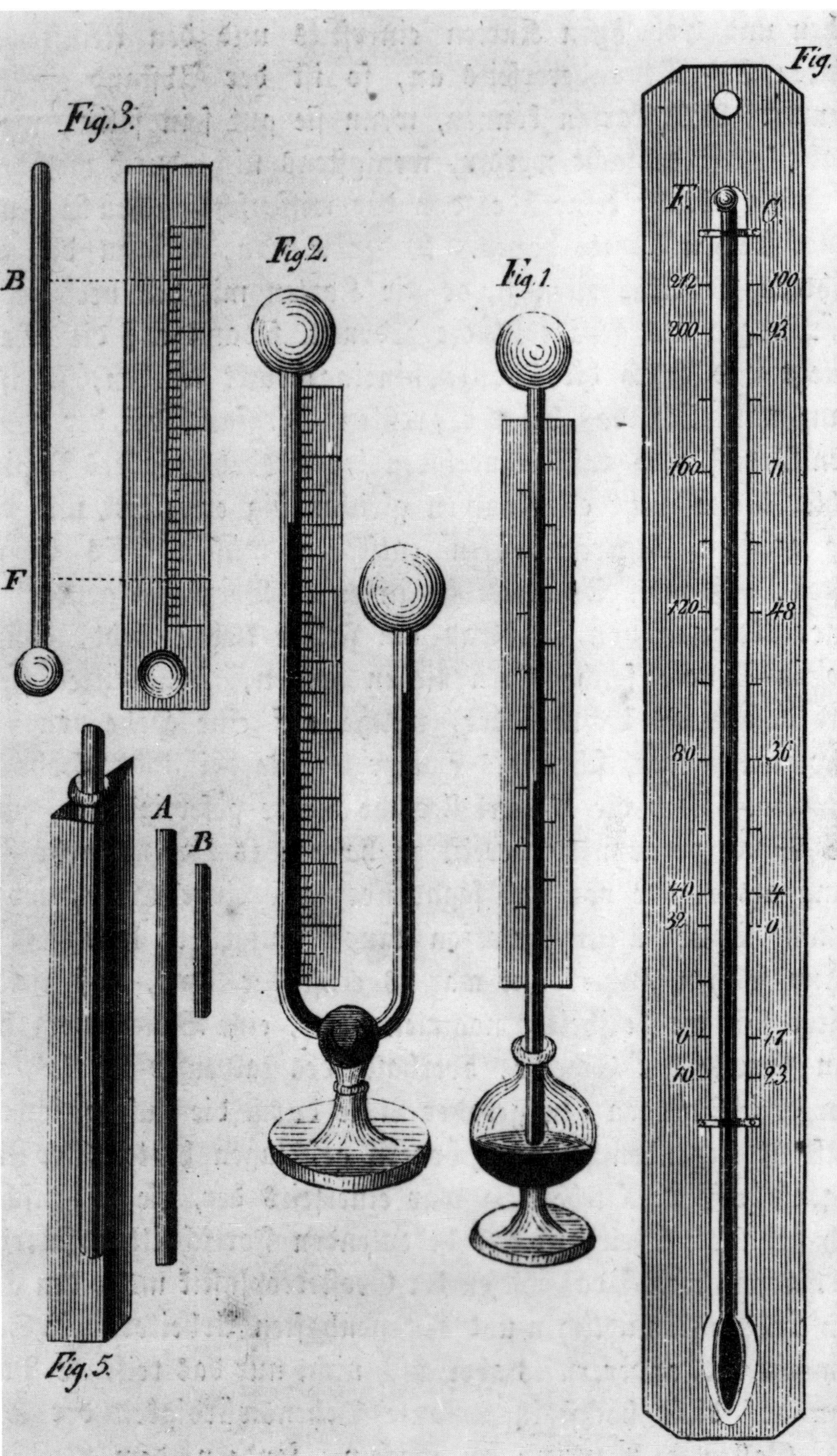

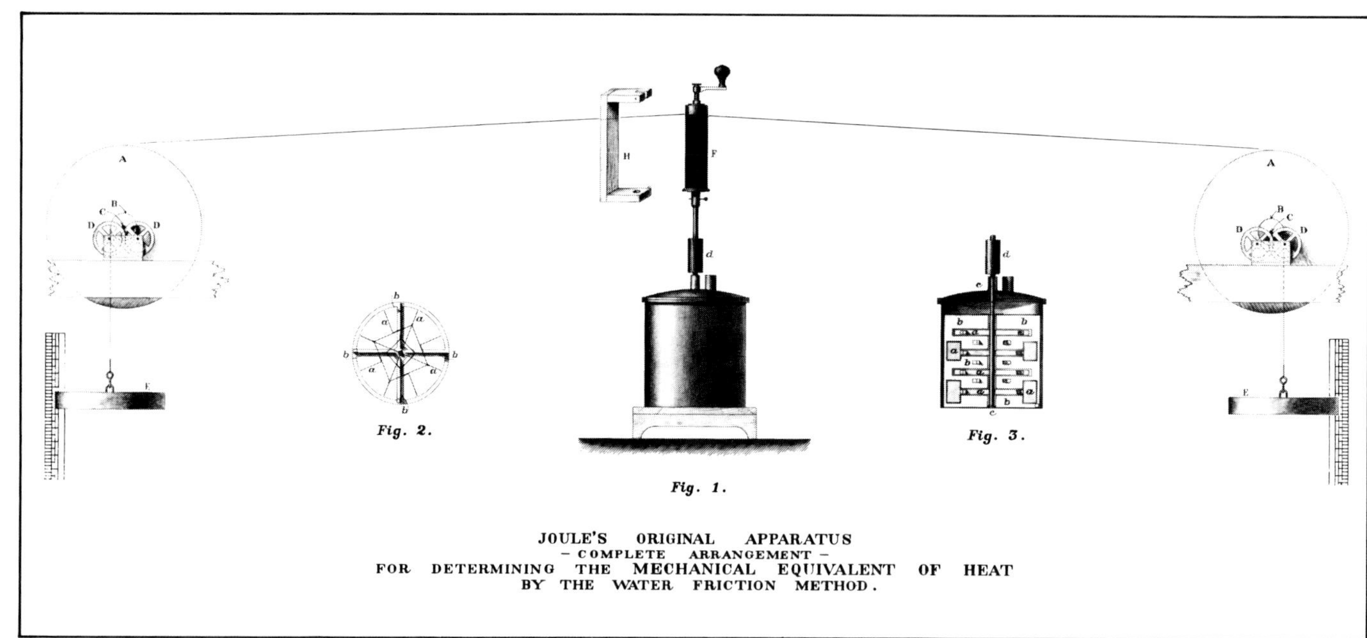

531. *James Prescott Joule's "crucial" experiment.* In the early 1840's, James Prescott Joule began his studies of the conversion of mechanical energy into heat that led to his determination of the mechanical equivalent of heat. Shown here is the apparatus he used. The weight at bottom left, marked *E*, fell along a measured path, doing work by turning the crank (*F*) that turned paddles in a water-filled pail beneath the crank. (A cross section of the pail, showing the paddles, is in Figure 3.) The amount of work done could be determined by measuring the rise in temperature of the water in the pail. This experiment proved that heat was not a substance, for if it were, it would not be created or destroyed by mechanical processes.

532. *Joule's pail.* The complex arrangement of paddles and baffles was necessary to keep the water from swirling in the pail. Joule had to ensure that the mechanical energy exerted by the falling weight was transformed into heat, rather than into mechanical energy again.

532

VI

The Marvels of Science

We are, today, so used to the pace of scientific and technological change that we expect immediate applications of scientific discoveries as a matter of course. Furthermore, our expectations are usually met. The transistor went from a scientific curiosity to an integral part of our radios and television sets in less than a decade. The discovery of the role of DNA (deoxyribonucleic acid) in heredity led, again in fewer than ten years, to the new field of genetic engineering. And so on. But, it was not always so. Until the nineteenth century, the relationship between science and society was almost entirely intellectual. The Copernican revolution was a revolution in ideas that had very little direct practical effect on the lives of Westerners for centuries thereafter. There was, of course, always the dream that knowledge of the laws of nature would permit man to control nature for man's benefit. We mentioned earlier that Sir Francis Bacon died of pneumonia contracted in his attempts to preserve a chicken by freezing it and thereby escape the age-old alternation of feast and famine caused by the overabundance of foods at certain seasons and their scarcity at others. Bacon's goal, we might simply add here, was to be achieved in the nineteenth century when mechanical pumps became available and made refrigeration practicable. But, when the nineteenth century began, Bacon's dream was just that, a dream that had come no closer to realization than when he had dreamed it in the seventeenth century. Life at the beginning of the century was much as it had been for centuries. It is worth looking at nineteenth century conditions in some detail so that we may appreciate how life was changed.

In its fundamentals in the home, life continued much as it had for ages. Homes were built of stone or brick, only rarely of wood in Europe for wood was becoming scarce. They were heated by fireplaces or by wood and coal stoves. They were illuminated by oil lamps. For the wealthy, there were primitive sanitary features, but for the generality of the population, the outdoor latrine was standard. Food production had been increased somewhat by the application of common sense to the problems of agriculture, but famine was still a constant threat. The potato famine in Ireland in the 1840's was an extreme and somewhat extraordinary event, but it reveals how close to the edge of starvation Western societies still were. Transportation remained what it had been for centuries. The horse provided land power, wind drove the ships across the seas. Communications were primitive. The printing press had evolved somewhat from Gutenberg's time, but printing remained a manual art. The French, during the French Revolution, had devised a signal telegraph that sent messages by positioning mechanical "arms" and permitted messages to be sent over fairly long distances rather rapidly as long as weather conditions permitted the stations to be visible from one another. Medicine was both crude and cruel. The causes of most diseases were unknown and there was no rational therapy that could be applied to most cases of illness. A sick person was better off observing the prescriptions of Hippocrates than allowing a contemporary physician to bleed, purge, and poultice him to death. Surgery was reserved for extreme cases, for there was no anesthesia and there were no antiseptic precautions. In short, the great scientific achievements of the seventeenth and eighteenth centuries had barely touched the life of the people. In 1834, when François Arago was asked in the French Chamber of Deputies why the government should support science, he could offer only the lightning rod as an example of how science had improved the life of the ordinary citizen.

All this was to change in the nineteenth century. Transportation was to be revolutionized, first by the steam engine and then, at the end of the century, by the internal combustion engine. Let us just note here a few of the more obvious effects of the steam engine on the daily life of Westerners. The steam engine made it possible to harvest enormous quantities of wheat in the heartland of America and ship it, via railroad and steamship, to Europe, in the process destroying much of the agrarian basis of European society and also providing vast new reserves of food so that European populations could grow and could avoid famine. Refrigerated steamships brought frozen beef from the Argentine and frozen lamb from Australia to peoples for whom meat had been a luxury. Meat became a standard part of the Western diet, thereby contributing to both population increase and an increase in cardiac seizures.

The steam engine and the railroad also revolutionized the economies and military organizations of Europe. With railroads, outlying agricultural districts could transport their products to cities that formerly had been too far away to absorb them and cities could be provisioned and grow to an extent never possible before. It was in the nineteenth century that the balance between agrarian and urban society began to tip in favor of the urban. Soldiers as well as cabbages could be transported rapidly and this fact gave certain advantages to those countries that built extensive railroad networks. Germany's willingness to gamble on war in 1914 was due in large part to the fact that Russia's railroad network was so primitive that a Russian army could not be put into the field for some weeks after a war was declared. The lesson had been learned in the nineteenth century, and any country that refused to industrialize and accept the steam revolution seemed doomed to decay and domination by more powerful nations.

The conquest of the land and the sea through steam power only whetted some men's appetite for the conquest of the air. This was to be the ultimate revolution in transport and nineteenth-century men were obsessed by it. The goal was to perfect guided flight either by making lighter-than-air balloons maneuverable or by providing mechanical power to a heavier-than-air craft. This was beyond the resources of the century, but, again, the pioneering work was done by the time the century ended. Here, as in so many other areas, the twentieth century was to reap the harvest of nineteenth-century efforts.

Somewhat less dramatically, but no less importantly, there was a communications revolution based on the telegraph and the power printing press. The world had always been parochial before the nineteenth century. News traveled so slowly that it was almost impossible to feel personally involved in events long since over when one first heard of them. What could one do, for example, for the victims of the Lisbon earthquake in 1755; by the time news of it reached London, most of its victims were past helping. Hands could be thrown up in horror, but that was about it. Now, news of catastrophes reached the public almost immediately and a new moral imperative was added to the Western conscience. Plague, famine, or natural catastrophe in distant parts of the world became the concern of those whose knowledge of these terrible events put them in a position to help. The Age of Anxiety was partially initiated by the new ability to worry on a global scale. Peoples, after all, could not callously be left to die when means were available to save them, and Western imperialism, at least in part, was impelled by this concern. This is the message of Rudyard Kipling's poem "The White Man's Burden." It was a burden imposed by the new systems of communication.

The other side of this coin, of course, was the added power that rapid communications gave to Western powers. The cartoon from *Punch* (Fig. 539) reveals how contemporaries saw the ability of great powers to react almost immediately to unfavorable political events throughout the world. Telegraphy would make the world safe for democracy!

The growth of telegraphy led to important advances in the generation of electricity and these, in turn, led to the domestication of electric power. The new electrical age was symbolized by the electric light and nineteenth-century people went to extraordinary lengths to celebrate this new factor in their lives. The electric light only dimly illuminated the future and the wonders that electricity would make possible, but no one doubted that life would be made wonderful by electricity. All the reader need do to appreciate what a change was accomplished by the ability to bring dependable, steady electrical power into the home is to look around and note the electrical appliances now taken for granted. The revolution, of which these are the consequence, began in the nineteenth century.

The most direct effect of science on the average man and woman came through the development of scientific medicine. Science helped create new diagnostic instruments like the gastroscope that permitted physicians to look into bodily cavities never before examined in living patients. X rays could detect pathological conditions long before they might otherwise have been detected and were particularly effective in the early diagnosis of the white plague of the century, tuberculosis. The germ theory of disease led both to better therapeutic measures and to the development of immunological procedures that helped to protect masses of people from diseases that had killed millions before. Again, all that the modern reader has to do to appreciate the extent of the revolution here started is to look around and realize how rare are such maladies as diphtheria, typhoid fever, smallpox, tuberculosis, and septicemia. Not all of these were conquered in the nineteenth century, but the foundations for the medicine that ultimately did wipe them out or reduce them to lesser importance were laid in that century.

With all its practical effects, we should not neglect to notice the continuance of the intellectual impact of science on Western civilization. It continued to erode the traditional foundations of society at an ever-increasing pace until, by the end of the century, it is proper to speak of the religion of science. All questions that were worth asking ought to be capable of scientific answers. Reality was that, and only that, which could be contained within the scientific tradition. Other questions and other answers were reactionary, out of date, useless, and detrimental to the progress of mankind. If only all people could acquire the scientific habit of mind, then all would finally come right with the world. Acolytes to the new religion could now start their training as children, through popular scientific kits that brought the principles of science directly into the home. If and when all mankind could be brought together under the mantle of science, the millennium would be achieved. It was a noble faith, but there was, already in the nineteenth century, a significant minority which doubted salvation of the human race through science. With all it had produced, had science really made life richer and better? What about the industrial worker who lived in grime and soot? What about the child whose life medicine had saved but who could find no place in an increasingly technological society? What about the affluent classes whose every material need was met by the production made possible by the new science but who found their personal lives increasingly meaningless and empty? These questions were asked, and it seems fitting to end this album on this note of dissent. Science had achieved wonders but was it the panacea for all man's ills? John Donne's words, written in the seventeenth century, still echoed through the Western mind. "The new philosophy calls all in doubt." Could the new philosophy create a better and more just world? Or would man use it to multiply the evils which he had, throughout his history, inflicted upon his fellows. Only the future would tell.

533

533. *Science and the world. A British view.* By the midnineteenth century, it was obvious that through science and technology, Western man had something the rest of the world lacked. It seemed only right to equate science and technology with "progress," and to show the rest of the world that progress could not be fought. The smugness and sense of righteous menace conveyed by this *Punch* cartoon were part of a new atmosphere that marked the beginning of the imperialist race.

534

534. *The telegraph.* Progress meant more than railroads. Certainly it included the revolution in communications that the nineteenth century achieved. The electric telegraph annihilated space and shrank the time required for communication between distant parts of the world to the fraction of a second that it took for an electric impulse to pass through a wire. With refinements, the telegraph, as seen here, could even print out its own message.

535

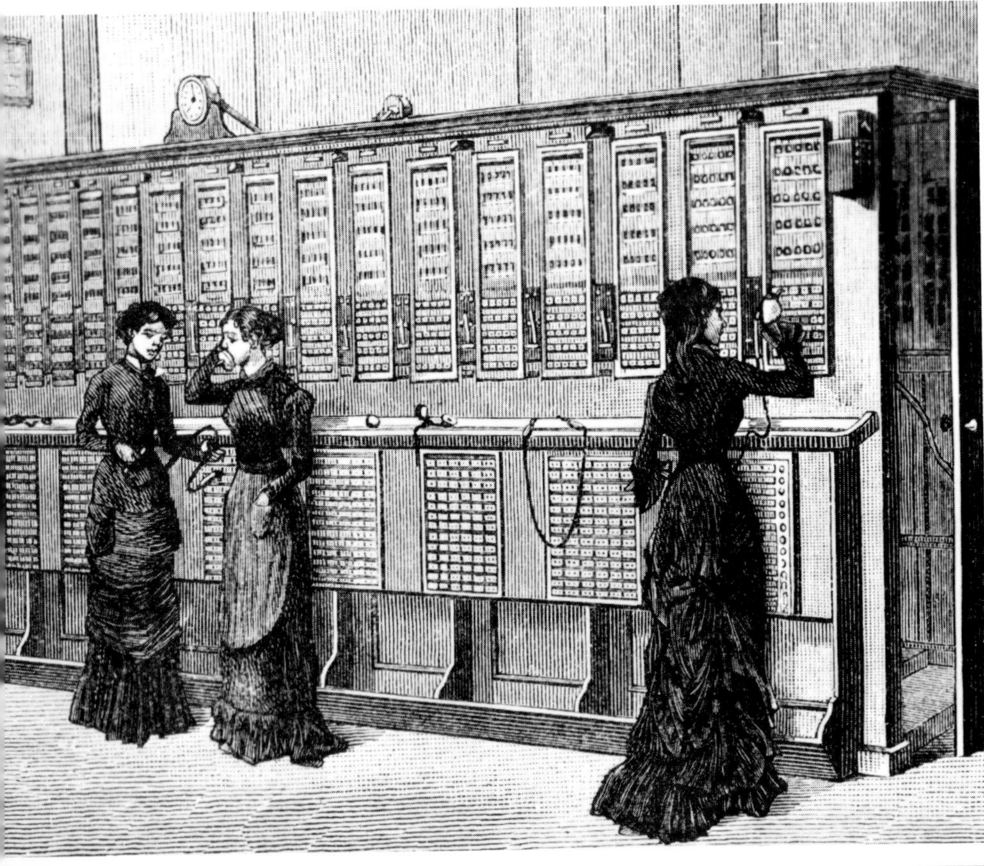

535. *Switchboard for a multiple telegraph.* Urban telegraphy stations became centers of communications for large regions. They required constant attendance to prevent wires from getting crossed and messages from being garbled. Hence, the creation of the switchboard that routed messages to the proper places.

536. *The terminal for a central telegraph station.* The requirements of telegraphic communication on a large scale created widespread telegraphic networks. The "nodes" of a network, as shown here, involved a respectably large number of circuits.

537. *The telegraphic center in Paris (1888).* As the demands on telegraphy multiplied, the size of the network had to increase. This gigantic installation was the result. It was, literally, the communications nerve center of Paris. It was eliminated by the invention of the telephone, which could be placed in individual homes and business premises and required no skill to operate.

538. *The battery for the Paris telegraph.* The power to run a telegraph station like the one in Paris still had to come from the chemical reactions of a galvanic battery. The size of such batteries, as shown here, was enormous. They would become obsolete only with the invention and introduction of efficient dynamos in the last decade of the nineteenth century.

536

537

538

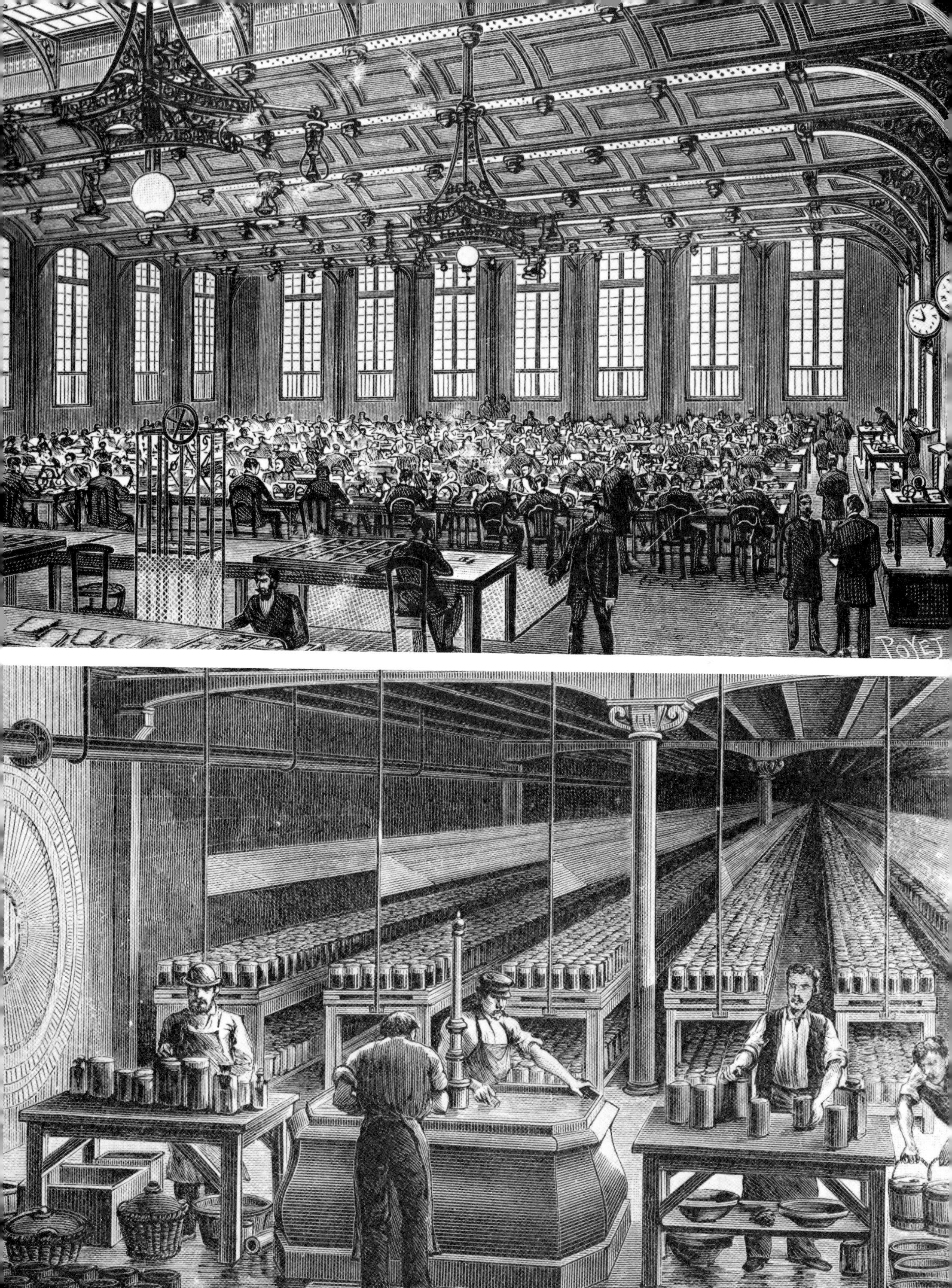
POYET

THE ATLANTIC TELEGRAPH—A BAD LOOK OUT FOR DESPOTISM.

JOHN BULL. "HOLD FAST, JONATHAN." JONATHAN. "ALL RIGHT, JOHNNY."

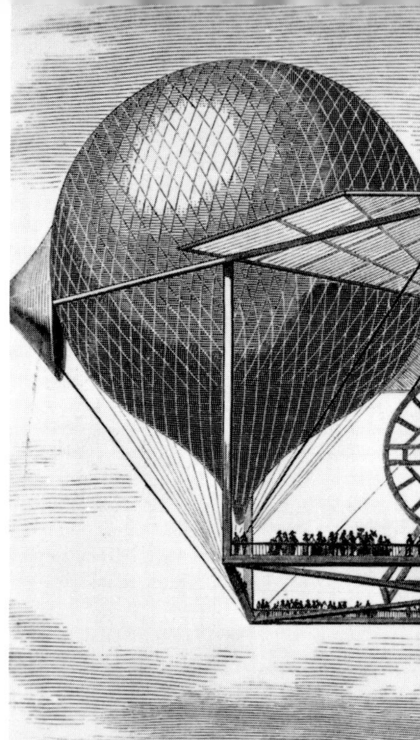

539 539. *Telegraphy and politics.* The laying of the Atlantic cable linked North America and England by telegraph. Its political impact was seen by *Punch* as inimical to any despotic government that might try to challenge the great Atlantic powers.

540. *Air transport by balloon.* The conquest of the air was an obsession of the nineteenth century. The triumph seemed so close and yet was so difficult to achieve, except in the flights of isolated balloonists. What was sought was directed flight that could be used for air transport. Mr. Petin designed the airship shown here and argued its merits in a public lecture at Paris in 1850. Vertical control was achieved by a pair of parachutes on each side; in each pair, the parachutes were placed one on top of the other and opened in opposite directions. Here the ship is shown rising, with the lower port parachute open to brake the ascent. On the descent, the upper parachute opened to brake the fall. Horizontal control was maintained by large plane "ailerons" that could be tilted up and down or from side to side. The motive power was provided by the wind. Petin's dream was grandiose, as can be seen by the number of passengers he envisioned; but his airship never got off the ground.

540

541

542

541. *Charvin's dirigible airship.* As the nineteenth century progressed, aerial navigation became more sophisticated. This 1869 design by Charvin for a propelled airship begins to look like the dirigibles of the twentieth century.

542. *A powered aircraft.* The alternative to the lighter-than-air dirigible was the power-driven, heavier-than-air airplane. The problem of what to use for power was "solved" in this aircraft by employing the newly discovered guncotton for fuel. The cotton, formed into pellets, drove a rotary engine that looked much like the modern Wankel rotary engine. This model, of course, did not fly.

543

544

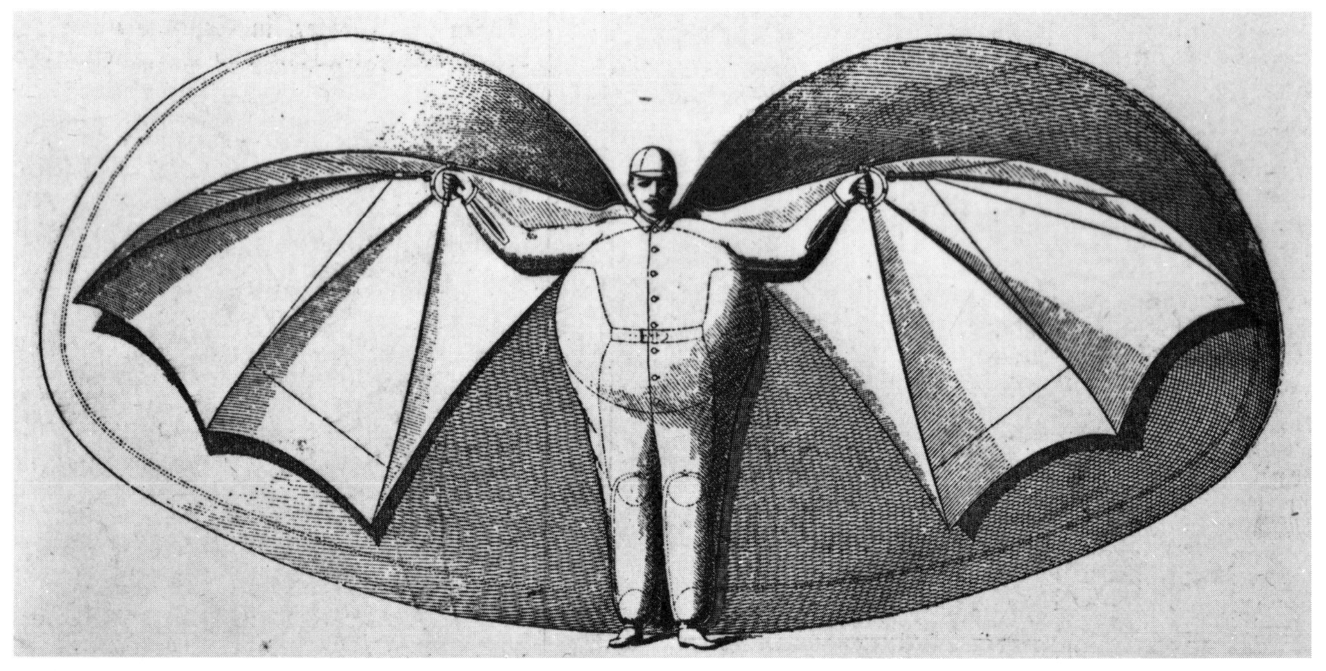

545

543. *An airship from the New World.* America was not to be left out of the aircraft invention sweepstakes. This bizarre machine was invented by W. J. Lewis of New York and was driven by the wind.

544. *A dirigible and personal airship.* By 1886, the problem of finding a suitable motor for a dirigible airship seemed insoluble. One result of this realization was the flying suit shown here. The flyer fit into what was essentially an inflatable suit, which provided him with the necessary lift. The flight path was controlled by flapping the wings.

545. *The death of an aeronaut.* On July 9, 1874, the ardent inventor and aeronaut, Guillaume de Groof put his flying machine to the ultimate test outside London. It failed, and he plunged to his death.

546. *Better things for better living through chemistry.* The spectacular advances of practical chemistry changed conditions of agricultural and industrial production in the nineteenth century. Chemical fertilizers, among which phosphoric acid figured prominently, seemed to promise unheard-of increases in agricultural production. These dreams seemed a fit subject for caricature. The captions read:

Upper left–The exhausted earth demands nourishment. Thanks to the new phosphoric acid fertilizer, this is both easy to do and cheap.

Lower left–M. de *, Secretary-General-Gentleman [a farmer], varnishing his coat of arms.

Center–The human race improved. Happy dwellers in the fields. For four sous of fertilizer per person, here is how they improve.

Upper right–The fertilizer of fertilizers. Yesterday there was nothing but stones in this field, and now there are so many millions of thousands of millions growing that we don't know where to put them.

Lower right–*Like the home of Robert Houdin* [a famous magician].

–Marvelous, to have asparagus in February!

–My friend, I just throw three drops of phosphoric acid on the table, and I get all the early vegetables.

547. *Science in the home.* Scientific advances provided a basis for advertising copy even a century ago. Maignen's filter protected against the "insidious horrors that lurk in the domestic water supply"; anti-calcaire helped the complexion and prevented terrifying domestic explosions.

548. *Promethean science unbound.* This version of the battle of the ancients and the moderns in *Punch* clearly awards the laurel to the moderns. With typical nineteenth-century hubris, the modern world is depicted as superior in every way even to Mount Olympus, the home of the gods.

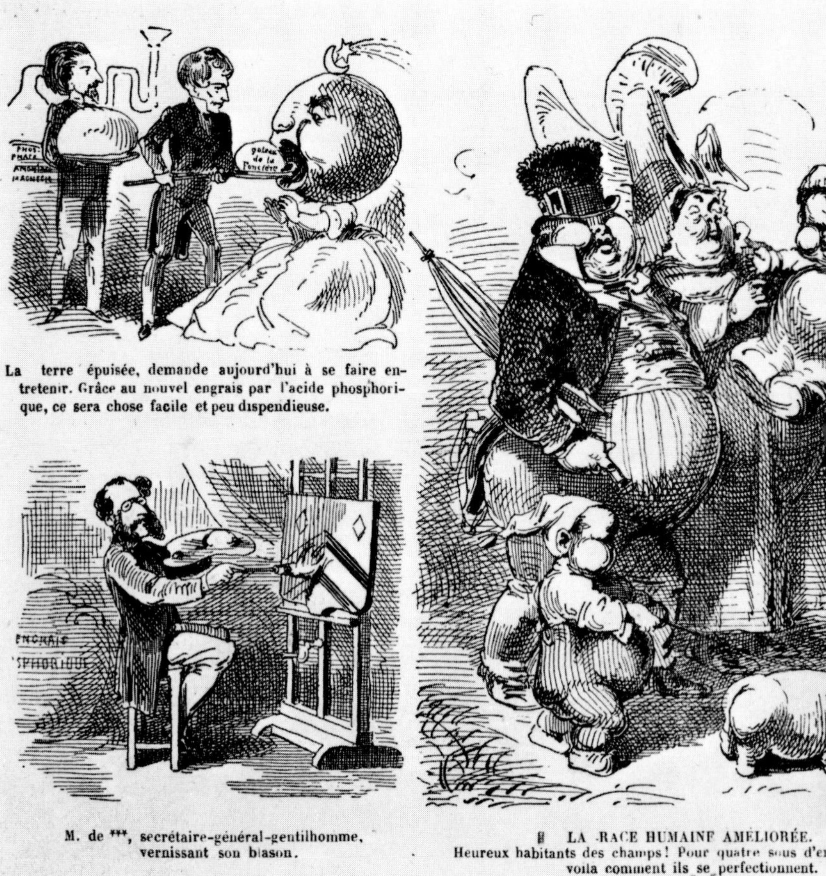
La terre épuisée, demande aujourd'hui à se faire entretenir. Grâce au nouvel engrais par l'acide phosphorique, ce sera chose facile et peu dispendieuse.
M. de ***, secrétaire-général-gentilhomme, vernissant son blason.
LA RACE HUMAINE AMÉLIORÉE.
Heureux habitants des champs! Pour quatre sous d'e
voila comment ils se perfectionnent.

547

PURE WATER

Look to the Water you Drink, and be Wary of Imperfect and Unsafe Filters.

MAIGNEN'S

PATENT

"Filtre-Rapide"

Is a certain protection against the insiduous horrors that lurk in the domestic water supply. It is guaranteed to remove all germs of Cholera, Typhoid, Diphtheria, and all causes of Blood Poisoning, such as dissolved Sewage, Lead, Zinc, Copper, or other noxious Metals.

The LANCET says,—"A better Filter could not be desired."

Differing in its construction and arrangements from all other Filters, the "Filtre-Rapide" is simplicity itself, easy to manage, and under perfect control, while its deodorising and purifying powers are in excess of any others. It can be taken to pieces, cleansed, and re-charged in a few minutes by the users. The Filters are made in various sizes and kinds to suit all requirements, but the principle is the same throughout, viz:—A large filtering surface, not liable to decay or disintegration,—pure Asbestos Cloth, specially woven from picked fibre, coated with Maignen's Patent Carbo-Calcis.

VISCOUNT WOLSELEY writes:

"I have much pleasure in stating that MAIGNEN'S *Patent* 'FILTRE-RAPIDE' *which was used by our Troops in the Expedition up the Nile, gave very satisfactory results indeed. It filtered the water both quickly and well, and was much valued by our Soldiers.*

To be Filled by Hand. *For attaching to the Service Pipes.*

"Main 'Filtre-Rapide.'"
Self-Filling, Easy to Fit, and Easy to Clean.
30/- 50/- 100/- 200/-.

Section of "Cottage" & Bijou D.

"Cottage 'Filtre-Rapide.'" (Brown Stoneware.)	"Bijou D 'Filtre-Rapide.'" (Decorated White Stoneware.)
15/- 25/- 35/- 55/- 70/- 105/-	25/- 40/- 60/- 105/-

See Pamphlet for description of other classes of "FILTRES-RAPIDES—"TABLE" (Glass), from 3/6—"EXPORT" (Enamelled Iron), from 12/6.—"TAP" (Rustless Metal), from 12/6,—"TOURIST" (White Metal) from 18s.,—"WATCH" (Pocket Filter) from 10/6.

Large Supply for Village and Town Water-Works. Eight different Classes approved by the Government for the Army, &c., &c.

SOFT WATER

IN EVERY HOME.

MAIGNEN'S

PATENT

"Anti-Calcaire"

Renders the Hardest Spring or Town Water like the Softest Rain Water.

Everybody will welcome "ANTI-CALCAIRE" so manifold are the uses, blessings, comforts, luxuries, needs, and applications of a real WATER-SOFTENER. It is a perfectly harmless powder which attacks, decomposes, and precipitates all the Carbonates of Lime and Magnesia, and the Sulphate of Lime, leaving nothing injurious in the water.

HARD WATER is cruelly hurtful, wasteful, and destructive. It ruins the complexion, makes the hair harsh and brittle, causes roughness and irritability of the skin, is a nuisance and an injury all round, and should be avoided in all toilet, bath, nursery, domestic, cooking, cleansing, gardening, medicinal, and mechanical uses.

"To soften Hard Water," says THE QUEEN, "use MAIGNEN'S Patent 'ANTI-CALCAIRE' Powder—quite harmless and inexpensive."

Explosion due to Incrustations, caused by Hard Water, preventable by "Anti-Calcaire." For Testimonials see Pamphlet.

"ANTI-CALCAIRE" is sold by Chemists, Ironmongers, Grocers, and Oilmen, in Tins, with printed directions for the Toilet and other domestic uses, at 2d., 6d., 1s., 2s., 3s. 6d., & 7s. each; also in Casks by the Cwt. or Ton, for Laundries, Steam Boilers, and Water-Works, by the Inventor. A Sample Tin and Pamphlet sent for 5 stamps.

Read Pamphlet, Post Free, "Water, Preventable Disease and Filtration."

P. A. MAIGNEN, Contractor to H.M. Government,

32, ST. MARY-AT-HILL, EASTCHEAP, E.C.; 15, PALL MALL, S.W. LONDON.

L'ENGRAIS DES ENGRAIS.

— Hier, il ne venait que des pierres dans ce terrain, à c'te'heure il y pousse des milliards de millions de milliards, qu'on ne sait plus où les mettre.

COMME CHEZ ROBERT-HOUDIN.

— C'est merveilleux! des asperges en fevrier!

— Mon cher, je n'ai qu'à verser sur la table trois gouttes d'acide phosphorique, et j'ai toutes les primeurs.

548

549. *A child's scientific laboratory*. The penetration of science into the intimate fabric of Western life is evident from this advertisement for an educational scientific kit. Toys like this first appeared in the 1890's and made it possible to recruit budding scientists among the very young.

549

550

551

552

550. *The marvels of electricity —rejuvenation.* The nineteenth century was the first electrical century. We have already seen some of the awe with which electricity was regarded. It also was supposed to recharge one's sexual batteries. Here a satirical artist depicts the visit of an old bridegroom, with his young bride, to his doctor for electrical stimulation.

551. *The apotheosis of electricity.* The nineteenth century worshipped strange new gods. In a ballet performed at Frankfurt-am-Main in 1891, electricity was raised almost to supernatural heights. Note the tribute to Volta and Galvani on the centerpiece.

552. *The electric light—an allegory.* This 1884 engraving illustrates how great was the impact of electrical illumination on the late nineteenth-century mind.

553. *The electrical future.* Electricity provided more than illumination: it also was seen as the energy source for the future. *Punch* illustrated its manifold uses for the twentieth century.

553

554

555

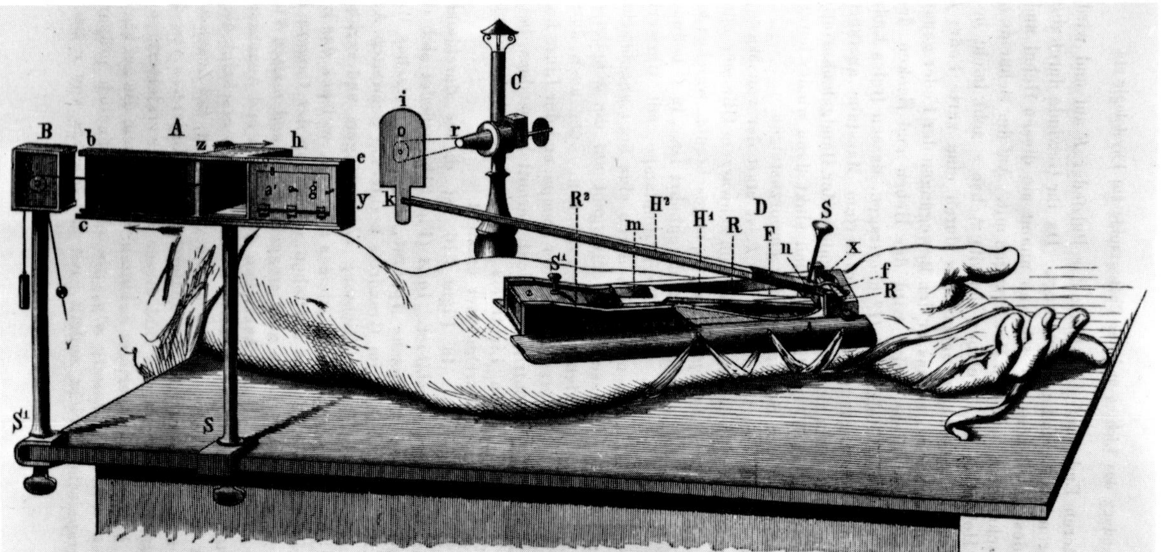

556

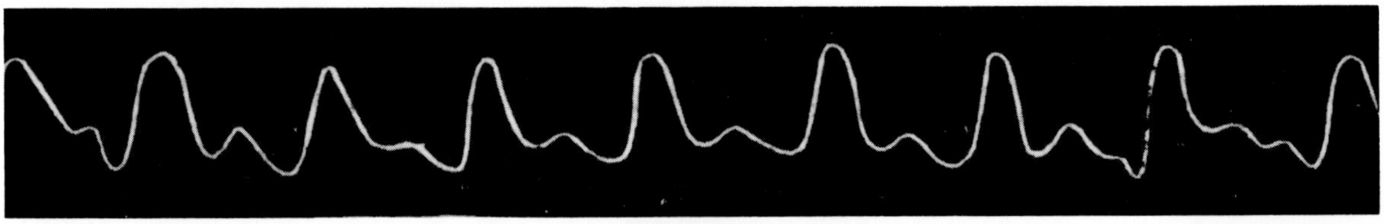

554. *Popular medicine at the beginning of the nineteenth century*. The scientific value of medicine around 1800 was not very high, and in the medicine encountered by the average person, it was very small. Here a "popular" doctor, using the crude air thermometer developed by the British natural philosopher John Leslie, attracts patients on a village street.

555. *Marey's sphygmograph*. The pulse had been used as a diagnostic aid since antiquity. With the sphygmograph, nineteenth-century technology devised a means for observing and preserving the record of it. A beam of light moved in response to the mechanism that detected the pulse, and this moving beam left a tracing on a photographic plate.

556. *A pulse tracing*.

557. *The gastroscope*. The ability to use electricity to create illumination made it possible to devise instruments for looking at internal parts of the body. The gastroscope, with an electric light at its end, could be lowered into the stomach, which could then be viewed.

558. *A gastroscopic examination*.

557

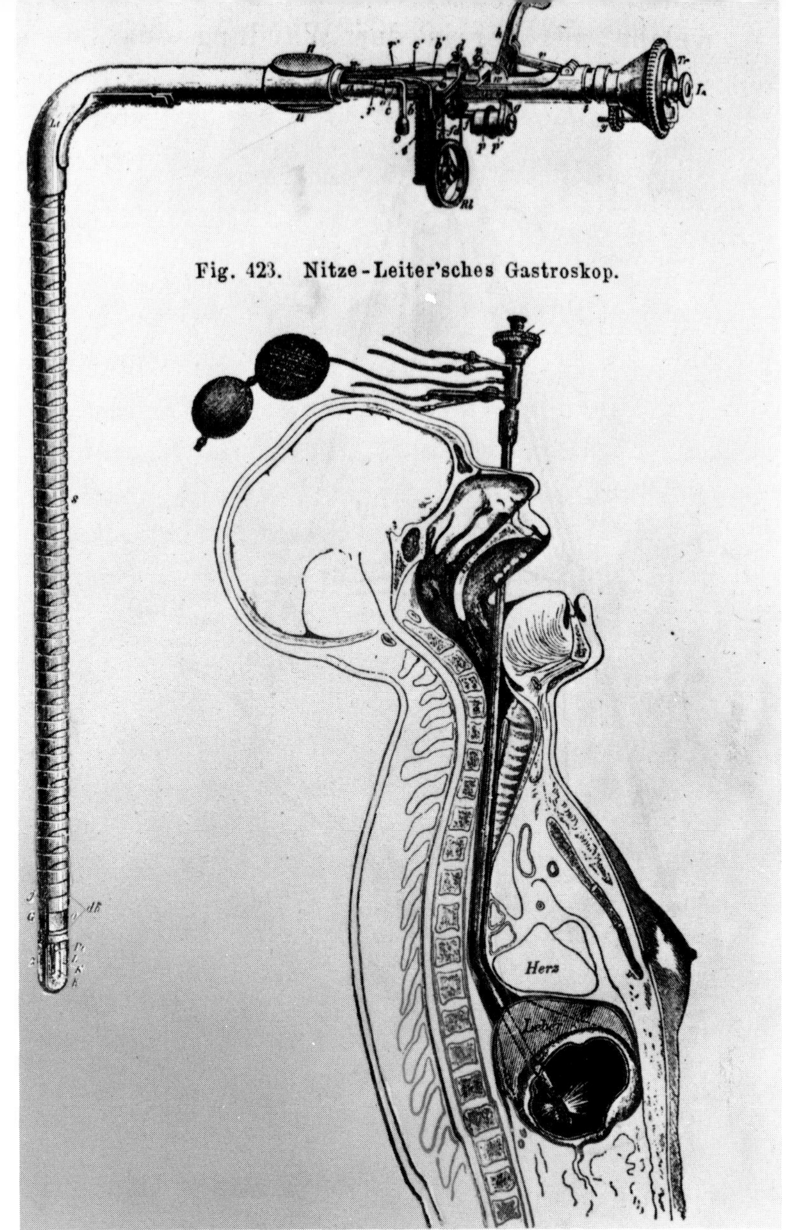

558

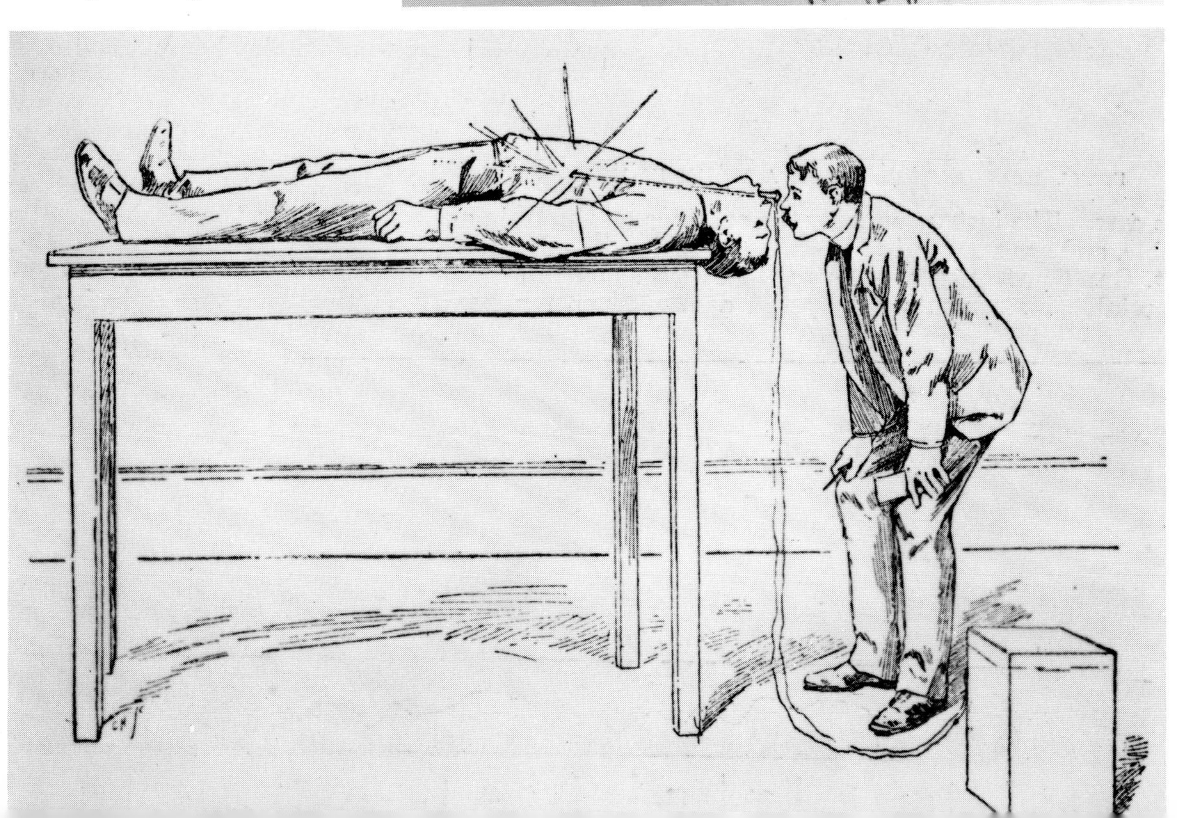

559

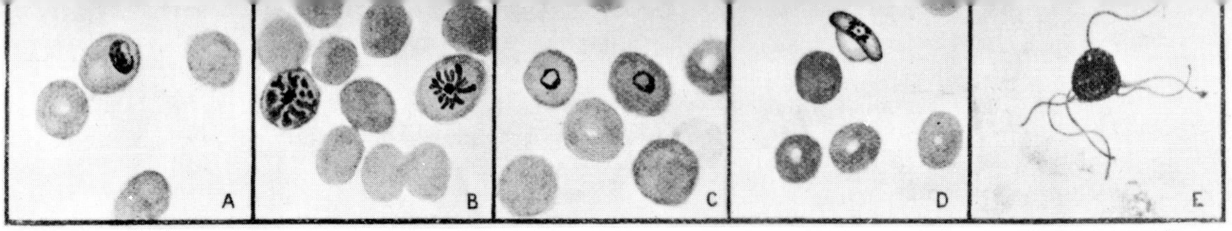

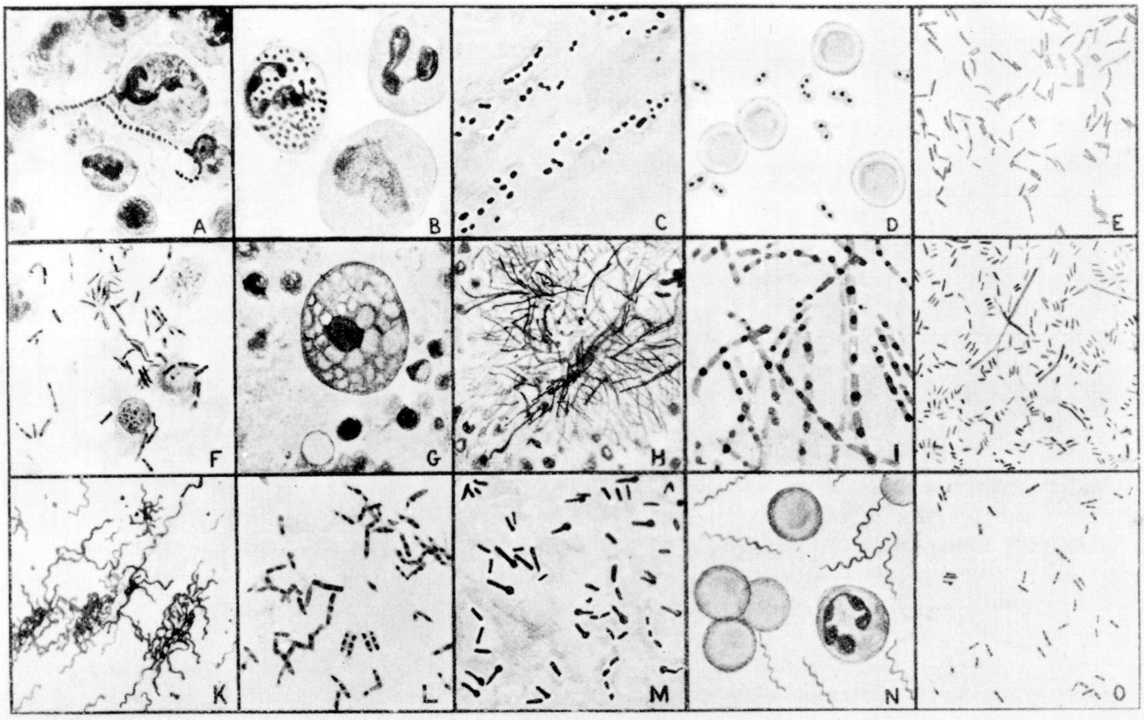

560

559. *Germs and medicine.* The major advance in nineteenth-century medicine was the germ theory of disease. The ability to see microorganisms and work with them made possible therapeutic and immunological steps that created a revolution in medicine. Shown here are some of the germs that had bedeviled mankind for millennia.

560. *Anesthesia.* The discovery of the anesthetic properties of ether and chloroform opened a new era in surgery. These substances replaced alcohol as a pain killer, as this cartoon of ether sniffers illustrates.

561. *The wonders of the new surgery.* Painless surgery opened new vistas for the surgeon, and for the cartoonist as well. The top four cartoons on the effects of ether show patients before operations, at left, and after, at right. At bottom left, a man is being cured of stupidity by having his empty head stuffed with straw. The young lady at bottom right is having her cold heart warmed by an ardent admirer.

561

Vor der Operation.

Nach der Operation.

Vor der Operation.

Nach der Operation.

Dummheit heilbar

Keine Untreue mehr.

562

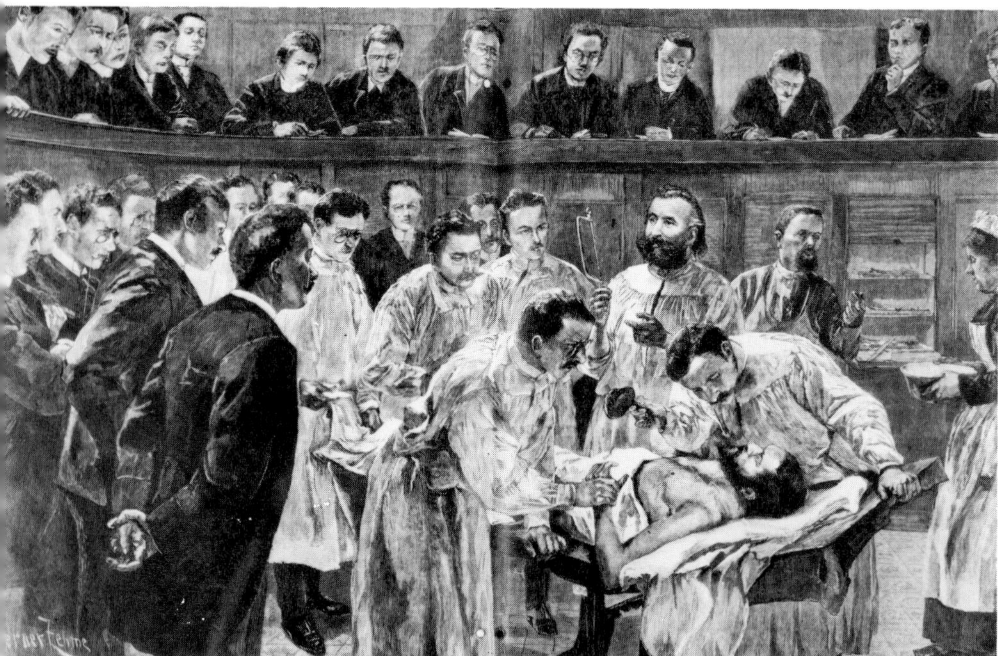

563

564

562. *The surgical amphitheater (1891).* The real surgical drama was played out in rooms such as this, where the master surgeon was assisted by a team of doctors and observed by a class of apprentice surgeons. What strikes the modern eye is the absence of aseptic or antiseptic precautions a generation after the value of these practices had been established.

563. *A vaccination station in Ireland.* It was in the nineteenth century that the first broad assault was made against diseases that had ravaged mankind for millennia. Public health measures, such as the establishment of this vaccination clinic in Ireland, reached the broad masses of the people for the first time in medical history.

564. *Inoculation against rabies.* Some diseases seemed to defy the finest microscopes and keenest eyes. Pasteur could not isolate the germ that caused rabies, and had to postulate the existence of a particle he could not see. Nevertheless, his experience in immunology with bacterial diseases allowed him to develop a vaccine against this dread disease. People bitten by mad dogs or other rabid animals could now hope to survive. This painting records the early inoculations given by Pasteur.

565. *Principle versus expediency.* The use of live animals in medical research quite rightly concerned people of high moral principle in the nineteenth century. The dilemma they faced was put rather starkly in this *Punch* cartoon.

ARGUMENTUM AD HOMINEM.
"Oh, Joseph! Teddy 's just been bitten by a strange Dog! Doctor says we 'd better take him over to Pasteur *at once!*"
"But, my Love, I 've just written and published a violent attack upon M. Pasteur, on the score of his cruelty to Rabbits! and at *your instigation,* too!"
"Oh, Heavens! Never mind the Rabbits *now!* What are all the Rabbits in the World compared to *our only Child!*"

565

566

375,000 Sold | in 5 Years.

THE LONDON GALVANIC GENERATOR

SIMPLE REMEDY FOR SIMPLE AILMENTS, NOW OFFERED TO THE WORLD BY THE PALL MALL ELECTRIC ASSOCIATION (LIMITED), 21, Holborn Viaduct, London, E.C.

WHICH CAN ALWAYS BE RELIED ON TO ALLAY LOCAL PAINS ARISING FROM

tomach, Liver, and Kidney Complaints; Debility, Heartburn, Rheumatism, Neuralgia, Weak Stomach, Dyspepsia, Aches and Pains, Weak Back, Malaria, Chills and Fever, Nervous Troubles, Indigestion, and other complications.

FRONT VIEW. THIS SIDE IS HANDSOMELY CARVED & RESEMBLES A LOCKET. IS HIGHLY POLISHED AND MADE OF A BEAUTIFUL MATERIAL RESEMBLING EBONY.

EXACT SIZE

The extraordinary popularity of this simple Galvanic arrangement, and the great demand that exists for it, has astonished even the Directors of the PALL MALL ELECTRIC ASSOCIATION themselves. THE LONDON GALVANIC GENERATOR was introduced merely to protect the public against the imposition practised by makers of rings, medals, toy batteries, and so-called generators, in which there is nothing galvanic or electric but the name. It is a thoroughly genuine article, constructed on accurate scientific principles, and the highly satisfactory reports that are constantly being received, as to its efficacy justify its designation as "a perfect marvel of curative science."

Sent Post Free on receipt of 5s. 6d., by the PALL MALL ELECTRIC ASSOCIATION, Limited, 21, Holborn Viaduct, London.

The following are specimens of the testimonials that are daily received. Hundreds of a similar character may be inspected, or will be sent to any address:

From Mrs. CLARKE, Wellington Street, Dolly Hill, near Bingley, Yorkshire.

"I am thankful to say that the Galvanic Generator I got from you has done me a great deal of good. Please send me another for my son, for which I enclose stamps value 5s. 6d."

From Mr. S. LEES, 13, Derby Street, Edgeley, Stockport.

"Enclosed are stamps, value 5s.6d., for another Galvanic Generator. My wife has improved much since receiving hers. This one is for a neighbour on our recommendation."

From Mr. JN. HOLMES, Clowne, near Chesterfield, Derbyshire.

"Please send me another Galvanic Generator, for which I enclose P.O. Order 5s. 6d. The former one received from you has cured both CRAMP and RHEUMATISM."

From Mr. P. HATTON, Garford, Abingdon, Berks.

"I enclose 11s. for two more of your Generators. The one I had cured my RHEUMATISM, and has also relieved WIND SPASMS, from which I suffered very much, so you see it shot at the pigeon and killed the crow. You may use this testimony for the benefit of others if you think fit."

WARRANTED FOR FIVE YEARS

BACK VIEW SHOWING THE SURFACE OF THE METALLIC BATTERY WHICH SHOULD BE WORN NEXT TO THE BODY.

LONDON GALVANIC GENERATOR. LONDON. PALL MALL ELECTRIC ASS'N 21 HOLBORN VIADUCT LONDON.

A Hundred-page Treatise, copiously illustrated, entitled "ELECTROPATHY; OR, DR. SCOTT'S GUIDE TO HEALTH." Post Free on application.

The Consulting Electrician of the Association (who has had many years experience in the application of Curative Electricity) attends daily for consultation (free) at the PRIVATE CONSULTING ROOMS, 21, Holborn Viaduct, E.C. An experienced lady is also in attendance. Residents at a distance should send for a PRIVATE ADVICE FORM.

NOTE THE ADDRESS—

THE PALL MALL ELECTRIC ASSOCIATION, LTD.
HOLBORN VIADUCT, LONDON, E.C.

567

566. *A strange "cure" for tuberculosis.* The many wonders of medicine made people impatient for cures to every disease. When no cure was forthcoming, they would grasp at any straw. One passing fad in Paris in 1891 was a "scientific cure" for tuberculosis that involved transfusing goat's blood into the patient. The results were catastrophic.

567. *The London galvanic generator.* The advances of scientific medicine created a public that tended to believe that science could do anything. As a result, quacks thrived. The galvanic generator was only one of dozens of pain relievers and cure-alls that gullible people bought in their search for health.

568. *Philosophy and the cosmos.* The nineteenth century was one of the most creative in the history of man. Science advanced by leaps and bounds. Man's total conquest of nature seemed almost at hand, and the millennium appeared to be just around the corner. Yet, underneath the bumptiousness and overweening pride, there was a strong current of pessimism and disillusion. Had man, by gaining the material world, lost his soul? The mood of this representation of philosophy and the cosmos is somber and melancholy. The picture appeared in the German periodical *Illustrirte Zeitung* in 1900, ushering in the new century on a peculiarly appropriate note.

Bibliography

The trustworthy literature on science in the nineteenth century is not large. Hence, a bibliography that will introduce the interested reader to works on various aspects of the history of nineteenth century science can be kept rather brief. The large gaps in any bibliography of nineteenth-century science are partly the result of the fact that there are no good works on so many of the subjects. For example, there is no profound and penetrating history of nineteenth-century anthropology that utilizes the primary sources and the manuscript resources that abound in libraries and private collections. Nor is there a decent history of scientific education. The history of instruments is also almost a virgin field.

The works listed and recommended here are those that will best serve the general reader. I have also tried to cite works that, themselves, contain extensive bibliographical aids so that a subject can be pursued in considerable depth if the reader so desires.

The best single work on the history of science (and of thought, in general) in the nineteenth century remains the four volumes composed by J. T. Merz. His *History of European Thought in the Nineteenth Century* (4th unaltered edition, Edinburgh–London, 1923-1950) is a remarkable work. Drawing only upon published sources, Merz produced a brilliant and coherent account of the development of all the sciences in this great century of scientific achievement. The first two volumes deal with the "hard" sciences and are a literal treasure trove of bibliographical information. The last two volumes deal with the social sciences and intellectual history and provide a valuable guide to the environment in which science thrived. A more modern work with similar scope is M. Capek, *The Philosophical Impact of Contemporary Physics*, (Princeton, 1961). Capek has read widely and his bibliography should be consulted by those interested in the progress of physics. The chapters devoted to the nineteenth century in Charles C. Gillispie, *The Edge of Objectivity* (Princeton, 1960), are both stimulating and controversial. Gillispie, too, has a fine bibliography. There are no modern histories of nineteenth-century astronomy that combine astronomical precision with historical insight. Among the better texts written by astronomers is A. Pannekoek, *A History of Astronomy* (New York, 1961). Agnes Clerke's volumes may still be read with profit, even though they are old. Her *Problems in Astrophysics* (London, 1903) and *The System of the Stars* (London, 1905) capture a good deal of the excitement that accompanied the birth of astrophysics in the last century.

The history of geology is still in its infancy. Charles C. Gillispie, *Genesis and Geology* (New York, 1959), is a pioneering work that beautifully describes the geological scene when the century began. The volume edited by Cecil Schneer, *Toward a History of Geology* (Cambridge, Mass., 1969), indicates what modern historians of geology are up to, and M. Rudnick, *The Meaning of Fossils* (London–New York, 1972), provides a detailed background of paleontology. William Coleman's work on Georges Cuvier, *Georges Cuvier, Zoologist* (Cambridge, Mass., 1964), is also well worth consulting.

The history of biology is also just getting started as a professional discipline and there are no good or trustworthy guides to the nineteenth century as a whole. William Coleman's *Biology in the Nineteenth Century: Problems of Form, Function and Transformation* (New York, 1971) is an interesting beginning, with a bibliography, that can start one on the journey into this fascinating topic, but guideposts soon vanish and reliance must be placed in the original sources themselves. Among these, of course, are the writings of Charles Darwin which are easily found today in any library. Care should be taken in attacking the "Darwin problem" to distinguish between the various editions that were produced during Darwin's lifetime. The sixth edition, for example, of *On the Origin of Species* is quite different from the first and contains many of Darwin's later and second thoughts on the problems of evolution and natural selection. It would be impossible in brief compass to list even a representative sample of the books on Darwin and the problem of evolution. I have found the following most useful. Sir Gavin de Beer, *Charles Darwin* (London, 1963), is by an eminent biologist who gets the biology right but is not always as sensitive as he might be to the social and intellectual currents impinging upon Darwin and his allies. Gertrude Himmelfarb's *Darwin and the Darwinian Revolution* (London, 1959) is a work written by an intellectual historian superbly sensitive to precisely those elements that de Beer neglects but not as good on the technical biology as could be hoped. Peter Vorzimmer, *Charles Darwin: The Years of Controversy* (Philadelphia, 1960), and H. Lewis McKinney, *Wallace and Natural Selection* (New Haven–London, 1971), are both good studies well worth reading.

The literature on the history of chemistry is extensive. For bibliography, volume 4 of J. R. Partington's otherwise unreadable text, *A History of Chemistry* (London, 1962–1964), is indispensable. Aaron Ihde's *The Development of Modern Chemistry* (New York, 1964) deals with many of the problems illustrated in this album. For John Dalton, see D. S. L. Cardwell, ed., *John Dalton and the Progress of Science* (Manchester–New York, 1968), and Frank Greenaway, *John Dalton and the Atom* (Ithaca, N.Y., 1966).

A number of works deal with the "imponderable fluids" with which the nineteenth century was fascinated when the century began. Among these are Wilson L. Scott, *The Conflict between Atomism and Conservation Theory, 1644–1860* (London–New York, 1970); and L. Pearce Williams, *Michael Faraday, A Biography* (London, 1965) and *The Origins of Field Theory* (New York, 1967), the latter written particularly for a popular audience. Mary Hesse's *Forces and Fields* (London, 1921) is a brilliant work that is definitely not for the beginner. Max Jammer has written three books on basic concepts in physics that are both difficult and fascinating. His *Concepts of Force* (Cambridge, Mass., 1957), *Concepts of Space* (Cambridge, Mass., 1954), and *Concepts of Mass* (Cambridge, Mass., 1961) will amply repay study of these basic ideas in the physical sciences.

For the history of technology, one can do no better than to refer to the massive work edited by Charles Singer *et al.*, *A History of Technology*, 5 volumes (Cambridge, England, 1954–1958).

Picture Sources

Citations have been made according to standard bibliographical form and in the shortest manner consistent with both precision and clarity. Thus, I have not used vol. for volume or p. for page, it seeming obvious that an italicized number indicates the number of the volume and a number following all other information, the page. The following abbreviations are used for those works cited most frequently.

A	*Astronomy* by James Ferguson, London, 1794
EM	*Electricity and Magnetism* by Amédée Guillemin, London, 1891
FN	*The Forces of Nature* by Amédée Guillemin, New York, 1872
G	*The Graphic* (London)
IL	*L'Illustration* (Paris)
ILN	*The Illustrated London News*
IZ	*Illustrirte Zeitung* (Leipzig)
L	*Das Licht* by S. T. Stein, 2 vols., Halle, 1886
M	*Der Mensch* by Johannes Ranke, 2 vols., Leipzig and Vienna, 1894
MS	*Les Merveilles de la Science* by Louis Figuier, 4 vols., Paris, 1867-70
P	*Punch* (London)
PA	*Popular Astronomy* by Camille Flammarion, New York, 1894
RV	*Le Règne végétal, Botanique générale, Atlas* by O. Reveil, A. Dupuis et al., 17 vols., Paris 1864-9
SFP	*A System of Familiar Philosophy* by A. Walker, London, 1802
SP	*Seminario Pintoresco Espanol* (Madrid)
SS	*The Silurian System* by Rodney I. Murchison, 2 vols., London, 1839
WM	*Weltall und Menschheit* by Hans Kraemer, 5 vols., Berlin, 1902-4
WT	*Wonders of the Telescope* by Anon., London, 1805
XIXJ	*Das XIX Jahrhundert in Wort und Bild* by Hans Kraemer, 4 vols., Berlin, 1900

1. *Science is Measurement* by Henry Stacey Marks, R.A., G *July-Dec.* (1879), 205. Reproduced by permission of the Royal Academy of Arts, London. **2.** IL, *18* (1851), 73. **3.** P, *2* (1842), 149. **4.** IL, *8* (1847), 413. **5.** British Crown Copyright. Science Museum, London. **6.** British Crown Copyright. Science Museum, London. **7.** *Babbage's Calculating Engines, Being a Collection of Papers Relating to Them; Their History and Construction* by H. P. Babbage, ed., London, 1889, after 342. **8.** British Crown Copyright. Science Museum, London. **9.** British Crown Copyright. Science Museum, London. **10.** IZ, *106* (1896), 277. **11.** ILN, *43* (1863), 597. **12.** Ibid. **13.** *Die Karikatur und Satire in der Medizin* by Eugene Hollander, Stuttgart, 1905, 345. **14.** ILN, *45* (1864), 301. **15.** G, *6* (1872), 185. **16.** ILN, *43* (1863), 269. **17.** IL, *10* (1847), 68. **18.** P, *117* (1899), 134. **19.** G, *44* (1891), 8. **20.** IZ, *88* (1887), 666. **21.** G, *59* (1899), 392. **22.** PA, 36. **23.** Ibid. **24.** Ibid., 37. **25.** IZ, *17* (1851), 205. **26.** MS, *4*, 721. **27.** IL, *15* (1850), 336. **28.** *Sky and Telescope*, I (1942), 3. **29.** From a tearsheet in possession of the author. **30.** *Baubericht über die technischen Anlagen für das königliche astrophysikalische Observatorium auf dem Telegraphenberg bei Potsdam* by P. Spieker, Berlin, 1879, Pl. 2. **31.** ILN, *39* (1861), 205. **32.** Ibid., *60* (1872), 61. **33.** IL, *66* (1875), 80. **34.** G, *36* (1885), 638. **35.** ILN, *83* (1883), 36. **36.** Ibid. **37.** G, *29* (1884), 32. **38.** Ibid. **39.** IZ, *4* (1845), 89. **40.** Ibid. **41.** Ibid. **42.** Ibid. **43.** Ibid. **44.** ILN, *45* (1864), 485. **45.** IZ, *82* (1884), 69. **46.** ILN, *85* (1884), 220. **47.** MS, *2*, 537. **48.** XIXJ, *1*, between 208-9. **49.** ILN, *21* (1852), 192. **50.** IZ, *95* (1890), 563. **51.** Ibid. **52.** IL, *65* (1875), 12. **53.** XIXJ, *3*, after 246. **54.** IZ, *101* (1893), 305. **55.** P, *93* (1887), 51. **56.** IL, *9* (1847), 108. **57.** Ibid., 109. **58.** Ibid. **59.** IL, *37* (1861), 136-7. **60.** IL, *10* (1847), 233. Reproduced by permission of Académie Nationale de Médécine, Paris. **61.** IL, *93* (1889), 9. **62.** Ibid., *27* (1856), 80. **63.** ILN, *53* (1868), 21. **64.** Ibid. **65.** IZ, *68* (1877), 111. **66.** *The Naturalist* by Henry Stacey Marks, R. A., G, *44* (1891), 517. **67.** MS, *1*, 669. **68.** Deutsches Museum, Munich. **69.** *The Discovery of Nature* by Albert Bettex, New York, 1965, 56. Photo: Albert Bettex. **70.** *Die Geschichte des chemischen Laboratoriums des Bayerischen Akademie der Wissenschaften in München* by Wilhem Prandtl, Weinheim: Verlag Chemie, 1952, 38. **71.** *La Nature, 1* (1873), 5. **72.** *Aperture, 15* (1970). Reproduced by permission of Aperture, Inc. **73.** IL, *77* (1881), 426. **74.** Ibid., *62* (1873), 72. **75.** G, *52* (1895), 419. **76.** IL, *61* (1873), 348. **77.** Ibid., *71* (1878), 45. **78.** G, *42* (1890), 625. Photo: Cadbury Schweppes Ltd. **79.** ILN, *66* (1875), 89. **80.** Ibid., *12* (1848), 98. **81.** Ibid. **82.** P, *93* (1887), 235. **83.** IL, *65* (1875), 257. **84.** Ibid., *16* (1850), 297. Reproduced by permission of Musée des Techniques—Conservatoire National des Arts et Métiers, Paris. **85.** Reproduced by permission of the Trustees of the British Museum. **86.** ILN, *28* (1856), 177. **87.** IZ, *76* (1881), 149. **88.** *The Science of Life* by Gordon Rattray Taylor, New York, 1963, 200-01. Reproduced by permission of the Trustees of the British Museum. **89.** IL, *76* (1880), 153. **90.** ILN, *54* (1869), 401. **91.** IL, *1* (1843), 233. **92.** P, *54* (1868), Punch's Almanack for 1868. **93.** IL, *18* (1851), 65. **94.** ILN, *70* (1877), 277. **95.** WT, Pl. 13. **96.** Ibid. 12. **97.** The Humanities Research Center, the University of Texas at Austin. Reproduced by permission of Mrs. Eileen Shorland. **98.** The Humanities Research Center, the University of Texas at Austin. Reproduced by permission of Mrs. Eileen Shorland. **99.** British Crown Copyright. Science Museum, London. **100.** ILN, *6* (1845), 253. **101.** Ibid. **102.** ILN, *21* (1852), 168. **103.** Ibid., *23* (1853), 307. **104.** IL, *66* (1875), 248. **105.** L, *2*, 20. **106.** Ibid., 21. **107.** *Some Firsts in Astronomical Photography* by Dorritt Hoffleit, Cambridge, Mass., 1950, 15. Photo: Sky Publishing Corp. **108.** Ibid., 18. Photo: Sky Publishing Corp. **109.** Ibid., 29. Photo: Sky Publishing Corp. **110.** Ibid., 63. Photo: Sky Publishing Corp. **111.** *Philosophical Transactions of the Royal Society*, 1811, 336, Pl. IV. Reproduced with acknowledgment to the Royal Society, London. **112.** Ibid., Pl. V. Reproduced with acknowledgment to the Royal Society,

London. **113.** IL, *39* (1862), 141. **114.** IL, *41* (1863), 96. **115.** Ibid. **116.** *The Discovery of Nature* by Albert Bettex, New York, 1965, 352. Reproduced by permission of Bibliothek der Eidgenossischen Technischen Hochshule, Zurich. Photo: Albert Bettex. **117.** *Annals of the Astronomical Observatory of Harvard College, 5* (1867), frontispiece. **118.** *A Selection of Photographs of Stars, Star-Clusters and Nebulae* by Isaac Roberts, London, 1893. Photo: Butler Library, Columbia University. **119.** *Publications of the Yerkes Observatory of the University of Chicago,* 2 (1903), Pl. 23. **120.** *A Selection of Photographs of Stars, Star-Clusters and Nebulae* by Isaac Roberts, London, 1893. Photo: Butler Library, Columbia University. **121.** *Publications of the Lick Observatory, 8* (1908), Pl. 59. Photo: Lick Observatory. **122.** Ibid., Pl. 47. Photo: Lick Observatory. **123.** *Atlas zum Lehrbuch des kosmischen Physik* by Johannes Müller, Braunschweig, 1875. **124.** XIXJ, *1,* between 464-5. **125.** SP, *3* (1838), 490. **126.** ILN, *33* (1858), 387. **127 a,b.** Ibid. **128.** Ibid. **129.** *Annals of the Astronomical Observatory of Harvard College, 3* (1862), Pl. 19. **130.** Ibid., Pl. 38. **131.** IL, *32* (1858), 272. **132.** WT, Pl. 16. **133.** Reproduced by permission of the Bibliothèque Nationale, Paris. Département des Estampes. **134.** IZ, *26* (1856), 101. **135.** Ibid., *23* (1854), 372. **136.** A, Pl. 15. **137.** P, *68* (1875), Punch's Almanack for 1875. **138.** IZ, 77 (1881), 454. **139.** G, *10* (1874), 600. **140.** IL, *68* (1876), 392. **141.** Ibid. **142.** Ibid., *65* (1875), 28. **143 a,b,c.** Ibid. **144.** Ibid. **145.** G, *10* (1874), 553. **146.** IL, *64* (1874), 420-1. **147.** *Atlas des Sonnensystems* by W. Valentiner, Lahr, 1884, Pl. 19. **148.** *The Sun by* Amédée Guillemin, New York, 1875, 105. **149.** WM, *1, 66.* **150.** SFP, *2,* Pl. 48, fig. 2. **151.** Ibid., Pl. 48, fig. 4. **152.** Ibid., Pl. 49, fig. 4. **153.** Ibid., Pl. 49, fig. 5. **154.** Ibid., Pl. 48, fig. 7. **155.** Ibid., Pl. 48, fig. 8. **156.** IL, *8* (1846), 156. **157.** Ibid. **158.** WT, Pl. 4. **159.** WM, *3,* between 252-3. **160.** PA, 439. **161.** WT, Pl. 6. **162.** ILN, *3* (1843), 293. **163.** WM, *3,* 247. **164.** WT, Pl. 5. **165.** IL, *42* (1863), 336. **166.** Ibid., *97* (1891), 6. **167 a,b,c.** Ibid., 7. **168.** Ibid., 30. **169.** Ibid., 7. **170.** IL, *92* (1888), 6. **171.** *Annals of the Lowell Observatory,* 3 (1905), Pl. 11. **172.** *La planète Mars,* 1659-1929 by E. M. Antoniadi, Paris, 1930, Pl. VI. **173.** *The Sky, 2* (1937), 19. **174.** WT, Pl.3. **175.** IL, *46* (1865), 192. **176.** *Royal Astronomical Society, Monthly Notices, 57* (1897), Pl. 6. **177.** *The Planet Venus,* 2nd ed., New York, opp. 109. Reproduced by permission of Union Observatory, Johannesburg, South Africa. **178.** WT, Pl.2. **179.** Reproduced by permission of Musée des Techniques, Conservatoire des Arts et Métiers, Paris. **180.** *Atlas des Sonnensystems* by W. Valentiner, Lahr, 1884, Pl. 20. **181.** *L'Astronomie pratique et les observatoires en Europe et en Amérique* by C. André and G. Rayet, 5 vols., Paris, 1874, *1,* 136. **182.** PA, 267. **183.** Ibid., 257. **184.** *The Sun* by Amédée Guillemin, New York, 1875, 219. **185.** A, Pl. 10. **186.** IL, *13* (1858), 175. **187.** P, *67* (1834), 98. **188.** *The Moon* by W. H. Pickering, New York, 1903, Pl. J. **189.** WT, Pl. 4. **190.** IZ, *23* (1854), 253. **191.** *The Moon* by W. H. Pickering, New York, 1903, Pl. H. **192.** IZ, *66* (1876), 461. **193.** Ibid. **194.** *The Moon* by James Nasmyth and James Carpenter, London, 1885, opp. 148. **195.** Ibid., opp. 184. **196.** *Mondkarte* by J. F. Julius Schmidt, Leipzig, 1892. **197.** WM, *3,* after 180. **198. a,b,c,d.** *The Moon* by W. H. Pickering, New York, 1903, end of vol. **199.** *The World Before the Deluge* by Louis Figuier, New York, 1869, opp. 74. **200 a,b, c,d,e,f.** IZ, *72* (1879), 108-9. **201.** A, Pl. 4. **202.** ILN, *18* (1851), 511. **203.** IL, *17* (1851), 213. **204.** *The Sea* by G. Hartwig, London, 1866, opp. 3. **205.** *Atlas zum Lehrbuch der kosmischen Physik* by J. Müller, Braunschweig, 1875, Pl. 40. **206.** One of series of lithographed plates in possession of the author. No attribution, date, or place of publication. **207 a,b,c.** *Philosophical Magazine,* 16 (1803), Pl. VI, VII, VIII. **208.** FN, 97. **209.** Ibid., 657. **210.** *Traité de l'Électricité et du Magnétisme, Atlas* by H. Becquerel, Paris, 1840. **211.** L, *2,* 139. **212.** WM, *1,* opp. 400. **213.** EM, Pl. 4. **214.** IZ, *16* (1851), 329. **215.** *The Life of Sir Rodney I. Murchison* by Archibald Geikie, 2 vols., London, 1875, *1,* opp. 309. **216.** IZ, *16* (1851), 144. **217.** *Vesuvius in Eruption,* 1817, by J. M. W. Turner. From the collection of Mr. and Mrs. Paul Mellon. **218.** ILN, *61* (1872), 553. **219.** WM, *1,* between 8-9. **220.** IZ, *16* (1851), 144. **221.** WM, *4,* after 220. **222.** IZ, *106* (1896), 274. **223.** ILN, *81* (1882), 400, **224.** Ibid., *111* (1897), 151. **225.** Ibid., *43* (1863), 8. **226.** Ibid., *63* (1873), 605. **227.** Ibid., *65* (1874), 457. **228.** WM, *4,* opp. 161. **229.** *Report on the Scientific Results of the Voyage of H.M.S. Challenger during the years 1873-76* by Sir C. Wyville Thomson, 40 vols., Edinburgh, 1880-95, *1,* first part, 1880, 201. Cited henceforth as *Report . . . Challenger.* **230.** *Report . . . Challenger, Botany, 2,* 1886, Pl. XV. **231.** Ibid., *Zoology, 1,* 1880, Pl. IX (Shore Fishes). **232.** Ibid., *4,* 1882, Pl. VIII (Deep Sea Medusae). **233.** Ibid., *8,* 1883, Pl. IX (Cirripedia). **234.** Ibid., *18,* 1887, Pl. XIII. **235.** Ibid., Pl. LIV. **236.** Ibid., Pl. CIX. **237.** Ibid., *22,* 1887, Pl. XIII. **238.** Ibid., *28,* 1888, Pl. XIV. **239.** IZ, *17* (1851), 204. **240.** WM, *1,* 137. **241.** *The Tertiary History of the Grand Cañon District* by Clarence E. Dutton, Washington, D.C., 1882. **242.** *A Delineation of the Strata of England and Wales with Part of Scotland* by William Smith, 1815. Photo: Butler Library, Columbia University. **243.** Ibid. Photo: Butler Library, Columbia University. **244.** *Siluria* by Sir R. I. Murchison, London, 1854, 330. **245.** Ibid., 328. **246.** *Geological Textbook* by Amos Eaton, Albany, 1830, opp. 19. **247.** Ibid., 23. **248.** *A Geological Map of the United States* by Jules Marcou, Boston, 1853, frontispiece. **249.** WM, *4,* 457. **250 a,b.** *La création et ses mystères dévoilés* by Antoine Snider, Paris. 1859. **251.** RV, *1,* Pl. 1. **252.** *The Jurassic Rocks of Britain* by Horace B. Woodward, London, 1893, 5 vols., *3,* 164. **253.** WM, *1,* 133. **254.** *Outlines of the Geology of England and Wales* by W. D. Conybeare and William Phillips, London, 1822, back of vol. **255.** SS, *1,* opp. 225. **256.** *The Life of Sir Rodney I. Murchison* by Archibald Geikie, 2 vols., London, 1875, *1,* 229. **257 a,b,c.** *Geological Textbook* by Amos Eaton, Albany, 1830, 28, 29, 30. **258.** *The Cretaceous Rocks of Britain* by A. J. Jukes-Brown, 3 vols., London, 1900, *1,* opp. 192. **259.** Ibid., 193. **260.** *Design in Nature* by J. Bell Pettigrew, 3 vols., London: Longman Group Ltd., 1908, *1,* 230. **261.** *The Jurassic Rocks of Britain* by C. Fox-Strangways, 5 vols., London, 1892, *1,* Pl. IV, opp. 272. **262.** *The World Before the Deluge* by Louis Figuier, New York, 1869, Pl. 33, opp. 371. **263.** Ibid., Pl. 30. **264.** *Reliquiae Diluvianae* by William Buckland, London, 1823, Pl. 20. **265.** WM, *1,* 161. **266.** *The Story of Creation* by E. Clodd, New York, 1887, 17. **267.** WM, *2,* 437. **268.** *Der Jura* by F. A. Quenstedt, Tübingen, 1858, 170. **269.** SS, *2,* Pl. 1. **270.** Ibid. **271.** *Der Jura* by F. A. Quenstedt, Tübingen, 1858, 232. **272.** IZ, *104* (1895), 153. **273.** IZ, *91* (1888), 75. **274.** WM, *1,* 177. **275.** IZ, *6* (1846), 16. **276.** Ibid., *74* (1880), 478. **277.** *Odontornithes* by Othniel Charles Marsh, Washington, D.C., 1880. **278.** IZ, *14* (1850), 256. **279.**

SP, *1* (1836), 61. **280.** Ibid., 2nd ser., *3* (1841), 185. **281.** Dinocerate by Othniel Charles Marsh, Washington, D.C., 1884. **282.** *P, 55* (1868), 272. **283.** *An Universal System of Natural History* by E. Sibly, 12 vols., n.p., n.d., but ca. 1800, *1*, frontispiece. **284.** Ibid., *1*, opp. 52. **285.** Ibid., *2*, opp. 18. **286.** Ibid., *2*, frontispiece. **287.** Ibid., opp. 152. **288.** Reproduced by permission of Edward Leigh, Cambridge, England. **289.** G, *26* (1882), 16. **290.** P, *73* (1877), 275. **291.** Ibid., *40* (1861), 213. **292.** *Charles Darwin and the Origin of Species* by Walter Karp, New York, 1968, 140. Reproduced courtesy of American Heritage Publishing Co., Inc. **293.** *Anthropogenie* by Ernst Haeckel, 2 vols., Leipzig, 1891, *1*, Tafel 1, between 272-3. **294 a,b.** Ibid., Tafel 8, 9, between 352-3. **295.** *Die Naturwunder der Tropenwelt Ceylon und Insulinde* by Ernst Haeckel n.p., n.d., frontispiece. **296.** P, *47* (1864), 239. **297.** *Der Neanderthaler* by Kurt Tackenberg, Bonn: Rudolf Habelt Verlag, 1956, Tafel 2, opp. 8. **298.** Ibid., 22. **299.** WM, *2*, 3. **300.** Ibid., 20. **301.** *Hundert Jahre Neanderthaler, 1856-1956* by G. H. R. von Koenigswald, ed., Cologne: Böhlau Verlag, 1958, Pl. 1, opp. 120. Copyrighted 1958 by the Wenner-Gren Foundation for Anthropological Research, Inc., New York. **302.** *Der Neanderthaler* by Kurt Tackenberg, Bonn: Rudolf Habelt Verlag, 1956, Tafel 5, opp. 40. **303.** SP, 2nd ser., *2* (1840), 17. **304.** WM, *1*, 5. **305.** *A Russian Mammoth Hunt* by V. M. Vasnetzoff. Reproduced by permission of the Russian Historical Museum, Moscow. **306.** *La fin du monde* by Camille Flammarion, Paris, 1893, 127, 129. **307.** *Geschichte der Physiologie* by K. Rothshuh, Berlin: Springer-Verlag, 1953, 100. **308.** *Anthropogenie* by Ernst Haeckel, 2 vols., Leipzig, 1891, frontispiece. **309.** *An Universal System of Natural History* by E. Sibly, 12 vols., n.p., n.d., but ca. 1800, *2*, opp. 343. **310 a, b.** Ibid., *1*, opp. 345, 347. **311.** *Design in Nature* by J. Bell Pettigrew, 3 vols., London: Longman Group Ltd., 1908, *1*, 396. **312.** *The Discovery of Nature* by A. Bettex, New York, 1965, 281. **313.** *Kalogynomia* by T. Bell, London, 1826, 23. **314.** Ibid. **315.** M, *1*, 20. **316.** Ibid., 28. **317.** Ibid., opp. 207. **318.** Ibid. **319.** Ibid., opp. 519. **320.** *Anatomy, Descriptive and Surgical* by Henry Gray, 15th ed., Philadelphia, 1901, 703. **321.** WM, *2*, between 24-5. **322.** *Die Charakteristik des Kopfes nach dem Entwicklungsgesetz desselben* by Robert Froriep, Berlin, 1845. **323 top, bottom.** *Essays on Physiognomy* by Johann Caspar Lavater, 10th ed., London, 1858, Pls. LXXVIII, LXXIX, between 496-7. **324.** G, *53* (1896), 17. **325.** *Phrenologische Bilder* by Gustave Scheve, Leipzig, 1874, 6-7. Photo: National Library of Medicine, Bethesda, Md. **326.** Reproduced by permission of the Clements C. Fry Collection, Yale Medical Library, New Haven, Conn. **327.** *The Expression of the Emotions in Man and Animals* by Charles Darwin, London, 1904, fig. 20. **328.** Ibid., Pl. III. **329.** *Essays on Physiognomy* by J. C. Lavater, 10th ed., London, 1858, Pl. LII, after 276. **330.** *Anthropométrie* by Adolph Quetelet, Brussels, 1871, 206. **331.** IZ, *108* (1897), 60. **332.** M, *1*, 389. **333.** Ibid., 393. **334.** WM, *2*, 31. **335.** IZ, *108* (1897), 61. **336.** IL, *95* (1890), 91. **337.** *A Manual of Anthropometry* by Charles Roberts, London, 1878, frontispiece. **338.** *Anthropométrie* by Adolph Quetelet, Brussels, 1871, 341. **339.** *A Manual of Anthropometry* by Charles Roberts, London, 1878, Pl. V, between 12-3 of "Report of the Anthropometric Committee." **340.** Ibid., Pl. IX. **341.** *History of the Earth* by Oliver Goldsmith, 4 vols., New York, 1830, *1*, opp. 239. **342.** *The Pre-Adamite* by A. Hoyle Lester, Philadelphia, 1875, frontispiece. **343.** *An Universal System of Natural History* by E. Sibly, 12 vols., n.p., n.d., *1*, opp. 97. **344.** Ibid., opp. 124. **345.** Ibid., opp. 143. **346.** Ibid., opp. 145. **347.** Ibid., 166. **348.** Ibid., opp. 220. **349.** *Report on the Voyages of the Adventure and Beagle* by Capt. Robert Fitzroy, London, 1839, frontispiece. **350.** *The Races of Man* by J. Deniker, London, 1901, 522. Photo: Prince Roland Bonaparte. Reproduced by permission of Princess Eugenie of Greece. **351.** Ibid., 574. **352.** Ibid., 79. **353.** Ibid., 82. Reproduced by permission of the Museum Nationale d'Histoire Naturelle, Paris. **354.** Ibid., 142. **355.** Ibid., 40. **356.** Ibid., 41. **357.** *Aus Insulinde* by Ernest Haeckel, Bonn, 1901, fig. 54, opp. 176. **358.** Ibid., fig. 55, opp. 192. **359.** M, *2*, between 166-7. Reproduced by permission of Veb Bibliographisches Institut, Leipzig. **360.** *History of the Earth* by Oliver Goldsmith, 4 vols., New York, 1830, *4*, Pl. 75, opp. 64. **361.** Ibid., Pl. 79, opp. 105. **362.** Ibid., *2*, Pl. 21, opp. 203. **363.** Ibid., Pl. 9, opp. 77. **364.** *An Universal System of Natural History* by E. Sibly, 12 vols., n.p., n.d., *3*, opp. 240. **365.** *Le Règne animal* by Georges Cuvier, 20 vols., *Mammifères, 2, Atlas,* Paris, 1834, Pl. 78. **366.** *Anthropogenie* by Ernst Haeckel, 2 vols., Leipzig, 1891, *1*, 579. **367.** IZ, *91* (1888), 541. **368.** Ibid. **369.** IZ, *115* (1900), 927. **370.** RV, *1*, Pl. 6. **371.** *Zoographie* by L. F. Jauffret, Paris, 1800, 3e carte de l'Atlas géographique. **372.** Ibid., 6e carte. **373.** Reproduced by permission of the Trustees of the British Museum, London. Photo: Albert Bettex. **374.** *Exotic Flora* by W. J. Hooker, 3 vols., Edinburgh, 1823-7, Pl. 13. **375.** Ibid., *3*, Pl. 209. **376.** Reproduced by permission of the Houghton Library, Harvard University. **377.** *Illustrationes Algarum* by A. Postels and F. Ruprecht, Weinheim, 1963 [orig. pub. 1840], frontispiece. **378.** Ibid., Pl. 35. **379.** *Voyage Pittoresque et Historique au Brésil,* 3 vols., Paris, 1834, 2e cahier, Pl. 1. **380.** RV, *3*, Pl. 18. **381.** *Pflanzengeographie* by A. F. W. Schimper, Jena, 1898, 184. **382.** Ibid., opp. 192. Reproduced by permission of the State Museum, the University of Nebraska. **383.** Ibid., opp. 284. **384.** Ibid., 292. **385.** Ibid., opp. 668. **386.** Ibid., end of vol., Karte 3. **387.** *Charles Darwin and the Origin of Species* by Walter Karp, New York, 1968, 49. Reproduced courtesy of American Heritage Publishing Co., Inc. **388.** *The Geographical Distribution of Animals* by Alfred Russell Wallace, 2 vols., New York, 1876, *1*, Pl. 4, opp. 251. **389.** Ibid., Pl. 13, opp. 455. **390.** Ibid., *2*, Pl. 20, opp. 136. **391.** *Biologie* by Gottfried Reinhold Treviranus, 5 vols., Göttingen, 1802-21, *1*, title page. **392.** *Vorlesungen über Pflanzenphysiologie* by Julius Sachs, Leipzig, 1882, 48. **393 a, b.** RV, *2*, Pl. 39, 40. **394.** Ibid., Pl. 43. **395.** *On the Origin of Species* by Charles Darwin, London, 1859, title page. **396.** RV, *1*, Pl. 14. **397.** *Iconographic Encyclopedia* by J. G. Heck, 2 vols., New York, 1851, *1*, Pl. B-6. **398.** Ibid., Pl. 1. **399 a, b.** *Natürliche Schöpfungsgeshichte* by Ernst Haeckel, 2 vols., Berlin, 1898, *2*, Pls. 14, 15. **400.** P, *75* (1878), 133. **401.** British Crown Copyright. Science Museum, London. **402.** *Optische Werkstaette* by Carl Zeiss, Jena, 1898, 39. **403.** Ibid., 96. **404.** *Die Infusionsthierchen als vollkommene Organismen* by Christian Ehrenberg, Leipzig, 1838, Pl. 51. **405.** Ibid., Pl. 26. **406.** Ibid., Pl. 23. **407.** *Design in Nature* by J. Bell Pettigrew, 3 vols., London: Longman Group Ltd., 1908, *1*, 174. **408.** Ibid., 175. **409.** *Microscopical Researches* by Theodor Schwann, Sydenham Society, London, 1847, opp. 227 and 265. **410.** *Studies in Microscopical Science, 4* (1886), Pl. 3. Photo: New York Botanical Garden Library. **411.** RV, *2*, Pl. 14. **412.** *Gesammelte Werke* by Robert Koch, 2 vols., Leipzig,

1912, *1*, Pl. 1. Reproduced by permission of Georg Thieme Verlag, Stuttgart. **413.** Ibid., Pl. 26, figs. 33, 34. Reproduced by permission of Georg Thieme Verlag, Stuttgart. **414.** *Archiv für Mikroskopische Anatomie, 21* (1882), Pl. 25. **415.** M, *1*, opp. 95. **416.** L, *1*, 182. **417.** *Atlas d'Anatomie Microscopique* by A. Donné, Paris, 1845, Pl. 1, fig. O. Photo: Special Collections, Butler Library, Columbia University. **418.** Ibid., fig. 3. Photo: Special Collections, Butler Library, Columbia University. **419.** Ibid., Pl. 11, fig. 44. Photo: Special Collections, Butler Library, Columbia University. **420.** Ibid., fig. 44bis. Photo: Special Collections, Butler Library, Columbia University. **421.** Ibid., Pl. 12, fig. 48. Photo: Special Collections, Butler Library, Columbia University. **422.** Ibid., Pl. 26, fig. 64. Photo: Special Collections, Butler Library, Columbia University. **423.** Ibid., Pl. 15, fig. 61. Photo: Special Collections, Butler Library, Columbia University. **424.** Ibid., fig. 62. Photo: Special Collections, Butler Library, Columbia University. **425.** Ibid., fig. 63. Photo: Special Collections, Butler Library, Columbia University. **426.** Ibid., Pl. 16, fig. 66. Photo: Special Collections, Butler Library, Columbia University. **427.** L, *1*, Pl. 3, opp. 322. Photo: Science Museum, London. **428.** Ibid., Pl. 6. Photo: Science Museum, London. **429.** *Anatomy, Descriptive and Surgical* by Henry Gray, 15th ed., Philadelphia, 1901, 836. **430.** Ibid., 1115. **431.** Ibid. **432.** *The Cell* by E. B. Wilson, New York, 1897, fig. 69. **433.** *Anthropogenie* by Ernst Haeckel, Leipzig, 1891, Pl. 2. **434.** *The Cell* by E. B. Wilson, New York, 1897, 299. **435 a, b.** *Illustrirtes Lexikon der Verfalschungen* by H. Klenske, Leipzig, 1879, 661 and 666. Photo: Butler Library, Columbia University. **436.** SP, *3* (1838), 578. **437.** P, *84* (1883), 75. **438.** FN, 55. **439.** SFP, *1*, Pl. 15. **440.** *Das Laboratorium*, 2 (1825), Pl. 8. **441.** *Laboratorium et Museum, 1* (1900-01), 67. **442.** British Crown Copyright. Science Museum, London. **443.** *Physisch-Chemische Abhandlungen in chronologischer Folge* by J. W. Ritter, Leipzig, 1806, Pl. 1, fig. 15. **444.** IZ, *88* (1887), 665. **445.** British Crown Copyright. Science Museum, London. **446.** *Cinquanténaire scientifique de M. Berthelot*, Paris, 1902, opp. 36. **447.** *Theoria Philosophiae naturalis* by R. J. Boscovich, English trans., Chicago, 1922, 42 [orig. pub. Venice, 1763]. **448 a, b.** *A New System of Chemical Philosophy* by John Dalton, 2 vols., 1808-1827, *1*, Pls. 3, 4. **449.** *Traité de chimie* by J. J. Berzelius, 8 vols., Paris, 1829, *1*, Pl. 2. **450.** *Tabelle, Exempel af nagra Dubbelsalter Sammansattning* by J. J. Berzelius, Stockholm, 1818. **451.** *Traité élémentaire de minéralogie* by F. S. Beudant, 2 vols., Paris, 1830-2, *1*, Pl. 2. **452.** *Krystallometrie* by J. F. C. Hessel, 2 vols., Leipzig, 1897, *2*, Pl. 11, figs. 338, 339. **453.** *Lehrbuch der organischen chemie* by August Kekulé, 2 vols., Erlangen, 1861-7, *1*, 162. **454.** Ibid., 160. **455.** Ibid., 143. **456.** Ibid., 164. **457.** Ibid., *2*, 515. **458.** *Revue universelle des Arts, 18* (1865), 318. **459.** *Lehrbuch der organischen chemie* by August Kekulé, 2 vols., Erlangen, 1861-7, *2*, 496. **460 a, b.** *Handbuch der stereochemie* by C. A. Bischoff, Frankfurt a. Main, 1894, 646-7. **461.** *Lehrbuch der organischen chemie* by August Kekulé, 2 vols., Erlangen, 1861-7, *2*, opp. 498. **462.** *Handbuch der stereochemie* by C. A. Bischoff, Frankfurt a. Main, 1894, 90-1. **463 a, b.** *L'Architecture du monde des atomes* by M. A. Gaudin, Paris, 1873, 24. **464.** Ibid., 52. **465.** Ibid., 82. **466.** *Handbuch der stereochemie* by C. A. Bischoff, Frankfurt a. Main, 1894, 169. **467.** *Abhandlung der Math. Phys. Klasses des König. Sachs. Ges. der Wiss., 14* (1888), 32. **468 a, b.** Ibid., 70, 71. **469.** *The Periodic System of Chemical Elements* by J. W. van Spronsen, Amsterdam: Elsevier Publishing Co., 1969, 98. **470 a,b.** (a) *The True Atomic Weights* by G. D. Hinrichs, St. Louis, Mo., 1894, Pl. 5, opp. 216; (b) Manuscript notebook, ca. 1865, in possession of author. **471.** *The Periodic System of Chemical Elements* by J. W. van Spronsen (Amsterdam: Elsevier Publishing Co., 1969), 141. **472.** British Crown Copyright. Science Museum, London. **473.** *Das genetische System der chemischen Elemente* by W. Preyer, Berlin, 1893, end of vol. **474.** SFP, *2*, Pl. 29, fig. 9. **475.** Ibid., Pl. 29, fig. 3. **476.** EM, 222. **477.** Ibid., Pl. 4. **478.** Ibid. Pl. 7. **479.** Ibid., 426. **480.** *Experimental Researches in Electricity* by Michael Faraday, 3 vols., London, 1839-55, *1*, Pl. 8. **481.** Reproduced by permission of the Museo Nazionale della Scienza e della Tecnica, Milan. **482.** MS, *1*, 641. **483.** Ibid., 673. **484.** British Crown Copyright. Science Museum, London. **485.** MS, *1*, 649. **486.** Ibid., 653. **487.** ILN, *17* (1850), 87. **488.** IL, *15* (1850), 384. **489.** Ibid. **490 a, b.** (a) MS, *1*, 713; (b) WM, 5, 213. **491.** *Experimental Researches in Electricity* by Michael Faraday, 3 vols., London, 1839-55, *2*, Pl. 4. **492.** *Exposé des nouvelles découvertes sur l'électricité et le magnétisme* by A. M. Ampère, Paris, 1822, 7. **493.** Ibid., 14. **494.** *Experimental Researches in Electricity* by Michael Faraday, 3 vols., London, 1839-55, *1*, Pl. 7. **495.** EM, 330. **496.** Ibid. **497.** Ibid., 332. **498.** British Crown Copyright. Science Museum, London. **499.** Photo: Deutsches Museum, Munich. **500.** Photo: Deutsches Museum, Munich. **501.** L, *1*, 137. **502.** MS, *1*, 733. **503.** *La Nature, 1* (1873), 249. **504.** *Experimental Researches on the Electric Discharge* by Warren de la Rue, London, 1880, Pl. 9. Reproduced with acknowledgment to the Royal Society, London. **505.** *The Discovery of Nature* by Albert Bettex, New York, 1965, 143. **506.** British Crown Copyright. Science Museum, London. **507.** British Crown Copyright. Science Museum, London. **508.** Lent to Science Museum, London, by the late Sir J. J. Thomson. **509.** IZ, *106* (1896), 172. **510.** Ibid., 146. **511.** IZ, *107* (1896), 231. **512.** ILN, *109* (1896), supp. to July 11 issue, after 64. **513.** G, *53* (1896), 208. **514.** XIXJ, *3*, 272. **515.** FN, 132. **516.** Ibid., 135. **517.** Ibid., 136. **518 a, b.** Ibid., 156-7. **519.** Ibid., 176-7. **520.** Ibid., 183. **521.** SFP, *2*, Pl. 38. **522.** IZ, *54* (1870), 353. **523.** FN, 197. **524.** WM, *5*, between 220-1. **525.** Lent to the Science Museum, London, by the Royal Society. **526.** *The Discovery of Nature* by Albert Bettex, New York, 1965, 116. Photo: Albert Bettex. **527.** L, *1*, 54. **528.** Ibid., 62. **529.** FN, 490. **530.** IZ, *14* (1850), 124. **531.** British Crown Copyright. Science Museum, London. **532.** British Crown Copyright. Science Museum, London. **533.** P, *25* (1853), 98-9. **534.** EM, Pl. 11. **535.** Ibid., 723. **536.** Ibid. **537.** IL, *92* (1888), 236. **538.** Ibid., 237. **539.** P, *35* (1858), 77. **540.** IL, *16* (1850), 149. **541.** Ibid., *44* (1864), 48. **542.** IZ, *8* (1847), 76. **543.** Ibid., *68* (1877), 81. **544.** Ibid., *87* (1886), 472. **545.** IL, *64* (1874), 56. **546.** Ibid., *43* (1864), 191. **547.** G, *34* (1886), 395. **548.** P, *76* (1879), Punch's Almanack for 1879. **549.** IZ, *97* (1891), 669. **550.** Lithograph in author's possession. **551.** IZ, *97* (1891), 100-01. **552.** Ibid., *83* (1884), 163. **553.** P, *82* (1882), Punch's Almanack for 1882. **554.** IL, *27* (1856), 241. **555.** L, *1*, 332. **556.** Ibid. **557.** Ibid., 423. **558.** ILN, *92* (1888), 377. **559.** *Design in Nature* by J. Bell Pettigrew, 3 vols. (London: Longman Group Ltd., 1908), *1*, 306. **560.** IZ, *8* (1847), 240. **561.** Ibid., 189. **562.** Ibid., *96* (1891), 556-7. **563.** G, *22* (1880), 620. **564.** Reproduced by permission of the Institut Pasteur, Paris. **565.** P, *97* (1889), 27. **566.** IL, *97* (1891), 213. **567.** G, *31* (1885), 95. **568.** IZ, *144* (1900), 952.

Index

The index of this book serves as a guide to both textual and pictorial material. The contents of the pictures themselves are indexed, in some cases, as explained in *A Note on Pictures*, on page xiii. The following system of reference is used:

187 reference to the text on page 187
187 reference to picture number 187

When the identification of a subject is not obvious, an explanation of its position appears in parentheses following the picture number: e.g. autoclave, used for sterilizing, **75** (left background).

I am grateful to Hope Hockenberry Yelich for preparing the Index.

L. P. Williams